Physical Chemistry and Its Biological Applications

Physical Chemistry and Its Biological Applications

Wallace S. Brey

University of Florida

Academic Press

New York
San Francisco
London
A Subsidiary of Harcourt Brace Jovanovich, Publishers

ACADEMIC PRESS, INC.
111 FIFTH AVENUE, NEW YORK, NEW YORK 10003

UNITED KINGDOM EDITION PUBLISHED BY
ACADEMIC PRESS, INC. (LONDON) LTD.
24/28 OVAL ROAD, LONDON NW1

ISBN: 0–12–133150–4
LIBRARY OF CONGRESS CATALOG CARD NUMBER: 77–91330

PRINTED IN THE UNITED STATES OF AMERICA

Preface

In writing this book, the aim has been to present and illustrate the basic principles of physical chemistry and to show how the methods of physical chemistry are being applied to increase our understanding of living systems. The reader should have some knowledge of organic chemistry and an acquaintance with calculus, but no very detailed mastery of either of these subjects is required.

The amount of material included is more than can be covered in a one-semester or two-quarter course and, for such a course, the instructor has considerable freedom in choosing the sections to be covered. Some suggestions for appropriate sequences will be found in a guide available to instructors. With modest additions of mathematical material by the teacher, the book should also be found suitable for a full-year major's course in physical chemistry. The author hopes that the presentation is sufficiently full and clear so that the book will be useful to biological scientists for self-study and reference.

My wife, Mary Louise, has provided invaluable aid in the preparation of the manuscript, as well as moral support during the period when it was being written. It is also a pleasure to acknowledge the help and cooperation of the Academic Press staff during the writing and production of the book.

Contents

13/MAGNETIC RESONANCE SPECTROSCOPY   494

14/PHOTOCHEMISTRY AND RADIATION CHEMISTRY   536

TABLE OF SYMBOLS AND ABBREVIATIONS   581

INDEX   583

One
States of
Matter

The differences we observe in the characteristics of the three states of matter—gas, liquid, and solid—depend upon the variation in the condition of aggregation of the molecules of which the matter is composed. In this chapter some of the principles governing transformation of one state of matter into another are considered. Structural models for gases and liquids are discussed, and the relationships between the macroscopic properties of these phases and the behavior and properties of individual molecules are examined, particularly from the viewpoint of the influence of forces between molecules.

1-1
MOLECULAR PICTURE
OF MATTER

From the properties of the gaseous state of matter, scientists have deduced a model in which the molecules are relatively far apart and are free to move almost independently of one another. This picture is embodied in the *kinetic theory*, which describes the molecules of a gas as separated particles in continuous motion. Each molecule travels in a straight line until it collides with another molecule or strikes the wall of the vessel in which it is confined. When the vessel is enlarged, molecular motion causes the gas to spread throughout all the newly accessible space; the application of external pressure, however, readily compresses the gas into a smaller volume, for the molecules have a relatively large amount of empty space between them.

In a liquid, the molecules are more restricted in their movement: They are able to roll past one another so that the liquid can flow, but it is only with considerable difficulty that they detach themselves from intimate association with other molecules in the bulk of the liquid, as they must do if the liquid is to be vaporized. In a solid, each molecule has a definitely assigned average position about which it vibrates; movement of the molecule away from its own small compartment, formed by neighboring molecules, is a comparatively unusual event. A crystalline solid is characterized by a relatively high degree of order in the arrangement of the atoms, ions, or molecules of which it is composed. In Figure 1-1 are some schematic representations of the possible molecular arrangements in various states of matter.

The state assumed by a particular sample of matter under a given set of conditions depends upon a balance between the kinetic energy of the molecules, on the one hand, and the sum of the intermolecular attractive forces plus the restraining effect, or pressure, imposed by the environment, on the other. The average kinetic energy per molecule in a group of molecules increases as the absolute temperature increases. In fact, a major part of the change associated with a rise in temperature is an increase in atomic and molecular motion resulting from the addition of energy: For gaseous molecules the velocity of translational motion increases, whereas in a solid the vibrational motion becomes greater in magnitude.

Let us imagine a quantity of matter at a temperature sufficiently low so that it is in the solid state, well below the melting point. We will now picture typical changes occurring as the temperature of the material is raised. The molecules acquire additional energy, which may for some solids lead to rotation of parts of the molecule or of the entire molecule

Figure 1-1
Molecular arrangements typical of (a) a crystalline solid, (b) a liquid or amorphous solid, and (c) a gas.

(a)

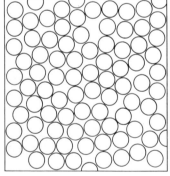

(b)

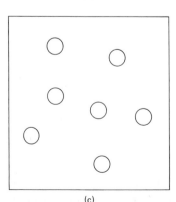

(c)

within its position in the solid. Eventually the amount of energy added will be sufficient to disrupt the solid structure, leading to fusion or melting to a liquid and permitting the molecules to move away from their localized positions. A further increase in the temperature of the liquid decreases its viscosity, a measure of its resistance to flow, and finally a temperature is reached at which another change of state occurs, the formation of a vapor. Vaporization requires energy sufficient not only to pull the molecules away from one another against the forces of attraction that hold them together in the liquid, but also to do the work of pushing back the atmosphere or the walls of the enclosing vessel to permit the large volume change associated with the conversion of liquid to gas.

1-2
PHASE DIAGRAMS

Each of the states of aggregation, or types of arrangement of molecules, a substance or mixture can assume is termed a *phase*. A sample of water may be in the vapor phase, in the liquid phase, in the solid phase which is commonly encountered as ice, or in one of a variety of other solid phases which appear at high pressures and which differ from ordinary ice by the manner in which water molecules are arranged in the crystalline pattern.

More generally, a phase can be described as a homogeneous portion of a material system set off from other phases by a boundary surface or discontinuity. The term "homogeneous" means that it is uniform in composition and structure throughout. A single phase may be entirely gaseous, or may comprise a single liquid layer, or may be a solid in which all the particles are in the same structural pattern or allotropic form. For liquid water with pieces of ice floating on the surface, two phases, liquid and solid, are present. If benzene and water, which do not dissolve very well in one another, are mixed together, there form two liquid phases separated by a visible boundary surface. If liquid water is mixed with a relatively large amount of sodium chloride so that not all the salt dissolves, the two phases present after equilibrium is reached are saturated aqueous salt solution and solid sodium chloride.

The behavior of a chemical substance under various conditions of temperature and pressure is often represented by a phase diagram. In this diagram are shown the ranges of conditions under which each of the several phases the substance can assume may exist as a stable form, as well as the more limited conditions under which equilibrium coexistence of two or more phases is possible.

At this point we illustrate the interpretation of a phase diagram for a relatively simple system containing only one substance; later, more complex systems will be considered. Figure 1-2, the phase diagram for the substance water in the lower pressure region, represents schematically the results of experimental determinations of the equilibrium relations of three phases of water in the absence of air or any other foreign material, as in a closed container. In order that the dis-

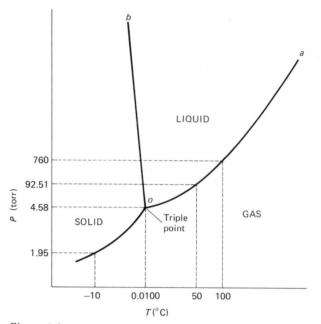

Figure 1-2
Schematic phase diagram for water in the region of relatively low pressure. One
torr, which is a pressure very nearly equal to 1 mm Hg, is defined as 1/760 atm.
At high pressures, several different crystalline forms of ice occur.

tinctive features can be more clearly seen, the diagram is not drawn
to scale.

Suppose that the temperature of a sample of water is 50°C. So long
as the pressure remains below 92.51 torr, the gas phase will continue
to be stable. If the pressure is momentarily increased to just above 92.51
torr, either by decreasing the container volume or by adding more
vapor, the vapor will partially condense to liquid until the loss of gas
is sufficient to restore the pressure to its equilibrium value at 50°C. If
the pressure is increased above 92.51 torr and kept there by external
means, the vapor will completely condense, for under these circumstances only liquid can exist. Thus there is only one pressure under
which the liquid and vapor of a pure substance can coexist permanently at a given temperature. Similar considerations apply to the
vapor and solid: At −10°C, for example, vapor can exist by itself at
pressures below 1.95 torr, while the solid alone exists at pressures exceeding this. Only at this one particular pressure can the two phases be
in equilibrium.

Water is rather unusual in that increasing pressure lowers the melting point of the solid phase; this trend is indicated by the inclination
of the line *ob* to the left. At about 1000 atmospheres (atm), the solid
phase and the liquid phase may be in equilibrium at −10°C. For most
substances, the solid phase is more dense than the liquid phase and
tends to be formed more readily at high pressures in accordance with

Le Chatelier's principle that, when the conditions of a system at equilibrium are altered, the equilibrium shifts in a direction so as to offset the stress of the change. The water system is distinctive in that the structure of the solid phase is less compact than is that of the liquid and therefore the solid is less dense.

A point of particular interest in Figure 1-2 is that designated o and known as the *triple point*. Only under the conditions represented by this point, a temperature of 0.0100°C and a pressure of 4.579 torr, is it possible for all three phases, solid, liquid, and gas, to be together in equilibrium. If the three phases are mixed together at any other temperature or pressure, one or two of them will disappear in order to establish equilibrium. The zero of temperature on the centigrade, or Celsius, scale was formerly defined as the freezing point of water; this *ice point* is not the same as the triple point but is the temperature at which ice and air-saturated water are in equilibrium under a total pressure of 1 atm. Under these conditions, most of the pressure of the vapor phase is contributed by air. Because the triple point is more precisely reproducible in the laboratory than the ice point, the temperature of the triple point by international agreement has been *defined* to be precisely 0.01°C.

The curve oa, along which the liquid phase exists in equilibrium with the vapor phase, has a definite termination point at 374°C and 218 atm. If the temperature exceeds this value, it is not possible to see a liquid–vapor interface. If, for example, water vapor in a container is initially at a temperature of 400°C and a pressure of 1 atm, the pressure can be raised at constant temperature to 250 atm, the temperature then lowered at constant pressure to 90°C, and the pressure finally lowered at constant temperature to 1 atm. The water is now clearly in the liquid state, but nowhere along this path will there have been an interface between the two phases. Indeed, we can not easily say at what stage the vapor became liquid.

The point a is termed the *critical point* of water. The temperature at this point, above which the vapor cannot be liquefied in the usual way, is the critical temperature, and the pressure is the critical pressure. Just as the triple point and the boiling point at 1 atm pressure depend upon the material, the critical temperature and pressure also vary widely from one substance to another. We shall consider further their significance in Section 1-7.

The phase diagram for the substance carbon dioxide is shown in Figure 1-3. In some respects, it resembles the diagram for water, but the triple point is at a pressure exceeding 1 atm, so that solid CO_2 sublimes rather than melts under ordinary external pressures. Further, the solid–liquid equilibrium curve goes to higher temperatures at higher pressures, a trend similar to that exhibited by most other substances for which the liquid phase is less dense than the solid phase.

Phase diagrams are also of great assistance in describing the behavior of systems having more than one constituent, for example, to display the conditions under which one substance is soluble in another; applications of this type will be described in Chapter 2.

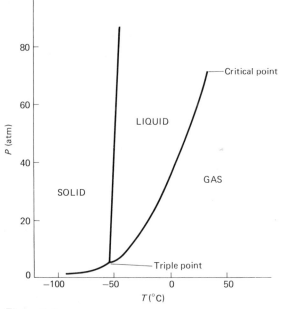

Figure 1-3
Phase diagram for carbon dioxide. The pressure at the triple point is 5.1 atm; this
is the lowest pressure at which the liquid phase is stable.

1-3
IDEAL GASES

We turn now to the description of properties of matter in the gaseous
state. The outstanding characteristic of a gas is the sensitivity of its
volume to changes in temperature and pressure. Experience has
shown that a generalized pattern, termed *ideal* or *perfect* gas behavior,
is approached or at least roughly approximated by most gases, according to which the volume varies in direct proportion to the absolute
temperature under constant pressure (Charles' law), and in inverse
proportion to the applied pressure at constant temperature (Boyle's
law):

$$V \sim T \tag{1-1}$$

$$V \sim \frac{1}{P} \tag{1-2}$$

These equations can be combined into one, useful for the prediction of
volume change when both temperature and pressure vary at the same
time:

$$V \sim \frac{T}{P} \tag{1-3}$$

Proportional variation implies that one quantity is equal to a constant numerical multiplier times the second quantity; thus the gas equation can be written with a constant of proportionality c:

$$V = \frac{cT}{P} \quad \text{or} \quad PV = cT \tag{1-4}$$

A further result of experiment is the conclusion that the numerical value of c varies directly with the number of gram-molecular weights, or moles, of gas in the sample being described, and is the same for samples of different gases containing the same number of moles. If c is accordingly set equal to nR, where n is the number of moles of gas and R is called the *gas constant*, the ideal gas law becomes

$$PV = nRT \tag{1-5}$$

Volumetric behavior in accord with this equation is shown in Figure 1-4 in three different ways.

It is often convenient to use an alternate form of the ideal gas equation in which each side of Equation (1-5) has been divided by n, the

Figure 1-4
Behavior of an ideal gas. (a) Ideal gas isotherms—variation of molar volume with pressure, at several constant temperatures. (b) Variation of gas pressure–molar volume product with pressure at several temperatures. (c) Variation of gas pressure–volume product with temperature for various amounts of gas.

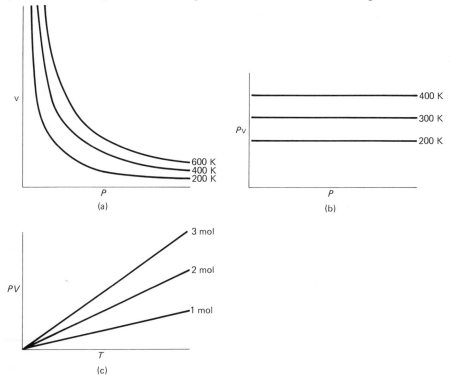

number of moles of gas, and in which v stands for the volume per mole of gas:

$$Pv = RT \qquad (1\text{-}6)$$

The ideal gas law, it must be remembered, is only a limiting law, in the sense that the volume variations of an actual gas approach those predicted by the equation more and more closely as the pressure is lowered and as the temperature is raised. However, it is to be emphasized that, under such conditions of sufficiently low pressure and sufficiently high temperature, a single numerical value of R does in fact represent gaseous behavior, independent of the particular gas being described.

Equation (1-5) is sometimes written in terms of the number of molecules, rather than the number of moles, of gas. If N' is the number of molecules, then

$$PV = N'kT \qquad (1\text{-}7)$$

Since N' is equal to n multiplied by N, Avogadro's number or 6.02 $\times 10^{23}$, k is equal to R divided by N. The gas constant per molecule, k in this equation, is often called the *Boltzmann constant*. Both R and k appear in many physical chemistry equations, and they have a significance wider than mere application in the calculation of gaseous volumes.

DIMENSIONAL NATURE OF THE GAS EQUATION

Let us now analyze the dimensional characteristics of the ideal gas equation. P, the pressure, is the ratio of a force to the area of the surface on which the force acts. The dimensions of P are force/area or force/(length)^2. V, the volume, has the dimensions of (length)^3. The product PV then has dimensions of (force)(length)^3/(length)^2 or (force)(length). Remembering that the force represented by the gas pressure and due to the collisions of the molecules with the walls of the containing vessel is exerted perpendicularly to these walls, we consider the force on any one particular side of the container, which, for simplicity, may be a rectangular box. If the dimensions of the box are a, b, and c, as in Figure 1-5, we let F be the force exerted on the face M, bounded by the edges of lengths b and c. The lengths in the denominator of the dimensional pressure expression above are then those of edges b and c. The lengths in the dimensional expression for

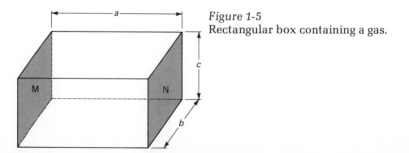

Figure 1-5
Rectangular box containing a gas.

volume are a, b, and c. Then the length that does not cancel out when pressure is multiplied by volume is a, the dimension of the box parallel to the direction of the force.

Imagine now that we start with a situation in which faces M and N are coincident with one another and are at the place occupied by face M in the diagram; let the absolute temperature of a hypothetical ideal gas between faces M and N be 0 K, at which the volume would be zero. As the temperature of this gas is allowed to rise to a final temperature T, the accompanying expansion will push face N away from face M; if a constant pressure, say atmospheric pressure, is maintained on the outside of face N, the work of expansion is done against a constant force. Any mechanical work can be expressed as the product of a force and the distance through which the force acts. Here the quantity of work is equal to the force Pbc that the gas exerts on the wall, multiplied by the distance at which the wall moves in a direction perpendicular to the force, and the work done is therefore $(Pbc)a$ or PV.

The product PV thus represents a quantity of work or energy associated with the gas because it occupies a portion of space not available for any other material substance. The dimensions, and therefore the units, of the PV product are those of work or energy, and R, which by the ideal gas equation equals PV/nT, has units corresponding to work per mole per degree.

NUMERICAL VALUES OF R

Although the dimensions of R are invariable, its numerical value depends upon the set of units in which the other quantities in the ideal gas equation are expressed. No problem is presented by the units of n, which is customarily given in gram-moles (g mol), nor by T, which is in kelvins (K), formerly called degrees, on the absolute temperature scale. However, the volume V can be specified in cubic centimeters (cm³), in liters, equivalent to cubic decimeters (dm³), or in cubic meters (m³), and the pressure P can be given in any of a variety of units.

In evaluating R numerically, it is convenient to remember that 1 mol of an ideal gas occupies a volume of 22,413.8 cm³ under standard conditions of 1 atm pressure and a temperature of 273.150 K. From these numbers,

$$R = \frac{(1 \text{ atm})(22{,}413.8 \text{ cm}^3)}{(1 \text{ mol})(273.150 \text{ K})} = 82.057 \text{ cm}^3 \text{ atm}/(\text{mol K})$$

For most calculations, it is sufficient to use for R values rounded to three significant figures, 82.1 cm³ atm/(mol K) or 0.0821 liter atm/(mol K.)

Since R represents a quantity of energy per mole per degree, it is often desirable to use a numerical value in terms of the usual energy units of ergs, joules (J) or calories (cal). For example, if all the factors entering into a calculation of energy are expressed in the *centimeter-gram-second* (cgs) system of units, the units of the quantity of energy are *ergs*. In the gas equation, the appropriate units are cubic centimeters for volume, and dynes (dyn) per square centimeter for pressure. One atmosphere can be converted to dynes per square centimeter by remembering that this pressure can support, against the force of gravity, a column of mercury 76 cm tall. The weight in grams of a column of mercury of this height and 1 cm² in cross section is equal to the density of mercury, 13.595 g/cm³, multiplied by the volume of the column, 76.00 cm³, or 1033.2 g. Multiplication by the numerical value of the acceleration due to gravity, 980.67 cm/sec² or 980.67 dyn/g, converts the mass to dynes force of gravity, and for R we set down the equation

$$R = \frac{(1033.2 \text{ g})(980.67 \text{ dyn/g})(22{,}413.8 \text{ cm}^3)}{(1 \text{ mol})(273.15 \text{ K})} = 8.314 \times 10^7 \text{ ergs/(mol K)}$$

In the internationally adopted SI (for Systeme International) set of units, length is expressed in meters and mass in kilograms (kg). The unit of force is the newton (N), equal to 1 kg m/sec² or 10⁵ dyn. The atmosphere as a unit of pressure is defined directly to be 101,325 N/m², and the torr, corresponding approximately to 1 mm of mercury, is 1/760 atm. If we multiply the value for 1 atm by the value of the gas constant in m³ atm/(mol K), or 82.057 cm³ atm/(mol K) $\times$ 10⁻⁶ m³/cm³, we find for the gas constant a value of 8.314 N m/(mol K). In SI units, the joule represents work or energy in newton meters, so that

$$R = 8.314 \text{ J/(mol K)}$$

This is consistent with the ratio of 10⁷ ergs in 1 J. From the "mechanical equivalent of heat," which is 4.184 J/cal,

$$R = \frac{8.314 \text{ J/(mol K)}}{4.184 \text{ J/cal}} = 1.987 \text{ cal/(mol K)}$$

This number can be remembered as approximately equal to 2.0.

Example: A sample of 0.250 g of a pure liquid is vaporized, and the vapor is collected in a buret over mercury as the confining liquid. The gas in the buret is at a pressure of 745 torr and a temperature of 60.0°C. Calculate the molecular weight of the substance if the volume of the vapor is 68.3 cm³.

Solution: If the weight of the sample in grams is represented by g and the molecular weight of the substance by M, the number of moles is

$$n = \frac{g}{M} = \frac{PV}{RT}$$

or

$$M = \frac{gRT}{PV}$$

Substituting the data given,

$$M = \frac{(0.250 \text{ g})[82.06 \text{ cm}^3 \text{ atm/(mol K)}](333.2 \text{ K})}{\left(\dfrac{745 \text{ torr}}{760 \text{ torr/atm}}\right)(68.3 \text{ cm}^3)} = 102 \text{ g/mol}$$

Example: In a respiration experiment, a bacterial culture consumed 22.7 cm³ of oxygen gas, measured at a pressure of 79.0 cm and a temperature of 32.8°C. Calculate the number of moles of oxygen used.

Solution: The measured quantities can be substituted directly into the ideal gas equation. Then

$$n = \frac{PV}{RT} = \frac{\left(\dfrac{79.0 \text{ cm}}{76.0 \text{ cm/atm}}\right)(22.7 \text{ cm}^3)}{[(82.06 \text{ cm}^3 \text{ atm/(mol K)}](306.0 \text{ K})}$$

$$= 9.40 \times 10^{-4} \text{ mol}$$

KINETIC THEORY
AND DERIVATION
OF THE GAS EQUATION

The *equation of state,* or equation relating P, V, and T for a sample of gas, can be derived from the following postulates of the kinetic theory:

(1) Each molecule of the gas is continuously in motion, traveling in a straight line until deflected by collision with another molecule or with the wall of the container. The average kinetic energy of all the molecules in a sample of gas is proportional to the absolute temperature; therefore, since the kinetic energy of a moving body varies directly with the square of the velocity of the body, the average value of the square of the molecular velocity is proportional to the absolute temperature.

(2) In the course of a large number of collisions, kinetic energy can be transferred from molecule to molecule, but there is no overall loss of kinetic energy from the group of molecules composing the gas. Thus the gas does not settle to the bottom of its container, as it would eventually if any energy were lost by a "frictional" effect upon collison of two molecules. Moving bodies consisting of more than one molecule cannot undergo frictionless collisons, because some of their kinetic energy is always converted into random motions of the separate molecules comprising the larger objects; but this restriction does not apply to collisions of single molecules.

(3) The gas exerts pressure by virtue of the force of the molecules striking the wall of the confining vessel.

The preceding description is valid for all gases, whether they obey the ideal gas equation exactly or only approximately. In addition, however, two stipulations, of the nature of approximations, are necessary to give an exact quantitative justification of the perfect gas equation:

(1) The molecules are so small that the space they actually occupy is negligible relative to the total volume V the bulk gas occupies. The diameter of a molecule of oxygen is about 3×10^{-8} cm [or 3 angstroms (Å)], and at 0°C and 1 atm pressure, the average distance between the centers of two molecules in any ideal gas is about 30 Å. Thus of a total volume of 22,414 cm³ for a mole of oxygen gas, only about 22 cm³ is actually occupied by matter, and the remainder is empty space. The approximation of negligible molecular volume is not valid when the total volume per mole is small because of high pressure or low temperature.

(2)   The molecules exert no forces of attraction upon one another. This is a good approximation when they are mostly far apart from one another, for the attractive forces drop off rapidly with distance, but it is invalid under the same conditions as those under which the previous approximation fails.

On the basis of these postulates and approximations, we can now derive the perfect gas equation. Suppose there are N' molecules of a gas contained within a rectangular box of dimensions a, b, and c. The box is placed with edges parallel to the axes of a system of rectangular coordinates, as shown in Figure 1-6. The molecules, each of mass m, are moving randomly in various directions and with various velocities. If the velocity of any one selected molecule is u, the components of this velocity parallel to the x, y, and z axes of the rectangular coordinate system can be designated u_x, u_y, and u_z, respectively. These components are related in magnitude to the velocity u by the equation

$$u^2 = u_x{}^2 + u_y{}^2 + u_z{}^2 \qquad (1\text{-}8)$$

The pressure exerted by the gas on one wall of the container is the same as that on any other wall; let us consider therefore one particular wall bounded by edges of length b and c and designated S. Pressure on surface S is produced by collisions of molecules with this wall; only u_y, the component of the velocity perpendicular to the surface, is effective, however. After each collision with the surface, a molecule rebounds with velocity unchanged in magnitude but reversed in direction; the total change in velocity per collision is therefore $2u_y$. The time required for the molecule to move from surface S to surface S' and then back to S is equal to $2a$, twice the distance between the surfaces divided by the velocity u_y. The number of collisions of a single molecule with surface S per second is the reciprocal of this time or $u_y/2a$, and the change in velocity per second is $2u_y(u_y/2a)$ or $u_y{}^2/a$. The force on surface S is the same as the force the surface exerts on the molecule to change its velocity; this force is equal to the mass of the molecule m times the acceleration, which is by definition the time rate of change in velocity:

$$F = m\frac{u_y{}^2}{a} \qquad (1\text{-}9)$$

The pressure is obtained by dividing the force by the area of S, which is bc:

$$P = \frac{mu_y{}^2}{abc} = \frac{mu_y{}^2}{V} \qquad (1\text{-}10)$$

The product abc is equal to V, the volume of the container.

Since the number of molecules in the box is N' rather than one, and we have been considering the effect of only one, the total pressure is equal to N' multiplied by the average pressure due to an individual molecule. If the molecules are identical in mass, the average pressure can be obtained from Equation (1-10) by replacing the square of the velocity of one molecule by the average of the square of the velocity

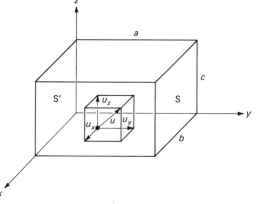

Figure 1-6
Gas molecule in a box.

for the group of molecules $\overline{u_y{}^2}$:

$$P = \frac{N'm\overline{u_y{}^2}}{V} \tag{1-11}$$

Since the motion of the molecules is entirely random, the components of velocity in the three directions are on the average equal, so that from Equation (1-8),

$$\overline{u_x{}^2} = \overline{u_y{}^2} = \overline{u_z{}^2} = \tfrac{1}{3}\overline{u^2} \tag{1-12}$$

and

$$PV = \tfrac{1}{3} N'm\overline{u^2} \tag{1-13}$$

The kinetic energy of the N' molecules is equal to $N'\tfrac{1}{2}m\overline{u^2}$, and therefore Equation (1-13) can be written

$$PV = \tfrac{2}{3}(N'\tfrac{1}{2}m\overline{u^2}) = \tfrac{2}{3}N'(\text{kinetic energy per molecule}) \tag{1-14}$$

The assumption that the kinetic energy of translational motion of molecules is proportional to the absolute temperature converts this equation to

$$PV = \tfrac{2}{3}N'c'T = \tfrac{2}{3}nNc'T \tag{1-15}$$

where N is Avogadro's number. Letting R represent the product of constants $2c'N/3$, one obtains the usual gas equation:

$$PV = nRT \tag{1-16}$$

1-4
MOLECULAR VELOCITIES

Let us consider Equation (1-13) applied to 1 mol of an ideal gas, which contains N molecules:

$$PV = \tfrac{1}{3}Nm\overline{u^2} \tag{1-17}$$

Since the product PV is also equal to RT for 1 mol, there results

$$\overline{u^2} = \frac{3RT}{Nm} = \frac{3kT}{m} = \frac{3RT}{M} \tag{1-18}$$

where M is the molecular weight. For oxygen at 300 K,

$$\sqrt{\overline{u^2}} = \sqrt{\frac{3(8.314 \text{ J/mol K})(300 \text{ K})}{0.032 \text{ kg/mol}}}$$
$$= 484 \text{ m/sec}$$

If a gas is allowed to escape through a very small hole—so small that no collisions between molecules occur within it—from a container into a region maintained at low pressure, the rate of effusion is proportional to the molecular velocity. Equation (1-18) is then the basis for the prediction that the effusion rate will be inversely proportional to the square root of the molecular weight, a prediction verified by Thomas Graham's observations of relative rates of effusion of different gases through the same hole.

It should be noted that the process of effusion is quite different from that of gaseous diffusion in which the average net motion of molecules is followed as they migrate through a region filled with gas, where their progress is impeded by frequent molecular collisions and the time required to traverse a given distance may be several orders of magnitude larger than would be calculated from the velocities of the individual molecules.

The quantity $\sqrt{\overline{u^2}}$, the square root of the mean of the squares of the velocities for a group of molecules, which we calculated above, is a type of average velocity known as the *root-mean-square* (rms) velocity. That it is necessary to consider an average velocity is a consequence of the fact that, in any sample of gas, the molecules have a wide range of kinetic energies. A few of them have very high veloc-

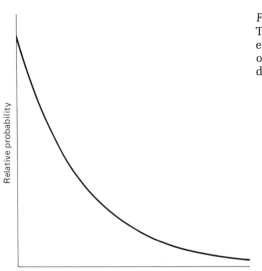

Figure 1-7
The probability that a molecular system has an energy in excess of the zero level as a function of that energy, according to the Boltzmann distribution for a single kind of energy.

Relative probability

Energy in excess of zero level

ities, a few others have very low velocities, and the great majority fall in an intermediate region.

The problem to be solved in establishing the distribution of velocities in a collection of gaseous molecules is that of the random distribution of a fixed total quantity of energy over a number of systems, the molecules, each of which may have a quantity of energy varying from none to a very great amount. If the various states in which a molecule can exist are equivalent except in the amount of energy possessed by the molecule, a statistical analysis developed by Boltzmann is applicable in the form

$$P(\epsilon) = e^{-\epsilon/kT} \qquad (1\text{-}19)$$

Here, $P(\epsilon)$ is the probability that a molecule has energy ϵ above the ground or base level, k is the Boltzmann constant defined earlier, and T is the absolute temperature. This distribution has the maximum probability value for the lowest energy, which is the quantity zero associated with the ground state, and decreases monotonically as the energy increases as shown in Figure 1-7.

The kinetic energy of a molecule is equal to $\frac{1}{2}mu^2$. However, as the total kinetic energy of the molecule increases, the number of ways in which the three independent components of the kinetic energy can combine to give the same total energy increases as the square of the magnitude of the energy. Combining this consideration with the Boltzmann distribution results in an expression for the probability that a molecule have a velocity u:

$$P(u) = Au^2e^{-mu^2/2kT} \qquad (1\text{-}20)$$

The constant A is independent of the velocity and is equal to $4\pi(m/2\pi kT)^{3/2}$. This equation describes what is known as the *Maxwell–Boltzmann* distribution of molecular velocities and is the basis for the typical distribution of velocities shown in Figure 1-8.

In Figure 1-8, several types of "average" velocity are indicated for one of the temperatures. The *most probable* magnitude u_{mp} of the velocity corresponds to the maximum in the curve and is obtained by setting the derivative of the probability with respect to velocity equal to zero. Solution of this equation yields

$$u_{mp} = \sqrt{\frac{2kT}{m}} = \sqrt{\frac{2RT}{M}} \qquad (1\text{-}21)$$

The *mean velocity* magnitude corresponds to the usual arithmetic average, in which the number of molecules in each small interval of velocity is multiplied by a weighting factor equal to that velocity, and the sum of all these expressions, which is the integral over all possible magnitudes of the velocity from zero to infinity, is divided by the total number of molecules, which is the integral of the distribution function over the same range:

$$\bar{u} = \frac{\int_0^\infty uAu^2e^{-mu^2/2kT}\,du}{\int_0^\infty Au^2e^{-mu^2/2kT}\,du} = \sqrt{\frac{8kT}{\pi m}} = \sqrt{\frac{8RT}{\pi M}} \qquad (1\text{-}22)$$

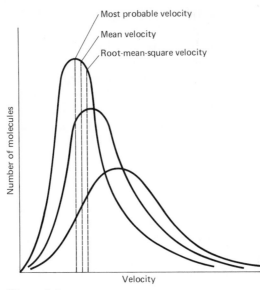

Figure 1-8
Distribution of molecular velocities. The three curves correspond to three different temperatures; the maximum shifts to higher velocities at higher temperatures.

The rms velocity, which we encountered above, can be calculated in a similar fashion, except that the weighting factor for each velocity interval in the numerator is the *square* of the velocity:

$$\sqrt{\overline{u^2}} = \frac{\int_0^\infty u^2 A u^2 e^{-mu^2/2kT}\,du}{\int_0^\infty A u^2 e^{-mu^2/2kT}\,du} = \sqrt{\frac{3kT}{m}} = \sqrt{\frac{3RT}{M}} \qquad (1\text{-}23)$$

1-5
GASEOUS MIXTURES;
MEASUREMENT OF GASES

We turn now to a description of ways in which compositions of gaseous mixtures can be described. The volumetric behavior of an ideal gas is independent of the chemical nature of the molecules comprising the gas. In a mixture of gases that separately behave ideally we expect to find no distinction among the different kinds of molecules. In fact, even if the separate components of a gaseous mixture deviate somewhat from ideality, the properties of the mixture are still found to be very nearly additive in the properties of the individual gaseous components. For example, suppose a mixture is composed of 0.10 mol of O_2, 0.30 mol of N_2, 0.02 mol of CO, and 0.08 mol of CO_2 and is present at 29.0°C in a container of 6.20 liters volume. The total number of moles in the mixture is 0.50; using $P = nRT/V$, the pressure is calculated to be 2.00 atm. One way of describing the mixture is in terms of the *partial pressures* of the four gases, that is, the pressures that each of the gases, individually, would exert if present in the same volume of 6.20 liters at the same temperature. For O_2, $p_{O_2} = n_{O_2}RT/V$

$= (0.10 \text{ mol})[0.08205 \text{ liter atm}/(\text{mol K})](302 \text{ K})/6.20 \text{ liter} = 0.40 \text{ atm};$ similarly for the other gases, $p_{N_2} = 1.20$ atm, $p_{CO} = 0.08$ atm, and $p_{CO_2} = 0.32$ atm. It is seen that the partial pressures add to give a value equal to the total pressure.

The partial pressure of a gas in a mixture is significant as an indication of the concentration of the substance to which it refers; for most purposes the concentration governs the behavior of a gaseous substance regardless of the presence of other materials in admixture.

An alternative treatment of the mixture described is a consideration of it as made up by combining the appropriate volumes of each of the components, taken separately under the same pressure of 2.00 atm as that of the mixture. These individual volumes are then the *partial volumes* of the components. In the example the partial volume of oxygen, $v_{O_2} = n_{O_2}RT/P$, is 1.24 liters; for N_2, CO, and CO_2, the partial volumes are, respectively, 3.72, 0.25, and 0.99 liter.

If a gaseous mixture is analyzed by absorbing one of its components in a liquid that dissolves or reacts chemically with that particular substance, the loss in volume of the gas mixture in this process corresponds to the partial volume of the component that is removed. Thus the analysis for CO_2 can be carried out for the mixture given above by passing the gas through NaOH solution. If the pressure of the mixture is 2 atm before and after this treatment, the volume will have been found to decrease from 6.20 liters to 5.21 liters.

A third way of specifying the concentration of a component of a mixture, suitable whether the mixture be gaseous, liquid, or solid, is in terms of its *mole fraction*. This is defined as the fraction that the number of moles of any component represents of the total number of moles of all the components of the mixture. Such a concentration scale is very easily applied to gases, since the mole fraction is equal to the volume fraction, or ratio of the partial volume to the total volume, as well as to the pressure fraction, or ratio of the partial pressure to the total pressure.

For O_2 in the mixture above, the mole fraction X_{O_2} is given by

$$X_{O_2} = \frac{0.10 \text{ mol}}{0.50 \text{ mol}} = \frac{0.40 \text{ atm}}{2.00 \text{ atm}} = \frac{1.24 \text{ liters}}{6.20 \text{ liters}} = 0.20$$

The mole fraction has no dimensions and no units; it is a pure number. Just as the sum of the partial pressures equals the total pressure and the sum of the partial volumes equals the total volume, so the sum of the mole fractions equals unity.

Often mole fraction values are converted to mole percent values by multiplying by 100. On the mole percent scale, the concentrations of all the components of a mixture total 100.

The volume of a gas sample can be measured in a buret, just as the volume of a liquid is measured in quantitative analytical work. However, to handle the gas conveniently, it is introduced into the buret through a stopcock at the top and confined by a liquid such as mercury kept in a leveling bulb attached to the bottom of the buret by a rubber hose. The leveling bulb can be raised or lowered to bring the gas to any desired pressure by changing the amount of mercury in the buret. The

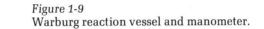

Figure 1-9
Warburg reaction vessel and manometer.

difference between atmospheric pressure and the pressure of the contents of the buret is indicated by the difference in levels of the mercury surface exposed to the two gas phases.

Biological investigations concerned with the respiration processes of living cells generally involve the measurement of small volumes of CO_2 produced and of O_2 consumed. One of the outfits used for this purpose, known as the *Warburg apparatus*, is shown diagrammatically in Figure 1-9. The organism to be studied is placed in the reaction bulb, and the whole apparatus is mounted on a support in a manner so that it can be shaken as a unit while the reaction vessel is immersed in a constant-temperature water bath.

At the bottom of the open-end U-tube manometer is a liquid reservoir, the volume of which is adjustable by means of a screw clamp. Initially, the screw clamp is adjusted to bring the liquid in the arm of the buret connected to the reaction vessel to a reference mark, and the liquid level in the limb open to the atmosphere is then read. As the process goes forward and O_2 is consumed, the liquid reservoir is readjusted to keep the level on the interior side at the reference mark, while readings are taken of the level on the atmospheric side. Thus the volume of the gas is kept constant, and the change in number of moles of gas can be calculated from the change in pressure observed on the manometer.

If it is desired to determine the rate of CO_2 evolution from the living organism, parallel determinations can be run in which only one reaction vessel contains alkali solution. The difference in the calculated change in the amount of gas in the two reaction vessels then represents the amount of CO_2 formed, since in the vessel containing alkali all the CO_2 is absorbed. However, this method is reliable only if the CO_2 content of the atmosphere in the vessel has no effect on the organism under investigation.

The Warburg apparatus is also suitable for measurement of the total dissolved CO_2 in blood. A sample of plasma can be placed in the vessel, excess acid added to convert all the carbonates and bicarbonates to CO_2, and the amount of gas evolved calculated from the increase in pressure.

The observed volumetric behavior of many gases deviates from that predicted by the ideal gas equation, and frequently it can be better represented by an equation containing additional arbitrary constants, such as the *equation of van der Waals,* which for n moles of gas is

$$\left(P + \frac{n^2 a}{V^2} \right)(V - nb) = nRT \qquad (1\text{-}24)$$

The quantities a and b are characteristic of the particular gas being described and must be determined by experimental measurements of molar volume over a range of pressures.

The term $n^2 a / V^2$ arises from the forces of intermolecular attraction that were neglected in the derivation of the ideal gas equation. The nature of these forces is discussed later in Section 1-8. Retardation of molecular motion by these forces leads to an effect sometimes described as an internal pressure. The result is a measured pressure P_m, the quantity described by the equation of state, which is smaller than the ideal pressure P_i, the quantity predicted by the ideal gas equation. Thus P_m is equal to P_i less the correction term, and, in Equation (1-24), the quantity in brackets which includes pressure corresponds to P_i.

The value of a is proportional to the forces between a pair of molecules, but the total effect for a gas is the summation over all the pairs of molecules. The significant forces are those that slow the molecules as they are about to strike the containing wall, and thus comprise the forces with which molecules in the interior of the gas attract those near the surface. The concentration of molecules in either region is proportional to n/V, and the number of molecular interactions then varies as the square of the molecular concentration, or n^2/V^2. To visualize this, we may suppose that there are two molecules in a certain volume at the surface and two in the corresponding volume in the interior of the gas; the number of interactions directed perpendicularly to the surface is four. If the concentration is then doubled, so that the number of molecules in each region becomes four, the number of interactions will increase to 16.

Space is occupied by the molecules themselves, for they are not actually mathematical points; therefore a certain part of the total volume V is not accessible to other molecules for free movement. This circumstance leads to the second van der Waals correction: The constant b represents the volume that, for each mole of gas, is effectively unavailable, and for n moles the volume nb is subtracted from the overall measured volume to obtain the ideal free volume.

The ease of condensation of a gas to a liquid is obviously related to the forces of attraction between the molecules, and a general parallelism is seen in Table 1-1 between the ratio of the van der Waals constants a/b and the boiling points of the substances listed. Also given

Table 1-1
Van der Waals constants, boiling points, and critical constants of various
materials

Substance	a (liter^2 atm/mol^2)	b (liter/mol)	Ratio a/b	Boiling point (°C at 760 torr)	T_c (°C)	P_c (atm)
H_2O	5.464	0.0305	179	100.0	374.1	218.3
C_6H_5Cl	25.43	0.1453	175	132	359.2	44.6
$n\text{-}C_8H_{18}$	37.32	0.2368	158	125.7	296	24.8
C_6H_6	18.00	0.1154	156	80.1	288.9	48.6
CCl_4	20.39	0.1383	147	76.6	283.1	45
$(CH_3)_2CO$	13.91	0.0994	140	56.2	235.5	47
SO_2	6.714	0.05636	119	−10.0	157.8	77.7
CH_3Cl	7.471	0.06483	115	−24.2	143.8	65.9
NH_3	4.170	0.03707	112	−33.4	132.5	112.5
HCl	3.667	0.04081	90	−84.9	51.4	82.1
N_2O	3.782	0.04415	86	−88.5	36.5	71.7
CO_2	3.592	0.04267	84	Sublimes	31	72.9
$CH_2{=}CH_2$	4.471	0.05714	78	−103.7	9.9	50.5
CH_4	2.253	0.04278	52.7	−164	−82.1	45.8
O_2	1.360	0.03183	42.7	−183.0	−118.4	50.1
Ar	1.345	0.03219	41.8	−185.7	−122.3	48
N_2	1.390	0.03913	35.5	−195.8	−147	33.5
H_2	0.2444	0.02661	9.18	−252.5	−239.9	12.8
He	0.03412	0.02370	14.40	−268.6	−267.9	2.26

in the table are the critical temperatures and pressures of the materials,
which are closely related to the van der Waals parameters, as discussed more fully in Section 1-7.

If the pressure is known and the volume occupied by a certain
amount of gas is to be calculated, the van der Waals equation is difficult
to solve directly, for it is an equation that is cubic in the volume. If,
however, the volume is known and the corresponding pressure is to be
found, the calculation is easily carried out as illustrated in the following problem.

Example: What pressure is exerted by 30.0 mol of CO_2 introduced into a vessel of
65.0 liters volume at 126.8°C?

Solution: The van der Waals equation is rearranged, and the constants for CO_2 are
substituted:

$$P = \frac{nRT}{V - nb} - \frac{n^2a}{V^2}$$

$$= \frac{(30.0 \text{ mol})[0.0821 \text{ liter atm/(mol K)}](400 \text{ K})}{65.0 \text{ liter} - (30.0 \text{ mol})(0.0427 \text{ liter/mol})}$$

$$- \frac{(30.0 \text{ mol})^2(3.59 \text{ liter}^2 \text{ atm/mol}^2)}{(65.0 \text{ liter})^2}$$

$$= 15.46 - 0.76 = 14.70 \text{ atm}$$

Both the corrections embodied in the van der Waals equation are
more significant when the concentration of molecules is large. This is
the situation when the temperature is low and the pressure is high.

Table 1-2
Observed and calculated gas pressures (in atmospheres)

Nitrogen								
Temperature		0°C				100°C		
P, observed	1.000	10.00	50.0	200.0	1.000	10.00	50.0	200.0
P, ideal	1.000	10.04	50.8	193.1	0.999	9.97	49.3	182.5
Carbon dioxide								
Temperature			30°C				100°C	
P, observed		50.0		200.0		50.0		200.0
P, ideal		71.6		504.4		56.6		335.3
P, van der Waals		51.8		227.6		49.2		198.6

*a*Values are in atmospheres.

Table 1-2 provides a comparison of the pressures found experimentally with those calculated from the perfect gas equation for two representative gases. For nitrogen, the van der Waals equation affords little improvement over the perfect gas equation in this range of conditions, but it is seen from the table that the van der Waals equation describes the behavior of carbon dioxide much better than does the perfect gas equation.

An alternative equation frequently used to describe the behavior of a real gas is a "power" series expansion in the pressure, which, because of its form, is referred to as the *virial equation*:

$$Pv = RT + BP + CP^2 + DP^3 + \cdots \tag{1-25}$$

The quantities B, C, D, and so on, are referred to as the virial coefficients and depend upon the temperature but not upon the pressure. The validity of this equation is based upon its success in describing experimental results with only a reasonable number of terms, very often only two or three.

Virial coefficients display the same information as that embodied in the van der Waals equation but in a different way. If, for example, the van der Waals equation is rearranged to give an expression parallel to Equation (1-25), the second virial coefficient B is found to be equivalent to $b - a/Pv$ and thus to represent the excess of the effect of volume occupied by the molecules themselves over that of the attractive forces.

In addition to the van der Waals equation and the virial equation, several other equations of state have been devised to describe the behavior of real gases as accurately as possible. These are principally of interest in engineering applications and are not discussed here.

1-7
CONTINUITY OF STATES; CORRESPONDING STATES

In considering further the behavior of a real gas as distinguished from that of an ideal gas, it is profitable to extend the examination of pressure–volume–temperature relations to temperatures low enough and to pressures high enough to lead to condensation to a liquid. Indeed, because the same kinds of forces that lead to deviations from ideal behavior are responsible for drawing the molecules together to form the liquid phase, it is possible to give at least a qualitative description of both liquid and vapor states of a substance in a single equation. The idea on which this approach is based is sometimes referred to as the principle of *continuity of states*.

As an example of the application of this principle, we cite carbon dioxide, for which volumetric data were obtained a century ago. Imagine a fixed amount of the substance, say 1 mol, placed in a cylinder fitted with a movable piston and immersed in a constant-temperature liquid bath. As the piston is moved into the cylinder to compress the gas, an increase in pressure accompanies the decrease in volume, and the results can be plotted, as in Figure 1-10, in the form of an *isotherm*, or constant-temperature curve.

If a typical isotherm, such as the curve from *a* to *d*, is followed, for instance, the initial segment from *a* to *b* corresponds to compression of the vapor, with an accompanying decrease in volume roughly similar to that expected from the ideal gas equation. As the point *b* is passed, the increase in pressure halts, although the volume continues to decrease. This reduction in volume *at constant pressure* occurs during presence of two phases, gas and liquid, in contact with one an-

Figure 1-10
Pressure–volume isotherms of carbon dioxide. Solid lines are experimental results, dashed lines are calculated from the van der Waals equation. Two phases coexist in the shaded region.

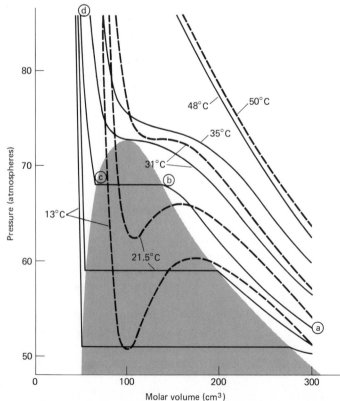

other. The interval bc represents the region of condensation of gas to liquid. The first drop of liquid appears at point b, the last bit of gas is condensed at point c, and further reduction in volume past c requires the application of increasingly larger pressures, consistent with the relatively low compressibility of the liquid.

If isotherms are followed at successively higher temperatures, one observes that the volume change on condensation becomes smaller and smaller until finally it disappears entirely. There is one particular isotherm, that at 304.3 K for carbon dioxide, which just touches the top of the two-phase region. The temperature of this isotherm is the critical temperature T_c discussed earlier. The pressure and volume approached as the curves bounding the two-phase region are followed toward the critical temperature are the critical pressure and the critical molar volume, P_c and v_c. As seen in Table 1-1, these critical constants are characteristic of particular substances.

At temperatures above the critical temperature the shape of an isotherm is similar to the hyperbola to be expected when the product PV is constant. Furthermore, in this temperature range, the contents of the cylinder can be compressed to the limit of the equipment without the appearance of an interface between the two phases.

Let us now investigate the behavior to be expected if the substance obeys the van der Waals equation in the transition to the liquid state. When the van der Waals equation is multiplied out and arranged in descending powers of the molar volume, it becomes

$$v^3 - v^2\left(b + \frac{RT}{P}\right) + v\frac{a}{P} - \frac{ab}{P} = 0 \qquad (1\text{-}26)$$

Since this equation contains the volume to the third power, there may be three roots, or values of the volume that satisfy the equation for any value of the pressure. Below the critical temperature, all three of the roots are real numbers, but above the critical temperature only one is real and two are imaginary. Several isotherms calculated from Equation (1-26) are shown in Figure 1-10 for comparison with the experimental results for carbon dioxide.

Several mathematical conditions apply at the critical point: (a) all three roots are equal, (b) the slope, or the first derivative of the function, is zero, and (c) since the critical point is an inflection point, the second derivative of the function is zero. Two of these conditions are sufficient to permit the equations to be solved to give the van der Waals parameters in terms of the critical constants: $a = 3P_c v_c^2$, $b = v_c/3$, and $R = 8P_c v_c/3T_c$. Using these relations with the form of the van der Waals equation applicable to one mole of gas, we find

$$\left(P + \frac{3P_c v_c^2}{v^2}\right)\left(v - \frac{v_c}{3}\right) = \frac{8P_c v_c}{3}\frac{T}{T_c} \qquad (1\text{-}27)$$

Division of each side of this equation by the product $P_c v_c$ leads to

$$\left[\frac{P}{P_c} + 3\left(\frac{v_c}{v}\right)^2\right]\left(\frac{v}{v_c} - \frac{1}{3}\right) = \frac{8}{3}\frac{T}{T_c} \qquad (1\text{-}28)$$

In this equation, temperature, pressure, and volume are involved as ratios of particular values to the corresponding critical values. It is convenient to define these ratios as reduced variables: $P_r = P/P_c$, $T_r = T/T_c$, and $V_r = v/v_c$. In terms of reduced variables, Equation (1-28) then becomes

$$\left(P_r + \frac{3}{V_r^2}\right)\left(V_r - \frac{1}{3}\right) = \frac{8T_r}{3} \tag{1-29}$$

If this equation is obeyed, two substances existing at the temperatures and pressures that are the same fraction of the critical values have the same reduced volume, which means that they have the same ratio of Pv to RT.

The equation just developed is an example of the principle of *corresponding states,* a generalization which says that different substances have similar properties when compared with each other at temperatures that are equal fractions of their respective critical temperatures, and pressures that are equal fractions of their critical pressures.

An empirical rule in accord with this principle is the statement that the boiling point of a liquid at 1 atm is about two-thirds of the critical temperature for the substance when each temperature is measured on the absolute scale.

1-8
INTERMOLECULAR FORCES

Atoms and molecules, as well as electrically charged species such as ions, are drawn together by a variety of nonchemical forces. By this term we means forces not directly associated with the presence of a covalent bond between atoms. Such forces are responsible for some of the deviations of real gases from ideal behavior, which have been described in earlier sections, and for the condensation of gases to liquids. Because of their relation to the behavior of nonideal gases, forces between neutral atoms or molecules are often called *van der Waals forces.* Although they may be between electrically neutral particles, physical or van der Waals forces are essentially electrical in origin. Despite the fact that an atom or molecule when viewed from a distance is electrically neutral, an imaginary observer close to the atom or molecule might find himself at any instant nearer the charge of one sign than to the charge of the other sign and thus feel predominantly the effect of the nearer charge.

Let us first examine the law of forces between two electric charges. Coulomb's equation, which describes the force between charges q_1 and q_2, is

$$F = \frac{q_1 q_2}{Dr^2} \tag{1-30}$$

where r is the distance between the two charges and D is the *dielectric constant,* a property of the medium separating the charges. For a vacuum, D is unity and, for a gaseous medium such as air, it is so slightly larger than unity that the difference can be neglected. In using this equation, the charges q_1 and q_2 carry their respective positive or negative signs in order that the force change sign, corresponding to a reversal in its direction, depending upon whether it is a force of repulsion between like charges or of attraction between unlike charges.

When a force of attraction or repulsion exists between two particles, there is necessarily a change in energy when the distance between the particles is altered. For example, if two particles bearing opposite electric charges are allowed to approach one another, they can in this process do work on the surroundings, and in the course of doing work they lose energy. Furthermore, if one now wishes to pull them apart, work must be done against the force of attraction that tends to hold them together.

In describing the energy of a system of neighboring charged particles, it is customary to take as the zero point the energy the group of particles would have if they were so far apart as to be completely independent of one another. If the particles bear electric charges of the same sign, their approach to one another corresponds to a gain in energy, and the energy of the less stable situation in which they are near one another is given a positive sign. If approach is facilitated by attractive forces between unlike charges, the energy of the combined system is less than the sum of the energies of the individual components, and the energy of the group of neighboring particles is said to be negative.

For a system of two ions in which the interaction is given by Coulomb's equation, the work required to pull apart the particles from distance r to an infinitely large distance is

$$w = -\frac{q_1 q_2}{Dr} \tag{1-31}$$

The energy of the system when the particles are at distance r apart, compared to the energy at infinite separation, is therefore

$$E = \frac{q_1 q_2}{Dr} \tag{1-32}$$

The large magnitude of the Coulomb attractive force between opposite charges is the principal factor that stabilizes the solid phase of a salt such as sodium chloride. In the sodium chloride crystal, each sodium ion is surrounded by six chloride ions and each chloride ion by six sodium ions, and the energy of each positive–negative interaction between a pair of nearest neighbors is substantially unaffected in magnitude by the presence of the other interactions. Each ion has more-distant neighbors of the same sign as the ion itself but, because the Coulomb force falls off with distance, the influence of these is outweighed by the attractive forces between nearest neighbors.

Although a molecule is neutral and has no net charge, it can be polarized; that is, there can be a separation of charge within the molecule. In this situation, the center of gravity of the negative charges carried by the electrons in a molecule is not at the location of the center of gravity of the positive charges on the atomic nuclei of the molecule. Polarization may exist permanently in a molecule or it can be induced by the approach of another particle which itself has either a charge or a dipole, and thus be only temporary.

The magnitude of the permanent separation of charge in a molecule can be expressed by a quantity known as the *dipole moment;* this is defined as the product of the amount of charge separated and the distance of separation. Since the charge on an electron is of the order of 10^{-10} electrostatic units (esu) and the distance between atoms is of the order of 10^{-8} cm, the unit of dipole moments is conveniently taken as 10^{-18} esu cm. This unit is termed the *debye,* honoring Peter Debye, a twentieth-century physical chemist who made major contributions to our understanding of the behavior of matter and of its interaction with light and other electromagnetic radiation. In Table 1-3 are listed some values of permanent dipole moments for typical simple molecules. Also included in the table are values, measured for the pure liquid substances, of another property, the dielectric constant, which was defined implicitly by Equation (1-30) and which is a measure of the effective polarity of the collection of molecules in the liquid phase.

We now turn to a consideration of the attractive forces associated with the existence of permanent or induced dipoles. Although our present concern is not primarily directed to ionic solutions in such solvents as water, ammonia, or alcohols, it should first be pointed out that solute ions interact strongly with the dipoles in solvent molecules.

In the water molecule, there exists a permanent dipole because the oxygen atom is more electronegative than hydrogen, drawing the bonding electrons closer to itself, and also because the unshared electrons on the oxygen tend to be in the region in space away from the hydrogen

Table 1-3
Polar properties of some common compounds

Substance	Dipole moment (debyes)	Dielectric constant of liquid[a]
N-Methylformamide	—	182
Nitrobenzene	3.99	34.8
Dimethyl sulfoxide	—	48.9
Acetonitrile	3.37	37.5
Formamide	3.37	110
Hydrogen cyanide	2.9	107
Acetone	2.72	20.7
Chloroethane	1.98	6.3 (170°)
Hydrogen fluoride	1.91	84 (0°)
Ethylenediamine	1.90	14.2
Water	1.85	78.5
Bromobenzene	1.70	5.4
Ethanol	1.68	24.3
Methanol	1.66	32.6
Sulfur dioxide	1.60	14.1 (20°)
Ammonia	1.47	16.9
Hydrogen chloride	1.03	4.6
Chloroform	1.15	4.8
Diethyl ether	1.15	4.3 (20°)
Hydrogen sulfide	0.93	9.1 (−78.5°)
Hydrogen bromide	0.80	7.0 (−85°)
Toluene	0.39	2.4
Benzene	0	2.27
Carbon tetrachloride	0	2.23

[a]Temperature is 25°C unless otherwise indicated.

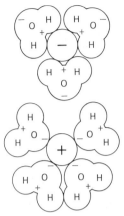

Figure 1-11
Solvation of ions by oriented water molecules.

atoms. When a positive ion, such as the sodium ion, is placed in water, it attracts to itself the negative ends of the water dipoles, whereas a negative ion attracts the positive ends of the dipoles. The resulting hydrated ions are diagramed in Figure 1-11. Maximum attractive force depends upon the presence of a particular orientation of the dipoles with respect to the ions; since thermal energy tends to destroy such a special arrangement, the effect of this type of ion–dipole force tends to decrease with increasing temperature.

In addition to orienting the permanent dipoles of the solvent, ions in solution also create or induce additional dipoles in the solvent molecules. The cation draws electrons toward itself, so that the induced dipole is in the same sense as the permanent dipole, while the anion repels electrons, and this too acts to reinforce the effect of the dipole already present in the solvent. Thus the induced dipole is always in the proper direction to contribute an attractive force. Since its direction is not fixed to the orientation of a rotating or diffusing molecule, its average energy of interaction with an ion is not dependent upon temperature. The magnitude of the induced dipole depends upon the ease with which the electron cloud of a molecule responds to the presence of an external electric field, a property called the *electric polarizability* and defined as the magnitude of the dipole produced by application of unit electric field to the molecule. The energy associated with either the ion–dipole or the ion–induced dipole force varies with the inverse fourth power of the distance between the ion and the dipole, so that these forces are somewhat shorter in range than ion–ion forces.

As we have already indicated, however, intermolecular electrostatic forces do not require the presence of ions. When two particles having permanent dipoles approach one another with appropriate orientation, either end to end or side to side, the attraction of unlike charges, which are nearer one another, predominates over the repulsion of like charges, which are farther apart:

$$
\begin{array}{cc}
A \cdots B \\
+ \qquad -
\end{array}
$$

$$
\begin{array}{cccc}
A \cdots B & A \cdots B \\
+ \quad - & + \quad -
\end{array}
$$

$$
\begin{array}{cc}
B \cdots A \\
- \qquad +
\end{array}
$$

These arrangements are preferred in a dipolar liquid, such as sulfur dioxide, but of course thermal agitation tends to disrupt them. To obtain the average interaction energy, then, it is necessary to average over all possible relative orientations of two dipolar molecules. In the absence of information about the distance between the centers of positive and negative charge in the dipole, the dipole is usually assumed to be located at a point. The equation that results for the interaction energy is

$$E = -\frac{2}{3kT}\frac{\mu_1^2\mu_2^2}{r_{12}^6} \tag{1-33}$$

where μ_1 and μ_2 are the two dipole moments, r_{12} is the distance between the dipoles, k is the Boltzmann constant, T is the absolute temperature, and D has been assumed to be unity.

In the same way that an ion can induce a dipole in a molecule—either a molecule that is nonpolar or one that already has some polarity—so too a permanent dipole can produce a temporary separation of charge in a nearby molecule. The direction of this induced dipole is again always that resulting in a net attractive interaction, and the result is consequently independent of temperature. As an example, Figure 1-12 shows two ways in which a water molecule can induce dipoles in a helium atom, an effect that contributes to the solubility of helium in water. The magnitude of the moment induced depends upon the polarizability of the molecule, symbolized by α, and so for a mixture of two polar substances 1 and 2 two temperature-independent terms are added to Equation (1-33) to give the total dipolar interaction energy:

$$E = -\frac{2}{3kT}\frac{\mu_1^2\mu_2^2}{r_{12}^6} - \frac{\mu_1^2\alpha_2}{r_{12}^6} - \frac{\mu_2^2\alpha_1}{r_{12}^6} \tag{1-34}$$

Dispersion force is the name given to yet another kind of molecular interaction, an attraction that exists between *any* two molecules, although both may be electrically neutral species without a permanent dipole. Because the electrons are in constant motion, chance may lead to a momentary unsymmetrical charge distribution. This in turn may affect the neighboring molecule, inducing an opposite electric dipole. Thus two molecules, such as two hydrogen molecules, which on an average over a period of time are nonpolar, may experience mutual attractive forces through synchronization of electronic motions. Such a mutual induction of dipoles is responsible for the fact that even the noble gases can be condensed if taken to sufficiently low temperature.

The energy of dispersion forces, which are sometimes called London forces in honor of the man who developed their theory, depends

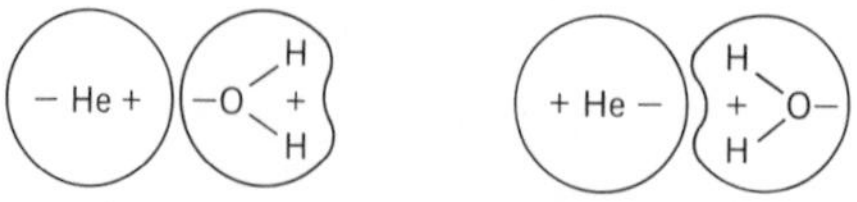

Figure 1-12
Attraction between dipole of water molecule and induced dipole in a helium atom.

upon the molecular polarizability and the inverse sixth power of the intermolecular separation:

$$E = - c \frac{\alpha_1 \alpha_2}{r_{12}{}^6} \tag{1-35}$$

The constant of proportionality c is approximately equal to $3I_1 I_2/2(I_1 + I_2)$, where I is the ionization potential of a molecule, the energy required to remove completely one electron from the molecule.

The term "dispersion force" arises because of the involvement of the molecular polarizability α, since this property of the molecules in a liquid also determines the extent to which the liquid can disperse into its various colored components a beam of light transmitted by the liquid. The phenomenon of optical dispersion will be discussed in Chapter 8.

Dispersion forces are relatively large between organic molecules containing double or triple bonds, and particularly between aromatic molecules. The tendency of nucleic acids to form a helical arrangement, although dictated in part by hydrogen-bonding interactions to be described later, is also strongly reinforced by the favorable energetics associated with tier-upon-tier stacking of the planar rings containing unsaturation electrons, the orbitals of which are directed outward from either side of the plane of the purine or the pyrimidine ring.

It is of interest to compare the magnitudes of the various contributions to the intermolecular attractions in some typical substances. In liquid carbon tetrachloride only dispersion forces can be present, but the total interaction energy is about 10 times as large as it is in hydrogen chloride aside from the effects of the hydrogen bond, which may also be present in HCl. The contributions at room temperature to the HCl interaction energy, omitting that of the hydrogen bond, are about 80 percent from dispersion, 15 percent from mutual orientation of dipoles, and less than 5 percent from the dipole-induced dipole effect. In methane where again there are only dispersion interactions, the energy of these is nearly as large as the total interaction energy in HCl. For hydrogen iodide the dispersion effect is more than four times greater than in HCl and constitutes practically the entire intermolecular interaction. For water, the dipole–dipole interaction is almost 10 times as strong as for HCl, but the dispersion effect is only about one-half as large.

Because the energy of both dispersion and dipole–dipole forces follows an inverse sixth-power dependence upon distance, these forces are much more short-range in spatial extent than Coulomb forces, and also of somewhat shorter range than ion–dipole forces. Upon very close approach of two molecules, the repulsions of electrons in one molecule for those in the other molecule become significantly large and must be explicitly considered. To describe this short-range repulsion, it is appropriate to use an inverse dependence of the energy upon a still higher power of the distance. An *empirical* relation often applied to the total molecular interaction is the Lennard-Jones "6-12" potential energy expression, where the first term represents the attractive interaction and the second term corresponds to the repulsive interaction:

$$E = -\frac{A}{r^6} + \frac{B}{r^{12}} \tag{1-36}$$

In concluding the discussion of van der Waals forces, it is worthwhile to point out that some solids are held together by such forces. Thus solid carbon dioxide and solid iodine are formed of molecules of the respective substances, packed in a regular arrangement but held together essentially by van der Waals forces. It is a characteristic of such solids that they sublime easily. Many organic crystals are also molecular solids stabilized by van der Waals forces. Particularly as the molecular weight becomes greater, the question of whether a substance exists as a liquid or solid depends as much upon aspects of molecular geometry that determine whether they can be readily packed together, and thus indirectly influence the magnitude of the attractive forces, as upon the specific nature of these forces. Furthermore, the molecular weight also plays an important part in determining the volatility of a material.

1-9
THE HYDROGEN BOND

In one situation, a molecular interaction arises primarily from dipole–dipole force, but the resulting bond differs from most of those described in Section 1-8 because it has a preferred spatial orientation and a higher degree of stability. This occurs when a hydrogen atom is attached to an electron-attracting atom, typically oxygen, fluorine, or nitrogen, so that the hydrogen is the positive end of an electric dipole and is then attracted to an atom at the negative end of another dipole. Because the hydrogen atom has only the bonding pair of electrons in its vicinity, there is little repulsion of the second atom by negative charge, and it can approach quite close to the hydrogen atom, so long as it stays opposite the electronegative atom to which the hydrogen is covalently bonded.

The energy associated with such a *hydrogen bond* may be as much as 10 kcal/mol, compared to 1 or 2 kcal/mol for a typical van der Waals interaction; it is, however, five to ten times less than that of an ordinary covalent interatomic bond.

The presence of hydrogen bonds is responsible for some unusual properties of liquids. For example, water has a boiling point about 142 K higher than the boiling point of hydrogen sulfide, a substance with a greater molecular weight, and hydrogen fluoride boils more than 100 K above hydrogen chloride. These liquids, as well as ammonia, are said to be associated—the individual molecules do not represent the true characteristic units of the liquid, but rather they are aggregated into clusters which have, however, only a very transitory existence. It appears that the small size of the elements of the first row—nitrogen, oxygen, fluorine—as well as their fairly strong tendency to attract negative charge, causes hydrogen bonding to be quite significant in determining the properties of their compounds.

A hydrogen bond is strongest when the hydrogen atom is collinear with the two electronegative atoms it links. However, the bond can be "bent" by perhaps 5 to 10° from linearity without loss of very much of its strength. Of course, it is only for the solid state, in which crystal structures can be determined, that we can obtain direct information about the geometry of the hydrogen bond. For systems of the type in which the hydrogen lies between two oxygen atoms, the oxygen–oxygen distance varies from 2.4 to 2.9 Å. A typical arrangement is

$$\underset{1.0\ \text{Å}}{\text{O}\text{---}\text{H}}\underset{1.7\ \text{Å}}{\text{-------O}}$$

where the shorter distance corresponds to an ordinary covalent bond and the longer distance corresponds to a hydrogen bond. In the lattice of ordinary ice, each oxygen atom is surrounded by four other oxygen atoms at the tetrahedral angle of 109° and at a distance of 2.76 Å; between the central oxygen and each of its neighbors is a hydrogen atom. Two of the hydrogens lie close to the central oxygen, forming with it a covalently bonded water molecule, and the other two hydrogens are closer to other oxygen atoms, forming part of their water molecules. Within the unit constituting a water molecule, the H—O—H bond angle is apparently about 105°, about the same as it is in the vapor phase, so that in ice the hydrogen bonds are slightly bent. The arrangement is indicated schematically in Figure 1-13, and the three-dimensional structure of ice is discussed further in Section 1-13.

The molecules of alcohols form, in the liquid phase, loose aggregates of fairly well-defined character as a result of intermolecular hydrogen bonding:

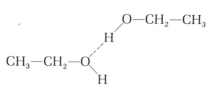

The size of the aggregates, as well as the fraction of the total number of

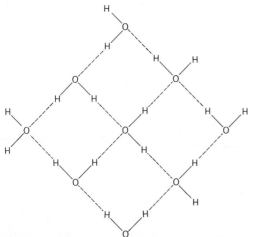

Figure 1-13
Schematic arrangement of hydrogen bonds between water molecules.

alcohol molecules involved in them depends upon the nature of the alcohol and the temperature. Other molecules that can act as both donors and acceptors of hydrogens include water, carboxylic acids, and primary and secondary amines. Chloroform, $CHCl_3$, can act as a hydrogen donor for, although the hydrogen is attached to a carbon atom, the carbon atom is much depleted of electron density by the three chlorine atoms. Typical hydrogen acceptor molecules are ketones, aldehydes, ethers, and tertiary amines. Intramolecular hydrogen bonds may occur when the geometry of a molecule permits a donor group and an acceptor group to approach one another sufficiently closely. An example is maleic acid, in which x-ray diffraction studies of the crystal indicate that oxygen atoms from the two carboxyl groups are quite close together, as would be consistent with the following structure:

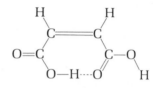

Several direct spectroscopic methods can be used to detect the presence of hydrogen bonding, and these will be discussed in Chapters 9 and 13. In the remainder of this chapter and in some subsequent sections, indirect effects of hydrogen bonding on liquid properties will be encountered. For discussing physical properties of liquids, it is often convenient to make a rough qualitative classification into those liquids that are *normal,* including those such as benzene, carbon tetrachloride, and paraffin hydrocarbons that have no permanent electric dipoles or only very small ones, and those that are *associated,* like alcohol and water, into definite if transient polymeric aggregates. The reader should be alert for examples of the consequences of hydrogen bonding on such properties as viscosity, volatility, and solubility.

Hydrogen bonds play a very important role in the structure and function of the components of living cells. Not only is the nature of water of critical significance in determining the nature of life processes, but also the structures of such macromolecules as proteins are governed by the existence of hydrogen bonds between neighboring amino acids. Likewise, the transfer of genetic information through nucleic acids is effected by a process in which matching, or complementary, base pairs are fitted together according to a pattern dictated by their ability to form hydrogen bonds to one another.

1-10
VAPOR PRESSURE

A liquid placed in a closed container having a volume larger than the liquid fills the free space with its vapor, and eventually the liquid comes into equilibrium with its vapor. This is not to say that the transfer of molecules from the liquid surface to the vapor space ceases, but rather that the molecular concentration in the vapor becomes great enough so that the rate of condensation of vapor molecules is equal to the rate of volatilization of liquid molecules. The pressure of the vapor when equilibrium has thus been attained is a function of the liquid substance and of the temperature and is termed the *vapor pressure* of the liquid.

Vapor pressure is a measure of the volatility of a liquid, of the ease with which it can be converted into a gas. It measures the *escaping tendency* of molecules from the liquid phase, and it is thus an inverse indication of the forces between molecules in the liquid.

The vapor pressure of a liquid is the one pressure for a given temperature at which liquid and gas can coexist, and on the phase diagram, like that of water discussed earlier, the curve that divides the gas region from the liquid region is exactly the vapor pressure–temperature curve for the liquid.

As the temperature increases, the tendency of molecules to fly off from the liquid as a vapor increases and the vapor pressure increases. The *boiling point* of a liquid is the temperature at which visible evolution of bubbles of vapor occurs at a particular value of the external pressure. The *normal* boiling point, which is commonly cited as the boiling point of a liquid, is the temperature at which ebullition occurs under an external pressure of 1 atm.

The mathematical form of the equation relating vapor pressure p and absolute temperature T is the same for liquids generally:

$$\ln p = -\frac{C_1}{T} + C_2 \tag{1-37}$$

In this equation, the symbol "ln" represents the *natural* logarithm, to the base e. The scale of logarithms often employed in numerical calculations is that in which the base is 10, and such logarithms are represented in this book by the symbol "log." A natural logarithm is always equal to 2.303 times the logarithm of the same number to the base 10, so that Equation (1-37) can be written as

$$2.303 \log p = -\frac{C_1}{T} + C_2 \tag{1-38}$$

Many equations describing natural phenomena are more simply written in terms of natural logarithms; both types of logarithms are used from time to time in this book.

It is found that both C_1 and C_2 vary from liquid to liquid, but that C_1 is related to the amount of energy required to vaporize a mole of liquid, the heat of vaporization ΔH_{vap}, by the simple relation

$$C_1 = \frac{\Delta H_{vap}}{R} \tag{1-39}$$

The heat of vaporization can also be regarded as the excess energy a mole of substance has as a vapor over that which it has as a liquid at the same temperature. The symbol Δ, delta, is frequently used, as it is here, to indicate a change or difference in some property associated with a physical or chemical change. ΔH in general is the amount of energy H the products of the change possess less the amount possessed by the reactants, and in this case the change is the process of vaporization. The quantity R is the gas constant, and it must be expressed of course in the same units of energy and for the same amount of material as the heat of vaporization.

Substitution of the value of C_1 in Equation (1-37) leads to

$$\ln p = -\frac{\Delta H_{vap}}{RT} + C_2 \qquad (1\text{-}40)$$

This equation can be applied to a given liquid at two temperatures, T_1 and T_2, at which the vapor pressures are p_1 and p_2, respectively:

$$\ln\frac{p_2}{p_1} = -\frac{\Delta H_{vap}}{R}\left(\frac{1}{T_2} - \frac{1}{T_1}\right) = \frac{\Delta H_{vap}}{R}\frac{T_2 - T_1}{T_2 T_1} \qquad (1\text{-}41)$$

This is the integrated form of what is termed the *Clausius–Clapeyron* equation, useful for predicting the vapor pressure at any temperature not too far removed from a temperature for which the vapor pressure is known, given also the heat of vaporization, or for calculating the heat of vaporization from any two vapor pressures measured at different temperatures. A differential form of the equation can be obtained by taking the derivative of each side of Equation (1-40) with respect to temperature:

$$\frac{d \ln p}{dT} = \frac{\Delta H_{vap}}{RT^2} \qquad (1\text{-}42)$$

Example: The heat of vaporization of water is 540 cal/g. Predict the vapor pressure in torr at 95.0°C.

Solution: The molar heat of vaporization is 540 cal/g multiplied by the molecular weight, 18.02 g/mol. The vapor pressure at the boiling point, 100°C, is 760 torr. Then

$$\log\frac{760}{p_{95}} = \frac{(540)(18.02)}{(2.303)(1.987)}\frac{(100.0 - 95.0)}{(373.2)(368.2)} = 0.0774$$

$$p_{95} = (760)/(1.195) = 636 \text{ torr}$$

The value obtained by experiment is 634 torr.

In Figure 1-14 are shown vapor pressure plots for several typical liquids. If Equation (1-40) is obeyed, the plot of log p against the reciprocal of the absolute temperature is a straight line with a slope of $-\Delta H_{vap}/2.303R$. This is evident upon comparison of the equation with the type equation of a straight line: $y = ax + b$, where a is the slope of the line or rise per unit distance on the scale of the abscissa, and b is the intercept or value of the ordinate y when x is equal to zero. The advantages of a straight-line plot are twofold: The graph provides a rapid means of taking into account the results of a series of measurements at different temperatures when the heat of vaporization is desired, and a straight line permits much more accurate interpolation or extrapolation of data to obtain values for temperatures at which measurements

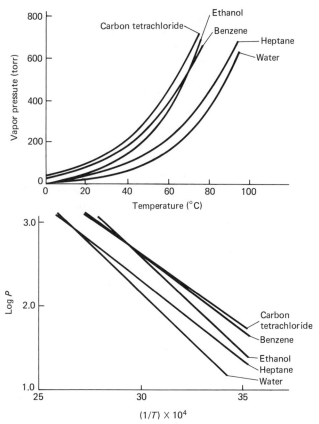

Figure 1-14
Vapor pressures of some common liquids.

have not been made than does a curved line.

Several different methods of vapor pressure measurement can be employed. In one method, the sample is placed in a vessel and subjected to a fixed external pressure, and the temperature of the vessel is raised until boiling is observed. As evaporation proceeds, the liquid is cooled because energy equivalent to the heat of vaporization is absorbed. Care must be exercised to strike a balance between the rate at which heat is supplied and the rate at which it is consumed by evaporation, so that a true equilibrium temperature is measured and superheating is avoided. In yet another method, a stream of inert gas is bubbled through the volatile liquid at a rate sufficiently slow so that the gas is saturated with vapor of the liquid. The partial pressure of the vapor in the gas mixture leaving the bubbler or saturator is then equal to the vapor pressure of the liquid.

Almost 100 years ago, Trouton pointed out the existence of some interesting regularities in the heats of vaporization of liquids. For many compounds, the ratio of the energy required to vaporize 1 mol of the liquid to the absolute temperature at which the liquid boils under 1 atm has a numerical value of about 20 to 21. A number of values of this ratio are shown in Table 1-4.

Deviations from Trouton's rule are seen to occur in both directions. A larger value of the heat of vaporization than expected occurs for associated liquids, such as water and alcohols. This may be interpreted as reflecting the additional amount of energy required to separate the molecules from the aggregates beyond that required to move the molecules apart against other van der Waals forces and to push back the atmosphere. A few values smaller than normal for Trouton's constant are also found, an example being that for acetic acid. For this substance, other experimental evidence indicates that the vapor exists mostly in the form of hydrogen-bonded dimers, as does the liquid:

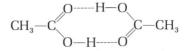

Thus, since dimers are being vaporized, Trouton's constant can better be calculated on the basis of a molecular weight higher than the formula weight for CH_3COOH, with the result that the proper heat of vaporization per mole of particles would be higher than that listed in Table 1-4. Somewhat similar in behavior is hydrogen fluoride, which in the vapor phase gives evidence of existing in polymers containing six hydrogen fluoride "molecules."

Solids also have well-defined vapor pressures, although they are generally smaller than those of liquids. Indeed, the vapor pressures of the solid phase of a given substance and the liquid phase of that substance must be identical at the melting point of the solid. The effect of temperature upon vapor pressure of a solid is similar to that upon vapor pressure of a liquid, and expressions such as Equations (1-40)

Table 1-4
Heat of vaporization and Trouton's constant for various liquids

Substance	Normal boiling point (K)	Heat of vaporization (kcal/mol)	Ratio $\Delta H_{vap}/T_b$
Ethanol	351.4	9.30	26.5
Water	373.2	9.71	26.0
Methanol	337.8	8.44	25.0
Ammonia	239.8	5.57	23.2
Sulfur dioxide	263.2	5.94	22.6
Mercury	630.4	13.61	21.6
Hydrogen sulfide	212.8	4.46	21.0
Benzene	353.3	7.36	20.8
Carbon tetrachloride	349.8	7.20	20.6
Cyclohexane	353.9	7.19	20.3
Chlorine	239.5	4.87	20.3
n-Hexane	341.9	6.90	20.2
Sodium	1172	23.68	20.2
Oxygen	90.2	1.630	18.1
Hydrogen fluoride	377	6.03	16.0
Acetic acid	390.9	5.83	14.9

to (1-42) apply, with substitution of the *heat of sublimation* for the heat of vaporization.

1-11
SURFACE TENSION

A droplet of liquid, such as a raindrop or a globule of mercury, tends to assume a spherical shape; this is a manifestation of the general tendency of a liquid to reach a condition in which *the surface area has the smallest possible value for a given volume of material.* The molecules in the interior of a condensed phase—solid or liquid—are attracted equally in all directions by the surrounding molecules, but those at the surface are subjected to an unbalanced force of attraction directed inward, for there are relatively few molecules above the surface to counteract the effect of those below the surface.

The inclination to minimize the surface area leads to a result equivalent to that which would be evident if the surface were under tension like a stretched membrane: a pull parallel to a plane tangent to the surface. This is the *surface tension,* defined quantitatively as the force in dynes perpendicular to a line 1 cm long in the surface.

The surface tension describes one aspect of the nature of surfaces, or interfaces, between phases. These interfaces are of great practical importance, for they contribute substantially to the structure and function of portions of living cells, and they represent the key to many catalytic processes, as we shall see in Chapters 11 and 12.

Imagine that a spherical droplet of liquid is distorted to a flattened shape so that a larger area of surface is established. *Work must be done in order to form the new surface,* and this work is equal to the force exerted on the surface multiplied by the distance through which the force acts. If the force is applied perpendicularly to an imaginary line of length l centimeters in the surface and pulls the line a distance x centimeters, a new area of extent lx is produced. The force necessary is γl, where γ is the surface tension, and the distance through which this force moves is x, so that the work done is γlx. The work required per unit area of new surface is thus γ ergs/cm². The surface tension is thus numerically equal to the surface energy per unit area of surface— the work supplied to produce the surface remains associated with that surface as *surface energy.*

It is possible to turn the spherical droplet of liquid "inside-out" and to imagine a bubble of gas at the lower end of a tube dipping into the liquid, as in Figure 1-15. The problem is now to relate the surface tension to the amount of extra pressure that must be exerted by the gas in the tube to maintain the bubble, aside from that required because the bubble is some distance beneath the surface. To do this, one can visualize a plane passing horizontally through the center of the spherical bubble and intersecting the gas–liquid surface in a circular line which serves as a hypothetical boundary between the two hemispheres of liquid surface. The surfaces above and below this line are exerting across it a contractile force equal to the length of the line multiplied by the surface tension, a force equal to $2\pi r\gamma$, where r is the radius of the sphere and therefore also of the circle. To keep the bubble from

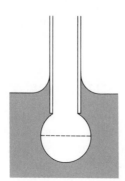

Figure 1-15
A bubble of gas immersed in a liquid at the end of a tube. The dashed line represents the intersection of a horizontal plane with the gas–liquid interface, as described in the text.

collapsing, there is exerted through the plane a force equal to the pressure of the gas multiplied by the area within the circle, or $P\pi r^2$. In a condition of equilibrium, which means that the size of the bubble is constant, these two forces are equal to one another:

$$2\pi r\gamma = P\pi r^2 \tag{1-43}$$

From this we see that P is equal to $2\gamma/r$, or

$$\gamma = \frac{Pr}{2} \tag{1-44}$$

Another result of the existence of surface tension is the rise of liquid in a small tube made of material that is wetted by the liquid. If the inside of the tube were wetted by the liquid but no *capillary rise* occurred, the surface of the liquid exposed to the air would then be larger than it is when the liquid is drawn up into the tube. The situations to be compared are shown in Figure 1-16. The effect of the liquid on the capillary wall upon the column of liquid rising in the capillary can be considered as localized in a circular line located where the meniscus meets the wall. The force that can be transmitted across unit length of this line is equal to the surface tension, the length of the line is $2\pi r$, where r is the radius of the tube, and the total lifting force is then $2\pi\gamma r$. The weight of—that is, the force of gravity pulling downward on—the column of liquid in the capillary is equal to its mass, which is the volume multiplied by the density of the liquid ρ times g, the acceleration due to gravity, or $\pi r^2 h\rho g$. Equating these two forces to satisfy the condition of equilibrium yields

$$2\pi\gamma r = \pi r^2 h\rho g$$

which rearranges to give an expression for the capillary rise h:

$$h = \frac{2\gamma}{\rho g r} \tag{1-45}$$

The measurement of capillary rise affords a means of evaluating the surface tension of a liquid. Care must be taken that the capillary employed is quite clean, so that it is thoroughly wetted. For absolute results, the dimensions of the capillary must be determined, but usually

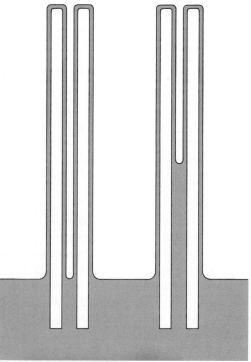

Figure 1-16
Capillary rise of a liquid, illustrating the
reduction of amount of liquid–gas surface.
The liquid is shown as wetting the entire
surface of the capillary tube.

values of the surface tension relative to a known value for a liquid
used for calibration of the tube are satisfactory. For measurements
in any one capillary of two liquids, 1 and 2, the simple proportion can
be applied:

$$\frac{\gamma_1}{\gamma_2} = \frac{h_1\rho_1}{h_2\rho_2} \tag{1-46}$$

When very accurate measurements are to be made, variation in radius of the capillary from one point to another becomes significant,
and the liquid level outside the tube must be adjusted so that in a series
of relative measurements the meniscus inside the tube is always at the
same place in relation to the tube.

The outer edge of the curved meniscus of a liquid in a tube may not
be precisely vertical. Another way of describing this situation is in
terms of the angle of contact θ between the liquid and the solid that
forms the tube wall, an angle represented in Figure 1-17. If the contact
angle in a capillary tube is not exactly zero, not all the force of the surface tension, but only its vertical component, is effective in lifting the

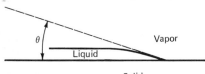

Figure 1-17
Diagram of the contact angle between a
liquid and a solid. If the liquid wets the
solid fully, as water does glass, $\theta = 0$.
If the liquid does not wet the solid at all,
as mercury on glass, $\theta = 180°$.

column of liquid, and it is necessary to multiply the maximum upward force by $\cos \theta$:

$$h = \frac{2\gamma \cos \theta}{\rho g r} \qquad (1\text{-}47)$$

Liquid mercury does not wet the surface of clean glass; and when a glass capillary is dipped into mercury, the liquid level inside the tube falls rather than rises. The mercury–glass interface is to be considered free surface and tends to become smaller. Alternatively one can explain this phenomenon by assigning a value of 180° to the contact angle θ; the cosine of this angle is -1, so that the value of h calculated from Equation (1-47) is negative.

Another method of measuring surface tension is by determining the force necessary to pull a platinum ring up through the surface of the liquid. This is conveniently done with the *du Nouy tensiometer,* which is commercially available. In Figure 1-18 is shown the shape of the liquid surface before the ring is detached from it. The wire ring is held by one film of liquid on the inside of the circle and by a second film on the outside. The average length of the two films is very nearly equal to the circumference of the ring, and the force that can be applied to the ring–liquid interface is thus equal to the product of the surface tension and twice the circumference. The tensiometer method is quite rapid and, with the aid of corrections which have been determined empirically to take into account the radius of the wire of which the ring is formed and the shape of the liquid lifted from the surface, accurate results can be obtained.

Surface tension decreases with increasing temperature and approaches zero in the vicinity of the critical temperature, at which the interface between the liquid and gas phases vanishes. The rate of change in surface tension of a liquid with temperature is related to the structure of the liquid, affording an indication of the existence, in polar or associated liquids such as water, of polymolecular aggregates which dissociate with an increase in temperature.

Table 1-5 includes surface tensions for several representative simple liquids. In Chapter 11, we shall encounter a variety of more complex

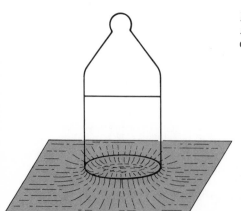

Figure 1-18
A du Nouy tensiometer ring lifting a quantity of liquid above the surface of the bulk liquid.

Table 1-5
Surface tensions for various liquids at 20°C

Liquid	Surface tension (dyn/cm)	Liquid	Surface tension (dyn/cm)
Water	72.7	Ethyl acetate	23.8
m-Cresol	37.5	Acetone	23.5
Benzene	28.9	Ethanol	22.5
Acetic acid	27.6	n-Hexane	18.4
Chloroform	27.2	Ethyl ether	17.0
Carbon tetrachloride	26.8		

systems, including many related to structures found in living organisms, in which surface tension plays an important role.

**1-12
VISCOSITY**

The viscosity of a fluid is a measure of the resistance of the fluid to flow. In the process of flow, the molecules comprising the fluid move past one another, and viscosity arises from what can be termed the "frictional" effects of relative motion. Both liquids and gases are viscous, but the molecules of a gas are much more nearly independent of one another, so that the viscosity of a gas is of much smaller magnitude than the viscosity of a liquid.

Directed flow necessarily occurs inside a tube, conduit, or channel of some sort. In order to maintain a given flow rate through a particular tube, a certain driving force must be applied at one end of the tube, and there is a corresponding drop in pressure along the tube from inlet end to exit end. In visualizing flow in a tube, it is important to remember that the molecules adjacent to the walls are practically stationary as compared to molecules in the interior, which may be moving at much greater velocities.

We shall discuss here flow through a circular tube, restricting our consideration to a flow pattern called *streamline* or *viscous* or *laminar*. The flowing fluid is viewed as being composed of concentric cylindrical layers, with all the molecules in any one layer moving with the same velocity, and with each molecule following a course parallel to the tube axis, without lateral motion. Each successive layer from the wall inward toward the axis is found to be moving faster than the one outside it, so that the velocity increases to a maximum at the center of the tube. Streamline flow is found in a given fluid–tube combination when the velocity is not too great; at higher velocities, the flow becomes *turbulent,* eddies are present, and the path of a given molecule is an irregular spiral.

To describe flow quantitatively, we imagine two small planar surfaces, each of area A, located within a flowing fluid and parallel to one another and to the axis of flow, each of the planes representing a surface of constant velocity. In order to maintain the velocity of flow at one plane greater than that at the second by an amount u, a force F must be exerted. The force is proportional to the area of the planes and to the difference in velocity between them, and inversely proportional to the distance r between the planes. The constant of proportionality depends upon the nature of the fluid and is designated as the coefficient of viscosity η of the fluid. These relations are expressed in the equation

$$F = \frac{\eta u A}{r} \tag{1-48}$$

For many simple liquids and almost all gases, the coefficient of viscosity is found to have a constant numerical value at a specified temperature, regardless of the stress applied to the fluid to cause it to flow. Such fluids are called *Newtonian*. For some systems containing molecules which are very large or are quite far from spherical in shape, the liquid structure is changed by the application of the force, and so the coefficient of viscosity varies with the rate of flow.

The unit of viscosity, the *poise* (P), is defined by Equation (1-48) with all quantities in cgs units, so that a liquid of unit viscosity requires a force of 1 dyn to cause two parallel planes of 1 cm² area 1 cm apart in the liquid to slide past one another at a relative velocity of 1 cm/sec. Most liquids have viscosities that are only a small fraction of a poise, so that viscosity coefficients are frequently given as multiples of a *centipoise* (cP), one one-hundredth of a poise, or of a *millipoise* (mP), one one-thousandth of a poise. The viscosity of water at room temperature is approximately 1 cP. In the SI system, the unit of viscosity is kg/(m sec), but this has not been given a special name.

A mathematical treatment of streamline flow in a cylindrical tube was carried out by J. L. M. Poiseuille, who showed that the volume V of liquid of viscosity coefficient η flowing through a tube in time t seconds is given by

$$V = \frac{\pi r^4 P}{8 \eta l} t \tag{1-49}$$

Here r is the radius of the tube, l is the length of the tube, and P is the driving pressure. Poiseuille's equation can be utilized to calculate the viscosity of a liquid from measurements of its flow rate through a capillary tube. However, it is difficult to measure accurately the dimensions of the tube, and so recourse is usually had to comparative measurements of several liquids in the same tube, using the viscosity of one of the liquids to establish the apparatus constant of the tube. With a device such as the Cannon–Fenske modification of the *Ostwald viscometer,* shown in Figure 1-19, the times for equal volumes of the unknown liquid and of the known liquid to pass through the same capillary can be measured. The capillary tube is placed in a nearly vertical position and is provided with reservoir bulbs which hold the liquid before and after passage through the tube. The liquid in the viscometer is drawn up until one of its levels is in the upper bulb. The liquid is then released, and the time interval recorded is that required for passage of the liquid meniscus from graduation *a* to graduation *b*, above and below the bulb, respectively. As the level of liquid falls in the upper bulb, the level in the lower bulb rises.

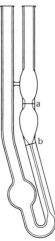

Figure 1-19
Cannon–Fenske modification of the Ostwald viscometer.

The driving pressure forcing a liquid through an Ostwald-type capillary is the product of the density of the liquid and the difference in liquid levels in the two arms of the U-tube. As a consequence, the volume of liquid introduced into the viscometer must be equal for each of a series of relative measurements and should be chosen so that the variation in the liquid head during the period of efflux is minimized by maintaining the lower liquid level in the widest part of the bulb. For the same capillary, all factors in the Poiseuille equation are identical except the viscosity, pressure, and time. If the liquid head difference is the same, the pressure is proportional to the liquid density. For two liquids, 1 and 2, the ratio of viscosity coefficients is

$$\frac{\eta_1}{\eta_2} = \frac{\rho_1 t_1}{\rho_2 t_2} \tag{1-50}$$

Several errors may enter into a viscosity determination by the Ostwald method; probably the most important is the kinetic energy error. If the linear velocity of the liquid flow is great, some of the potential energy of the liquid head goes into acceleration of the liquid and not all is available for overcoming frictional resistance. The time of flow is then greater than that calculated from the Poiseuille equation.

In Table 1-6 are given the viscosities of several common liquids at various temperatures. When liquids of the same type are compared,

Table 1-6
Viscosity coefficients of liquids (in centipoises)

Substance	20°C	30°C	40°C	50°C	60°C
Water	1.009	0.800	0.654	0.549	0.470
Carbon tetrachloride	0.969	0.843	0.739	0.651	0.585
Methyl alcohol	0.593	0.515	0.449	0.395	0.349
Ethyl alcohol	1.200	1.003	0.831	0.701	0.591
Benzene	0.647	0.561	0.492	0.436	0.389
Aniline	4.40	3.15	2.37	1.85	1.51
Mercury	1.55	1.50	1.45	1.41	1.37
n-Octane	0.542	0.483	0.433	—	0.355

it is seen that the viscosity coefficient increases with increasing molecular size. It is noted also that associated liquids have, in general, viscosity coefficients larger than those of normal liquids.

For any liquid, the viscosity is found to diminish with increasing temperature. Often the temperature dependence is discussed in terms of the fluidity ϕ, which is defined to be $1/\eta$, the reciprocal of the viscosity, and the variation in the fluidity with temperature is given by an equation very similar in mathematical form to that describing the variation in vapor pressure with temperature:

$$\ln \phi = -\frac{\Delta E}{RT} + \ln A \qquad (1\text{-}51)$$

where ΔE and A vary in magnitude from one liquid to another. An equation equivalent to Equation (1-51) can be obtained by taking the antilogarithm of each side:

$$\phi = Ae^{-\Delta E/RT} \qquad (1\text{-}52)$$

Inspection of these equations shows that the dimensions of ΔE must be those of energy; the physical significance of ΔE is that it is somehow related to the amount of energy required to move a molecule from its location in the liquid and so to produce flow. The effect of temperature is important for circulation in the human body: a 1°C rise in temperature of the body during fever causes a decrease of about 3 percent in blood viscosity and makes the circulation accordingly easier.

For liquid mixtures, the presence of significant attractive interactions between molecules of different components is often reflected in a viscosity larger than that expected on the basis of additivity, but the detailed interpretation of such results is fraught with difficulty because of uncertainty about the role played by the shapes of individual molecules. Measurements of the effect of macromolecules on the viscosity of a solvent have, however, been quite successfully applied to determination of the size and shape of the macromolecules.

1-13
STRUCTURE OF LIQUIDS

In discussing the behavior of liquids, we have emphasized the effect of forces of molecular interaction as reflected in such properties as vapor pressure, viscosity, and surface tension. Because the molecules in a liquid are always relatively close together, these forces are of much greater significance than in the gas phase. However, it is much more difficult to systematize or to predict quantitatively the properties of liquids, such as density, viscosity, and vapor pressure, than it is to handle the correlation of gaseous properties. For liquids, a more detailed model is required, which takes into account such characteristics of the molecules as their size, their shape, and the anisotropic nature of their attractive forces, and it has not yet been possible to work out a single scheme incorporating all these details.

One approximate description of the liquid phase was presented in Section 1-7, in which the continuity of behavior from the gaseous to the liquid state was emphasized. This model can be extended somewhat by postulating that in a liquid there are scattered about some interstices about the size of a single molecule. It is proposed that these *holes* move about through the mostly continuous liquid somewhat as molecules fly about through the mostly empty gas phase. A hole in a liquid moves by virtue of being filled by molecular diffusion, which in turn creates another hole adjacent to the original one. In this view, a liquid is somewhat like a gas "turned inside-out."

An alternative basis for a model of liquid structure is the molecular arrangement of the corresponding solid. The liquid resembles a solid in the intimate approach of molecules to one another, and the densities of the two phases are roughly equal under a variety of conditions. To understand the model of a liquid that describes it as a slightly disordered solid, we describe briefly and qualitatively some aspects of the nature of molecular order in a solid.

Solids are characterized, as mentioned earlier, by the assignment of each atom to a local site, or small cavity formed by its neighbors. Since its neighbors are fixed in location as well, it has a definite relationship to each of them. If the pattern of sites is ordered and regular and repeated many times throughout the solid, the solid is said to be a crystal. Other solids are amorphous, or without a definite recurring pattern, although the individual particles are localized; these solids are often referred to as glasses and regarded as supercooled liquids.

In some elemental solids, individual atoms are the basic units of which the solid is formed. In other solids, molecules are present in substantially the same form as they are in the vapor, as mentioned in Section 1-8 for carbon dioxide and iodine. Salts such as sodium chloride consist of positive and negative ions, held very tightly by electrostatic forces, but with no unit corresponding to a single "molecule" distinguishable in the crystal. There are also covalent crystals, in which the atoms are linked by shared electron pairs into extensive chains, sheets, or three-dimensional networks. As exemplified by the diamond, in which each carbon atom is linked to four other carbon atoms symmetrically placed around it, these are hard and high-melting, and their structure has little bearing on the problem of the nature of liquids. Rather it is molecular crystals of which the structure is relevant.

In Figure 1-1a is represented a two-dimensional section through a typical crystalline arrangement, such as can be found in an atomic or molecular solid. Each circle represents one of the atoms in an atomic solid, or one of the molecules in a molecular solid, and directional forces are so small that the component units are packed together as closely as geometry permits. It is seen that each unit in the pattern has six nearest neighbors, all at a single distance. There are six second-nearest neighbors, all equivalent and again all at a single distance. The six third-nearest neighbors are all equivalent and all at twice the distance of the nearest neighbors. The pattern can be followed on out to greater and greater distances and, in principle, one can characterize the nature of the whole array by specifying the number of neighbors at each successive distance. The nature of such arrangements in crystals is determined from results of x-ray diffraction measurements, briefly described in Chapter 12, which give direct information about the repeat distances as well as the directions associated with these distances. In the comparison of a liquid with a crystal, it is natural to ask what kind of diffraction results are obtained for the liquid. The answer is that the short-range or nearest-neighbor distances are about the same, but that *the longer-range order has been reduced* and the larger repeat distances, sharply distinguished from one another in the solid, tend to merge into a continuous sequence.

Both the hole model and the disordered solid model of liquids have deficiencies when attempts are made to apply them quantitatively, and the elucidation of liquid structure remains a challenge to physical chemistry.

A particularly significant and perplexing problem in liquid theory is how to describe the structure of water and how to account quantitatively for its properties. Of numerous models that have been proposed, no one is completely successful. In our discussion here it is only possible to indicate the salient features of a few of these models. A full evaluation of each model would require the understanding of a variety of the tools of thermodynamics and spectroscopy that will be encountered later in this book, as well as acquaintance with some advanced theoretical and mathematical techniques.

As a beginning for this discussion, we again turn to the solid phase corresponding to the liquid in which we are interested. Ordinary ice—the form usually encountered in human experience—has a density lower than water at the same temperature as a result of the rather open structure of the solid, dictated by the geometry of the bonds, described in Section 1-9. A diagram of the resulting three-dimensional structure is given in Figure 1-20; examination of this shows characteristic six-membered rings of oxygen atoms, with a hydrogen atom located between each pair of oxygen atoms. One view of liquid water is that it is derived from ice by small modifications, such as the bending of the hydrogen bond angles, perhaps accompanied by a change from the six-membered rings of the solid to five- or seven-membered rings.

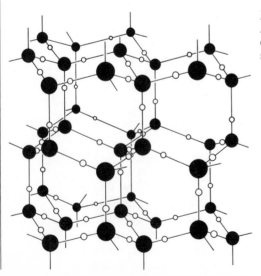

Figure 1-20
A portion of the lattice of an ice crystal. The open circles represent hydrogen atoms, the solid circles, oxygen atoms.

Another sort of theory of water structure is based upon a *mixture* model. Perhaps the earliest model of this type envisaged regions having a structure much like ice—they have been called icebergs—floating in a disordered liquid having relatively few hydrogen bonds. This concept was modified by H. S. Frank and W.-Y. Wen, who took into account the impossibility of the continuous existence of any such icelike regions, since physical measurements give no direct indication of them, and pictured instead "flickering clusters"—regions in which the water molecules are joined by hydrogen bonds, but which are formed suddenly, have a very short lifetime, and disappear suddenly. The formation of a cluster is considered to be a *cooperative* process in which a great many hydrogen atoms move very quickly into place to form hydrogen bonds. It is proposed that the cooperative effect results from the fact that, when a hydrogen atom enters into a hydrogen bond, the oxygen to which it is covalently bonded becomes even more negative than it otherwise would be and thus attracts another hydrogen atom to form a hydrogen bond. Because of a "zipper" effect running through the liquid, the probability of formation of hydrogen bonds in sequence is thus much greater than the probability of random formation.

Another version of the mixture model was developed by G. Nemethy and H. A. Scheraga. Emphasis is on the equilibrium number of hydrogen bonds that exist in the liquid, a number related to cluster size, for water molecules on the outside of a cluster have but two or three rather than four hydrogen bonds. Increasing temperature decreases somewhat the number of hydrogen bonds and the size of the cluster. It is calculated that at 20°C there are about 57 molecules to the cluster on the average, with 46 percent of the total possible hydrogen bonds present.

An important aspect of the structure of water is the effect of solutes on the arrangements of the water molecules. In general, nonpolar solutes tend to increase the amount of order of water and are termed *structure formers,* but ionic materials reduce the amount of order and are termed *structure breakers.* We shall return to the discussion of aqueous solutions in later chapters.

EXERCISES

1. A mixture of gases containing 24.0 g of CO_2, 15.5 g of N_2, and 100 g of O_2 is placed in a container of 50 liters volume at a temperature of 23°C. Calculate the mole fraction, partial volume, and partial pressure of each of the three components of the mixture.

2. Estimate the rms velocity of ethane molecules at 100 K and at 300 K.

3. The surface tension of a liquid is measured in a du Nouy tensiometer, and a force of 125 dyn is found to be required to pull the ring from the surface. In a calibration run, using the same ring, a force of 520 dyn is required to pull the ring from water at 25°C. Calculate the surface tension of the liquid.

4. Find in the literature a phase diagram for the element sulfur and describe the phase changes that occur as the pressure is varied at a constant temperature of 120°C, and those occurring as the temperature is varied at a constant pressure of 1 atm, assuming that equilibrium is maintained. By inspection of the diagram, determine the maximum number of phases of sulfur that can coexist in equilibrium with one another.

5. A sample of 0.2281 g of a pure organic liquid is converted to vapor and found to occupy a volume of 28.96 ml at 28.0°C and a pressure of 755 torr. What is the molecular weight of the liquid?

6. Calculate the volume occupied by 25 g of ammonia at 350 K and 100 torr pressure, assuming the gas to be ideal.

7. What mass in kilograms could be lifted by a 20,000-liter balloon filled with helium at 1 atm pressure and 27°C? How many average-weight students could ride in the balloon?

8. Calculate the total molecular kinetic energy, in joules and in calories, in 56.0 g of ethylene gas at 15°C.

9. Determine the numerical value of the gas constant to be used when the pressure is expressed in torr and the volume is expressed in cubic meters.

10. Show how the virial coefficient B in Equation (1-25) can be evaluated in terms of the van der Waals constants.

11. A hydrocarbon is found by analysis to contain 91.30 percent C and 8.70 percent H. A sample of 0.460 g vaporizes to yield 227 cm³ of gas at 200°C and 650 torr. What is the empirical formula of the compound?

12. Calculate the numerical ratio between the most probable velocity, the mean velocity, and the rms velocity of the molecules in a gas sample.

13. Hydrogen molecules are found to leak through a pinhole in the shell of a space capsule at the rate of 0.22×10^{-6} mol/hr. Under the same conditions, how rapidly will molecules of water vapor escape through the same hole?

14. Predict the position of the liquid meniscus within a clean glass tube of 0.50 mm inside diameter dipping into a mercury surface at room temperature.

15. From the data on vapor pressure of ice at −10°C and at the triple point given on pages 4 and 5, estimate the amount of energy required to sublime 1 g of ice to vapor.

16. Describe the sequence of physical changes that occurs as water vapor at −0.10°C and 1.0 torr is subjected to successively higher pressures up to 20 atm.

17. In the van Slyke method of determination of amino nitrogen which is alpha to a —COO— unit in proteins and peptides, nitrous acid is reacted with the sample, generating a molecule of gaseous nitrogen for each such nitrogen atom. A 1.04-g sample of peptide evolves an amount of nitrogen gas which occupies 45.2 cm³ at 25°C and 750 torr. Find the weight percent of α-amino nitrogen in the sample.

18. Calculate the density in grams per liter of diethyl ether vapor at a pressure of 700 torr and a temperature of 37°C.

19. A liquid of density 0.825 g/cm³ requires 132.2 sec to flow through a viscosity tube. Water requires 78.3 sec at 25°C to flow through the same tube. Calculate the viscosity of the liquid.

20. A sample of nitrogen gas is collected in a 50.0-ml gas buret at 25°C. The bottom of the buret is connected to a reservoir of liquid water to keep the gas from escaping, and the external surface of the water, exposed to 1 atm pressure of air, is at a level 68.0 cm below the meniscus in the buret which is exactly at the 50.0-ml mark. How many milligrams of nitrogen are there in the sample?

21. Using the constants in Table 1-1, calculate the pressures the van der Waals equation predicts for 1 mol of nitrogen at 0°C and at 100°C when the ideal pressure is 50.0 atm. Compare the results with the observations cited in Table 1-2.

22. Assume that the energy of the dispersion interaction between two molecules at a distance of 2.5 Å between centers is the same as the Coulombic interaction energy between two ions at the same distance. How would the interaction energies compare with one another if the pairs of molecules and ions were each separated by distances of 5 Å?

23. What fraction of the total volume of an ideal gas would be occupied by molecules of 2 Å diameter at an external pressure of 50 atm and a temperature of 100°C?

24. A sample of air expired by a human subject is collected. After 100.0 cm³ of the mixture is passed repeatedly through sodium hydroxide solution, the volume occupied is 94.7 cm³. Calculate the percent by

volume, mole fraction, and percent by weight of CO_2 in the expired air.

25. At 50°C, a volume of 40.0 liters of inert gas under a pressure of 1 atm is bubbled through liquid bromobenzene of which the vapor pressure is 16.96 torr. Assuming that the gas becomes saturated with bromobenzene vapor, how much liquid is evaporated?

26. The vapor pressure of chloroform, $CHCl_3$, is found to be represented by the equation

$$\log p = -\frac{1638}{T} + 7.735$$

where T is the temperature in kelvins. Estimate the heat of vaporization of the liquid in the temperature range over which this equation is valid.

27. Estimate the molar volume of water at the critical point from the van der Waals constants and the critical temperature and pressure given in Table 1-1.

28. Certain of the liquids in Table 1-3 have dielectric constants that are disproportionately large as compared to the molecular dipole moment. Formulate a possible explanation.

29. The vapor pressure of ethanol is 78.8 torr at 30°C and 222.2 torr at 50°C. What is the heat of vaporization of ethanol in this temperature range?

30. A volume of 3.25 liters of argon at 1 atm pressure is saturated with the vapor of pyridine at 38.0°C. If 0.530 g of pyridine is vaporized, what is the vapor pressure of liquid pyridine at this temperature?

REFERENCES

Books

Farrington Daniels and Robert A. Alberty, *Physical Chemistry,* 4th ed., Wiley, New York, 1975. Chapter 9 contains a more detailed account of the kinetic theory of gases.

Dale Driesbach, *Liquids and Solutions,* Houghton-Mifflin, Boston, 1966. The first nine chapters present an introductory, well-illustrated description of models of the liquid state and intermolecular forces.

D. Eisenberg and W. Kauzmann, *The Structure and Properties of Water,* Oxford Univ. Press, London and New York, 1969. A thorough discussion, at an intermediate level, of both structure and methods for its investigation, of solid, liquid, and gaseous water.

Joel H. Hildebrand, *An Introduction to Kinetic Theory,* Reinhold, New York, 1963. An elementary account.

R. A. Horne, Ed., *Water and Aqueous Solutions,* Wiley-Interscience, New York, 1972. Rather advanced, but Chapters 9 and 10 give good accounts of the models of liquid water.

Walter Kauzmann, *Kinetic Theory of Gases,* W. A. Benjamin, Menlo Park, Calif., 1966. Detailed accounts, beginning at an elementary level, but including more advanced mathematical details of equations of state,

molecular collisions, and distribution of molecular velocities.

Harold J. Morowitz, *Entropy for Biologists,* Academic Press, New York, 1970. Contains a good section on the kinetic theory of gases.

G. C. Pimental and A. L. McClellan, *The Hydrogen Bond,* Freeman, San Francisco, 1960. A comprehensive but readable account, including both the characteristics of hydrogen bonds and the methods by which they are investigated.

Journal Articles

Robert E. Apfel, "The Tensile Strength of Liquids," *Sci. Am.* **227,** 58 (December 1972).

A. F. M. Barton, "The Description of Dynamic Liquid Structure," *J. Chem. Educ.* **50,** 119 (1973).

R. M. J. Cotterill, "Glassy Metals," *Am. Sci.* **64,** 430 (1976).

Earl Frieden, "Non-Covalent Interactions. Key to Biological Flexibility and Specificity," *J. Chem. Educ.* **52,** 754 (1975).

Julian H. Gibbs, "Elementary Derivation of the Boltzmann Distribution Law," *J. Chem. Educ.* **48,** 542 (1971).

A. L. McClellan, "The Significance of Hydrogen Bonds in Biological Structures," *J. Chem. Educ.* **44,** 547 (1967).

A. C. Norris, "SI Units in Physico-Chemical Calculations," *J. Chem. Educ.* **48,** 797 (1971).

Peter Oesper, "The History of the Warburg Apparatus," *J. Chem. Educ.* **41,** 294 (1964).

J. Bevan Ott, J. Rex Goates, and H. Tracy Hall, "Comparisons of Equations of State in Effectively Describing *PVT* Relations," *J. Chem. Educ.* **48,** 515 (1971).

L. Petrakis, "A Metric (SI) Energy Scale: Conversions and Comparisons," *J. Chem. Educ.* **51,** 459 (1974).

F. L. Pilar, "The Critical Temperature: A Necessary Consequence of Gas Non-Ideality," *J. Chem. Educ.* **44,** 284 (1967).

F. L. Swinton, "The Triple Point of Water," *J. Chem. Educ.* **44,** 541 (1967).

Two
Solutions of Nonelectrolytes

The term *solution* can be applied to any homogeneous mixture. "Mixture" implies that there are present several different chemical substances which to a large degree retain their chemical identity, and that the ratio in which these substances are present can be varied over a fairly wide range. "Homogeneity" means that the individual molecules, ions, or atoms of each constituent are dispersed randomly throughout the mixture and that the solution is a single phase. Gaseous mixtures, which were considered in Chapter 1, are all solutions, for the molecules of each component substance are uniformly distributed throughout the entire volume of the mixture. Solutions most frequently encountered by chemists and biologists are in the liquid state, consisting of a gas, liquid, or solid dispersed in a liquid medium.

In this chapter, we are concerned with solutions in the liquid phase and with gaseous mixtures in equilibrium with such solutions. Measured macroscopic properties of solutions of nonelectrolytes are discussed, with primary attention to the equilibrium composition of coexisting phases, and some experimental results are interpreted in terms of intermolecular interactions. Because of their rather special properties, solutions in which appreciable concentrations of ions are present will be considered later in Chapters 5–7.

2-1
CONCENTRATION SCALES

Because a solution is a mixture for which the composition can vary over wide limits, the description of a specific solution must include a statement of its composition. The most fundamental and most widely applicable method of expressing concentrations is in terms of the *mole fraction*. This scale, which was employed in Chapter 1 for gaseous mixtures, gives a direct indication of the relative number of molecules of each component compared to the total number present. The whole range of concentration of a two-component system from unit mole fraction of the first component to unit mole fraction of the second component can be represented graphically on a single scale. It is also conveniently applied to multicomponent systems. Directly related to the mole fraction concentration value is the *mole percent* value, which is simply 100 times the mole fraction.

For some solutions it is convenient to designate one material as the *solvent* and the other material or materials as *solutes*. The solvent can be defined as the substance whose physical state is preserved when the solution is formed. However, there may be several components of a solution for which the physical state when pure is the same as the physical state of the solution; for instance, in a mixture of acetone and water, either component can be regarded as the solvent, and the substance present in the larger amount can arbitrarily be taken as the solvent.

Several concentration scales are designed to indicate primarily the amount of solute present in a solution. On one type of scale it is a given volume of solution for which the amount of solute is cited. Typical units are grams per liter, or moles per liter, or equivalent weights—for some particular reaction types—per liter. Of these, the last two constitute the *molarity* and *normality* scales, respectively. Expressing concentration on a volume-of-solution basis has the advantage, especially convenient in quantitative analytical chemistry, that a desired amount of solute can be obtained by measuring out a certain volume of solution from a buret or pipet. However, the density of a liquid changes with temperature, so that any concentration scale based on volume varies slightly with variation in temperature, and an aqueous solution that is 1.00 molar at 5°C is only about 0.96 molar at 95°C.

Another type of concentration scale which, like mole fraction, is independent of temperature, is often employed by physical chemists. This is the *molality,* which is defined as the number of moles of solute present along with each kilogram of solvent.

The distinctions between the molality (m) and molarity (M) scales should be kept carefully in mind. For example, a liter of 1.000 M sucrose solution can be prepared by weighing out 342 g, or 1 mol, of the solid solute, placing it in a volumetric flask, and adding water until the total volume of the solution is 1000 ml. The density of the resulting solution is 1.129 g/ml so that a liter of solution weighs 1129 g. The amount of water added, then, was 1129 less 342, or 787 g. Thus the molality of the 1.000 M solution is 1/0.787 or 1.271. If instead it were desired to prepare a 1.000 m solution, 1000 g of water would be added to 342 g of sucrose.

For dilute aqueous solutions, the molarity and molality are approximately the same, since a liter of dilute solution does contain about 1000 g of water. For concentrated solutions, as well as for solvents other than water, this is not likely to be true. An extreme case is that of solutions in mercury, for which a 0.1 m solution is approximately 1.3 M.

Example: An aqueous solution contains 12.00 percent by weight sucrose. The solution density is 1.0465 g/cm³. Calculate the mole fraction, molarity, and molality of the solution in sucrose.

Solution: Take as a basis 1 liter of solution, which has a total weight of 1046.5 g. Of this, 12.00 percent or 125.6 g is sucrose and the remainder, 920.9 g, is water. The weight of sucrose corresponds to 0.367 mol, so that the solution is 0.367 M. The molality is the number of moles of solute divided by the number of kilograms of water, $0.367/0.9209 = 0.399$ m. The mole fraction is $0.367/[0.367 + (920.9/18.020)]$, or 0.0071.

2-2
IDEALITY OF SOLUTIONS
RAOULT'S LAW
AND THE IDEAL SOLUTION

In discussing gases and their properties, it was convenient to set up a standard, the ideal gas, to which real gases approach and then to describe real gases in terms of deviations from this standard. A similar procedure is of value in considering the properties of liquid solutions: A solution is characterized by its conformity to, or divergence from, ideality.

The ideality of a solution is judged by the escaping tendency of the components from the solution. The partial vapor pressure of a component from a solution measures the tendency of the material to leave the solution—the pressure driving that material to escape into another phase. In an ideal solution, the partial vapor pressure of each component is equal to the product of the mole fraction of that component in the mixture and the vapor pressure of the pure substance at the same temperature. This principle of the ideal solution is known as *Raoult's law:*

$$p_i = X_i p_1^0 \qquad (2\text{-}1)$$

In this equation, p_i is the vapor pressure of component i from a solution in which its mole fraction is X_i, and p_i^0 is the vapor pressure of pure i at the same temperature as that at which the vapor pressure of the solution is measured.

It is a corollary of this definition of the ideal solution that the molecules of a substance upon entering an ideal solution are not affected by the change in medium except that their concentration is reduced: The only change in vapor pressure is the decrease caused by dilution of the material by other molecules. One way of describing this situation is to say that the rate of evaporation of the given material from the surface is proportional to the fraction of the surface occupied by molecules of this material, and this fraction is in turn proportional to the mole fraction of the material in the liquid mixture. The rate of condensation of gas to liquid is proportional to the pressure of the gas; the equilibrium pressure, or vapor pressure, is that at which the rate of condensation equals the rate of evaporation, and thus the equilibrium pressure decreases in direct proportion as the mole fraction of material in the liquid phase decreases.

It must be clearly borne in mind that the absence of intermolecular forces is *not* a characteristic of an ideal solution in the same way that it is of an ideal gas. In any liquid there are, of necessity, intermolecular forces of considerable magnitude; ideality only demands that the molecules in solution are in the same environment there as they would be in the exclusive company of their own kind in the pure liquid. As a consequence of this requirement, fully ideal solutions are commonly formed only by two materials containing molecules of similar size, shape, and polarity, such as benzene and toluene, chloroform and carbon tetrachloride, n-hexane and n-heptane, or ethyl bromide and ethyl iodide.

When two liquids mix to form an ideal solution, there is no volume change on mixing, so that the volume of the solution is equal to the sum of the separate volumes of the constituents. Physical properties of the solution such as the viscosity are additive in the respective properties of the constituents of the mixture. Moreover, there is no energy change, that is, no heat of mixing. If a gas dissolves in a liquid to form an ideal solution, the heat change is merely that which would occur if the same gas were condensed to form a liquid, and the evolution of heat can be evidenced by a rise in temperature. It is only rarely that a solid dissolves in any appreciable concentration in a liquid to form an ideal solution, but the heat change for such a process is equal to the heat of fusion of the solid—the amount of heat absorbed when the solid is converted into a liquid.

DEVIATIONS FROM IDEALITY

Mere dissimilarity leads both components of a binary nonideal mixture to have vapor pressures greater than those predicted from Raoult's law. Suppose one substance, for example ethanol, with large intermolecular forces is mixed with another substance, such as benzene, with smaller intermolecular forces. The effect on the ethanol is a lessening of the strong forces by an increase in distance between molecules, and thus the ethanol molecules are freer to evaporate from the solution than when surrounded by other ethanol molecules. The molecules of benzene are not, however, held to one another very strongly. Located between ethanol molecules with large mutual attractive forces, they can be "squeezed out" of the liquid phase and thus are vaporized more readily than from pure liquid benzene. As a consequence, it is found that the vapor pressures of both ethanol and benzene are greater from the solution than would be expected by reference to the vapor pressures of the pure liquids; the solution is said to show positive deviations from Raoult's law.

Intimately related to the question of the ideality of solutions is the problem of the extent to which two materials are miscible with each other. Indeed, the extreme of positive deviations from ideality is insolubility. As one might suppose on the basis that similar materials produce more nearly ideal solutions, solubility can be predicted to some extent on the basis that "like dissolves like." To a considerable degree "likeness" is measured by polarity, and it is also measured quite well, aside from interactions corresponding to complex formation which are specific to particular solvent—solute combinations, by the magnitude of the van der Waals forces in each component.

The formation of complexes or compounds between molecules of the solvent and molecules of the solute tends to produce negative deviations from Raoult's law and to increase solubility where otherwise miscibility would be limited by the dissimilarity of the components. In many cases, it is the existence of a hydrogen bond that makes two unlike materials soluble in one another. Acetone acts as a proton acceptor, forming hydrogen bonds with chloroform to yield solutions with vapor pressures less than the ideal, as described in Section 2-3. The solubility of the alcohols in water can be compared with the solubility of the corresponding hydrocarbons in water which, although measurable, is quite small. The introduction of a hydroxyl group confers some solubility; the solubility of the alcohol is greater the lower the molecular weight of the alcohol, and therefore the larger the fraction of the molecular character represented by the hydroxyl group. The hydroxyl group affords the opportunity for hydrogen bond formation with the solvent, as well as contributing polarity to the solute. Dimethyl sulfoxide is a good solvent for materials containing —NH— groups, principally by virtue of its capacity to act as an acceptor in hydrogen bond formation.

2-3
MISCIBLE LIQUID PAIRS

In this section, the liquid–vapor equilibria of pairs of substances miscible in the liquid phase over the complete range of concentration from one pure material to the other are considered. Of course, some of the binary systems included may not be completely miscible at some temperatures or pressures other than those dealt with here, but behavior under such conditions is treated later in Section 2-5.

VAPOR PRESSURES
OF MIXTURES

In an ideal binary system, where both components are volatile, the vapor pressure relations are as shown in Figure 2-1. Each of the liquids exerts a partial vapor pressure as given by Raoult's law:

$$p_A = X_A p_A^0 = \frac{n_A}{n_A + n_B} p_A^0 \qquad (2\text{-}2)$$

$$p_B = X_B p_B^0 = \frac{n_B}{n_A + n_B} p_B^0 \qquad (2\text{-}3)$$

The p's are partial vapor pressures and the n's are numbers of moles. The total vapor pressure of a solution is the sum of the partial vapor pressures of the components, whether or not the solution is ideal. In the ideal case, the total vapor pressure curve is linear in the mole fraction of either component of a binary mixture, as in Figure 2-1.

Consider now the vapor in equilibrium with a liquid mixture. Let the mole fractions of A and B in the vapor be Y_A and Y_B, respectively.

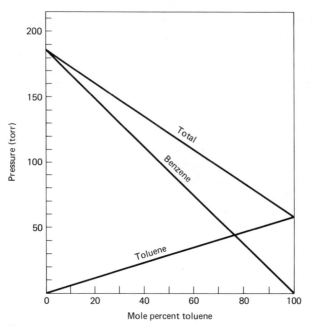

Figure 2-1
Partial vapor pressures of benzene and of toluene, and total vapor pressures of
the mixture, for solutions of benzene and toluene at 40°C. Composition varies
from pure benzene on the left to pure toluene on the right.

These are the ratios of the partial vapor pressures of the components
to the total vapor pressure of the solution:

$$Y_A = \frac{p_A}{p_A + p_B} = \frac{X_A p_A^0}{X_A p_A^0 + X_B p_B^0} \tag{2-4}$$

$$Y_B = \frac{p_B}{p_A + p_B} = \frac{X_B p_B^0}{X_A p_A^0 + X_B p_B^0} \tag{2-5}$$

Only if p_A^0 and p_B^0 are equal, which means that the component liquids
are equally volatile, does Y_A equal X_A and Y_B equal X_B. In general, a
vapor in equilibrium with an ideal binary liquid mixture is richer than
the liquid in the component that has the higher vapor pressure when
pure.

Example: At 40°C pure benzene and pure toluene have vapor pressures of 183.6
and 59.1 torr, respectively. A liquid solution contains 40.0 percent benzene by weight.
What is the composition of vapor in equilibrium with the solution if the solution is
ideal?

Solution: As a basis for calculation, any desired amount of material may be chosen.
Using 100.0 g of liquid as the basis, we calculate that there is 40.0/78.1 or 0.512 mol
of benzene to 60.0/92.1 or 0.651 mol of toluene in the liquid. The mole fraction of
benzene in the liquid is then 0.512/(0.512 + 0.651) or 0.440, and the mole fraction of
toluene is 0.560. The partial vapor pressure of benzene is (0.440)(183.6 torr) or 80.8 torr,
and that of toluene (0.560)(59.1 torr), or 33.1 torr. The total vapor pressure is the sum of
these two partial vapor pressures, or 113.9 torr. The mole fraction of benzene in the
vapor is 80.8/113.9 or 0.709, and the mole fraction of toluene is 0.291. The weight of

benzene in the vapor is a fraction $(0.709)(78.1)/[(0.709)(78.1) + (0.291)(92.1)]$ of the total weight of vapor or 67.4 percent.

By a series of calculations for an ideal system such as the calculation just given, or by experimental determination of the vapor–liquid equilibrium for a nonideal system, it is possible to obtain a series of values of corresponding compositions of liquid and vapor, from which a phase diagram can be constructed. Because there are three variables, temperature, pressure, and composition, and only two variables can be shown on a single two-dimensional plot in a plane, one of the three must be held constant when the diagram is represented on paper.

Figure 2-2 is the vapor–liquid composition diagram for the benzene–toluene system at 40.0°C. At pressures below those on the "vapor" curve, only vapor exists. At pressures above those on the "liquid" curve, only liquid exists. Between these two curves, two phases co-exist. Since the two phases must be at the same pressure to be in equilibrium with one another, the compositions of coexisting phases lie at the ends of a horizontal line joining the two curves. Such a line joining phases that can exist in equilibrium with one another is called a *tie line,* and several examples are shown crossing the two-phase region. If a mixture is made up with the overall composition and at the pressure corresponding to point b, it cannot continue to exist as one phase, but rather it separates into two phases, one of which is liquid of composition c, and the other of which is vapor of composition a.

A tie line affords a method for easily estimating the relative amounts of the two phases present at equilibrium. If, as on line ac in Figure 2-2,

Figure 2-2
Liquid–vapor composition diagram for solutions of benzene and toluene at 40°C. The horizontal lines cross the two-phase region, and each joins a particular liquid composition with the composition of the vapor in equilibrium with that liquid.

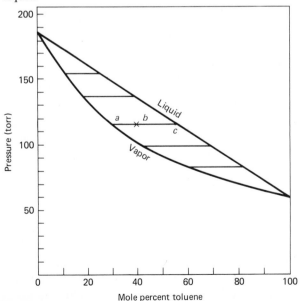

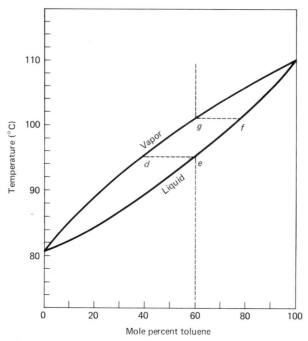

Figure 2-3
Boiling point diagram for benzene–toluene mixtures at 1 atm pressure. The lines
de and *fg* are tie lines.

b represents the overall composition of the system, the amount of the
phase having the composition *a* is proportional to the distance *bc*, and
the amount of the phase having the composition *c* is proportional to
the distance *ab*.

Example: A mixture of 46.2 g of benzene and 53.8 g of toluene is kept at 40.0°C under
a pressure of 114.0 torr. What is the distribution of material between the liquid and
vapor phases?

Solution: From the previous example, this mixture lies in the two-phase region, for
which at this temperature and pressure the liquid composition is always 40.0 percent
benzene by weight and the vapor composition is 67.4 percent benzene by weight.
The difference between mixture and vapor compositions is 67.4 − 46.2 or 21.2 percent. The difference between mixture and liquid compositions is 46.2 − 40.0 or 6.2
percent. Thus 21.2/(21.2 + 6.2) or 0.77 of the mixture exists as liquid at equilibrium.
This corresponds to 77 g, leaving 23 g as vapor.

Because the phase relations of a system of this kind also depend
upon temperature, it is useful to construct a diagram in which temperature is varied and the pressure is held constant. Many such diagrams are possible, for there is one corresponding to each value of the
pressure. In Figure 2-3 is a representative "boiling point" diagram for
the benzene–toluene system at a pressure of 1 atm. As in the vapor
pressure diagram, the region between the two curves is a region of coexistence of two phases, a region which on the graph we can imagine

to be covered by a whole series of horizontal—and therefore constant-temperature—tie lines linking points corresponding to the compositions of the coexisting phases. A solution of composition corresponding to 60 mole percent toluene can be increased in temperature until point e is reached. At this temperature, it begins to boil, with the first vapor coming off having a composition corresponding to point d. Since the vapor being formed is richer in benzene than is the original liquid, the remaining liquid becomes increasingly richer in toluene, and its boiling point rises, following the liquid curve to the right. The composition of the last drop of liquid to vaporize is that of point f, and the vapor in equilibrium with this liquid has the same composition as the original liquid. At temperatures above that at which the vertical line from the point corresponding to 60 mole percent toluene crosses the vapor curve, the equilibrium state of the mixture is a single phase, the vapor, this phase has a composition the same as that of the original liquid, and no more liquid phase is present.

If the deviations of a mixed system from Raoult's law are fairly large in magnitude, there is the possibility that at some composition of the

Figure 2-4
Vapor pressure of mixtures of ethanol and carbon tetrachloride at 65°C.

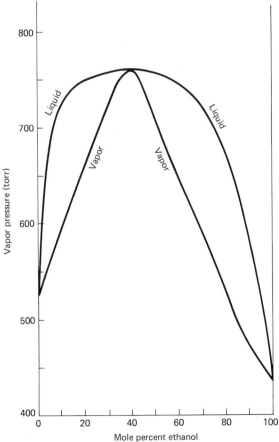

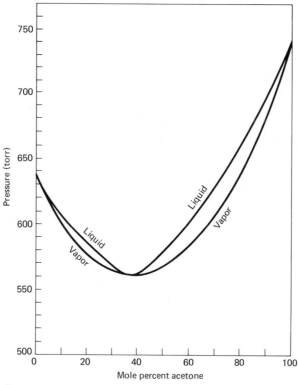

Figure 2-5
Vapor pressure of mixtures of acetone and chloroform at 55°C.

mixture the vapor pressure of the solution will pass through either a maximum value or a minimum value, as illustrated for the systems carbon tetrachloride–ethanol and acetone–chloroform by the vapor pressure diagrams in Figures 2-4 and 2-5, respectively. At a vapor pressure extreme of this sort, the composition of vapor in equilibrium with the liquid is the same as the composition of the liquid. Such a point is termed an *azeotropic point,* since the liquid vaporizes without change in composition. Some examples of systems that form constant-

Table 2-1

Constant-boiling binary liquid mixtures at one atmosphere pressure

Component A	Component B	Boiling point of A (°C)	Boiling point of B (°C)	Weight percent of B in azeotrope	Boiling point of azeotrope (°C)
Water	Hydrogen chloride	100.0	−83.7	20.2	110
Water	Nitric acid	100.0	86	68	120.5
Chloroform	Methyl acetate	61.2	57.1	23	64.8
Acetic acid	Pyridine	118.1	115.3	65	139.7
Benzene	Methyl alcohol	80.2	64.7	40	58.3
Ethyl alcohol	n-Hexane	78.3	69.0	79	58.7
Water	Ethyl alcohol	100.0	78.3	95.6	78.1
Water	Isopropyl ether	100.0	69.0	96.4	61.4

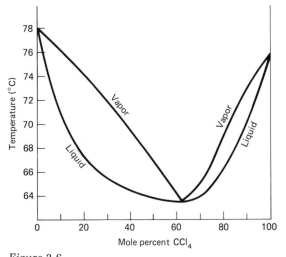

Figure 2-6
Boiling point diagram for ethanol–carbon tetrachloride mixtures.

boiling mixtures are listed in Table 2-1. For a system that, at constant temperature, has a maximum vapor pressure value at a composition intermediate between the limits of the two pure substances, there is a solution of intermediate composition which has a minimum boiling point at any fixed pressure. Likewise, a series of solutions of two substances that has a minimum on the vapor pressure curve is found to have a maximum on the boiling point–composition curve. Boiling point curves corresponding to the vapor pressure curves in Figures 2-4 and 2-5 are shown in Figures 2-6 and 2-7, respectively.

Figure 2-7
Boiling point diagram for acetone–chloroform mixtures.

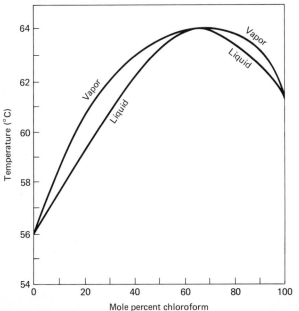

DISTILLATION

The difference in volatility between two liquids can be employed as a means of separating or analyzing a mixture of the two liquids. The process of separation by partial, selective vaporization is known as *fractional distillation*. Of course, one partial vaporization produces only an enrichment of the vapor in the more volatile material and never brings about complete separation of the ingredients of the mixture. However, the process of vaporization and condensation can be repeated any necessary number of times in order to achieve the desired degree of separation of the liquids.

Consider how this process can be applied to a mixture of benzene and toluene, starting with liquid of composition X_4 as indicated in Figure 2-8. When this mixture is heated, the first vapor to come off has the composition Y_4. This vapor can be completely condensed to a liquid, still of the same composition as Y_4, at point X_5. This liquid, if boiled, will initially give off vapor of composition Y_5, which condenses to liquid at point X_6. One must remember, however, that as each vapor portion is removed from the liquid the material remaining in the liquid phase becomes richer in toluene, the less volatile component. The vapor produced from the liquid residue thus has a higher toluene content than the vapor produced when the solution first begins to boil. The greater the amount of vapor taken off, the greater is the concentration of toluene in the residual liquid as well as in the vapor being removed. When the liquid has all volatilized, the vapor composition is obviously the same as the original liquid composition. In distillation,

Figure 2-8
Distillation diagram for benzene–toluene mixtures.

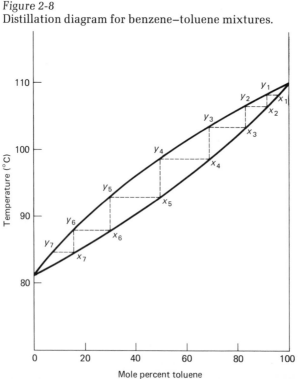

Mole percent toluene

then, separation can be achieved only by processes involving partial evaporation of the liquid or partial condensation of the vapor.

As shown in Figure 2-8, vapor–liquid equilibrium curves can be followed by a succession of complete volatilizations and partial condensations from composition X_4 toward the toluene side of the diagram. The greater the number of such steps carried out, the closer the composition of the liquid is to pure toluene.

Each fraction obtained by boiling off a series of portions of a liquid or by condensing portions of a vapor can be redistilled or refractionated to obtain additional separation, and this process can be repeated a number of times. In practice, however, a tremendously large total number of manipulative operations, involving handling of the various fractions, would be required to accomplish a useful separation by this method. What is usually done as an alternative to this cumbersome procedure is to operate a fractionating column. The column is a vertical tube which can have a diameter of anywhere from the few millimeters of a laboratory-scale glass tube to the many feet of an industrial-scale tower. The column is filled with packing or regularly spaced horizontal trays or perforated dividers, so arranged that vapor from boiling liquid rises through it, meeting and mixing with condensed liquid descending through the column. As the two phases pass one another, a portion of the vapor condenses and a portion of the liquid evaporates. In each of these steps, the vapor becomes richer in the more volatile component and the liquid becomes richer in the less volatile component, so that the more volatile material is carried to the top of the column, where it can be removed as distillate, and the less volatile material is carried toward the bottom of the column, where it concentrates in the still pot.

It must be observed that, for a system in which an azeotropic composition is present, the best that can be achieved in a fractional distillation, no matter how efficient the column, is the separation of one pure component on the one hand, and the azeotropic mixture on the other. This limitation can sometimes be circumvented by adding a third substance to a binary mixture to change the phase relations, or by redistilling at an altered pressure at which the azeotropic composition is also altered.

2-4
SOLUTIONS OF GASES IN LIQUIDS

When a gas dissolves in a liquid, the molecules in the resulting solution are all in a "liquid" environment, just as if two mutually soluble liquids had been mixed together. The principal difference in the treatment of gas-in-liquid solutions from that of liquid–liquid mixtures, which have been discussed in terms of vapor pressures of volatile materials from the liquid, is that we can think in terms of the solubility of the gas rather than of its volatility. The reason is in part that at the temperature of interest there is no "liquid" behavior of the pure substance to which we can refer its behavior in solution. There is no difference in the equilibrium principles that apply to the two types of solutions, but it is true that the extent of deviation from ideality is likely to be greater when the two components of the mixture are more dissimilar in physical properties.

The amount of gas that dissolves in a liquid solvent depends upon the particular combination of gas and solvent, upon the pressure of the gas above the solution, and upon the temperature of the system. Each component of a mixture of gases dissolves according to its own partial pressure in the mixture, pretty much independently of the other gases.

For a system consisting of the liquid solution of a gas and the same gas in the vapor phase, so long as it approaches ideality, the change in concentration of the solute gas is paralleled by a proportionate change in concentration of the gas in the vapor phase, as expressed by the equilibrium equation

$$K_i = \frac{\text{concentration of } i \text{ in vapor}}{\text{concentration of } i \text{ in solution}} = \frac{p_i}{X_i} \qquad (2\text{-}6)$$

Here the partial pressure of the gas in the vapor phase p_i has been chosen to measure its concentration in the vapor, and the mole fraction in solution X_i to measure its concentration in the liquid phase. This equation is a statement of *Henry's law,* and the constant of proportionality between pressure in the vapor and concentration in the liquid is known as the Henry's law constant. The existence of this proportionality is the basis for the practice, often followed by workers in physiology, of describing the concentration of solute gas in terms of its "tension," that is, its vapor pressure. The statement that the carbon dioxide tension of the blood is 45 mm means that the blood contains the concentration of carbon dioxide that corresponds to a partial vapor pressure of 45 mm Hg.

Example: Under a partial pressure of 760 torr of hydrogen, 100 cm³ of water at 40°C dissolves 0.0731 millimole of hydrogen. How much hydrogen will be dissolved by 200 cm³ of water under a hydrogen partial pressure of 25 torr?

Solution: The amount dissolved is proportional to the amount of solvent and to the partial pressure. Since the concentration is quite small, the mole fraction is directly proportional to the number of moles. Therefore the amount dissolved will be (0.0731)(200/100)(25/760) or 0.00481 mmol of hydrogen.

Henry's law expresses the same ratio, that of vapor pressure to concentration, as Raoult's law, and thus the two must be related. If Henry's law applies all the way to the pure solute material, for which X_i is unity, then K_i must be numerically equal to p_i^0, and Henry's law is identical to Raoult's law. However, Henry's law usually does not apply over the complete solution composition range, K_i has a value other than p_i^0, and the law is less restrictive than Raoult's law. It seems reasonable to expect that solutions containing substances that, when pure, are in different physical states under the same conditions, as a liquid and a gas, constitute at best "partially ideal" systems which follow Henry's law over a limited range of concentration. It is often

Table 2-2
Solubilities of gases in water[a,b]

Gas	0°C	10°C	20°C	30°C	40°C
CO_2	18.02	12.69	9.26	7.19	5.78
H_2	0.226	0.207	0.192	0.184	0.173
N_2O	13.5	9.26	6.45	4.76	4.07
C_2H_4	2.70	1.81	1.33	1.00	0.81
O_2	0.518	0.405	0.329	0.279	0.248
CH_4	0.588	0.442	0.351	0.293	0.254
H_2S	49.2	35.9	27.3	21.6	17.7

[a]Reciprocal Henry's law constants, with pressures expressed in torr and concentrations in mole fraction. All values are to be multiplied by 10^{-7}.
[b]Values recalculated from *Handbook of Chemistry and Physics*, Chemical Rubber Publishing Company, Cleveland, Ohio.

said that Henry's law applies to solutes, that is, to materials present in low concentration, while Raoult's law applies to the solvent, that is, to a material for which the mole fraction is not much less than unity.

Experimental results show that Henry's law holds within a few percent for many gaseous materials dissolved in water when the gas pressure is not more than a few atmospheres. If the gas is one that combines chemically with water, as do such substances as the hydrogen halides and ammonia, deviations from Henry's law may be quite large. In most cases an increase in temperature decreases the solubility of a gas and raises the pressure of the gas in equilibrium with a solution of any given concentration. This change is parallel to the increase in vapor pressure of a pure liquid with temperature, for the process in which a gas goes into solution in a liquid is energetically and physically much like that in which the gas condenses to the liquid state: Heat is given off in the solution process, and the molecules going into solution become surrounded closely by other molecules as they do in condensing to pure liquid. In Table 2-2 are given solubility constants for several common gases in water at various temperatures. The solubilities of gases in other solvents are of course different from their solubilities in water. The solubilities of many common gases, such as nitrogen, oxygen, and carbon dioxide, are two to ten times as great in ethanol, acetone, or benzene as in water.

The presence of dissolved solids, particularly of electrolytes, reduces the solubility of gases in water, except for those few special cases in which the gas happens to react chemically with the solute. The diminution in solubility of the gas is an example of what is termed "salting out." The amount of salting-out is nearly independent of the nature of the gas; this is explained by assuming that the ions of the electrolyte combine with molecules of solvent, reducing the amount of water left free and thus reducing the amount available to act as a solvent for the gas. The effect is similar to that in which organic liquids are made less soluble in water by the addition of a salt such as sodium chloride, or that in which proteins are selectively precipitated by addition of high concentrations of ammonium sulfate.

Let us now apply some of the considerations of gas solubility to the question of the solubility of oxygen, nitrogen, and carbon dioxide in blood. The air inhaled into the lungs contains, on a dry basis omitting the variable amounts of water vapor, about 21 percent oxygen, 79 percent nitrogen, and usually 0.04 percent carbon dioxide. In the lungs, oxygen is lost from the air to the blood and carbon dioxide is removed from the blood, passing into the air; this exchange occurs in small chambers, or alveoli, containing air and surrounded by a very thin membrane bathed in the stream of blood passing through the lungs.

As a result of the exchange of gases occurring in the lung, the average composition of the alveolar air on the dry basis becomes about 15 percent oxygen, 80 percent nitrogen, and 5 to 6 percent carbon dioxide. The alveolar air is almost saturated with water vapor, which exerts a partial pressure of about 45 torr. The remaining gases together contribute a pressure of 715 torr, of which oxygen accounts for approximately 107 torr, nitrogen about 570 torr, and carbon dioxide perhaps 38 torr. The salting-out effect of the solutes in the plasma and the low water content of the blood corpuscles reduce the expected physical solubility of gases in whole blood to about 92 percent of that in water. With this consideration, one can calculate the expected solubility in whole blood to be 0.31 cm^3 of oxygen, 0.83 cm^3 of nitrogen, and 2.4 cm^3 of carbon dioxide, measured at standard conditions, per 100 cm^3 of blood.

These volumes can be compared with the volumes of gases actually found by pumping out a sample of blood, again in units of cubic centimeters of gas per 100 cm^3 of blood:

	O_2	N_2	CO_2
Arterial blood	19.5	0.9	49.7
Venous blood	13–14	0.9	55–58

It is seen that much greater amounts of carbon dioxide and oxygen are present in the blood than can be explained by simple physical solubility in the liquid. The largest portion must be present in some chemically combined form, although the chemical bonds are relatively easily broken since the gas can be removed from the blood by reducing the pressure. Oxygen is chiefly carried in combination with hemoglobin in the red blood cells. Saturation is reached at about 20 cm^3 of oxygen per 100 cm^3 of blood at an oxygen partial pressure of 150 torr. The oxygen content of arterial blood corresponds to a tension of about 100 torr, and that of venous blood to about 40 torr. Inspection of oxygen-tension values shows that the proportionality of amount of gas in the liquid phase to partial pressure does not hold very well at moderate pressures and fails as saturation is approached; this is characteristic of a state of chemical combination rather than of physical solution. The carbon dioxide in excess of that physically dissolved is carried both in the plasma and in the corpuscles, principally in the form of bicarbonate ion as well as partly in combination with $-NH_2$ groups of proteins.

2-5
LIQUID MIXTURES SHOWING LIMITED SOLUBILITY

When two liquids that are immiscible are present together, each exerts its own vapor pressure, and the total vapor pressure is the sum of the two pure-liquid vapor pressures. Even if each of the liquids is slightly soluble in the other, the situation is essentially the same, because the small concentration of solute does not appreciably lower the vapor pressure of the predominant liquid, and the condition of equal vapor pressure of any component from two phases at equilibrium requires that the material at low concentration in one phase have the same vapor pressure that it does from the other phase.

As a consequence of this situation, the total vapor pressure is independent of the amounts of the two materials present and is larger than the vapor pressure of either of the substances, so that the mixture distils at a temperature below the boiling point of either of the component liquids. This phenomenon is utilized in the process of steam distillation of organic materials. For example, if the boiling point of a water-insoluble substance is so high that there is danger of decomposition during conventional distillation, an impure sample can be purified by passing a current of steam through it. Both water and the water-insoluble material as a single vapor phase are carried over into the condensate, where they separate into two liquid layers.

In a category intermediate between pairs of liquids that are almost completely insoluble in one another, such as water and bromobenzene, and pairs that are miscible in all proportions, such as water and methyl alcohol, are mixtures that are mutually soluble over a limited range of concentration. At a given temperature, at least within a wide region of variation in temperature, there is a range of intermediate composition in which two phases are present.

As an example of such a partially miscible liquid pair, consider the system n-butyl alcohol–water at 25°C. The alcohol dissolves in water up to a concentration of 7.35 weight percent alcohol, while water dissolves in the alcohol up to a concentration of 20.27 weight percent water. If mixtures with overall compositions between these limits are prepared, they are found to separate into two layers, one a saturated solution of alcohol in water, the other a saturated solution of water in alcohol. Thus no homogeneous liquid phase can exist at this temperature containing only n-butyl alcohol and water and having more than 7.35 percent but less than 79.73 percent alcohol.

The solubility of partially miscible liquids depends upon temperature. Usually the solubility increases with temperature, evidently as a result of increasing thermal motion of the individual molecules. A typical case is represented in Figure 2-9, which is a phase diagram for the system n-hexane–methyl alcohol. In this diagram, the region outside the curve ACB includes those combinations of temperature and

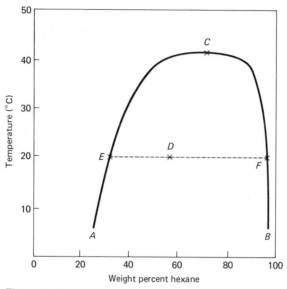

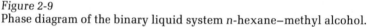

Figure 2-9
Phase diagram of the binary liquid system n-hexane–methyl alcohol.

composition for which there exists but one liquid phase. At temperatures exceeding that corresponding to point *C*, which is termed the *consolute* point or *critical solution temperature*, the liquids are completely miscible.

Any mixture of methyl alcohol and n-hexane for which the composition and temperature fall inside curve *ACB* consists of two phases; the compositions of the two phases are given by the two points on the curve at the same temperature as that of the mixture. Points *E* and *F*, for example, give the compositions of the two liquid layers into which a mixture of composition *D* will separate at 20°C. Further, the ratio of the amount of layer of composition *E* to the amount of layer of composition *F* is equal to the ratio of the distance *DF* to the distance *DE*. Any horizontal line drawn from one limb of the curve to the other is a tie line, for it joins together points corresponding to the compositions of two phases that can be in equilibrium with each other. In any heterogeneous equilibrium, it is possible to apply the tie-line ratio rule, as illustrated here for one system, to prediction of the relative amounts of the two phases for a mixture of any arbitrary overall composition.

Example: A mixture of 50 g of methyl alcohol and 50 g of n-hexane is warmed to 35°C. What phases are present and in what amounts?

Solution: The compositions of the phases are independent of the overall amount of each component; at 35°C the hexane content of the two layers is 43.6 and 91.2 percent by weight. Using the tie-line rule, one finds the alcohol-rich layer to be a fraction (91.2 − 50.0)/(91.2 − 43.6) or 0.866 of the total weight, and the hydrocarbon-rich layer to be a fraction (50.0 − 43.6)/(91.2 − 43.6) or 0.134 of the total weight, 100 g. The weight of the alcohol-rich layer is 86.6 g, and the weight of the hydrocarbon-rich layer is 13.4 g.

Table 2-3
Critical solution temperatures of partially miscible liquid pairs

Components		Upper consolute point (°C)	Lower consolute point (°C)
Nicotine	Water	208	61
3-Methylpiperidine	Water	235	57
Triethylamine	Water	—	18
Ethylene glycol	Lauryl alcohol	135	—
Glycerol	Acetone	96	—
Phenol	Water	66	—
Aniline	Hexane	60	—
Perfluoro-n-heptane	n-Heptane	50	—
Methanol	Cyclohexane	49	—
Carbon disulfide	Methanol	41	—

Some binary liquid systems have lower consolute temperatures: Below a certain temperature the two liquids are completely miscible, and above this temperature they are only partly miscible. This behavior is an indication of compound formation between the two materials, and the effect of increasing temperature is to break up the complex between the two different types of molecules. A few systems have both upper and lower consolute temperatures. In Table 2-3 are listed upper and lower critical solution temperatures for a few representative liquid pairs.

Some partially miscible liquid pairs show increasing solubility as the temperature is raised but, at least at atmospheric pressure, never reach an upper consolute temperature because the liquid vaporizes first. Examples of such systems are aniline and water, butyl alcohol and water, and ethyl acetate and water. A schematic version of the phase diagram displayed by a liquid pair of this type is shown in Figure 2-10. At temperatures below the line ABC, two liquid phases coexist over the middle range of composition. At temperature T_b, the

Figure 2-10
Typical phase diagram for a partially miscible binary liquid system which boils below the temperature of complete miscibility.

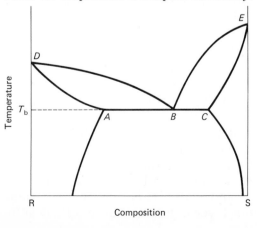

vapor pressure of each phase becomes equal to 1 atm, and vapor at this pressure and of composition B can coexist with the two liquid phases of compositions A and C.

2-6
DISTRIBUTION OF
A SOLUTE BETWEEN
IMMISCIBLE SOLVENTS

A three-component system consisting of a solute present simultaneously in two immiscible or practically immiscible solvents represents an important type of phase equilibrium. The question of the relative concentrations of a solute material in several liquid layers is of significance in problems involving extraction by an immiscible solvent and in the consideration of the distribution of materials in different parts of the living organism. Ether, for example, tends to concentrate in the brain and nerve tissue, which are rich in fatty materials, rather than in the blood, which is primarily an aqueous medium.

NERNST DISTRIBUTION
EQUATION

Let us assume that there are two immiscible liquid phases each containing some quantity of a solute material, and that each of these solutions is in equilibrium with a common vapor phase. Despite the fact that the solute can be a material of low volatility and thus may have a very low vapor pressure, there will be some finite, numerically expressible, even if very small, pressure of solute vapor in the common vapor phase. In order that the two liquid phases may coexist, each must have a solute concentration that corresponds to the same solute vapor pressure; otherwise, solute would transfer from one phase to the other. If Henry's law applies to both solutions, at equilibrium there results the equation

$$p = \frac{X_1}{k_1} = \frac{X_2}{k_2}$$

in which the subscripts 1 and 2 refer to the two different liquid phases, and k is the reciprocal Henry's-law constant. Rearranging yields

$$\frac{X_2}{X_1} = \frac{k_2}{k_1} = K_D \tag{2-7}$$

This is known as the *Nernst distribution equation* and K_D is called the *Nernst distribution coefficient*.

The concentrations employed in the distribution equation can be expressed on any scale. For dilute solutions, K_D does not depend in numerical value on the particular concentration scale used, but for more concentrated solutions, its value varies from one scale to another. In Table 2-4 are given values, for dilute solutions, of the distribution coefficients for a variety of solute types between solvent pairs consisting of an organic solvent and water. It is seen that for some solutes the affinity for the aqueous layer is greater, but for others solu-

Table 2-4
Liquid–liquid distribution coefficients at 25°C, for all concentrations expressed in moles per liter.

Solute	Solvent A	Solvent B	$K_D = \dfrac{\text{(concn. in B)}}{\text{(concn. in A)}}$
Iodine	Water	Carbon tetrachloride	85
Iodine	Water	Chloroform	131
Bromine	Water	Carbon tetrachloride	26
Bromine	Water	Bromoform	64
Trimethylamine	Ethyl ether	Water	1.52
Mercuric chloride	Benzene	Water	12
Caffeine	Chloroform	Water	14
Caffeine	Water	Ethyl ether	7.2
Acetylsalicylic acid	Water	Ethyl ether	4.7
Acetylsalicylic acid	Xylene	Water	4.2

[a]Values are from *International Critical Tables*, Vol. III, McGraw-Hill, New York, 1928.

bility in the organic solvent is greater than in water.

Treatment of a solution by a solvent not miscible with the solution can serve to extract preferentially one solute from among several, or it can serve to concentrate a solute that has a greater solubility in the added solvent. For example, iodine can be extracted from aqueous solution by carbon tetrachloride, in which it is highly soluble; the iodine is removed from other solutes—in particular, polar or ionic materials—which may be in the water and at the same time is put into a much more concentrated solution. Materials present in the blood in small concentrations can be separated from other constituents of the blood, and can sometimes be concentrated, by extraction with a suitable nonaqueous solvent. Procedures of selective extraction and concentration are applied both on a laboratory scale and in chemical processing on an industrial scale. One selective extraction is frequently insufficient to bring about the desired separation, but the process can be repeated many times in an apparatus permitting countercurrent flow of the two liquid phases with repeated contacts between the phases.

There is a significant limitation of the Nernst equation which must be borne in mind in quantitative applications: The equation refers only to the concentration of the *same molecular species* in both phases. If there is a change in the molecular condition in transfer from one phase to another, the Nernst equation does not apply to the total amount of solute in each phase. This change may be the solvolysis of the solute, as in the hydrolysis of bromine in an aqueous phase, or it may be an association or dissociation process. For example, when acetic acid is distributed between water and benzene, it exists in water primarily as CH_3COOH—in say 1 N solution, only a small fraction is ionized—but in benzene it is primarily $(CH_3COOH)_2$. The system benezene–water–acetic acid involves, then, two equilibria. In the benzene layer,

$$2CH_3COOH \text{ (benzene)} \rightleftharpoons (CH_3COOH)_2 \text{ (benzene)}$$

and between the two layers,

$$2CH_3COOH \text{ (water)} \rightleftharpoons 2CH_3COOH \text{ (benzene)}$$

Adding these two equations together yields an equation for the overall equilibrium:

$$2CH_3COOH \text{ (water)} \Longrightarrow (CH_3COOH)_2 \text{ (benzene)} \qquad (2\text{-}8)$$

The equilibrium constant corresponding to this overall equation is given by

$$K' = \frac{[(CH_3COOH)_2]_{\text{benzene}}}{[CH_3COOH]^2_{\text{water}}} \qquad (2\text{-}9)$$

Since the concentration of single CH_3COOH molecules in the benzene layer is negligible, the molarity of the double molecules in this layer is simply one-half the total molar concentration of CH_3COOH calculated from the customary formula weight.

2-7
COLLIGATIVE PROPERTIES— VAPOR PRESSURE LOWERING

In a solution in which the solutes are nonvolatile, only the solvent contributes measurably to the vapor pressure. However, the presence of the solute reduces, by the mere effect of diluting the solvent, the vapor pressure of the solvent. When the solution is reasonably dilute, perhaps 0.1 m or less, it behaves very nearly ideally, since the environment of a solvent molecule is not greatly changed by the presence of the relatively small number of solute molecules. The magnitude of the vapor pressure lowering is then almost independent of the specific composition of the solute or solutes and is dependent only on the total concentration of solute particles.

The vapor pressure lowering brought about by the presence of nonvolatile solutes has associated with it consequences for other properties of the solution related to phase equilibria: The boiling point of the solution is higher than that of the solvent, the freezing point of the solution is lower than that of the solvent, and the solution displays an osmotic pressure.

The vapor pressure lowering, freezing point depression, boiling point elevation, and osmotic pressure of a solution are all mechanistically "bound together"; as a consequence they are termed *colligative* properties. Since for dilute solutions the magnitude of any one of these properties is independent of the nature of the solute material although fixed by the solute concentration, the term "colligative" has also come to be applied to any *non-substance-specific* property.

If any one colligative property of a solution has been measured, the other properties can be calculated accurately from the single experimental result. Determination of these properties affords an important method for establishing the molecular weight of a nonelectrolyte. Later, we will see that colligative properties of solutions containing ions supply significant information about the structure of such solutions.

In setting down equations to represent the lowering of the vapor pressure of the solvent by the solute, we shall denote the solvent by the subscript 1 and the solute by the subscript 2. If more than one solute is present, the concentration on a mole basis of solute to be employed is the sum of the concentrations of the individual molecular types.

If a solution is sufficiently dilute that Raoult's law applies,

$$p = X_1 p_1^0 \qquad (2\text{-}10)$$

Here p is the vapor pressure of the solution, practically equal to the vapor pressure of the solvent from the solution, and p_1^0 is the vapor pressure of the pure solvent at the same temperature. The vapor pressure lowering is $p_1^0 - p$, which by Equation (2-10) can be written

$$p_1^0 - p = p_1^0 - X_1 p_1^0 = p_1^0(1 - X_1) \qquad (2\text{-}11)$$

The sum of the mole fractions of all the components of a solution is equal to unity, and accordingly the quantity $1 - X_1$ must be equal to the sum of the mole fractions of all the solutes, which is represented as X_2. Equation (2-11) thus becomes

$$\Delta p = p_1^0 - p = p_1^0 X_2 \qquad (2\text{-}12)$$

From this equation, the fraction that the vapor pressure lowering represents of the original vapor pressure, termed the *relative vapor pressure lowering*, is equal to the mole fraction of solutes:

$$\frac{\Delta p}{p_1^0} = \frac{p_1^0 - p}{p_1^0} = X_2 \qquad (2\text{-}13)$$

Using Equation (2-13), one can calculate the molecular weight of a solute from the vapor pressure lowering produced by a known concentration of solute.

The definition of the molality of a solution is embodied in the equation

$$m = \frac{w_2/M_2}{w_1/1000} = \frac{1000 w_2}{w_1 M_2} \qquad (2\text{-}14)$$

where M represents molecular weight and w is the number of grams of a component in the solution. The mole fraction of solute is

$$X_2 = \frac{w_2/M_2}{w_1/M_1 + w_2/M_2} \qquad (2\text{-}15)$$

In dilute solutions, w_2/M_2 is much smaller than w_1/M_1 and thus can be neglected in the denominator. From Equation (2-14), $w_2/w_1 M_2$ can be replaced by $m/1000$, leading to

$$X_2 \approx \frac{m M_1}{1000} \qquad (2\text{-}16)$$

Combining this with Equation (2-13) results in

$$\frac{p_1^0 - p}{p_1^0} = \frac{m M_1}{1000} \qquad (2\text{-}17)$$

Example: A portion of 1.25 g of a solute dissolved in 100 g of ethanol lowers the vapor pressure at 70°C from 542.50 to 540.77 torr. Calculate the molecular weight of the solute.

Solution: The vapor pressure lowering is 542.50 − 540.77 or 1.73 torr. The relative lowering is 1.73 divided by 542.5. Using Equation (2-15),

$$\frac{1.73}{542.5} = \frac{1.25/M_2}{100/46.1 + 1.25/M_2} = \frac{1.25}{2.17M_2 + 1.25}$$

from which M_2 equals 181.

The direct measurement pressure of vapor in equilibrium with a liquid to any high degree of accuracy is beset with difficulty. Therefore, if a small difference in vapor pressure between solvent and dilute solution is to be determined, relative rather than direct procedures are employed. Usually a comparison is made with a solution of known vapor pressure. When water is the solvent, solutions of potassium chloride can be employed as reference solutions, since the vapor pressures of aqueous potassium chloride solutions have been very accurately determined.

The solution for which the vapor pressure is to be measured by the method of comparison is placed in a closed container—a vacuum desiccator is often convenient—together with a reference solution of which the vapor pressure is close to that of the unknown. The container is evacuated to facilitate vaporization and, after a period of time, the two solutions reach equilibrium by transfer of solvent through the gas phase.

2-8
FREEZING POINT DEPRESSION AND BOILING POINT ELEVATION

In this section we consider two colligative properties of dilute solutions that involve phase equilibria. The first of these is depression of the freezing point of a pure solvent by a solute. It is assumed that the solution of interest is sufficiently dilute and of such a nature that neither solute nor a solution of the two components comes out as the temperature is lowered, so that the solid that appears on freezing is pure solvent.

The condition for equilibrium between solid and solution, parallel to that for any other phase equilibrium, is that the vapor pressure of the solvent from solution be equal to the vapor pressure of the pure solid solvent. The vapor pressure of solvent in each phase varies according to an equation of the form of Equation (1-40):

$$\ln p = -\frac{\Delta H}{RT} + C$$

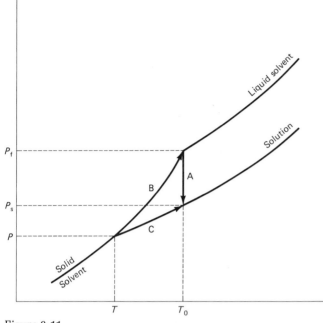

Figure 2-11
Depression of the freezing point of a liquid by the presence of a solute.

For the solid, ΔH refers to the heat of sublimation, whereas for the liquid it represents the heat of vaporization. Since the heat of sublimation is always larger than the heat of vaporization by an amount equal to the energy required to melt the solid, at any given temperature the vapor pressure curve for the solid has a greater slope than does that for the liquid. As a result, the vapor pressure curve of the solid approaches that for the liquid from below as the temperature is raised toward the melting point. The vapor pressure curve for the solvent from solution lies below the curve for the pure solvent. These three curves are shown for a typical system in Figure 2-11. Because of their relationships, the intersection corresponding to the freezing point of the solution is at a lower temperature than that for the pure solvent.

For an ideal solution, which obeys Raoult's law, the vapor pressure lowering along arrow A in Figure 2-11 is described by the equation

$$\frac{p_s}{p_f} = X_1 = 1 - X_2 \tag{2-18}$$

where X_2 is the total of the mole fractions of all the solutes. To establish the temperature T at which the solid and solution curves intersect, Equation (1-41) is applied to each equilibrium from this temperature up to the freezing point of the pure solvent T_0. For the solution, it is assumed that the heat of vaporization of the solvent is the same as it is from the pure liquid:

$$\ln \frac{p_s}{p} = \frac{\Delta H_{vap}}{R} \frac{T_0 - T}{T_0 T} \tag{2-19}$$

Table 2-5
Molal freezing point depression
constants

Solvent	Freezing point (°C)	K_f
Water	0.00	1.86
Benzene	5.45	5.09
Nitrobenzene	5.82	8.1
Cyclohexane	6.55	20.0
1,4-Dioxane	11.8	4.63
Acetic acid	16.7	3.9
t-Butyl alcohol	25.1	8.37
Phenol	40.9	7.40
Naphthalene	80.1	6.90
Camphor	178.4	40.0

For the solid,

$$\ln \frac{p_f}{p} = \frac{\Delta H_s}{R} \frac{T_0 - T}{T_0 T} \tag{2-20}$$

where ΔH_s is the heat of sublimation of solid solvent. Subtracting Equation (2-20) from Equation (2-19), and using the fact that the difference between the heat of sublimation and the heat of vaporization is equal to the heat of fusion ΔH_f one obtains

$$\ln \frac{p_s}{p_f} = \frac{-\Delta H_f}{R} \frac{T_0 - T}{T_0 T} \tag{2-21}$$

Combining Equations (2-18) and (2-21) leads to

$$\ln (1 - X_2) = \frac{-\Delta H_f}{R} \frac{T_0 - T}{T_0 T} \tag{2-22}$$

Two simplifications are possible because the solution under consideration has been required to be dilute. First, since the value of the natural logarithm of a number close to unity is equal to the amount by which the number differs from unity, the quantity $-X_2$ can be substituted for $\ln(1 - X_2)$. Second, because the freezing point depression is very small compared to the value of perhaps 200 to 500 K for T_0, $T_0 T$ in the denominator can be replaced by T_0^2. With $T_0 - T$ symbolized by ΔT_f, Equation (2-22) becomes

$$X_2 = \frac{(\Delta H_f)(\Delta T_f)}{R T_0^2} \tag{2-23}$$

In working with colligative properties, it is customary to express the concentration of the solution on the molal scale rather than on the mole fraction scale. Use of the approximation of Equation (2-16) for dilute solutions, and rearrangement, leads to

$$\Delta T_f = \frac{R T_0^2 M_1}{1000 \, \Delta H_f} m = K_f m \tag{2-24}$$

The quantity K_f is the *molal freezing point depression constant.* Since it contains only universal constants and properties of the solvent, K_f is characteristic of the solvent alone and independent of the solute; values for some common solvents are listed in Table 2-5.

The depression of the freezing point can be employed as a means of determining the total molal concentration of solute or solutes in a solution. This affords, as well, a path to the molecular weight of the solute, which can be calculated from the weight composition of the solution and the measured molality. In performing a molecular-weight determination by this method, the solvent is chosen to afford a reasonably large temperature change for a given concentration.

Example: The heat of fusion of carbon tetrachloride is 670 cal/mol, and the freezing point is $-24°C$. Calculate the molal freezing point depression constant.

Solution: On the absolute temperature scale, the freezing point is 249°. The molecular weight of CCl_4 is 154. Then

$$K_f = \frac{RT_0^2 M_1}{1000\, \Delta H_f} = \frac{(1.987)(249)^2(154)}{(1000)(670)} = 28.3$$

Example: What is the molecular weight of a material if 2.35 g lowers the freezing point of 130.0 g of benzene by 0.369°?

Solution: The molality of the solution is 0.369/5.09 or 0.0725. The number of moles of solute is then (0.0725)(130.0/1000) or 0.00943. The molecular weight is then 2.35/0.00943 = 249.

Since the best results for freezing point measurements to be used for molecular-weight calculations are obtained if the solution is dilute, it is desirable to be able to measure accurately temperature changes of a few tenths of a degree. The customary method for this is by use of a Beckmann thermometer, which is a mercury thermometer with a large bulb and a small-diameter capillary tube, along side of which there is a scale graduated in units of 0.01° and readable to 0.001°. A mercury reservoir attached to the capillary permits any desired temperature from perhaps -15 to 150°C to be brought onto the scale, which itself covers an interval of only five to six degrees.

We turn now to the other colligative property involving the temperature of a phase transition, elevation of the boiling point. The lowering of the vapor pressure of the solvent by the solute requires that a solution be raised to a higher temperature to reach the boiling point than the temperature at which the pure solvent boils. In this statement it is assumed of course that the solute is essentially nonvolatile, contributing no measurable amount of pressure to the total vapor pressure of the solution.

Elevation of the boiling point for a dilute solution can be treated quantitatively in much the same way as the freezing point depression. Raoult's law and the Clausius–Clapeyron equation are applied to the equilibrium between liquid solvent and solvent in the vapor phase to yield the equation

$$\ln \frac{p_a}{p} = \frac{\Delta H_{vap}}{R} \frac{T - T_0}{T_0 T} \tag{2-25}$$

where p_a is the vapor pressure at which boiling is to occur, usually atmospheric pressure, and p is the vapor pressure of the solvent from the solution at the temperature at which pure solvent boils at the same external pressure. This equation can be simplified, as was done for the corresponding equation for the freezing point depression, by assuming the solution to be dilute. The result is

$$\Delta T_b = \left(\frac{R T_0^2 M_1}{1000 \, \Delta H_{vap}} \right) m = K_b m \tag{2-26}$$

The quantity in parentheses, represented by K_b, is the *molal boiling point elevation constant;* it contains only universal constants and quantities characteristic of the solvent, and thus the magnitude of the boiling point elevation, like the magnitude of the freezing point depression, depends only on the solvent employed and the molal concentration of the solute and not upon the nature of the solute.

Table 2-6 lists some molal boiling point elevation constants for typical solvents. Because these values are smaller than freezing point depression constants, relative errors are in general larger in the determination of molal concentration from boiling point elevation. Furthermore, many substances, particularly biochemical molecules, tend to undergo decomposition more readily at the higher temperatures required for boiling point determinations.

An instrument of rather recent development used to measure the boiling point elevation of a solution is the *vapor pressure osmometer.* In this apparatus, two thermistor beads are mounted at the ends of probes leading into a closed chamber saturated with the solvent vapor. Thermistors sense the temperature, as described in Section 3-6. On one bead is placed a drop of pure solvent, and on the other a drop of solution. The amount by which the temperature of the two thermistors differs is a measure of the heat released by condensation of the solvent into the solution, and thus it is an indirect measure of the vapor pressure difference between the solvent and the solution. Although the instrument is called an osmometer, it does not measure directly the osmotic pressure, but rather the osmotic pressure can be calculated from the results of the experiment.

Table 2-6
Molal boiling point elevation constants

Solvent	Boiling point (°C)	K_b
Nitrobenzene	210.8	5.24
Acetic acid	117.9	2.53
1,4-Dioxane	101.3	3.27
Water	100.0	0.51
t-Butyl alcohol	82.4	1.75
Cyclohexane	81.4	2.79
Benzene	80.1	2.53
Ethanol	78.3	1.16
Carbon tetrachloride	76.8	4.88
Methanol	64.7	0.79
Chloroform	61.2	3.61
Acetone	56.2	1.71

2-9
OSMOTIC PRESSURE

If a quantity of sucrose is placed in a beaker or flask and the container is then filled with water, the solid dissolves in the water to produce, initially, a concentrated solution near the bottom of the vessel. If the container is allowed to stand undisturbed for a period of time, however, the concentration of the solution is found to be uniform throughout, for the sugar has become uniformly distributed throughout the solvent. Another way of describing the effect is to say that the molecules of sucrose *diffuse* from the region where they dissolve to all points in the solution, just as the molecules of a gas tend to fill any region available to them, although the diffusion of a solute in a liquid solution is usually very much slower than the diffusion of gaseous molecules. Diffusion corresponds to the random motion of the molecules; when there is a concentration gradient present, the effect of diffusion is to transport material from the region of higher concentration to the region of lower concentration.

Let us now imagine a system consisting of a concentrated sucrose solution placed on one side of a barrier such as a parchment membrane through which sugar molecules cannot pass, and on the other side of the membrane a quantity of dilute solution or pure solvent. Because sugar cannot penetrate the membrane, the process of sugar diffusion is not available to equalize the concentration. The tendency to equalization of concentration on the two sides of the membrane is nevertheless present. If the membrane is permeable to water molecules, it is these that pass through, from the phase more concentrated in water to the phase less concentrated in water, and thus more concentrated in sugar, until equalization of the concentration has been accomplished.

The process of diffusion of a solvent through a semipermeable membrane from a solution of higher solvent concentration to a solution of lower solvent concentration is known as *osmosis*. Some of the mechanisms by which a particular kind of membrane or barrier can pass solvent molecules while holding back solute molecules will be considered in Chapter 12, but it is important to point out that the one significant characteristic of an osmotic barrier is its selective permeability, without regard to the mechanism of selection. Indeed, osmotic transport through the vapor phase can be realized in the laboratory between solutions of differing concentrations exposed to the same vapor phase, as described on page 74 for the measurement of vapor pressure differences.

CALCULATION
AND MEASUREMENT OF
OSMOTIC PRESSURE

The process of osmosis can be halted by applying a pressure to the concentrated solution sufficiently great so that solvent molecules cannot pass through the barrier into this solution. If the barrier is a membrane, we tend to view this pressure as simply a mechanical one, driving the solvent molecules back. However, the effect of pressure can be interpreted in another and more general way. The key to osmosis is the decrease in the escaping tendency, or vapor pressure, of the solvent resulting from the reduction in solvent concentration on introduction of the solute. Both theory and experiment show that a large increase in external pressure effects a slight increase in the vapor pressure of any liquid. Thus one should regard the action of applying pressure to the more concentrated solution as that of raising the vapor pressure, and thus the escaping tendency of the solvent, from that solution until it is equal to the escaping tendency from the more dilute solution.

The effect of external pressure on the vapor pressure is described by the equation

$$\ln \frac{p''}{p'} = \frac{v_1}{RT}(P'' - P') \tag{2-27}$$

The vapor pressure p'' is that found under external pressure P'', and p' is the vapor pressure under pressure P'. The quantity v_1 is the volume occupied by 1 mol of solvent. For a solution, the osmotic pressure is defined to be the pressure that is just sufficient, when applied to the solution, to prevent passage of pure solvent into the solution through a perfectly semipermeable membrane. From Equation (2-27) this pressure, equal to $P'' - P'$ and symbolized by π, is predicted to be

$$\pi = \frac{RT}{v_1} \ln \frac{p_0}{p} \tag{2-28}$$

The vapor pressure of the pure solvent is p_0 and the vapor pressure of the solution is p; the excess pressure on the solution must be enough to raise p to p_0. The osmotic pressure of a solution is a colligative property, just as are the freezing point depression and the boiling point elevation. Equation (2-28) relates *exactly* the two colligative properties, vapor pressure lowering and osmotic pressure.

Example: Given that the vapor pressure of pure water is 26.739 torr at 27°C and the vapor pressure of a sucrose solution at the same temperature is 26.409 torr, calculate the osmotic pressure of the solution.

Solution:

$$\pi = \frac{[0.0821 \text{ liter atm}/(\text{mol K})](300.2 \text{ K})}{18.02 \text{ g/mol}/(996.5 \text{ g/liter})} \ln \frac{26.739}{26.409} = (1363)(0.01242) = 16.93 \text{ atm}$$

The osmotic pressure of a solution of known concentration can be directly predicted without recourse to knowledge of the vapor pressure lowering if the solution can be assumed to be dilute and ideal. If Raoult's law applies,

$$\ln \frac{p_0}{p} = \ln \frac{1}{X_1} = -\ln(1 - X_2) \approx X_2 = \frac{n_2}{n_1} \tag{2-29}$$

If this result is substituted in Equation (2-28), there is obtained

$$\pi = \frac{RT}{v_1} \frac{n_2}{n_1} = \frac{n_2 RT}{V_{solv}} \tag{2-30}$$

where the product of the molar volume of the solvent v_1 and the number of moles of solvent n_1 is equal to the total volume of solvent that contains n_2 moles of solute. For aqueous solutions, the number of moles of solute per liter of solvent is practically the same as the number of moles of solute per kilogram of solvent, so that Equation (2-30) becomes

$$\pi = mRT \qquad (2\text{-}31)$$

where m is the molality. This equation is known as the *Morse equation* and was developed as an empirical generalization before it was derived.

For dilute solutions, the volume of the solution is almost the same as the volume of solvent, and thus Equation (2-30) can be written as

$$\pi = \frac{n_2 RT}{V_{\text{soln}}} = cRT \qquad (2\text{-}32)$$

In this equation, which is known as the *van't Hoff equation,* c is the concentration of solute in the solution in moles per liter. The equation was originally proposed by van't Hoff on an empirical basis to correlate his experimental results for osmotic pressure measurements. It is apparent that the equation is formally like the ideal gas equation, but there is no direct mechanistic connection between the two.

Example: A solution of 171 g of sucrose in 1000 g of water has a density of 1.0575 g/ml at 20°C. Estimate the osmotic pressure by each of the approximate equations.

Solution: By the Morse equation, since the molecular weight of sucrose is 342,

$$\pi = \frac{171}{342}(0.0821)(293) = 12.0 \text{ atm}$$

To calculate the molarity, note that the total weight of a liter is 1057.5 g, of which a fraction 171/1171 is sucrose. Then

$$c = 1057.5 \times \frac{171}{1171} \times \frac{1}{342} = 0.452$$

By the van't Hoff equation,

$$\pi = 0.452(0.0821)(293) = 10.9 \text{ atm}$$

(These results can be compared with the experimental value of 12.86 atm.)

In order to carry out an experimental measurement of osmotic pressure that yields meaningful results, it is necessary to have a perfectly semipermeable membrane, one that passes the solvent but is entirely impervious to the solute. Many membranes available approach this requirement but do not really meet it; they appear semipermeable only because the solvent diffuses through at a much greater rate than the solute. There is, for example, a simple demonstration often used to show osmotic effects in which a parchment membrane is placed over the end of a glass tube, the tube is filled with sugar solution, and the film-covered end of the tube is immersed in solvent. The liquid level then rises slowly in the vertical glass tube, but, if it is observed for a period of time, the level reaches a maximum and then begins to fall, for the solute diffuses through the membrane into the solvent, reducing the osmotic pressure difference between the two phases. Further, the mechanical pressure to which the membrane is subjected by the column of solution may cause it to stretch, resulting in an increase in permeability.

In quantitative measurements, stretching is avoided by supporting the membrane on a grid or porous plate. Membranes have also been made of inorganic substances, such as cupric ferrocyanide or metal silicates, by precipitating the salt in the pores of a porous earthenware form which provides the mechanical strength. The most common type of film is a synthetic one, derived from solubilized cellulose by precipitation from an organic solvent in the shape desired. An example is collodion, which is cellulose nitrate dissolved in a mixture of alcohol and ether and from which a film can be deposited by evaporation of the solvent.

Because of the difficulty of obtaining a suitable membrane that is impermeable to small solute molecules, direct measurements of osmotic pressure are not often made on solutions in which the solute has a molecular weight of less than 1000. The osmotic pressure, when expressed for such solutions, is usually calculated from the experimental results for another of the colligative properties. In Chapter 12, there will be a further account of the use of osmotic pressure measurements in the study of macromolecules.

OSMOTIC EFFECTS
IN LIVING ORGANISMS

We turn now to some applications of the concept of osmosis in the discussion of fluid transfer in living systems. In considering these effects, the problem of the permeability of the membrane in question to the solutes in the fluids must be dealt with. Since some solutes can pass through a membrane of the sort found in a living organism, say the membrane surrounding a red blood cell, the osmotic pressure of the solution does not represent a measure of the tendency for solvent transfer. There is defined instead the *tone* of a solution, which is the effective osmotic pressure of the solution with respect to any particular membrane. Although the osmotic pressure is a colligative property of a solution, the tone is decidedly not, since it depends specifically on the nature of the solutes present. Only those solutes that are restrained by the membrane under consideration contribute to the tone of the solution.

Suppose pairs of aqueous solutions are placed on opposite sides of a membrane that is permeable to water and glucose but impermeable to sucrose and phenol. Various possible pairs are illustrated in Figure 2-12. In each case shown, the two solutions are *isosmotic* with one another, or have the same osmotic pressure. In case (a) the solutions are also *isotonic*, having the same tone, for the concentration of non-diffusible solute is the same on both sides of the membrane. In case (b) the solution on the right side is hypotonic, or of lower tone, while the solution on the left is hypertonic, or of higher tone. Only the solution on the left has any nondiffusible solute. In order to reach equilibrium,

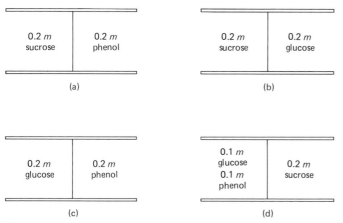

Figure 2-12
Illustrations of tone and osmotic pressure of solutions.

both water and glucose will migrate through the membrane toward the left. In case (c) the solution on the right is hypertonic, because only it contains nondiffusible solute. To reach equilibrium, water and glucose will migrate through the membrane toward the right. In case (d) the solution on the right is again hypertonic, for it contains twice the concentration of nondiffusible solute that the solution on the left contains. Equilibrium can be achieved by transfer of some of the water and some of the glucose from left to right.

If a section of plant tissue is examined under the microscope, the cells of which it is composed are seen to be enclosed in a rigid wall of cellulose. Normally the cells are in a condition of turgor, pressing against the supporting wall. If the tissue is placed in a hypertonic medium, water diffuses out of the cells and the protoplasm can be seen to shrink away from the wall. If placed in a hypotonic solution, the cells gain water and may indeed rupture the cellulose wall. The effective osmotic pressure of the cell contents can be determined by placing the tissue in solutions of varying concentration and observing what solution is isotonic with the cell contents.

The permeability of the walls of red blood cells, or erythrocytes, can be studied conveniently, since a suspension of the cells can be prepared and observed in various aqueous solutions. The cell contents are found to be isotonic with 0.9 percent sodium chloride solution; the cell wall seems impermeable to the sodium ion. In a hypotonic solution, the cells absorb water, change from a disk shape to a spherical shape, and finally undergo *hemolysis,* or leakage of the hemoglobin from within the cell into the solution. In hypertonic solution, the cells lose liquid and undergo *plasmolysis;* they can be seen to shrink, and they settle out of suspension.

The walls of human red corpuscles are found to be permeable to monosaccharides, urea, creatinine, H_3O^+, OH^-, Cl^-, HCO_3^-, and SCN^-. They are permeable to a limited degree to glycerin, Zn^{2+}, and K^+. They are practically impermeable to disaccharides, amino acids, and Na^+, as well as to many higher-molecular-weight materials.

The transfer of fluids within the body is governed in part by osmotic effects. The blood plasma is composed of about 7.5 percent proteins by weight, along with about 0.5 percent other organic compounds and 1.0 percent inorganic salts in water. The molality of the solutes averages 0.33, and the freezing point of the plasma is about $-0.56°C$. Of the osmotic pressure of 7.7 atm at 37°C, most is contributed by the inorganic salts; because of the high molecular weight of the proteins, they probably contribute only 30 to 40 torr.

The walls of the capillaries through which the blood flows are apparently permeable to all the principal constituents of the plasma (water, inorganic ions, glucose, amino acids, urea, and so on) except the proteins, so that it is only the proteins that determine the tone with respect to the capillary wall "membranes." Any decrease in the concentration of plasma proteins, such as may be caused by starvation, damage to the kidneys, or by infiltration into the tissues by leakage resulting from the presence of certain toxins, lowers the tone of the blood, so that water flows out of the bloodstream, increasing the volume of interstitial fluid outside the capillaries. This can be overcome by intravenous injection of concentrated blood plasma.

In normal circulatory function, there is a greater concentration of proteins in the blood than in the interstitial fluid bathing the capillaries and other body tissues. This would be expected to cause osmotic transfer to fluid in the bloodstream through the capillary walls. However, the heart acts as a pump to supply pressure to force the blood through the capillaries. At the inlet, or arterial, end of the capillary, this pressure is sufficient to overcome the effective osmotic pressure, and water and diffusible solutes are forced out of the bloodstream into the tissues. The pressure drop in passage of the blood through the capillaries reduces the hydrostatic pressure so that at the outlet, or venous, end of the capillary it is smaller than the difference in osmotic pressures of the interstitial fluid and blood, and liquid is transferred into the capillary.

The kidneys normally regulate the salt concentration of the blood. If they do not function properly in salt elimination, the freezing point of the serum may fall as low as a degree below zero. If an individual loses water from the tissues and bloodstream as a result of fever or another cause, the fluids become more concentrated in protein than normal. In this case the blood osmotic pressure may be so high that the kidneys are unable to excrete water and dissolved diffusible solutes. For filtration through the kidneys, the blood pressure must exceed the osmotic pressure of the blood; failure of the kidneys may be caused by low blood pressure as well as by high osmotic pressure.

2-10
PARTIAL MOLAR VOLUME

When the properties of a mixture, such as the volume occupied by a given amount of material, are measured, they may or may not be equal in value to the sum of the properties of the individual constituents, weighted according to the fraction of each in the mixture. Very often the interactions between components cause deviations from additive behavior. For example, if one mixes 884 cm³ of water with 279 g of solid potassium chloride, which has a volume of 140 cm³, the resulting solution has a volume of 1000 cm³ rather than the additive volume of 1024 cm³.

How shall we express the properties of nonideal solutions? The conventional method, and one that is convenient because it permits ready comparison of the behavior of a component of the solution with that of the corresponding pure substance, is to specify the contribution of each constituent as a *partial molar quantity*. For example, the partial molar volume of a component in a mixture is defined as the ratio of the change in volume of the mixture to the number of moles of that component added, when so small an amount of the component is added that the composition is essentially unchanged and when the temperature, pressure, and numbers of moles of the other constituents all remain unchanged. The partial molar volume of component i is thus equal to the partial derivative of the volume with respect to the number of moles of component i:

$$\overline{V}_i = \left(\frac{\partial V}{\partial n_i}\right)_{T,P,\text{other } n_i\text{'s}} \tag{2-33}$$

The partial molar volume of a solute in aqueous solution is conveniently evaluated by plotting the volume of solution of the given solute that contains 1000 g of water against the molality, and measuring the slope of a tangent to the curve at the desired concentration. Since molality is calculated as the number of moles of solute in a constant amount, 1000 g, of solvent, the change in molality is equal to the change in number of moles for a fixed amount of solvent, as demanded by the definition of partial molar volume. Figure 2-13 is a plot of this type for solutions of ethanol in water. Calculation of the partial molar volume is shown for a concentration of 2 m, at which the slope of the tangent is equal to 53 cm³/mol. This can be compared with the molar volume of pure ethanol, which is 58.3 cm³. The method of drawing tangents to the plot of volume against concentration is, however, not very accurate. Several other, more complicated but more accurate, methods of calculation have been worked out and can be found in the literature.

A partial molar volume can be found for every component in a mixture, solvent as well as solute. For a binary mixture, the partial molar volumes of the two components are related by the equation

$$\frac{d\overline{V}_1}{d\overline{V}_2} = -\frac{n_2}{n_1} \tag{2-34}$$

where n_i represents the number of moles of the ith component. For an ideal solution, rarely found, the partial molar volume of each component is independent of concentration and equal to the molar volume of the pure component. In many nonideal mixtures, the partial molar volumes vary regularly and smoothly from one composition extreme to the other.

The partial molar volumes of proteins in aqueous solution are usually quite well behaved. For example, data in the literature for aqueous

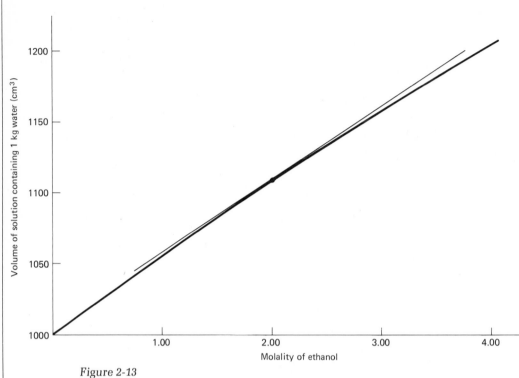

Figure 2-13
Volume of aqueous solutions of ethanol.

solutions of hemoglobin indicate that the partial molar volume of the protein is very nearly independent of concentration over the range from 8 to 42 g of hemoglobin per 100 cm³ of solution. Under these conditions, the partial molar volume is equal to a quantity called the *apparent partial molar volume,* defined by the equation

$$\bar{V}_{app} = \frac{V_{total} - v_1^0 n_1}{n_2} \tag{2-35}$$

In this definition, the solvent has been assumed to have the same molar volume v_1^0 that it does as a pure substance, and all deviations from ideality are attributed to the solute. The apparent partial molar volume can be quickly calculated from experimental results.

We will have occasion later to apply the idea of partial molar volumes in the discussion of buoyancy effects of the solvent on colloidal solutes in Chapter 12. Other types of partial molar quantities will be introduced in Chapters 3 and 4.

EXERCISES

1. It is desired to make an exactly 2 M solution of glucose. Describe the experimental procedure. How would it be modified to make an exactly 2 m solution?

2. Determine the volume of each material required to prepare 50 g of solution containing 0.35 mole fraction benzene in carbon tetrachloride. The density of

benzene is 0.879 and that of carbon tetrachloride is 1.594.

3. The density of an aqueous solution of glycerol containing 30.0 percent glycerol by weight is 1.0747. Calculate the molarity, molality, and mole fraction of glycerol in the solution.

4. The vapor pressure of heptane is 302 torr at 70°C; that of hexane is 796 torr. Calculate the mole fraction and weight percent composition of the vapor in equilibrium with an equimolar solution of these compounds at 70°C, assuming the solution is ideal.

5. What weight of p-aminonitrobenzene would decrease the freezing point of 20.0 g of benzene by 0.250°?

6. The heat of fusion of cyclohexanol is 419 cal/mol, and the freezing point is +25.5°C. Estimate the molal freezing point depression constant from these data.

7. Express the relative vapor pressure lowering of a dilute aqueous solution as a function of the molality of solute.

8. A mixture of 50 g of benzene and 50 g of toluene is kept at 93°C under a pressure of 760 torr. What phase or phases are present? If more than one is present, what are the relative amounts?

9. A quantity of 1.28 g of solute raises the boiling point of benzene by 0.137°. What is the molecular weight of the solute?

10. The freezing point of an aqueous solution of glucose is −0.63°C. Calculate the boiling point of the solution and its osmotic pressure at 25°C.

11. What is the vapor pressure at 100°C of an aqueous solution containing 2.50 g of a nonvolatile solute of molecular weight 150 in 500 g of water?

12. From the data in Table 2-2, calculate the amount of ethylene that will dissolve in 1 liter of water under a partial pressure of 100 torr at 30°C.

13. A mixture of 2.00 mol of methanol and 1.00 mol of n-hexane is maintained at 30°C. What phases are present—specify their compositions—and in what amounts?

14. An aqueous solution contains 0.30 g of caffeine in 50 ml. To this is added 20 ml of ether, and the system is allowed to come to equilibrium at 25°C. How much caffeine remains in the water? Compare with this the amount remaining after the aqueous solution is extracted with two successive 10-ml portions of ether.

15. Referring to Figure 2-10, describe the changes that occur when an equimolar mixture of the two components in the liquid phase at room temperature is heated until the entire sample has been vaporized.

16. From handbook data on the densities of methanol–water solutions at 20°C, estimate the partial molar volume of CH_3OH in a solution containing 25 percent by weight of alcohol.

17. A solution of 3.00 g of protein in 250 cm³ of water has an osmotic pressure of 176 torr at 25°C. Estimate the molecular weight of the protein.

18. Calculate the vapor pressure and osmotic pressure of a 0.150 m solution of lactose in water at 27°C. The vapor pressure of pure water at this temperature is 26.74 torr.

19. The freezing point of a solution in water is −1.78°C. Calculate the osmotic pressure of the solution at 25°C.

20. The osmotic pressure of an aqueous solution is 9.25 atm at 20°C. What is the vapor pressure of the solution if the vapor pressure of pure water is 17.535 torr and the solute is nonvolatile?

21. A solution of 20 g of glycerol in 200 g of water has a density of 1.020 g/cm³ at 20°C. Estimate the osmotic pressure at this temperature by each of the two approximate equations.

REFERENCES

Books

Dale Driesbach, *Liquids and Solutions,* Houghton Mifflin, Boston, 1966. Chapters 10,

11, and 12 cover solutions of nonelectrolytes, including a thermodynamic treatment appropriate after the reader has studied Chapters 3 and 4 in this text.

J. R. Overton, "Determination of Osmotic Pressure," in *Techniques of Chemistry,* A. Weissberger and B. W. Rossiter, Eds., Vol 1, Part V, Chapter VI, Wiley-Interscience, New York, 1971.

Evald L. Skau and Jett C. Arthur, Jr., "Determination of Melting and Freezing Temperatures," in *Techniques of Chemistry,* A. Weissberger and B. W. Rossiter, Eds., Vol. 1, Part V, Chapter III, Wiley-Interscience, New York, 1971.

A. G. Williamson, *An Introduction to Non-Electrolyte Solutions,* Wiley, New York, 1966. An elementary overview and an account of experimental methods, in addition to a more advanced thermodynamic and theoretical treatment.

Journal Articles

Pat Gaskins, "Countercurrent Distribution: Separation and Purification of Multigram Samples," *Am. Lab.,* p. 73 (October 1973).

Robert H. Goldsmith, "Solubility Relationships of Drugs and Their Metabolites," *J. Chem. Educ.* **51,** 272 (1974).

Three
Thermodynamics: First Law and Thermochemistry

The subject of thermodynamics—literally, "movement of heat"—deals with energy in its various forms, which include thermal, chemical, electrical, and mechanical, with the restrictions on the transformation of one type of energy into other types, and with the relation of energy changes to physical and chemical changes. It is concerned with the manner in which energy is supplied for or produced by processes of metabolism in living organisms, with the means for producing heat and work from fuels, and, by no means least, with the prediction of whether or not it is energetically possible for a certain chemical reaction to occur. Although we shall meet applications of thermodynamics in many of the subsequent sections of this book, the basic principles are outlined in this chapter and in Chapter 4.

3-1
ENERGY, WORK, AND HEAT

When a force acts upon a material body and accelerates the body—it may act upon a stationary body and set it in motion, for example—*work* is done upon the body and kinetic energy is imparted to it. Energy and work are two aspects of the same thing: Energy can, within certain limitations, be converted into work, while it is also possible for work to be converted into energy. Indeed, energy is best defined as *the capacity to do work*, or that which may be produced from or converted into work.

Heat is the term applied to energy transferred from one place to another because of a temperature difference between the two locations. When heat is taken up by a sample of material, the effect is observed as a rise in the temperature of that sample. We saw in Chapter 1 that thermal energy consists of the energy of random motion of the atoms and molecules composing matter. Matter in which this motion is more intense is at a higher temperature than material in which it is less intense, and the energy of the motion tends to become distributed as extensively and uniformly as possible—the result is the flow of heat down a temperature gradient. The concepts heat and work are related to energy in the process of transfer: Work represents the transformation of energy into concerted action of some sort, that is, some movement with a special orientation due to a force in a particular direction, but heat flow leads only to greater disorder of the motions of atoms and molecules.

Energy associated with matter can be classified—according to the manner in which it is related to the matter—as kinetic or potential energy. *Kinetic energy* is possessed by a body moving through space, which is able to do work as its velocity is decreased, as by striking another body. A body falling toward the earth's surface is acted upon by the force of gravity, which does work upon it; and this work becomes the kinetic energy of its downward motion, which can be reconverted into work when the falling body lands on an object. Whether the kinetic energy actually is converted into work or becomes dissipated as heat depends upon circumstances, but at least it is possible to visualize a device suitable for transforming it into work.

Potential energy, too, is a measure of capacity to perform work, not, in this case, because the body is moving and can be deprived of its kinetic energy by stopping it, but rather because it is located in a field of force and the force can be permitted to act upon it and to set it in motion. The force field can be that of gravity, that of an electric charge, or that of valence forces which draw atoms to one another. Consider as an example of the effect of the gravitational force field a body of water located in a reservoir at an elevation above sea level. In flowing down to sea level the water can do work—it can turn the blades of a turbine which drives an electric generator—and therefore, so long as it remains in the reservoir, the water is said to possess a certain positive potential energy with respect to sea level. This energy was of course supplied to it at some time in the past, possibly by solar radiation which raised it from the sea to the higher elevation. As an example of the second type of force field, there is the case of a positively charged ion placed in a gas space between two metal plates which bear positive and negative electric charges. The positive ion is drawn by electrostatic forces toward the negative plate and thus is accelerated. The electric energy associated with the electric potential difference—or voltage gradient—is then converted into the kinetic energy of the moving particle. Representing the third type of force field is the field acting on an atom bound into a molecule by covalent forces; the atom has a potential energy that is negative in sign with respect to the energy of the same atom set entirely free from bonding and removed from the influence of other atoms. If the atom is excited by thermal energy to vibrate about its equilibrium position, when it is removed from equilibrium it possesses positive potential with respect to its equilibrium position, to which it tends spontaneously to return.

The measurement of energy turns out always to involve the consideration of two factors: One of these is a *potential* or "driving force" factor and the second a *capacity* factor. Multiplying these two factors together yields an energy quantity. For example, thermal energy can

Table 3-1

Types of energy and typical units

Potential factor	Capacity factor	Energy type
Temperature Degrees	Heat capacity Calories per degree	Thermal Calories
Force Newtons	Distance Meters	Mechanical Joules
Potential Volts	Charge Coulombs	Electrical Joules
Pressure Dynes per (centimeter)^2 Newtons per (meter)^2	Volume (Centimeters)^3 (Meters)^3	Pressure-volume Ergs Joules
Surface tension Dynes per centimeter	Surface area (Centimeters)^2	Surface Ergs
(Velocity of light)^2 (Centimeters per second)^2	Mass Grams	"Mass" Ergs
Osmotic pressure Atmospheres	Solution volume Liters	Osmotic Liter atmospheres
(Velocity)^2 (Centimeters per second)^2	Mass Grams	Kinetic ($E = \frac{1}{2}mv^2$) Ergs

be measured by allowing a body to absorb the energy and measuring the increase in temperature of the body; a knowledge of the heat capacity of the body is also necessary in order to calculate the amount of heat energy taken up by the body, however. The heat capacity represents merely the heat absorbed by the body for each single degree rise in temperature. The temperature is a potential factor, for the temperature difference measures a driving force or a difference in level of potential; the heat capacity is the corresponding capacity factor. Common units for these factors for various sorts of energy are shown in Table 3-1.

We do not usually know the total amount of energy associated with any sample of matter, but rather we are concerned with *energy changes* occurring as the sample of matter undergoes physical and chemical changes. In order that one can think clearly about these changes in energy, it is necessary to define precisely that portion of the universe which is under consideration. That is, it is necessary to set up the limits or boundaries of a *thermodynamic system*. For each problem under study, the thermodynamic system may be defined in the most convenient manner for the particular situation. The system may include a certain group of molecules and whatever space they occupy: It may be a gram of ice, or a mole of ideal gas, or 10^{20} molecules of ethanol. It may be a living cell or it may be an entire organism. If the temperature of a system defined in this manner changes, the volume it occupies of course also changes, and the boundaries of a system defined to contain a certain amount of matter are flexible. It may be more convenient under other circumstances to define a system as a certain region of space and all the matter that happens at any time to lie within that region. For example, one might consider as a system the portion of gas flowing through a tube that at any time lies within a certain 1-foot length of that tube. The amount of material in this system may vary from time to time as the temperature or the pressure changes. Although the latter type of system is often useful in engineering work, for our purposes in chemistry we usually consider a system to include a fixed amount of matter rather than a fixed portion of space.

3-2
EQUIVALENCE
OF ENERGY FORMS—
FIRST LAW
OF THERMODYNAMICS

Although energy can be changed from one form to another, it cannot arise out of nothing nor can it disappear into nothing. This is a statement of the principle of conservation of energy, sometimes called the *first law of thermodynamics*. There are other ways in which this principle can be stated. For example, one can say: When energy in one form is lost, an exactly equivalent amount of energy appears in some other form. This simply means that the same number of units of the new kind of energy is always produced by the disappearance of a given number of units of the original kind of energy. An example of the application of the first law in this form is the statement of the so-called "mechanical equivalent of heat": 4.184 J of electrical or mechanical work is the equivalent of 1 cal of heat and thus, when it is completely transformed into heat, always produces exactly 1 cal of heat energy. Electrical work can be converted into heat by allowing current to flow through a resistance, and mechanical work can drive a stirrer in a liquid, raising the temperature of the liquid by friction-generated heat. The quantities of work and heat associated with these changes can be measured, and by this procedure the first law of thermodynamics is experimentally confirmed.

The principle of conservation of energy can be stated as a law governing the behavior of a thermodynamic *system*: The only way the amount of energy in a system can change is by the passage of energy through the boundaries of the system. It is possible to classify all energy in transit into or out of a system as either heat or work. We shall represent heat absorbed by a system by the symbol q, understanding that a numerical value of q with a negative sign corresponds to a loss of heat from the system. Work done on the system will be represented by w, and a negative value of w indicates that work is done by the system upon its surroundings. If the system undergoes a change, such as the compression of a gas, or the melting of a solid, or a chemical reaction, the internal energy becomes greater, because of the heat absorbed by the system and because of the work done on the system.

$$\Delta E = q + w \tag{3-1}$$

On the left side of this equation, the symbol delta represents "change of" or "increase in," and thus ΔE stands for the increase in *internal energy* and is a positive quantity for a process in which energy is given to the system and a negative quantity for a process in which the system undergoes a loss in energy. The internal energy E can be thought of as a measure of the energy resident in the atoms and molecules composing the system; however, no attempt is made to evaluate the magnitude of E, and it is defined solely in terms of changes in its value such as given by the first-law expression, Equation (3-1).

For an incremental change, this equation can be written

$$dE = \delta q + \delta w \tag{3-2}$$

The special symbol δ is used, instead of the usual d denoting a differential, because the magnitudes of heat and work for a process depend upon how the process is carried out, and so "dq" or "dw" could not be integrated from one state to another without specific knowledge of the pathway followed.

In the treatment of many simple processes, it is sufficient to take as the only work involved in the process the work the system does in expanding to push back the surroundings or the work done by the surroundings in compressing the system. This circumstance is of frequent occurrence in reactions carried out in the laboratory. However, it does not apply, for instance, to a chemical process conducted in an electrochemical cell in a manner so as to produce an electric current which can do electrical work. Types of work other than pressure–volume work can later be included, if necessary, by extending the simple equations.

Under the special assumption, then, that the only kind of work involved is pressure–volume work, w is obviously equal to zero for any process carried out in a closed vessel which is rigid so that no change can occur in the volume of the contents. Then Equation (3-1) becomes

$$\Delta E = (q)_V \tag{3-3}$$

Here V specifies the quantity held constant during the process being considered; this notation is an example of a common convention in thermodynamics by which a subscript such as P, V, T, or C indicates that the quantity pressure, volume, temperature, or concentration, respectively, is to be held constant during the event being described.

As shown in Chapter 1 for a gas, the product PV represents the energy associated with the system because the material constituting it prevents other material at pressure P from pushing into and occupying a portion of space of volume V. With this in mind, thermodynamicists have defined a quantity called the *enthalpy*, represented by the symbol H, and equal to the sum of the internal energy and the pressure–volume product:

$$H = E + PV \tag{3-4}$$

The symbol H derives from the former name of this property, the "heat content."

We now describe the application of the defining relationship, Equation (3-4), to changes from one set of conditions to another. Suppose a sample of material undergoes a change in volume and pressure, and perhaps also in temperature. The initial state of the material is designated by the subscript 1, and the condition after the change has occurred by the subscript 2. To find how much the enthalpy has increased, one subtracts the initial enthalpy from the final enthalpy:

$$\Delta H = H_2 - H_1 = (E_2 + P_2 V_2) - (E_1 + P_1 V_1) = \Delta E + \Delta(PV) \quad (3\text{-}5)$$

Equation (3-5) is a very general and very important equation. It applies to any process. It has a special form for *processes in which the pressure is held constant*; P_1 and P_2 are identical, and each is represented by the symbol P:

$$\Delta H = \Delta E + P \, \Delta V \quad (3\text{-}6)$$

Let us now consider, as we did above for ΔE, the special circumstance in which the only sort of work that can be done is pressure–volume work. Then the work done on the system is $-P \, \Delta V$, and the appropriate form of Equation (3-1) for a constant-pressure process is

$$\Delta E = (q)_P - P \, \Delta V \quad (3\text{-}7)$$

If this expression is substituted for ΔE in Equation (3-6), the result is

$$\Delta H = (q)_P - P \, \Delta V + P \, \Delta V = (q)_P \quad (3\text{-}8)$$

At least one reason for the particular way in which enthalpy is defined should now be evident: *For a process carried out at constant pressure, the heat absorbed is equal to the increase in enthalpy.* This statement can be compared with that embodied in Equation (3-3), which is: *For a process carried out at constant volume, the heat absorbed is equal to the increase in internal energy.*

3-3
SOME ISOTHERMAL
PHYSICAL CHANGES

PHASE CHANGES

A simple type of physical process, the vaporization of a liquid, affords us a first example of the application of the thermodynamic quantities that have been defined. This change can be readily carried out at constant pressure, for example by boiling the liquid at atmospheric pressure. The heat absorbed in converting a mole of liquid to vapor at constant temperature, or isothermally, is equal to the increase in enthalpy accompanying vaporization, with the proviso that the change is truly one at constant pressure.

Two contributions account for the energy required for the conversion of a liquid to vapor. Energy must be supplied to pull apart the molecules against the attractive forces that hold them together in the liquid state, and this energy becomes part of the internal energy E. In addition to this, pressure–volume work, equal in magnitude to the pressure multiplied by the volume change in going from liquid to vapor, must be done to push back the surroundings, such as the air if vaporization occurs against the atmosphere. Thus,

$$\Delta H_{vap} = \Delta E_{vap} + P \, \Delta V \quad (3\text{-}9)$$

Because the process is at constant pressure, ΔH equals q, so that

$$(q)_P = \Delta E_{vap} - w \qquad (3\text{-}10)$$

which is simply an expression of the first law.

When the process of vaporization is carried out as we have just considered it, it is an example of a *reversible* process, provided the temperature and pressure are exactly uniform over the system throughout the process. In reality, such a process would never get anywhere, because at least a very small temperature gradient is necessary in order for heat to be transferred from the surroundings to the liquid and at least a very small pressure gradient to force back the surroundings. However, we can imagine, as a limit to be approached by a real process occurring more and more slowly, a sequence in which a system passes through a continuous series of states, each of which represents equilibrium. In a *reversible* liquid–vapor transformation, for example, an infinitesimal decrease in pressure at any stage in the process would cause vaporization to occur, whereas an infinitesimal increase would suffice to send the system moving in the opposite direction, producing condensation.

Vaporization of a liquid can be carried out in many other ways than by a reversible path. Instead of opposing the vapor by a constant pressure, we could suddenly open the vessel containing the liquid to an empty chamber, allowing the vaporization to occur freely into a vacuum. In this pathway, no work is done by the vapor, and consequently q is smaller than for the constant-pressure process. However, if the same amount of liquid is converted into vapor, and the final temperature, pressure, and therefore volume are the same as for the reversible process, values of the quantities ΔH and ΔE will each be identical to the values for the constant-pressure process. This follows from the fact that the magnitudes of H and E are determined by the state of a system with no regard to the pathway by which the system has reached that state. Quantities such as H and E, the values of which are uniquely fixed when the composition, temperature, and pressure of a system are specified, independent of the history of the system, are termed thermodynamic *properties*.

Turning from the process of interconversion of a liquid and vapor, we now examine briefly other types of phase change. The conversion of a solid to a vapor, sublimation, is quite similar to the vaporization of a liquid, and a similar equation applies for the heat of sublimation:

$$(q)_P = \Delta H_{sub} = \Delta E_{sub} + P\,\Delta V \qquad (3\text{-}11)$$

Phase transformations between a solid and a liquid or between two different solid forms usually involve so little increase or decrease in volume that they can be treated as if ΔH were identical with ΔE.

VOLUME CHANGES OF AN IDEAL GAS

Another important kind of isothermal process is that in which the volume and pressure of a certain quantity of a gas undergo changes.

For simplicity, we deal only with the behavior of an ideal gas. It is possible to show, by methods of thermodynamics based on the second law, which is introduced in Chapter 4, that *the internal energy of a gas that obeys the equation of state of an ideal gas is independent of pressure and volume so long as the temperature is constant.* Thus for any isothermal change in pressure and volume of an ideal gas, ΔE is equal to zero. From the first law, it then follows that

$$q = -w \qquad (3\text{-}12)$$

This is equivalent to the statement that any heat absorbed is converted into the work required when the gas expands against its surroundings. Furthermore, the enthalpy change in the isothermal expansion of an *ideal* gas is found easily from Equation (3-5): Since the product PV has a value independent of P and V, and since ΔE is zero, ΔH is also equal to zero.

The isothermal expansion of a gas from volume V_1 and pressure P_1 to volume V_2 and pressure V_2 can occur in any one of various ways. Suppose that the gas is confined in a cylinder with a movable, weightless, frictionless piston. Expansion can be produced by suddenly pulling out the piston so that the gas expands into the newly created empty space. The pressure opposing the expansion of the gas is thus zero, and no work is done by the gas. Therefore no heat is absorbed in such a free expansion or expansion into a vacuum. A second arrangement that can be employed is one in which the piston is restrained by an external constant force so that the pressure is at all stages in the expansion equal to a fixed value, P_2; the gas can spontaneously expand against this pressure until its own pressure becomes equal to P_2. The amount of work done and of heat absorbed then corresponds to the product of the change in volume and the constant pressure P_2.

Still a third possible path of the expansion process is one in which the pressure exerted by the piston on the gas is always equal to the pressure of the gas itself, which from the ideal gas equation is nRT/V. The pressure of the gas and the pressure exerted by the piston thus both decrease as the expansion proceeds and the volume increases. At any stage in this expansion, an infinitesimal increase in the pressure of the piston would suffice to reverse the direction of the change and cause the volume of the gas to decrease. Of course, the expansion would not actually make any progress under the conditions of equal internal and external pressure, but again, as for the liquid–vapor phase change, the hypothetical reversible process is to be considered as a limit that any actual process can be made to approach.

The reversible process is important because the amount of work done can never be greater in magnitude than that done when the process is carried out reversibly; the work for the reversible process is thus the *maximum work*. Since the pressure varies throughout this pathway, in calculating the work as the product of the pressure and the change in volume, one must write the product as $-P\,dV$ and integrate, or sum, over all stages of the change from beginning to end:

$$-w_{\max} = \int_{V_1}^{V_2} P\,dV \qquad (3\text{-}13)$$

If the gas is ideal, the functional dependence of P on V is given by $P = nRT/V$, and the integral can be evaluated:

$$-w_{max} = \int_{V_1}^{V_2} nRT \frac{dV}{V} = nRT \ln V \Big|_{V_1}^{V_2} = nRT \ln \frac{V_2}{V_1} \qquad (3\text{-}14)$$

In view of the inverse proportionality of the pressure of an ideal gas to its volume, an equation equivalent to Equation (3-14) is

$$-w_{max} = nRT \ln \frac{P_1}{P_2} \qquad (3\text{-}15)$$

For an isothermal change of an ideal gas, the heat absorbed is equal to minus the work done, as stated in Equation (3-12). Consequently Equations (3-14) and (3-15) also give expressions for q_{rev}, the particular value of the heat absorbed when the process is carried out reversibly.

One of the important ideas in the preceding discussion of the amounts of heat and work associated with changes in a system is that the values of these two quantities depend upon how a process is carried out. This is not true for the thermodynamic *properties* ΔE and ΔH, the values of which depend only on where the process starts and where it ends and not on the path between. We can show the distinction by an analogy with an automobile journey: The road distance between New York and San Francisco depends upon the particular route the traveler chooses to follow, and so it is like q and w. However, the traveler from New York to San Francisco ends his journey some 3° of latitude farther south and 48.5° of longitude farther west, regardless of whether he goes by way of Chicago or by way of New Orleans—like ΔE and ΔH, the changes in latitude and longitude depend only on the initial and final location. Furthermore, just as any location of the traveler on the earth's surface corresponds to a definite latitude and longitude, so any state of a system corresponds to specific values of E and of H.

3-4
HEAT CAPACITY

To the extent that it is not consumed in an endothermic process such as a chemical reaction or phase change and is not used to do work on the surroundings, heat absorbed by a system appears as a rise in temperature of the system. The heat capacity of a system is defined as the ratio, $\delta q/\delta t$, of the heat absorbed by the system to the increase in temperature of that system. If the heat capacity C has a value that is independent of temperature, it can be calculated by dividing the amount of heat absorbed q by the corresponding rise in temperature ΔT:

$$C = \frac{q}{\Delta T} \qquad (3\text{-}16)$$

The more material there is in a system, the greater is its heat capacity. For this reason the heat capacity is termed an extensive property of the system: It increases with increase in the extent of the system.

The *specific heat capacity,* which is the value for 1 g of a substance, and the *molar heat capacity,* equal to the specific heat capacity multiplied by the molecular weight, are both intensive properties, which means that they have values independent of the amount of material but characteristic of its composition.

The amount of heat absorbed by a system for a given increase in temperature depends upon the exact path followed during the rise in temperature, and it is customary to distinguish the heat capacity values for the particular paths of constant volume and of constant pressure. For the constant-volume process, the heat absorbed is equal to the increase in internal energy, and for the constant-pressure process, it is equal to the increase in enthalpy. Accordingly, the heat capacities at constant volume C_V and at constant pressure C_P are given by the equations

$$C_V = \frac{dE}{dT} \tag{3-17}$$

$$C_P = \frac{dH}{dT} \tag{3-18}$$

To calculate the amount of heat absorbed for an incremental change in temperature, one can write these equations as

$$dE = C_V\, dT \tag{3-19}$$

$$dH = C_P\, dT \tag{3-20}$$

Integration of these equations over a range of temperature leads to expressions for the internal energy change and the enthalpy change from temperature T_1 to temperature T_2:

$$E_2 - E_1 = \Delta E = \int_{T_1}^{T_2} C_V\, dT \tag{3-21}$$

$$H_2 - H_1 = \Delta H = \int_{T_1}^{T_2} C_P\, dT \tag{3-22}$$

If the variation in heat capacity with temperature is known in analytical form, the expression can be substituted in the appropriate one of these equations and the integration explicitly carried out.

Example: Determine the enthalpy change required to increase the temperature of 2.20 g of CO_2 gas from 300 to 400 K if the molar heat capacity at constant pressure is given by the equation

$$c_P = 6.396 + 10.100 \times 10^{-3}T - 3.405 \times 10^{-6}T^2$$

Solution: The general equation for the enthalpy change of 1 mol between any two temperatures is

$$\Delta_H = \int_{T_1}^{T_2} c_P\, dT = \left[6.396T + 10.100 \times 10^{-3}\frac{T^2}{2} - 3.405 \times 10^{-6}\frac{T^3}{3} \right]_{T_1}^{T_2}$$

$$= 6.396(T_2 - T_1) + 5.050 \times 10^{-3}(T_2{}^2 - T_1{}^2)$$

$$- 1.135 \times 10^{-6}(T_2{}^3 - T_1{}^3)$$

Evaluating the enthalpy change for 1 mol from 300 to 400 K:

$$\Delta H = 639.6 + 5.05 \times 10^{-3}(16 - 9) \times 10^4 - 1.135 \times 10^{-6}(64 - 27) \times 10^6$$
$$= 951.1 \text{ cal/mol}$$

For the given amount of material, which is 2.20/44.0 or 1/20 mol, the heat required is 47.6 cal.

Most materials expand on warming at constant pressure, and therefore work must be done against the surroundings, as well as possibly to overcome intermolecular forces of attraction. In consequence C_P is almost always larger than C_V. For ideal gases, it is possible to deduce the magnitude of the difference between C_P and C_V. Ideal behavior requires that intermolecular forces be negligible, so that no work is required to overcome them as the gas expands. The work of expansion against the surroundings $P \Delta V$ is the only work that need be done, and the equation of state shows that this is $R \Delta T$ calories per mole. For unit temperature increase, the work is therefore equal to R, and the molar heat capacity difference is

$$c_p - c_V = R = 1.987 \text{ cal/(mol deg)} \tag{3-23}$$

where we have used small capitals to symbolize molar quantities.

Because the energy absorbed by a material as it is warmed is utilized for the most part in setting the atomic or molecular constituents into translational, rotational, and vibrational motions, the molar heat capacity often gives valuable information about atomic and molecular behavior in the material. We therefore turn now to an examination of the magnitudes of the contributions of these several kinds of motion to the heat capacity.

For a monatomic gas the only change in energy with a change in temperature at constant volume is the change in translational energy of the molecules. From Equation (1-23) the average value of the square of the molecular velocity $\overline{u^2}$ is $3RT/M$; the kinetic energy of translation of one molecule is $\frac{1}{2}mu^2$, which corresponds to $\frac{1}{2}M\overline{u^2}$ or $\frac{3}{2}RT$ for a mole of material. The rate at which this changes with temperature is $\frac{3}{2}R$ or 2.98 cal/(mol deg). The observed values of c_V for monatomic gases are found to be quite close to this number, and of course the values of c_P are then about $\frac{3}{2}R + R$, or 4.97 cal/(mol deg).

Gaseous molecules composed of two or more atoms may have rotational energy in addition to translational energy. If the molecule is linear, whether it be a diatomic molecule such as hydrogen, oxygen, or nitrogen, or one of certain polyatomic molecules such as carbon dioxide or acetylene, the rotational motion consists of turning end over end and can be resolved into two independent rotations. Each of these rotations is about one of two axes which are perpendicular to one another and to the axis of the molecule, and which pass through the center of mass of the molecule. For such linear molecules, rotation about the third axis does not move any appreciable mass and accordingly can be disregarded. For most linear molecules, it is found that the rotational energy is two-thirds of the translational energy. Since there are two independent modes of rotation and three independent components of translational motion, one parallel to each of the Cartesian axes, the generalization can be made that each independent kind of molecular energy has a magnitude of $\frac{1}{2}RT$ per mole, and therefore a heat capacity contribution of $\frac{1}{2}R$ per mole per degree. This generalization is known as the *principle of equipartition of energy*.

The experimental results for the heat capacities at constant volume for diatomic molecules are thus about 5 cal/(mol deg). Nonlinear molecules, such as water, can rotate about any or all of three mutually perpendicular axes, these axes, too, passing through the center of mass of the molecule. Once more each mode of rotation contributes about $\frac{1}{2}R$ to the molar heat capacity, following the equipartition principle, and the heat capacity at constant volume is about 6 cal/(mol deg).

In addition to translational and rotational contributions, there is evidently a third contribution to the heat capacities of gases. For example, we find at room temperature values of the molar heat capacity for the chlorine molecule, Cl_2, of about 1 cal/mol deg more than can be accounted for by translation and rotation, and for the carbon dioxide molecule, CO_2, values almost 2 cal more. What is involved here is the energy of vibration—the energy of the movement of atoms in a molecule with respect to one another. In a diatomic molecule, vibration represents a stretching of the interatomic bond, as the atoms move apart from one another, followed by a shrinking of the bond as they approach one another, a cyclic sequence repeated at a very high frequency. For N_2 or H_2 or CO or many other diatomic molecules aside from the halogens, the bond between the atoms is so strong that vibration is not occurring in any measurable fraction of molecules at room temperature, but only becomes significant at higher temperatures.

The general statement of the equipartition principle is that there is a contribution of one unit of $\frac{1}{2}RT$ to the energy for each "squared" term that enters into the expression for the energy. If the energy of the atom is kinetic, its magnitude is expressed by a term proportional to the square of the velocity, of the form $mv^2/2$. If the energy is potential, its magnitude is proportional to the square of the displacement from the equilibrium position, the position for which the potential energy is taken to be zero, and there is a term of the type $k(\Delta x)^2/2$, where k is a measure of the force required to produce deformation. Since vibration of the atoms in a molecule includes both kinetic and potential energies, each degree of vibration that is fully active contributes two units of $\frac{1}{2}RT$ to the energy or two units of $\frac{1}{2}R$ to the heat capacity.

For a diatomic molecule, the contribution of one vibrational degree to the heat capacity is therefore expected to be about 2 cal/(mol deg). However, even for halogen molecules, this contribution is not fully in evidence at room temperature and is only approached as a limit at high temperatures.

In molecules having more than two atoms, vibrations representing bending motions are possible, along with those in which valence bonds are stretched; some vibrational modes are active and contribute to the heat capacity in the vicinity of room temperature, although others become active only at quite high temperatures. Each fully active mode, like a fully active diatomic molecular stretching vibration, contributes two units of $\frac{1}{2}R$ to the heat capacity.

For polyatomic molecules, it is possible to predict the maximum value of the heat capacity that is approached at high temperatures and corresponds to the situation in which all rotational and vibrational modes are active. The starting point is the rule that, for a collection of N bodies, the location of the bodies at any time can be described by three coordinates for each, or a total of $3N$ coordinates. For a molecule containing N atoms, there are thus $3N$ variables available to describe the molecular behavior. Separation of translational motion of the molecule corresponds to using three coordinates out of the total of $3N$ as locators of the center of mass of the molecule. If the molecule is linear, two more coordinates are taken to describe rotation by specifying the orientation in space of the molecular axis with respect to some frame of reference. For a nonlinear molecule, three coordinates are associated with rotation. The *remainder* of the coordinates, $3N - 5$ or $3N - 6$, correspond to vibrations—that is, to distortions from the equilibrium, most stable, shape of the molecule.

To illustrate, we first list the number of coordinates of each type for a linear triatomic molecule, such as carbon dioxide, and the possible heat capacity contributions of each kind of energy if fully excited:

Translational	$3 \times \frac{1}{2}R = 3$ cal/(mol K)
Rotational	$2 \times \frac{1}{2}R = 2$
Vibrational	$[(3 \times 3) - 5] \times R = 8$
Total	13 cal/(mol K)

For a nonlinear triatomic molecule such as water, the contributions are

Translational	$3 \times \frac{1}{2}R = 3$ cal/(mol K)
Rotational	$3 \times \frac{1}{2}R = 3$
Vibrational	$[(3 \times 3) - 6] \times R = 6$
Total	12 cal/(mol K)

By similar arguments, one can predict for ammonia an upper limit of heat capacity of 18 cal/(mol K). These predictions can be compared

Table 3-2

Heat capacities of gases at constant pressure in cal/(mol K)

Gas	300 K	500 K	700 K
He	4.968	4.968	4.968
H_2	6.895	6.993	7.035
HCl	6.96	7.00	7.17
CO	6.97	7.12	7.45
N_2	6.961	7.070	7.351
O_2	7.019	7.429	7.885
Cl_2	8.119	8.624	8.821
CO_2	8.89	10.66	11.85
H_2O	8.03	8.42	8.96
CH_4	8.55	11.13	13.91
HC$\equiv$CH	10.53	12.97	14.37
NH_3	8.53	10.04	11.54

with the observed heat capacities at constant volume, obtained by subtracting R from the molar heat capacities at constant pressure given in Table 3-2. The fact, already mentioned, that the experimental values are always smaller than those predicted is an indication that the principle of equipartition of energy is not obeyed; in other words, not all the possible kinds of molecular motion are really being produced by the absorption of energy.

3-5
ENERGY CHANGES
IN CHEMICAL REACTIONS

Each chemical reaction has associated with it an energy change which is characteristic of the reaction and proportional to the amounts of the materials consumed or produced in the reaction. If the only work done is pressure–volume work, and the reaction is carried out at constant volume, then the amount of heat absorbed in the reaction is equal to the increase in internal energy ΔE. If the reaction is carried out at constant pressure, the amount of heat absorbed is equal to the enthalpy change ΔH and, because constant-pressure conditions are common since many reactions are carried out rather readily at atmospheric pressure, the value of ΔH is referred to as the *heat of reaction*. In this section we discuss for the most part changes in enthalpy, but the relation of these to changes in internal energy is also briefly considered.

THERMOCHEMICAL
EQUATIONS

A thermochemical equation represents the relation between energy change and the amount of reaction occurring, for some particular set of reaction conditions. For example, one can write

$$C_{12}H_{22}O_{11} \text{ (sucrose)} + H_2O(l) \longrightarrow C_6H_{12}O_6 \text{ (glucose)}$$
$$+ C_6H_{12}O_6 \text{ (fructose)} \qquad \Delta H_{25°C} = -4.8 \text{ kcal} \qquad (3\text{-}24)$$

This represents the experimental result that, at a temperature of 25°C and at constant pressure, the hydrolysis of 1 mol of sucrose by liquid water to form 1 mol of glucose and 1 mol of fructose is accompanied by the evolution of 4.8 kilocalories of energy. Now the entire course of the reaction was not necessarily at 25°C—the specification of temperature applies only to the initial and final states of the process, and in practice simply requires that the reactants begin at 25°C and the products end at 25°C and that all heat changes along the way are included. Note that the negative sign denotes a reaction in which the products have *less* enthalpy than the reactants, and thus the reaction is an exothermic one.

As another example of a thermochemical equation consider that for the oxidation of benzoic acid by gaseous oxygen:

$$C_6H_5COOH(s) + 7\tfrac{1}{2}O_2(g, 1 \text{ atm}) \longrightarrow 7CO_2(g, 1 \text{ atm})$$
$$+ 3H_2O(l) \qquad \Delta H_{25°C} = -771.6 \text{ kcal} \qquad (3\text{-}25)$$

This equation describes the reaction in which 1 mol of benzoic acid, in solid form as indicated by the "s" in parentheses, and $7\frac{1}{2}$ mol of gaseous molecular oxygen at 1 atm react to produce 7 mol of gaseous carbon dioxide at 1 atm pressure plus 3 mol of liquid water. Again, the temperature subscript attached to the symbol ΔH indicates that the reactants are initially at 25° and that there are included in the process all changes that would be required to bring the products back to 25° at the conclusion of the reaction.

In addition to strictly chemical processes, various physical changes together with the accompanying enthalpy changes are conveniently represented by thermochemical equations. For example, the sublimation of iodine can be described by the equation

$$I_2(s) \longrightarrow I_2(g) \qquad \Delta H_{25°C} = 7.44 \text{ kcal} \qquad (3\text{-}26)$$

Similar equations can be written for enthalpies of fusion, enthalpies of vaporization, or enthalpies of transition from one crystalline form to another:

$$S \text{ (rhombic)} \longrightarrow S \text{ (monoclinic)} \qquad \Delta H_{25°C} = 80 \text{ cal} \qquad (3\text{-}27)$$

A type of process in which there has been quite a bit of interest in recent years is that in which a molecule, usually one of high molecular weight such as a protein or synthetic polymer, undergoes a conformational change in solution. For instance, the enthalpy value associated with the change in which the enzyme lysozyme is unfolded in aqueous solution at pH 1.0 so that it loses its activity has been measured:

$$\text{lysozyme (native)} \longrightarrow \text{lysozyme (denatured)} \qquad \Delta H_{45°C} = 55 \text{ kcal} \quad (3\text{-}28)$$

The enthalpy value quoted refers of course to the denaturation of 1 mol of the enzyme, a quantity of about 13,000 g.

Heats of solution can be similarly treated. When several components mix to form an ideal solution, the change in enthalpy during the process is zero, but such behavior is the exception rather than the rule. The enthalpy change when a nonideal solution is formed depends not only on the amount of solute dissolved but also on the final composition of the solution. The concentration can be specified on a molar or molal scale, but it is satisfactory to define both the amount of solute *and* the concentration by simply stating the amount of each component going into the mixture. Thus the equation

$$H_2SO_4(l) + 4H_2O(l) \longrightarrow H_2SO_4(4H_2O) \qquad \Delta H_{25°C} = -12.92 \text{ kcal} \quad (3\text{-}29)$$

describes the process in which 1 mol of liquid sulfuric acid is mixed with 4 mol of water. The amount of heat evolved when 1 mol of sulfuric acid is dissolved in more water, say 10 mol, is greater:

$$H_2SO_4(l) + 10H_2O(l) \longrightarrow H_2SO_4(10H_2O) \qquad \Delta H_{25°C} = -16.02 \text{ kcal} \quad (3\text{-}30)$$

The heat of dilution is the enthalpy change per mole of solute present when solvent is added to a solution of a given concentration to dilute it to a specified lower value. Thus subtraction of Equation (3-29) from Equation (3-30) yields

$$H_2SO_4(4H_2O) + 6H_2O \longrightarrow H_2SO_4(10H_2O) \qquad \Delta H_{25°C} = -3.10 \text{ kcal} \quad (3\text{-}31)$$

The resulting enthalpy change is the heat of dilution of the acid solution from a concentration containing 1 mol of acid to 4 mol of water to one containing 1 mol of acid to 10 mol of water.

The illustration just given shows a simple case in which two thermochemical equations are algebraically combined to obtain the energy change for a third process. If the equations are added to obtain the desired equation, then the ΔH values are added in like fashion to give ΔH for the resultant. If one equation is subtracted from the other, then the ΔH values are subtracted in the same way.

Example: The heats of hydrogenation of a number of olefins were determined experimentally by G. B. Kistiakowsky and co-workers [*J. Am. Chem. Soc.* **57**, 876 (1935)]. The value for butene-1 was found to be -30.3 kcal/mol, and that for *trans*-butene-2, -27.6 kcal/mol. Calculate the heat of isomerization of butene-1 to *trans*-butene-2.

Solution: Write out the equations corresponding to the given heats of reaction:

$$CH_2{=}CHCH_2CH_3 + H_2 \longrightarrow CH_3CH_2CH_2CH_3 \qquad \Delta H = -30.3 \text{ kcal}$$
$$trans\text{-}CH_3CH{=}CHCH_3 + H_2 \longrightarrow CH_3CH_2CH_2CH_3 \qquad \Delta H = -27.6 \text{ kcal}$$

The equation for the desired reaction is obtained by subtracting the second equation from the first:

$$CH_2{=}CHCH_2CH_3 \longrightarrow trans\text{-}CH_3CH{=}CHCH_3$$

The corresponding value of ΔH is $-30.3 - (-27.6)$ or -2.7 kcal.

If more than two equations are to be combined, the prescription is simply to add or subtract the ΔH values just as the equations are added or subtracted.

ENTHALPIES OF FORMATION AND COMBUSTION

Certain types of heat of reaction have been given special names. One of these is the *heat of formation,* or enthalpy of formation, which is defined as the value of ΔH when 1 mol of the compound is formed from the elements in their standard states. An element is said to be in its standard state when it is at 1 atm pressure in the physical form that is stable at the pressure and temperature specified. Usually heats of formation are given for a temperature of 25°C. The standard state thus described is ambiguous for a few elements, such as carbon which can exist either as diamond or graphite under these conditions, and the basis chosen for the tabulation of values of heats of formation must then be stated. For Table 3-3, graphite has been taken as the standard state for carbon.

When it is stated that the enthalpy of formation of liquid water at 25°C is 68.32 kcal, it is possible to write down directly the corresponding thermochemical equation, for the "formation" value is understood to refer to 1 mol of water:

$$H_2(g, 1 \text{ atm}) + \tfrac{1}{2}O_2(g, 1 \text{ atm}) \longrightarrow H_2O(l) \qquad \Delta H_{25°C} = -68.32 \text{ kcal} \quad (3\text{-}32)$$

Enthalpies of reaction are often computed from tabulated values of enthalpies of formation by taking the difference between the enthalpies of formation of all the products and the enthalpies of formation of all the reactants. An alternative method is to subtract the *enthalpies* of the reactants from the enthalpies of the products but, unfortunately,

Table 3-3
Enthalpies of formation of representative compounds at 25°C in kcal/mol[a]

Name and state	Formula	ΔH_f	Name and state	Formula	ΔH_f
Sodium chloride(s)	NaCl	−98.23	Methanol(l)	CH_3OH	−57.04
Carbon monoxide(g)	CO	−26.42	Ethanol(l)	CH_3CH_2OH	−66.20
Carbon dioxide(g)	CO_2	−94.05	Acetic acid(l)	CH_3COOH	−115.73
Water(l)	H_2O	−68.32	Benzoic acid(s)	C_6H_5COOH	−91.7
Water(g)	H_2O	−57.80	Acetaldehyde(g)	CH_3CHO	−39.68
Sulfur dioxide(g)	SO_2	−70.96	Urea(s)	NH_2CONH_2	−79.58
Aluminum oxide(s)	Al_2O_3	−399.09	Glycerol(l)	$C_3H_8O_3$	−159.8
Ammonia(g)	NH_3	−11.02	D-Glucose(s)	$C_6H_{12}O_6$	−304.6
Methane(g)	CH_4	−17.89	Sucrose(s)	$C_{12}H_{22}O_{11}$	−531.1
Ethane(g)	C_2H_6	−20.24	Myristic acid(s)	$C_{13}H_{27}COOH$	−200.1
Propane(g)	C_3H_8	−24.82	Trimyristin(s, α)	$C_{45}H_{86}O_6$	−503
n-Octane(l)	C_8H_{18}	−59.74	Adenine(s)	$C_6H_5N_5$	23.2
Cyclohexane(l)	C_6H_{12}	−37.34	L-Glutamic acid(s)	$C_5H_9O_4N$	−241.2
Ethylene(g)	C_2H_4	12.50	Glycine(s)	NH_2CH_2COOH	−128.4
Butene-1(g)	C_4H_8	−0.03	Glycylglycine(s)	$C_4H_8O_3N_2$	−178.12
Benzene(l)	C_6H_6	12.3	Fumaric acid(s)	$C_4H_4O_4$	−193.8
			Malic acid(s)	$C_4H_6O_5$	−264.2

[a]Values for carbon-containing compounds are based on graphite as the standard state.

absolute values of the enthalpies are not known. Since the heats of formation are all measured with respect to the same reference condition, that of the elements in the standard state, the enthalpies of the elements in that state simply cancel out when the difference between formation enthalpies is calculated.

Example: Calculate the enthalpy change of the following reaction from the enthalpies of formation given in Table 3-3:

$$HOOC-CH=CH-COOH(s) + H_2O(l) \longrightarrow HOOC-CH_2-CHOH-COOH(s)$$

Solution: $\Delta H = \Delta H_f(\text{malic acid}) - \Delta H_f(\text{fumaric acid}) - \Delta H_f(\text{water})$
$= -264.2 - (-193.8) - (-68.3) = -2.1 \text{ kcal}$

The *heat of combustion* refers to another specialized type of reaction enthalpy. It is the value of ΔH for the reaction in which 1 mol of substance is completely oxidized. It is usually for an organic compound that one is interested in the heat of combustion, and "complete oxidation" for such a compound corresponds to conversion of carbon to carbon dioxide gas, hydrogen to liquid water, and nitrogen to the free element. Many heats of formation are calculated from experimental determination of the heats of combustion, since the combustion reaction is relatively easy to carry out.

Example: The value of the heat of combustion of tyrosine is given in tables of thermodynamic data as 1070.2 kcal/mol. Calculate the heat of formation.

Solution: The formula of tyrosine is

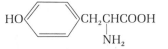

Using the empirical formula, we write the combustion equation:

$$C_9H_{11}O_3N(s) + 10\tfrac{1}{4}O_2(g) \longrightarrow 9CO_2(g) + \tfrac{1}{2}N_2(g) + 5\tfrac{1}{2}H_2O(l)$$

The heat of combustion is equal to the heats of formation of compounds on the right side of the equation less the heat of formation of tyrosine:

$$\Delta H_{comb} = 9(-94.05) + 5\tfrac{1}{2}(-68.32) - \Delta H_f(tyrosine) = -1070.2 \text{ kcal}$$

From this,

$$\Delta H_f = -1222.2 + 1070.2 = -152.0 \text{ kcal}$$

The heat of combustion is also significant as an indication of the energy that can be obtained by complete oxidation of a given material. For example, the energies obtained from foodstuffs when they are ingested by the living organism are related to the heats of combustion; on the average, the food calorific value of fats is 9 kcal/g, and that of proteins and carbohydrates, 4 kcal/g.

RELATION OF INTERNAL ENERGY CHANGE TO ENTHALPY CHANGE

It is often desired to relate the change in internal energy ΔE to the value of ΔH. For instance, when a reaction is carried out in a closed, rigid vessel such as a bomb calorimeter, the process is one at constant volume and the heat change measured is equal to ΔE. The two quantities are related by the equation

$$\Delta H = \Delta E + \Delta(PV) \tag{3-5}$$

Since the solids and liquids among the reactants and products have relatively small volumes and any changes in their volumes are quite negligible, the one circumstance under which ΔH differs appreciably from ΔE is that in which the number of moles of gaseous products differs from the number of moles of gaseous reactants. From the ideal gas equation, $\Delta(PV)$ is then simply (ΔnRT) where Δn is the change in number of moles of gas during the reaction.

As an illustration, consider the combustion reaction of sucrose, for which the thermochemical equation is

$$C_{12}H_{22}O_{11}(s) + 12O_2(g) \longrightarrow 12CO_2(g) + 11H_2O(l) \qquad \Delta H_{25°C} = -1349.6 \text{ kcal} \tag{3-33}$$

Since there is no change in the number of gaseous molecules, there being 12 mol of gas in the products for the reaction as written and 12 mol of gas in the reactants, the value of the enthalpy change is the same as the internal energy change. In the following reaction, there is a decrease of 1 mol in gases, and $\Delta H = \Delta E - RT$:

$$2NO(g) + O_2(g) \longrightarrow 2NO_2(g) \tag{3-34}$$

Example: The enthalpy of formation of aspartic acid is -232.64 kcal/mol, and that of β-alanine is -134.72 kcal/mol. Calculate ΔE for the decarboxylation of aspartic acid at 298 K.

Solution: The reaction is

$$HOOCCH_2CH(NH_2)COOH(s) \longrightarrow HOOCCH_2CH_2NH_2(s) + CO_2(g)$$

The enthalpy change is calculated from the difference in enthalpies of formation:

$$\Delta H = -134.72 - 94.05 - (-232.64) = +3.87 \text{ kcal}$$

Because the number of moles of gas increases by 1, $\Delta H = \Delta E + RT$. Therefore

$$\Delta E = 3.87 - (1.987)(298)/1000 = 3.87 - 0.59 = 3.28 \text{ kcal}$$

EFFECT OF TEMPERATURE ON ΔH and ΔE

The energy change in a reaction, whether that reaction is a chemical or a physical process, almost always depends in value upon the temperature. If the value is known for a particular temperature, however, it is possible to calculate the energy change at some other temperature by taking into consideration the heat capacities of the materials consumed and produced in the reaction. The general rule is: The energy of reaction is more positive at temperature T_2 than at temperature T_1 by the amount of energy required to raise the products from T_1 to T_2 less the amount of energy required to raise the reactants from T_1 to T_2. This is illustrated in Figure 3-1.

For the present, we consider that the heat capacities of all the materials involved in the process are themselves independent of temperature, although this is often not exactly correct. Then the enthalpy

Figure 3-1
Diagram of relation between heat of reaction and heat capacity difference between products and reactants. ΔH_{T_2} is larger than ΔH_{T_1}, in this illustration, by an amount $(T_2 - T_1)[C_P(\text{products}) - C_P(\text{reactants})]$.

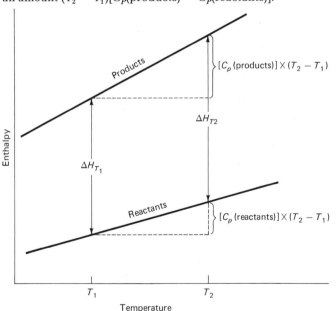

increase in each of the reactants or products can be calculated by simply multiplying its heat capacity at constant pressure by the relevant temperature interval. Consequently, the enthalpy change for the process at temperature T_2 is related to that at temperature T_1 by the equation

$$\Delta H_{T_2} - \Delta H_{T_1} = C_{p,\text{prod}}(T_2 - T_1) - C_{P,\text{react}}(T_2 - T_1)$$
$$= (C_{P,\text{prod}} - C_{P,\text{react}})(T_2 - T_1) = \Delta C_P(T_2 - T_1) \tag{3-35}$$

The quantity ΔC_P is the excess of the heat capacity, at constant pressure, of the products over that of the reactants.

The situation can also be summarized qualitatively: If it takes more energy to heat the products than it does to heat the reactants, an endothermic reaction requires more energy input at a higher temperature than at a lower temperature, whereas an exothermic reaction yields less energy at higher temperature than at lower temperature.

Example: The enthalpy change for the reaction $2H_2(g) + O_2(g) \longrightarrow 2H_2O(g)$ is -115.6 kcal at 300 K. Calculate the value at 500 K. The average values of the heat capacities at constant pressure over the range from 300 to 500 K are 6.95 cal/(mol deg) for H_2, 7.21 for O_2, and 8.21 for H_2O.

Solution: The change in heat capacity is $\Delta C_P = 2(8.21) - 2(6.95) - 7.21 = -4.69$ cal/(mol deg). Therefore $\Delta H_{500} = \Delta H_{300} - 4.69(200)/1000$ kcal $= -116.5$ kcal.

If the heat capacities of the individual reactants and products depend upon temperature, then it is likely that the value of ΔC_P or ΔC_V also varies with temperature. Then the differential equation equivalent to Equation (3-35) is

$$d(\Delta H) = \Delta C_P \, dT \tag{3-36}$$

The difference in enthalpy change between two given temperatures can be evaluated by integrating this equation between the two temperatures:

$$\Delta H_{T_2} - \Delta H_{T_1} = \int_{T_1}^{T_2} \Delta C_P \, dT \tag{3-37}$$

In order to carry out this integration explicitly, the algebraic form of the heat capacity difference is first substituted in this equation.

3-6
CALORIMETRY AND THERMAL ANALYSIS

Determination of the amount of heat used or produced in a chemical reaction or physical change is usually carried out by measuring the change in temperature of a system for which the heat capacity is known. For example, a chemical reaction can be carried out in a calorimeter, an insulated vessel of known heat capacity, and the change in temperature of the calorimeter and contents measured.

Because the accurate determination of temperature is a very important part of calorimetric work, as well as of other experimental procedures, it is appropriate to describe some of the methods of measuring temperature that can be employed.

The most familiar type of thermometer employs the expansion of a liquid as an indication of change in temperature. The accuracy is limited by the ratio of the smallest detectable expansion to the total volume of the liquid. The ratio can be made smaller and the accuracy improved by decreasing the bore of the glass capillary in which the liquid rises or by increasing the volume of the liquid in the thermometer bulb. However, there are problems associated with the fabrication of uniform small-bore capillaries and with their fragility, and, the larger the reservoir, the longer it takes for the sensing liquid to come to thermal equilibrium. The Beckmann thermometers used in measuring boiling point elevation or freezing point depression as described in Chapter 2 represent about the practical limit of sensitivity of mercury-in-glass thermometers.

Another principle of temperature measurement involves determination for a metal, such as platinum, of the electrical resistance, which increases with increasing temperature. Since the change with temperature is relatively small, a very sensitive measuring device is required—usually based upon a Wheatstone bridge such as is described in Chapter 5 for measurement of the resistance of conducting solutions. The resistance thermometer is convenient because the sensing element can be fairly small; it can be located some distance away from the measuring circuitry and can even be placed inside a living organism.

When two dissimilar metals are placed in contact with one another, a potential which varies with temperature is developed at the junction. This effect is applied in another type of device called a *thermocouple,* which consists of a junction situated at the point where the temperature is to be measured. A second junction, which must exist somewhere in order to complete the electric circuit, is located at a site where the temperature is known, frequently in an ice water bath in a Dewar flask. A potentiometer circuit is used to measure the difference in potential between the two junctions, which is an indication of the temperature difference between them. Typical pairs of metals used together in thermocouples include iron and the alloy constantan, and the two alloys chromel and alumel.

The potential difference developed at a thermocouple is of the order of 10^{-5} volts per degree, an amount too small to permit precise measurements in the range of fractions of degrees. However, the sensitivity can be increased by connecting a group of thermocouples in series to form a thermopile. In this arrangement, junctions are located alternately at the place where the temperature is to be measured and at the reference point, and the potential produced by the thermopile is the sum of the potentials produced by the individual couples. A thermistor is a recently developed device for measuring temperature, based on a semiconductor element. A mixture of oxides, the thermistor undergoes a change in resistance with temperature, which is about ten times that of a platinum thermometer.

In an *adiabatic* calorimeter, an effort is made to keep all the energy produced in a reaction within the calorimetric vessel itself, so that

Figure 3-2
Calorimeter for determination of heats of reaction using a Dewar flask.

none is lost to the surroundings. Of course, this goal is never quite attainable in practice, so that corrections for heat loss must be made.

A relatively crude form of calorimeter, suitable for large-scale liquid-phase reactions, can be easily set up in the laboratory by using an ordinary Dewar flask as a reaction vessel. One possible arrangement of the apparatus is shown in Figure 3-2. In order to correct for exchange of heat with the surroundings, a series of readings is taken on each of the two thermometers at time intervals of, say, a minute for a period of time before the solutions are mixed, and readings on the thermometer immersed in the reaction mixture are taken for some period after mixing. The sets of readings before and after mixing are each then extrapolated to a point halfway between the time of mixing and the time of maximum temperature. The difference between the extrapolated temperatures corresponds closely to the true rise in temperature that would have occurred as a result of reaction in the absence of any heat exchange with the surroundings. This treatment of data is illustrated for an exothermic reaction in Figure 3-3. It is of course necessary to stir the reaction mixture gently but thoroughly to ensure complete mixing and uniform temperature. The effective heat capacity of the calorimeter can be found by introducing a known amount of heat, either by carrying out a standard reaction or by passing a known amount of electric current through a resistance immersed in the calorimeter. The precision of this method is usually of the order of 1 percent or better.

Much more elaborate devices have been constructed, and are now available commercially, to give results of higher sensitivity and precision. One of the directions of extension of the adiabatic principle has been to surround the calorimetric vessel by a bath maintained by controlled heating at the same temperature as the vessel. This procedure has been carried out automatically by comparing the two temperatures by means of a thermopile with one junction attached to the calorimeter and the other attached to the vessel containing the bath fluid, and then providing a servomechanism to control the input of heat so as to null the potential difference and therefore also eliminate the temperature difference.

A less commonly used approach is to operate a calorimeter at a fixed temperature, isothermally rather than adiabatically. Heat loss is then either eliminated or takes place at a constant rate. For example, the Bunsen ice calorimeter determines the change in volume produced by the melting of ice at 0°C, and from this value is calculated the amount of ice melted and thus the amount of heat produced.

The methods described above, together with other modern developments in calorimetric techniques, have made possible the detection of very small amounts of heat and thus greatly extended the potential of calorimetry in the study of biochemical reactions. The aim of the investigations is not always to obtain quantitative thermodynamic data, but often merely to demonstrate that a reaction is occurring or to monitor the overall metabolic rate of a collection of living cells.

The typical microcalorimetric design includes two calorimeters, both mounted in a single heat sink, such as a block of metal. One of the calorimeters serves as a blank against which the calorimeter containing the sample under study can be compared.

Types of processes that have been studied by these methods include conformational transitions of macromolecules, the coupling of substrate to enzyme, the progress of enzyme-catalyzed reactions, hydrolysis and ionization reactions, and rates of bacterial and blood-cell metabolism.

Figure 3-3
Time–temperature curves in a calorimetric measurement, showing extrapolation to obtain true temperature rise.

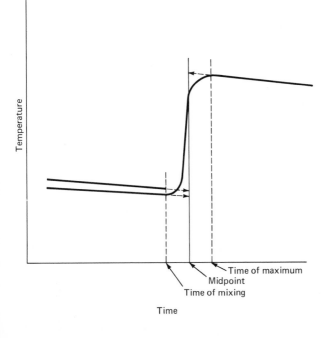

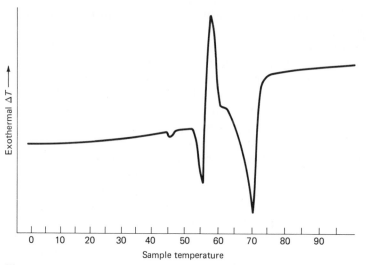

Figure 3-4
Differential thermogram of tristearin, 0.00531 g sample, heated at a rate of 4°C per minute. Reprinted with permission from E. M. Barrall and J. C. Guffy, p. 4 of *Oriented Fluids and Liquid Crystals,* American Chemical Society, Washington, D.C., 1967. Copyright by the American Chemical Society.

Closely related to the twin-calorimeter techniques of microcalorimetry is the procedure known as *differential thermal analysis.* Two samples, one an inert reference and the other the substance under study, are subjected to the same thermal process; that is, heat is supplied to each at the same rate. For an endothermic process, the temperature lags below that of the reference, whereas for an exothermic process in the sample, the temperature rises above that of the reference.

A differential thermal analysis has been carried out on tristearin, the triglyceride of octadecanoic acid. This material is of particular interest because of its relationship to some of the components of cell membranes, and because of its rather unusual phase behavior. A plot of the results is shown in Figure 3-4. A small endothermic effect is observed at 47°C, which is attributed to a phase transition in the solid from an orthorhombic crystal structure to a hexagonal crystal. A second endothermic phase transition occurs at 57°, corresponding to conversion of the hexagonal form to a triclinic form. However, this transition apparently includes a strongly exothermic part which follows at about 60°. The strongly endothermic process near 70° is the melting of the triclinic solid to the liquid form.

One version of differential thermal analysis is called *differential scanning calorimetry.* In this procedure, the temperatures of the sample and reference are automatically kept equal to one another within 0.01°C, and the power required to do this is measured as a region of temperature is "scanned" according to a definite program.

An example of the results obtained by this procedure is the heat of transition of bovine pancreatic ribonuclease A [T. Y. Tsong, R. P. Hearn, D. P. Wrathall, and J. M. Sturtevant, *Biochemistry* **9**, 2666

(1970)]. This protein undergoes a transition at a temperature varying from 30°C at pH 0.3 to 61°C at pH values greater than 7. The protein concentration used was 0.5 percent by weight, and the ΔH value found for the transition was 87.5 ± 0.5 kcal/mol.

EXERCISES

1. Calculate the energy in joules required to heat 25 g of helium gas at constant pressure from 25°C to 250°C.

2. A quantity of 48.0 g of oxygen gas, which may be assumed ideal, is expanded reversibly and at a constant temperature of 77°C from a pressure of 6.48 atm to a pressure of 2.16 atm. What are the values of q, w, ΔE, and ΔH in joules for this process?

3. The same process is carried out as in Exercise 2, except that the expansion occurs suddenly into a larger vessel rather than reversibly. Calculate the same quantities for this change as for the process in Exercise 2.

4. How much work is done against the atmosphere when a mole of carbon tetrachloride is vaporized into the air at its normal boiling point?

5. Calculate the heats of combustion of D-glucose, of urea, and of n-octane at 25°C from the enthalpies of formation at that temperature.

6. Calculate the heat of formation of n-butane from the data on page 104 and the value for butene-1 in Table 3-3.

7. A volume of 50 ml of 0.52 M hydrochloric acid is allowed to react in a calorimeter with 50 ml of 0.50 M sodium hydroxide solution. Calculate the heat of reaction for the formation of 1 mol of water by neutralization of the strong acid by the strong base. The effective heat capacity of the calorimeter is 16 cal/deg, the initial temperature of the calorimeter and sodium hydroxide solution is 25.02°C, the initial temperature of the acid is 24.90°C, and the final temperature of the mixture is 27.94°C. Assume the heat capacities of the aqueous solutions are the same as the heat capacity of water.

8. The heat of formation of α-D-glucose in aqueous solution is -301.88 kcal/mol. The enthalpy change on conversion of α-D-glucose to β-D-mannose in aqueous solution is 2360 cal/mol. Calculate the enthalpy of formation of β-D-mannose.

9. Calculate the enthalpy change at 25°C for each of the following reactions:
   (a) 2 mol gaseous ammonia + 1 mol gaseous carbon dioxide $\longrightarrow$ 1 mol solid urea + 1 mol liquid water
   (b) 1 mol ethylene + 1 mol hydrogen $\longrightarrow$ 1 mol ethane
   (c) 1 mol liquid ethanol $\longrightarrow$ 1 mol acetaldehyde gas + 1 mol hydrogen

10. Obtain a value of the enthalpy change at 298 K for the reaction

$$\alpha\text{-D-glucose(s)} \longrightarrow \beta\text{-D-glucose(s)}$$

from the following data:

$$\alpha\text{-D-glucose(aq)} \longrightarrow \beta\text{-D-glucose(aq)}$$
$$\Delta H_{298} = -1162 \text{ J}$$

The heat of solution is 10716 J/mol for solid α-D-glucose and 4680 J/mol for β-D-glucose.

11. The enthalpy of vaporization of benzene at its normal boiling point of 353 K is 7500 cal/mol. Calculate the value of ΔE for the vaporization process.

12. Determine the enthalpy change for the following reaction at 25°C:

$$CaCO_3(s) + 2HCl(aq) \longrightarrow CaCl_2(aq) + H_2O(l) + CO_2(g)$$

Known enthalpy changes are

$$CaCO_3(s) \longrightarrow$$
$$CaO(s) + CO_2(g) \qquad \Delta H_{298} = +42.5 \text{ kcal}$$
$$Ca(OH)_2(s) \longrightarrow$$
$$CaO(s) + H_2O(l) \qquad \Delta H_{298} = +15.6 \text{ kcal}$$
$$Ca(OH)_2(s) + 2HCl(aq) \longrightarrow CaCl_2(aq) + 2H_2O(l)$$
$$\Delta H_{298} = -37.8 \text{ kcal}$$

13. Predict the maximum value of the heat capacity at constant volume of each of the following molecules, assuming all degrees of freedom contribute fully to the energy: acetylene, ethylene, sulfur dioxide, nitric oxide, fluorine. Express your answer in J/(mol K).

14. The heat of combustion of glycine (α-aminoacetic acid) at 25°C is −232.5 kcal/mol. Calculate the heat of formation at 25°C, in calories and in joules.

15. Calculate the value of ΔH at 500 K from the value at 298 K and the average heat capacity for each substance involved as estimated from Table 3-2, for the following reaction:

$$2CO(g) + O_2(g) \longrightarrow 2CO_2(g)$$

REFERENCES

Books

H. D. Brown, Ed., *Biochemical Microcalorimetry,* Academic Press, New York, 1969. Descriptions of various aspects of experimental techniques and applications.

Paul D. Garn, *Thermoanalytical Methods of Investigation,* Academic Press, New York, 1965. Accounts of differential thermal analysis, thermogravimetric analysis, and related fields.

Irving M. Klotz, *Introduction to Chemical Thermodynamics,* W. A. Benjamin, Menlo Park, Calif., 1964. Chapters 1 to 6 give a more extensive treatment of material related to the first law, along with problems.

R. T. Sanderson, *Chemical Bonds and Bond Energy,* 2nd ed., Academic Press, New York, 1976. Heats of formation of compounds are correlated in terms of molecular parameters, and empirical correlations are used to estimate "bond energies."

D. R. Stull, E. F. Westrum, and G. C. Sinke, *The Chemical Thermodynamics of Organic Compounds,* Wiley, New York, 1969. A collection of useful thermodynamic data.

J. M. Sturtevant, "Calorimetry," in *Techniques of Chemistry,* A. Weissberger and B. W. Rossiter, Eds., Vol. 1, Part V, Chapter VII, Wiley, New York, 1971. Experimental methods.

Jürg Waser, *Basic Chemical Thermodynamics,* W. A. Benjamin, Menlo Park, Calif., 1966. Chapters 1 and 2 expand on the material in this chapter and provide more problems.

Bernhard Wunderlich, "Differential Thermal Analysis," in *Techniques of Chemistry,* A. Weissberger and B. W. Rossiter, Eds., Vol. 1, Part V, Chapter VIII, Wiley, New York, 1971. An account of experimental methods and their applicability.

Journal Articles

Bruce Cassel, "Recent Developments in Quantitative Thermal Analysis," *Am. Lab.,* p. 9 (January 1975).

Eric S. Cheney, "U.S. Energy Resources: Limits and Future Outlook," *Am. Sci.* **62,** 14 (1974).

Katherine F. Daly, "Applications of Thermal Analysis to Pharmaceutical Compounds and Related Materials," *Am. Lab.,* p. 57 (January 1975).

Mansel Davies, "Studies of Molecular Interactions in Organic Crystals," *J. Chem. Educ.* **48,** 591 (1971).

Joseph B. Dence, "Heat Capacity and the Equipartition Theorem," *J. Chem. Educ.* **49,** 798 (1972).

G. K. Estok, "Temperature Conversions and the New IPTS-68 Temperature Scale," *J. Chem. Educ.* **50,** 495 (1973).

Niel D. Jespersen, "A Thermochemical Study of the Hydrolysis of Urea by Urease," *J. Am. Chem. Soc.* **97,** 1662 (1975).

A. H. Kalantar, "Nonideal Gases and Elementary Thermodynamics," *J. Chem. Educ.* **43,** 477 (1966).

J. Kirschbaum, "Biological Oxidations and Energy Conservation," *J. Chem. Educ.* **45,** 28 (1968).

Ralph Roberts, "Energy Sources and Conversion Techniques," *Am. Sci.* **61,** 66 (1973).

Chauncey Starr, "Energy and Power," *Sci. Am.* **225,** 36 (September 1971).

V. A. Tucker, "The Energetic Cost of Moving About," *Am. Sci.* **63,** 413 (1975).

W. W. Wendlandt, "Thermal Analysis Techniques," *J. Chem. Educ.* **49,** A671 (1972).

P. G. Wright, "Quantities of Work in Thermodynamic Equations," *J. Chem. Educ.* **46,** 380 (1969).

Four
Thermodynamics: Second Law and Equilibrium

In Chapter 3 we explored some of the ramifications of the principle of conservation of energy and some of the methods of handling the bookkeeping concerned with an energy balance. In this chapter we consider the limitations on the possibility of converting thermal energy into work. Let us restate the essence of the first law of thermodynamics: If a certain amount of thermal energy disappears because it is converted into work, it is possible to calculate precisely the work done from the amount of thermal energy lost; if a certain amount of work is converted into thermal energy, it is possible to calculate precisely the heat produced from a knowledge of the work that has been done. What is not included in the first law is a statement of the conditions under which such energy conversions can take place. It is found by experience that the second type of process—conversion of work to heat—is possible without any limitation, but that the first type of conversion—heat to work—is restricted in amount, even *in principle*. This experience is embodied in the generalization known as the *second law of thermodynamics*.

This chapter also includes an examination of the connection between restrictions on energy conversion and the tendency for a physical or chemical change to proceed in such a way that equilibrium is approached. In the course of detailed calculations involving the new thermodynamic functions to be introduced, the reader should not lose sight of the ultimate aim of the application of thermodynamics to chemistry and biochemistry, that of predicting the conditions under which a system will be in equilibrium by employing only the results of calorimetric measurements without directly determining any equilibrium composition. Thus one can sometimes exclude, as possible chemical or biochemical mechanisms, proposed processes which contradict the predictions of thermodynamics.

4-1
THE TENDENCY FOR SPONTANEOUS CHANGE

The principle designated as the second law of thermodynamics can be stated in various ways, but each version expresses two ideas: the tendency of a system to approach a state of *maximum randomness* or a condition of minimum order; and the resulting *decrease in availability of energy for doing work,* a decrease associated with all spontaneous changes.

Let us consider some simple examples of spontaneous processes. We assert confidently that, if left to itself, water runs downhill. All that is required for this loss of positional potential energy is an unobstructed channel in which the water can flow. It is also reasonable to state that a watch spring never winds itself tighter but tends to uncoil. Sugar dissolved in water never concentrates itself at one place in the solution but rather diffuses spontaneously throughout the liquid to yield a uniform concentration if originally there was a concentration gradient. A gas never of its own accord compresses itself into a smaller volume but always expands into the space available to it. Heat flows from a body at higher temperature to a body at lower temperature. All these spontaneously occurring changes have common characteristics:

(1) They can be made to do work, for if properly harnessed they can be used to give a driving force to some mechanical operation.
(2) Even if they are not accompanied by the performance of work, when they occur there is a *net loss in capacity to do work,* as illustrated by the expanded gas which, without the intervention of some outside agency, can never undergo the same expansion again, even if it has expanded into a vacuum and therefore has done no work.
(3) They represent a change toward a more random arrangement with a lesser degree of orientation of molecules and smaller constraints upon their motions, as shown by the expanded gas and the diffused solute in which the molecules are no longer confined to so small a portion of space as before these processes occurred.

To summarize then, the second law can be stated in the form: Spontaneous changes are those that alter a system in the direction of maximum probability and make the energy in the system increasingly less available for doing work. This rule is only a statement of experience and cannot be "derived" from more basic principles. It is the description of a course of events, the occurrence of which has a very high degree of statistical probability; but because of its statistical nature, it applies only to systems large enough to constitute collections of many molecules and not to events involving only one or a few atoms or molecules.

It is convenient to describe how far away a particular state of a system is from the equilibrium condition of that system by specifying the value of a thermodynamic property. At this point we introduce one property that can be applied in this way: the *entropy*, represented by the symbol S. One way of defining the entropy is based on dividing the energy E of the system into two parts. One part, represented by the symbol A, called the Helmholtz free energy, is energy that can be converted into work through some process at constant temperature. The energy that cannot be converted into useful work by any possible isothermal process is given by the product of temperature and entropy, TS. Thus,

$$E = A + TS \qquad (4\text{-}1)$$

For a change occurring at constant temperature, this becomes

$$\Delta E = \Delta A + T \, \Delta S \qquad (4\text{-}2)$$

Based on the first law, this is equivalent to

$$\Delta E = w_{max} + q_{rev} \qquad (4\text{-}3)$$

This leads to an experimental method of determining ΔS for a process:

$$\Delta S = \frac{q_{rev}}{T} \qquad (4\text{-}4)$$

In words, the entropy change of a system during any process is equal to the heat that would be absorbed by the system if the process were carried out reversibly, divided by the temperature at which the process occurs. Of course, if the temperature changes during the process, the relationship must be applied successively to each small increment of the process, during which the temperature can be considered practically constant, and one writes

$$dS = \frac{\delta q_{rev}}{T} \qquad (4\text{-}5)$$

Let us examine the physical meaning associated with the entropy function. It is a measure of the extent of "run-down-ness," or disorder, or randomness in a system. For example, the entropy of a gas increases as the volume of the given amount of gas increases at constant temperature, because the molecules are freer and less limited in their motion. The melting of a crystalline solid is another instance of an increase of entropy: In this process the order of the atoms or molecules in the solid is at least partially lost.

Since the second law describes as the most probable state a condition of maximum randomness or disorder, it can alternatively be stated: *The entropy of the universe tends to a maximum.* Of course we can apply this statement only if we consider *all* the systems involved in a change. As an illustration, a hypothetical experiment can be imagined, involving a system consisting of two blocks, perhaps of metal, which are in contact with one another but isolated from the rest of the universe. One block is at temperature T_2 and the other at a slightly *lower* temperature T_1. Suppose that a spontaneous flow of heat occurs: From our intuitive feeling for temperature, we know that the block at the higher temperature T_2 will lose heat to the block at the lower temperature T_1. We make the further assumption that the amount of heat transferred in the process under discussion is so small that it does not affect the temperature of the two blocks.

In this process, the entropy change of the first block is q/T_2, where q is the amount of heat transferred, and that of the second block is q/T_1. Because T_2 is larger than T_1, the loss in entropy of the first block is smaller than the gain in entropy of the second block. Accordingly, the entropy for the entire system consisting of the two blocks increases during the process, and the total ΔS is positive. If the temperature difference between the two blocks is imagined to become smaller and smaller until the two blocks are virtually at the same temperature, the increase in entropy during the process will also grow smaller. In the limit of equal temperatures, there will be no net increase in entropy, but neither will there any longer be a transfer of heat; this limit corresponds to the equilibrium condition for the system of two isolated blocks.

Extending these ideas to cover processes of any type, we can state the second law in yet another way: For any spontaneous process, the total entropy change summed over all the systems affected by the process is greater than zero, whereas for an isolated group of systems at equilibrium, the total entropy change that would occur for an infinitesimal displacement in any direction from equilibrium is zero and for any slightly larger displacement is negative. If the second law is applied to an *isolated* system, one perfectly insulated from its surroundings so that no heat can pass its boundaries, then the only entropy change is that of the system itself, and the requirement for a spontaneous process is

$$\Delta S_{system} > 0 \qquad (4\text{-}6)$$

Most chemical reactions in which we are interested take place in a container—a beaker, a reactor, or a living organism—which is in some sort of thermal contact with the environment. Under these circumstances, the changes of entropy both in the environment and the reacting system must be added together to determine whether or not the reaction will occur spontaneously.

The entropy of a system is a thermodynamic property of the system. Like the other properties, internal energy and enthalpy, encountered previously, it has a value that is independent of the history of the system and which is uniquely fixed when the temperature, pressure, volume, and composition of the system are specified. Furthermore, the magnitude of the entropy change in a process is independent of whether the process is carried out reversibly or irreversibly. It is therefore possible to calculate the entropy change for any process occurring from a particular initial state to a particular final state by applying Equation (4-4) to the process, using the heat change *that would occur if the process were carried out reversibly.*

It is interesting to attempt to philosophize about how the ability of living systems to develop highly organized structures can be reconciled with the second law. This ability would seem to contradict the general principle that the world is "running down," or at least to represent a highly improbable state of affairs. However, in assessing the net entropy changes associated with the processes of growth and development, one must remember to include the entropy increase involved in the consumption of foods used by the living organism and in the conversion of these food materials to simpler molecules.

4-3
ENTROPY CHANGES IN ISOTHERMAL PHYSICAL PROCESSES

For the change in volume of an ideal gas maintained at constant temperature, the heat absorbed when the process is reversible is obtained from Equation (3-14) or (3-15):

$$q_{rev} = nRT \ln \frac{V_2}{V_1} = nRT \ln \frac{P_1}{P_2} \qquad (4\text{-}7)$$

In this equation P_1 and V_1 represent the initial pressure and volume and P_2 and V_2 represent the conditions at the end of the process. The change in entropy is found simply by dividing by the absolute temperature:

$$\Delta S = S_2 - S_1 = nR \ln \frac{V_2}{V_1} = nR \ln \frac{P_1}{P_2} \qquad (4\text{-}8)$$

Example: A quantity of 0.250 mol of nitrogen is initially confined in a vessel at 5.20 atm pressure. A valve is opened, allowing the gas to expand into a previously evacuated vessel such that the final pressure is 1.30 atm. The initial and final temperatures are both 300 K. Calculate the change in entropy of the gas.

Solution: Although the process is irreversible and no work is done, the change in entropy is the same as the change in entropy for a reversible process. If the gas is assumed to behave ideally,

$$\Delta S = (0.250 \text{ mol})[8.314 \text{ J/(mol K)}]2.303 \log \frac{5.20 \text{ atm}}{1.30 \text{ atm}}$$

$$= 4.787 \log 4 = +2.882 \text{ J/K}$$

If one is dealing with a gaseous mixture, and the mixture is ideal, the entropy of each component can be calculated from its partial pressure, without regard to the presence of the other components.

For a phase change at constant temperature and pressure, the value of q_{rev} is simply equal to ΔH, the enthalpy change for the phase transition. Thus for any reversible phase change, the value of ΔS for the sample can be calculated from the equation

$$\Delta S = \frac{\Delta H}{T} \tag{4-9}$$

The qualification that the phase change be *reversible* is important. Whether the change is actually carried out reversibly depends upon the conditions under which it occurs. Thus the vaporization of water at 1 atm pressure and 100.0°C is reversible, because these conditions represent a combination of temperature and pressure at which equilibrium exists. If water is vaporized at 100.0°C to form gas at a pressure of less than 1 atm, there is an increase in entropy of the water greater than $\Delta H/T$ and the process is spontaneous; if the pressure of the vapor is greater than 1 atm, the change in entropy of the water is less than $\Delta H/T$ and the process is not spontaneous.

Example: For carbon tetrachloride, the heat of vaporization is 7140 cal/mol and the normal boiling point is 77°C. Calculate the entropy of vaporization per mole of carbon tetrachloride when the liquid is vaporized at 77°C against a pressure of 76 torr.

Solution: The reversible entropy of vaporization is 7140 cal/mol divided by the absolute temperature, which is 350 K, or 20.4 cal/(mol K). Reversible vaporization would produce the vapor at 1 atm pressure. It is then necessary to calculate the further entropy change when the vapor is expanded from 1.000 to 0.100 atm, which is 1.987 $\ln(P_1/P_2) = 4.576 \log 10 = 4.6$ cal/(mol K). Adding these two changes in entropy gives a net change of 25.0 cal/(mol K).

Parallel to the change in entropy found for the change in pressure of a gas is the change in entropy of a component of a solution when the concentration is changed. On the mole fraction concentration scale, the difference in entropy for n moles of component i in an ideal solution when concentration 1 is changed to concentration 2 is given by

$$\Delta S_i = S_2 - S_1 = -n_i R \ln \frac{X_2}{X_1} \tag{4-10}$$

It should be easy to see that, qualitatively, an increase in concentration decreases the entropy, since the molecules are confined to a smaller region in space, whereas dilution increases the region available to each molecule and therefore corresponds to an entropy increase.

If the *entropy of solution* is defined as the entropy change of 1 mol of pure substance when it is mixed with another substance or other substances to form a solution of mole fraction X_i, then for an ideal solution,

$$\Delta S_{soln} = s_i - s_i^0 = -R \ln X_i \tag{4-11}$$

The quantity s_i^0, the *standard state* entropy, refers to the entropy when the mole fraction is unity, in other words, when the substance is pure.

Since it takes more than one material to make a solution, the entropy of mixing of the several components is

$$\Delta S_{\text{mix}} = -\sum_i n_i R \ln X_i \qquad (4\text{-}12)$$

It is also possible to express, at least approximately, the relation of entropy to concentration on the molarity scale:

$$s_i = s_i^0 - R \ln c_i \qquad (4\text{-}13)$$

Here, s_i^0 has a different meaning than in Equation (4-11): It is the entropy per mole of the substance when the concentration is 1 mol/liter.

Example: A solution of 0.002 M N-acetylglucosamine is on one side of a cell membrane which is normally impermeable to this solute. What change in entropy per mole is associated with the transfer of NAG across the membrane to a solution of 0.008 M concentration?

Solution: The difference in entropy is taken as the difference between the expression of Equation (4-13) for the two solutions: $\Delta S = s^0 - R \ln 0.008 - (s^0 - R \ln 0.002)$ $= -R \ln (0.008/0.002) = -(2.303)(1.987) \log 4 = (-4.576)(0.602) = -2.75$ cal/mol K.

If a solution is not ideal, there may be a change in entropy when it is formed which is in addition to the sum of the changes for the several components as given by the ideal equations above. The term "regular solution" has been applied by J. H. Hildebrand to mixtures for which there is neither a particular development of order in the solution, such as the formation of clusters or complexes, nor a breaking down of aggregates which might exist in the pure liquid. As a more rigorous criterion, we can state that a regular solution is one in which the only entropy change observed upon formation of the solution is the ideal entropy of mixing.

4-4
ENTROPY CHANGES IN CHEMICAL REACTIONS

It is possible to treat the entropy change in a chemical reaction in the same way in which enthalpy changes were dealt with earlier: The change occurring during the course of the reaction is set equal to the entropy of the products less the entropy of the reactants. Table 4-1 lists the entropies of a number of elements and compounds at the customary temperature of 25°C.

A significant difference in the use of entropy values from the use of enthalpy values arises from the possibility of obtaining and tabulating, from procedures to be explained in Section 4-5, absolute entropies rather than merely differences from an arbitrary zero. A second important point is that values of the entropy depend upon concentration, and it is therefore necessary to specify the pressure or concentration of the material for which an entropy is quoted. The values in the table

Table 4-1
Standard entropies at 25°C

Substance	Formula	S^0 (cal/mol K)
Graphite(s)	C	1.361
Oxygen(g)	O_2	49.00
Hydrogen(g)	H_2	31.21
Nitrogen(g)	N_2	45.77
Carbon monoxide(g)	CO	47.30
Carbon dioxide(g)	CO_2	51.06
Water(l)	H_2O	16.71
Water(g)	H_2O	45.105
Ammonia(g)	NH_3	46.03
Sodium chloride(s)	NaCl	17.3
Ethylene(g)	$CH_2{=}CH_2$	52.45
Ethane(g)	CH_3CH_3	54.85
Benzene(l)	C_6H_6	41.9
Cyclohexane(l)	C_6H_{12}	71.28
Ethanol(l)	CH_3CH_2OH	38.49
Glycerol(l)	$CH_2OH{-}CHOH{-}CH_2OH$	48.9
Acetone(l)	$(CH_3)_2C{=}O$	47.5
Acetic acid(l)	CH_3COOH	38.2
Ethyl acetate(l)	$CH_3COOCH_2CH_3$	62.8
Succinnic acid(s)	$COOH{-}CH_2{-}CH_2{-}COOH$	42.0
Fumaric acid(s)	$COOH{-}CH{=}CH{-}COOH$	39.7
Urea(s)	NH_2CONH_2	25.00
Glycine(s)	NH_2CH_2COOH	24.74
DL-Leucine(s)	$C_6H_{13}O_2N$	26.1
L-Glutamic acid(s)	$C_5H_9O_4N$	45.0
L-Tyrosine(s)	$C_9H_{11}O_3N$	53.0
Glycylglycine(s)	$C_4H_8O_3N_2$	45.4
DL-Leucylglycine(s)	$C_8H_{16}O_3N_2$	67.2

are *standard-state* entropies—symbolized by S^0—which means that each refers to the substance in the standard state which is defined to be the pure substance at 1 atm pressure.

Example: Calculate the entropy change for the following reaction at 25°C:

$$2 \text{ glycine(s)} \longrightarrow \text{glycylglycine(s)} + H_2O(l)$$

Solution: From Table 4-1, we find that the entropies of water and glycylglycine are 16.71 and 45.4 cal/(mol K), respectively, and the entropy of glycine is 24.74 cal/(mol K). The entropy change for the reaction is then

$$\Delta S = 45.4 + 16.7 - 2(24.7) = 12.7 \text{ cal/K}$$

4-5
DEPENDENCE OF ENTROPY ON TEMPERATURE

In order to calculate how much the entropy increases with an increase in temperature it is necessary to know only the heat capacity of the system being considered. The relationship

$$dS = \frac{\delta q_{rev}}{T} = \frac{C\,dT}{T} \qquad (4\text{-}14)$$

can be applied to any process in which the temperature is changed, using $C\,dT$ as the increment of heat absorbed in the reversible process during an incremental change in temperature dT.

If a process starts at temperature T_1 and the temperature is changed continuously to a final value T_2, Equation (4-14) can be integrated after inserting an explicit expression for the heat capacity in terms of temperature. In many cases, it is adequate to assume that C is independent of temperature, and the result then is

$$S_2 - S_1 = C \ln \frac{T_2}{T_1} \qquad (4\text{-}15)$$

The value of the heat capacity used in Equation (4-14) or (4-15) depends upon the circumstances of the change. The two commonly encountered types of processes are changes at constant volume, for which C_V is used, and changes during which the pressure is held constant, for which the value of C_P must be used.

Example: One mole of oxygen gas is heated at constant pressure from 300 K to 330 K. The average molar heat capacity C_P over this interval is 7.15 cal/K. Calculate the entropy change.

Solution: The result is obtained simply by substituting the heat capacity value and temperatures in Equation (4-15):

$$S_{330} - S_{300} = (7.15)(2.303) \log \frac{330}{300} = 0.682 \text{ cal/(mol K)}$$

The calculation of the dependence of entropy upon temperature is of particular interest because the results can be used to calculate "absolute" values of the entropies of chemical substances with the aid of a principle often termed the *third law of thermodynamics*. The third law states: The entropy of a perfect crystal at the absolute zero of temperature is equal to zero. All that is needed to calculate the entropy of a substance at any temperature, provided the substance exists as a suitable crystal at low temperature, is that the heat capacity be measured from very low temperatures up to the given temperature and that the enthalpy changes of any phase transitions in this range be known or measured. The amount by which the entropy at any finite temperature exceeds that at zero can then be calculated, and this quantity is equal to the entropy of the substance at the given temperature.

4-6
SOME APPLICATIONS OF THE ENTROPY FUNCTION; ENTROPY AND PROBABILITY

VAPORIZATION OF LIQUIDS

In Chapter 1, it was pointed out that, for many liquids, the enthalpy of vaporization follows the rule that its magnitude is about 21 times the temperature of the normal boiling point of the liquid on the absolute temperature scale:

$$\frac{\Delta H_{vap}}{T} = 21 \text{ cal/K} \tag{4-16}$$

Examination of this generalization, often called Trouton's rule, in the light of our knowledge of entropy changes enables us to reformulate it as follows: The molar entropy of vaporization of a normal liquid at the boiling point under atmospheric pressure is 21 cal/K.

Exceptions to the rule attributed to changes in the degree of association of the molecules can also be discussed in terms of the idea of entropy. If a polymeric liquid goes to a monomeric gas, there is a greater freedom of material in the gas phase beyond that resulting from the difference in physical state, and therefore the entropy of vaporization is greater than the usual value.

Liquids that boil at temperatures much below room temperature have values of the Trouton's constant somewhat smaller than 21. If the increase in entropy on vaporization is primarily a result of a decrease in concentration of molecules, then the entropy change on going to the vapor is smaller when the concentration of the vapor is greater, and thus it is smaller the lower the temperature if the pressure is always 1 atm. Trouton's constant should therefore be smaller for low-boiling materials if evaluated at the normal boiling point. As a consequence of these considerations, J. H. Hildebrand suggested that entropies of vaporization be compared, not at the normal boiling point, but at temperatures that lead to equal molecular concentrations in the vapor phase, and the consistency of the values for nonassociated materials is indeed then much better.

ENTROPY OF SOLUTION IN WATER

In any attempt to set up a model for the structure of water, one of the phenomena that must be considered is the change in thermodynamic properties which occurs when materials go into solution in water. Suitable measurements of solubility combined with results for the temperature dependence of solubility permit the entropy change accompanying the solution process to be determined.

For nonpolar solutes, the entropy of solution is generally more negative than the value for an ideal solution. This corresponds to an increase in some kind of molecular order. For example, methane dissolves in water with an excess negative entropy change over that for an ideal process almost twice as great as when it dissolves in hexane. Does this extra decrease in entropy correspond to the ordering associated with the formation of some structure involving a particular mode of binding of methane to water? Probably not, because the forces of attraction between methane molecules and water molecules appear to be quite small. Neither is there any reason to believe that the methane molecules dissolved in water form aggregates among themselves because of a strong mutual attraction. There remains the possibility of some increase in the structure of the water. Exactly what this increased structure corresponds to in molecular terms or how it comes about is by no means certain. One model of water views the liquid as containing large but transient clusters of molecules. These clusters have a very high order, but the order vanishes in a very short time by rearrangement of hydrogen bonds. A possible explanation of the effect of foreign molecules like methane is that they keep the clusters intact for a longer period of time by getting in the way of the formation, by molecules at the periphery of the cluster, of new hydrogen bonds with water molecules not in the cluster.

Some small polar molecules show effects similar to those of methane when their behavior on solution in water is compared to that on dissolution in other solvents. Figure 4-1 shows values of the function $\Delta H - T \Delta S$ plotted against ΔH for the solution of a number of polar molecules in several solvents. With water as solvent, the value of this function, the meaning of which is discussed more fully in Section 4-7, is more positive than would be expected from the ΔH value by comparison with the behavior of the same solute in other solvents, and thus the ΔS value is more negative for solution in water.

When ions of small size and large charge dissolve in water, there is no doubt about the explanation for the accompanying negative entropy change. Each ion orients about itself a spherical shell of some four to six water molecules which are relatively rigidly held in place. Of course, the rearrangement of molecules near the ion to provide the hydration shell also disrupts the structure a bit further out from the ion, increasing the entropy and partially offsetting the effect of the very pronounced local order. Larger ions, such as Br^-, I^-, or Cs^+, have in contrast a marked structure-breaking effect as judged by the positive entropy change associated with formation of their solutions. These

Figure 4-1
Relation of entropy and enthalpy of solution for several solutes in water and in other solvents. Open circles represent water as solvent; solid symbols represent the solvents dimethyl sulfoxide ($\triangle$), acetonitrile ($\square$), nitromethane ($\blacktriangle$), formamide ($\bullet$), and methanol ($\blacksquare$). Redrawn with permission from B. G. Cox, A. J. Parker, and W. E. Waghouse, *J. Am. Chem. Soc.* **95**, 1011 (1973). Copyright by the American Chemical Society.

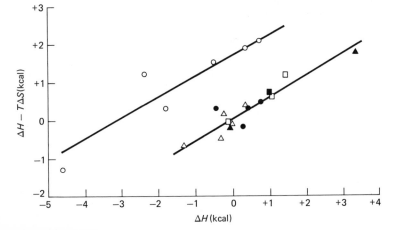

ions interfere with the normal water arrangement without introducing a high degree of order in their immediate vicinity.

ENTROPY AND PROBABILITY

The concept of entropy was related in Section 4-2 to the degree of disorder or randomness. It can also be connected with the idea of probability. The essential feature is that a state of perfect order can be achieved in only one way, whereas the number of ways in which states of disorder can be attained increases with the extent of the disorder. In consequence, the more disordered a state, the more probable is its occurrence.

To illustrate this, we consider a series of objects that can be oriented in two ways. Each member of the series can be assigned to either of two classes, let us say "up" or "down." The objects could be bits in a computer, magnetic moments of electrons, or just disks with one side red and the other side black. Suppose there are six of these "up-down" units in the series. This system can exist in various states; each state will be described by the total number of units up. For example, the state with all units up can be represented as ↑↑↑↑↑↑. This state is the most ordered one possible and has zero entropy. A state with five units up and one down can be attained in six different ways, each corresponding to one of the six units selected for the down orientation. Two examples are ↑↑↑↑↑↓ and ↑↑↑↓↑↑. The state with four up and two down can be achieved by any of 15 different combinations. The number of possibilities is calculated by multiplying the six different possibilities that can be selected for the first unit to be down by the five different choices that remain for a second unit to be down after the first one has been chosen, and dividing by two to eliminate duplicate states. A few of the 15 arrangements are ↑↑↑↑↓↓, ↑↑↑↓↑↓, ↑↑↓↑↑↓, . . . , ↓↑↓↑↑↑, and ↓↓↑↑↑↑.

The state of maximum probability is that of three up and three down, which can be obtained in $6 \cdot 5 \cdot 4/3 \cdot 2 \cdot 1$ or 20 different ways. This is the state that can be termed random, and it is the state with the maximum entropy. The state of two up and four down is quite equivalent in probability to the state of four up and two down, and so on.

Look at the same series of possibilities from a different point of view. Suppose that we repeatedly choose series of six units each, in random sequence as to up or down orientation. If we did this often enough, each of the various possible combinations of up and down would appear the same number of times. There would be 1 state with none up and all down for every 6 states with one up and five down for every 15 states with two up and four down, and so forth. The probability of a state with all up is one out of $1 + 6 + 15 + 20 + 15 + 6 + 1$ or 1 out of 64. If a statistical analysis is carried out for a system containing a great many units rather than just six, the conclusion is reached that the entropy of any state of the system is given by the equation

$$S = k \ln W \tag{4-17}$$

where W is the probability of that state and k is a constant which, on the molecular scale, turns out to be the Boltzmann constant.

THE STRUCTURE OF ICE

Ice is an example of a material that appears to violate the third law because it retains some entropy or disorder at the absolute zero of temperature. The statistical or probability interpretation of entropy has been applied to explain this and at the same time to draw conclusions about the structure of ice in its normal crystalline form, Ice I.

Of course a first suspicion about a material that does not have zero entropy at absolute zero is that it contains impurities, because the presence of an impurity always corresponds to an entropy of mixing of the components. This is not the case for ice, however, at least not in the sense of the presence of chemical impurities, for very pure samples have been studied. Linus Pauling pointed out that the residual entropy of ice is probably related to the circumstance that each hydrogen atom, lying as it does between two oxygens, can be closer to one of the two oxygen atoms than to the other. This corresponds to the existence of a covalent bond to one oxygen, the nearer one, and a hydrogen bond to the oxygen that is farther away. Each hydrogen atom has two possible sites at which it can be located, one closer to one oxygen and the other closer to the other oxygen, limited by the requirement that each oxygen atom has two nearer neighbors corresponding to two covalent bonds. For an ice crystal there are a great many possible ways in which the hydrogen atoms can be distributed over the available pairs of sites, each pattern of locations being energetically equivalent and equally probable. No one of the patterns can be said to correspond to a unique arrangement of perfect order.

The significance of this arrangement is pointed up by considering some of the alternatives. One is a structure in which each hydrogen is suspended halfway between two oxygen atoms, with the bonds being neither purely covalent nor purely of the hydrogen bond type. Since this structure assigns a definite position to each hydrogen atom, the entropy would be zero. Another possibility is one in which all the water molecules are arranged in a particular sense in the crystal—the pairs of covalent bonds to a single oxygen thus would all project in the same direction from all the oxygen atoms in the ice crystal. The existence of the residual low-temperature entropy excludes both of these proposals.

Pauling's quantitative estimate of the residual entropy of the ice structure is based on the following arguments: In a mole of water molecules there are $2N$ hydrogen atoms. If each atom can occupy either of two sites without any restriction, there are 2^{2N} possible arrangements. Around a given nucleus of oxygen, there are 2^4 ways of arranging the four hydrogen atoms, but only 6 of these ways yield a water molecule. Thus the total number of allowed arrangements corresponds to the fraction $(\frac{6}{16})^N$ of the 2^{2N} total. Multiplying the fraction by the total yields the result $(\frac{3}{2})^N$ for the number of allowed arrangements. The entropy contribution calculated from this number by use of Equation

(4-17) is 0.805 cal/(mol K), compared to the experimental value of 0.82. Thus the entropy values support Pauling's proposal for the locations of hydrogen atoms in ice.

4-7
THE FREE ENERGY FUNCTION AND ITS SIGNIFICANCE

In previous sections, the entropy was employed to measure the extent to which the energy of a system is unavailable for doing work at constant temperature and to characterize the tendency for a system to change toward the equilibrium condition. However, as we have pointed out, it is necessary to consider the contributions to the entropy change of both the system and the surroundings when using the criterion of positive entropy change for a spontaneous reaction.

It is possible to concentrate attention on changes within the system itself by employing another thermodynamic function defined by the equation

$$G = H - TS \tag{4-18}$$

This function is called the *Gibbs free energy*, or more often merely the *free energy*; in this book we symbolize it by the letter G; the letter F has also been used quite frequently to represent the same quantity.

We can think of the enthalpy as composed of two parts, the Gibbs free energy and the isothermally unavailable energy TS:

$$H = G + TS \tag{4-19}$$

Because the quantities H, S, and T are all thermodynamic properties, the free energy, too, is a thermodynamic property or function of the state of a system, which means that the magnitude of the change in free energy during a process is independent of the pathway of the process and depends only upon the conditions of the system in the initial and final states.

The equation for the change in free energy for a constant-temperature process proceeding from state 1 to state 2 is

$$\Delta G = G_2 - G_1 = (H_2 - H_1) - T(S_2 - S_1) = \Delta H - T\,\Delta S \tag{4-20}$$

If the process is carried out so that the pressure is also constant, the enthalpy change is equal to $\Delta E + P\,\Delta V$. Thus,

$$(\Delta G)_{T,P} = \Delta E + P\,\Delta V - T\,\Delta S \tag{4-21}$$

Now the value of ΔE can be expressed by the first-law equation, using the fact that maximum work is obtained when the process is carried out reversibly:

$$(\Delta G)_{T,P} = q_{\text{rev}} + w_{\text{max}} + P\,\Delta V - T\,\Delta S \tag{4-22}$$

But it was stated above that ΔS is equal to q_{rev}/T, so that the first and last terms on the right-hand side of this equation are equal and can be canceled.

With signs changed, Equation (4-22) thus becomes

$$-(\Delta G)_{T,P} = -w_{\max} - P\,\Delta V \qquad (4\text{-}23)$$

Thus the decrease in free energy of the *system itself* during a change at constant temperature and pressure is equal to the maximum work that can be done by the system in the course of the change *less* the pressure–volume work that must be done against the atmosphere; this difference is called the *net work.*

FREE ENERGY CHANGE AS A CRITERION FOR EQUILIBRIUM

If all the work done in a process is of the pressure–volume type, the value of $-w_{\max}$ at constant pressure is just $P\,\Delta V$ and the net work is zero. Thus if a process is to be spontaneous, which means that the magnitude of the work that can be done is less than $-w_{\max}$, ΔG must be negative.

An alternative proof of this result proceeds by considering the entropy changes resulting from the interchange of heat between a system and its surroundings. For a spontaneous process, there is a net increase in entropy, and thus

$$\Delta S_{\text{system}} + \Delta S_{\text{surroundings}} > 0 \qquad (4\text{-}24)$$

But $\Delta S_{\text{surroundings}}$ is equal to $q_{\text{surroundings}}/T$, which is equal to $-q_{\text{system}}/T$, which in turn is equal to $-\Delta H_{\text{system}}/T$. Therefore

$$\Delta S_{\text{system}} - \frac{\Delta H_{\text{system}}}{T} > 0 \qquad (4\text{-}25)$$

On the basis of the definition of the isothermal free energy change, Equation (4-20), this is equivalent to

$$-\Delta G_{\text{system}} > 0 \qquad (4\text{-}26)$$

To reiterate: *For a change at constant pressure and temperature to occur spontaneously, the free energy change of the system must be negative.* Spontaneous chemical processes are those that take the system in the direction of a free energy minimum; the situation at the minimum corresponds to an equilibrium condition. Reactions tending away from equilibrium, reactions that therefore lead to an increase in the free energy of the system, must have some external driving force if they are to take place.

FREE ENERGY CHANGES FOR PHYSICAL PROCESSES AT CONSTANT TEMPERATURE

For processes in which the temperature does not change, the free energy change can be calculated from Equation (4-20). For a physical change, such as a phase change, at constant temperature and pressure, ΔH is equal to $T\,\Delta S$, so that ΔG is equal to zero. It is therefore possible to make the very important statement that *the free energy per mole of any material in one phase is the same as the free energy per mole in any other phase with which the first phase is in equilibrium.* Processes such as the melting of a pure solid or the vaporization of a pure liquid, if conducted at a fixed pressure such as atmospheric pressure, necessarily fall into the category of constant-temperature processes.

If the phases concerned in an equilibrium are mixed phases, the same principle of equal free energy applies for each component. If two phases in which the free energy of some component differs are in contact, then there will be a spontaneous transfer of that component from the phase in which its free energy is greater to the phase in which its free energy is less.

For an ideal gas, the value of the enthalpy change for any alteration in pressure and volume at constant temperature is equal to zero. Consequently ΔG is equal to $-T\Delta S$. Since the change in entropy by Equation (4-8) for expansion from volume V_1 to volume V_2 is $nR \ln(V_2/V_1)$, the free energy change is

$$G_2 - G_1 = \Delta G = -nRT \ln \frac{V_2}{V_1} = nRT \ln \frac{P_2}{P_1} \qquad (4\text{-}27)$$

In a special form of this equation, G_P^0 is defined as the *standard free energy*, the value when the pressure is 1 atm; thus, using small capitals to denote the values for 1 mol,

$$G_2 - G_P^0 = RT \ln \frac{P_2}{1}$$

or

$$G = G_P^0 + RT \ln P \qquad (4\text{-}28)$$

This form of the free energy equation is a general expression of the dependence of the molar free energy G upon the pressure.

Example: Liquid water is vaporized at 100°C and 76 torr pressure. What is the free energy change per mole?

Solution: This vaporization process is not one at equilibrium, for the vapor pressure of water at 100°C is 760 torr. The process can, however, be divided into two reversible steps: vaporization at 760 torr, followed by isothermal, reversible expansion to 76 torr. The free energy change is the sum of the changes for the two steps:

$$\Delta G = 0 + RT \ln(76/760) = -(2.303)(8.314)(373) \log 10$$
$$= -7.14 \text{ kJ/mol}$$

As one would expect for a vaporization at less than the equilibrium pressure, the process is spontaneous.

Consider now the free energy of a component in a solution. It is the same, per mole of the component, as the free energy of the same material in the vapor phase in equilibrium with the solution. If the solution obeys Henry's law, as described in Section 2-4, the pressure of the material in the vapor is equal to a constant times the mole fraction of the material in the solution:

$$P = KX \qquad (4\text{-}29)$$

Substitution of this in Equation (4-28) leads to

$$\begin{aligned} G_{soln} = G_{vapor} &= G_P^0 + RT \ln KX \\ &= G_P^0 + RT \ln K + RT \ln X \\ &= G_X^0 + RT \ln X \end{aligned} \qquad (4\text{-}30)$$

Inspection of this equation shows that G_X^0 is the value of the free energy per mole when the mole fraction of the material in question is unity, that is, when the material is pure. If the molar concentration c is employed instead of the mole fraction, a parallel equation can be written:

$$G = G_c^0 + RT \ln c \qquad (4\text{-}31)$$

The constant in this equation, G_c^0, is the value of the free energy when the molar concentration is unity. Equations (4-28), (4-30), and (4-31) all represent a similar dependence of free energy on concentration, differing only in the scale of concentration used and the corresponding value for the standard-state free energy. The most convenient of the three can be chosen for use in a particular problem.

FREE ENERGY CHANGES FOR CHEMICAL REACTIONS

A very significant application of the free energy function is in the evaluation of the driving force for a chemical reaction in order to answer the question of how far a reaction may proceed without the intervention of an outside agency to supply energy. It is therefore a matter of great importance to be able to calculate the free energy change for a chemical reaction.

Either of two general procedures can be employed in utilizing data from the literature for this calculation. One method is to combine data for the enthalpy change occurring in the reaction with information on the entropy change in order to obtain the change in free energy, using the relation $\Delta G = \Delta H - T \Delta S$. As an alternative, the free energy change can be calculated as the difference between the free energies of formation of the products and the free energies of formation of the reactants, using values from a compilation such as Table 4-2. The definition of the free energy of formation of a compound and the ways in

Table 4-2
Standard free energies of formation of compounds at 25°C

Compound	Free energy of formation (kcal/mol)	Compound	Free energy of formation (kcal/mol)
Carbon monoxide(g)	−32.81	Cyclohexane(l)	6.37
Carbon dioxide(g)	−94.26	Ethanol(l)	−41.63
Water(l)	−56.69	Glycerol(l)	−114.6
Water(g)	−54.64	Acetic acid(l)	−93.08
Ammonia(g)	−3.98	Urea(s)	−47.12
Sodium chloride(s)	−91.79	α,β-D-Glucose(s)	−219.16
Ethylene(g)	16.28	Glycine(s)	−90.27
Ethane(g)	−7.86	Glycylglycine(s)	−117.25
Butene-1(g)	17.09	DL-Leucine(s)	−85.7
Benzene(l)	29.76		

which values of the free energy of formation are used parallel the definition and manipulation of the enthalpy of formation discussed in Chapter 3.

Example: In the example on page 122 the value for the entropy change for the reaction of 2 mol of glycine to form glycylglycine and liquid water was found to be 12.7 cal/mol K. Combine this value with enthalpy of formation values to find the standard free energy change for the reaction.

Solution: The enthalpy change is found by subtracting twice the enthalpy of formation of glycine from the sum of the enthalpies of formation of glycylglycine and of water:

$$\Delta H^0 = -178.12 - 68.32 - 2(-128.4) = +10.4 \text{ kcal}$$

The free energy change is this difference less the temperature multiplied by the entropy change:

$$\Delta G^0 = +10.4 - (298)(12.7)/1000 = +6.6 \text{ kcal}$$

The result can be checked by using values from Table 4-2:

$$\Delta G^0 = -117.25 - 56.69 - 2(-90.27) = +6.6 \text{ kcal}$$

Since the free energies of materials depend upon concentrations, the free energy change in a chemical reaction varies with the concentration or pressure of each material involved in the reaction. For a general reaction involving gaseous materials,

$$aA + bB \longrightarrow mM + nN \tag{4-32}$$

the free energy change is given by

$$
\begin{aligned}
\Delta G &= mG_M + nG_N - aG_A - bG_B \\
&= m(G_M^0 + RT \ln P_M) + n(G_N^0 + RT \ln P_N) \\
&\quad - a(G_A^0 + RT \ln P_A) - b(G_B^0 + RT \ln P_B) \\
&= mG_M^0 + nG_N^0 - aG_A^0 - bG_B^0 + RT \ln \frac{(P_M)^m (P_N)^n}{(P_A)^a (P_B)^b} \\
&= \Delta G_P^0 + RT \ln Q_P
\end{aligned}
\tag{4-33}
$$

The quantity ΔG_P^0 is referred to as the *standard* free energy change; it is equal to the free energy change when each reactant and each product is present in its standard state of 1 atm pressure. The quotient Q_P expresses the relationship of the pressures in any chosen set of reaction conditions to those in the standard states. The expression for Q_P can be set up for any given situation by multiplying together the pressures of the products, each raised to a power equal to the coefficient of the respective species in the equation for the reaction as written, and dividing by the product of the pressures of the reactants, each likewise raised to the power equal to the number of molecules appearing in the equation for the corresponding species.

If one wishes to express the concentrations of reactants and products on the molar or mole fraction scale instead of the partial pressure scale, analogous equations can be applied:

$$\Delta G = \Delta G_c^0 + RT \ln Q_c \tag{4-34}$$

$$\Delta G = \Delta G_X^0 + RT \ln Q_X \qquad (4\text{-}35)$$

These equations imply standard states of unit molar concentration or unit mole fraction, respectively, and the quotients are set up just as described for Equation (4-33) but with values expressed on the appropriate concentration scale.

Example: For the reaction

$$H_2(g) + I_2(g) \longrightarrow 2HI(g)$$

the value of ΔG^0 at 25°C is −3.1 kcal. Calculate ΔG when hydrogen at 0.1 atm and iodine at 0.1 atm react to form hydrogen iodide at 10 atm, all at 25°C.

Solution:

$$\Delta G = \Delta G^0 + (2.303)(1.987)(298) \log \frac{P_{HI}{}^2}{P_{H_2} P_{I_2}}$$

$$= -3100 + 1364 \log \frac{10^2}{0.1^2}$$

$$= -3100 + 4(1364) = +2360 \text{ cal}$$

Since the free energy change is positive for the given conditions, hydrogen and iodine at these pressures do not combine to form more hydrogen iodide at a pressure of 10 atm, but if any reaction occurs spontaneously it will be the reverse process, dissociation of hydrogen iodide into the elements.

It should be emphasized at this point that we have made the assumption that the systems with which we are dealing behave ideally: that the gases follow the ideal gas equation, and that the solutions follow Henry's law. If this is not a valid approximation, the same mathematical relationships can be used to calculate the free energy change, but an effective concentration, called the activity, must be introduced in place of the stoichiometric concentration. Section 4-11 describes this approach for nonideal systems. In Chapter 5, the rather large deviations from ideal behavior which are encountered for solutions of electrolytes will be discussed.

Many of the reactions occurring in aqueous solutions that are of interest in biochemical applications involve materials which are weak electrolytes. Molecules containing such groups as the carboxylate group and the phosphate group exist in ionic form to an extent determined by the pH of the solution and by other factors such as the concentration of inert salts. These equilibria will be considered in detail in Chapters 5 and 6, but it is appropriate to point out here that, for convenience, free energies of materials taking part in biological reactions are often specified for conditions near actual physiological conditions, and most often for a pH value of 7. Thus the symbols $G^{0\prime}$ and $\Delta G^{0\prime}$ refer to the standard free energy and the standard free energy change, respectively, under these conditions. A true equilibrium expression for this situation can involve only one species, either the neutral molecule or the ion formed from it by the ionization process. For the standard state, the total concentration of all such forms of material, whatever the state of ionization, can be included. However, if the change in concentration from the standard state involves a change in the ionization equilibrium, then allowance for this must be specifically made when expressions such as Equation (4-34) are applied.

Example: Determine whether malic acid at 0.300 M concentration will be spontaneously dehydrated to fumaric acid at 0.0100 M concentration in an aqueous solution of pH 7 at 25°C in the presence of a suitable catalyst.

Solution: The equation for the reaction involving acids is

$$\text{HOOC—CHOH—CH}_2\text{—COOH} \longrightarrow \text{HOOC—CH=CH—COOH} + \text{H}_2\text{O}$$

However, at this pH, the carboxylate groups are almost completely ionized, so that the reaction is essentially

$$(\text{OOC—CHOH—CH}_2\text{—COO})^{2-} \longrightarrow (\text{OOC—CH=CH—COO})^{2-} + \text{H}_2\text{O}$$

Free energies of formation for the fumarate and malate ions at unit concentration in aqueous solution at pH 7 are -144.41 and -201.98 kcal/mol, respectively, and the value for water is -56.69. The standard free energy change is then

$$\Delta G^0 = -144.41 - 56.69 + 201.98 = +0.88 \text{ kcal/mol}$$

For the given concentrations, the free energy change is

$$\Delta G = \Delta G^0 + RT\frac{[\text{fumarate}]}{[\text{malate}]}$$
$$= +0.88 + (1.987)(2.303)(298)(\log \tfrac{1}{30})/1000$$
$$= +0.88 - 2.01 = -1.13 \text{ kcal/mol}$$

Thus the reaction is a spontaneous one for the given concentrations, although it is not spontaneous for equal concentrations of malate and fumarate.

4-8
CHEMICAL EQUILIBRIUM

Many chemical reactions are reversible: As soon as some of the products are formed from the reactants, the products react to regenerate some of the reactants and the reaction then never goes to completion. The state of equilibrium for a reversible reaction system is one in which the ratio of the concentrations of the materials involved is such that no further change in overall composition occurs so long as the temperature and pressure remain unchanged. This equilibrium is a dynamic one, for it does not result from the cessation of reaction but from a condition of equal rates of the two opposing reactions. Furthermore, the same equilibrium condition can be approached from either direction. Consider, for example, the reaction of nitrogen dioxide, NO_2, to form nitrogen tetroxide, N_2O_4. If initially pure NO_2 is present, it combines to form some N_2O_4. As the concentration of NO_2 falls, the rate at which it reacts also falls, because the rate of a reaction is proportional to the concentration of the reactant. As more N_2O_4 is formed, the rate at which it dissociates increases. Eventually a condition is reached, the equilibrium condition, in which the rate of combination of NO_2 is equal to the rate of dissociation of N_2O_4, and there is then no further net change in the composition of the system.

If initially pure N_2O_4 is present, it dissociates into NO_2, with the rate of reaction declining as the concentration declines. NO_2 recom-

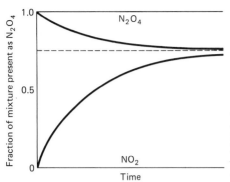

Figure 4-2
Approach to an equilibrium mixture
from either direction, for the reaction
$N_2O_4 \rightleftharpoons 2NO_2$. The exact value of the
equilibrium composition of the mixture
varies with the temperature.

bines more and more rapidly until its rate of reaction equals the rate
of reaction of the N_2O_4. For a given container in which the initial gases
are placed, the same equilibrium is reached whether two moles of NO_2
were initially present or one mole of N_2O_4 was initially present. Figure
4-2 represents the approach from either direction to the equilibrium
for the reaction

$$2NO_2 \rightleftharpoons N_2O_4 \tag{4-36}$$

We discussed above the requirement that the condition of equilibrium correspond to a free energy minimum of a system. This is
illustrated schematically in Figure 4-3. If we wish to express the
consequences of this condition quantitatively, we need only observe
that the free energy curve is flat for an infinitesimal distance on either
side of the equilibrium point. Thus for any very small change occurring in the vicinity of the equilibrium composition, ΔG is equal to zero.
Now ΔG is related to the standard free energy change by the equation

$$\Delta G = \Delta G^0 + RT \ln Q \tag{4-37}$$

To find the particular numerical value of the ratio of products to reactants Q_{equil} that applies at equilibrium, one can simply set ΔG equal
to zero:

$$\Delta G^0 + RT \ln Q_{equil} = 0 \tag{4-38}$$

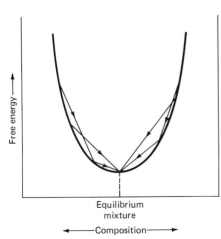

Figure 4-3
Schematic representation of the variation
of the free energy near a minimum
corresponding to equilibrium. The
changes corresponding to the arrows are
spontaneous.

But Q_{equil} is just the quantity chemists define to be the equilibrium constant K, and thus

$$\Delta G^0 = -RT \ln K \qquad (4\text{-}39)$$

This is an extremely important equation: It permits calculation from purely thermal data of the point of equilibrium in a chemical system, one of the most widely useful applications of chemical thermodynamics.

In the application of Equation (4-39), the equilibrium constant can be expressed as a ratio of pressures, a ratio of molar concentrations, or a ratio of mole fractions, denoted respectively by K_P, K_c, or K_X, depending upon the concentration scale used to define the standard states for which ΔG^0 is measured. In this connection, it should also be noted that a reactant or product that is a pure liquid or a pure solid is considered to have unit concentration, whatever the concentration scale employed. Another way of phrasing this convention is to state that the concentrations of these materials, which do not change, are incorporated into the numerical value of the equilibrium constant.

Example: Calculate the equilibrium constant for the dehydration of malate to fumarate in aqueous solution of pH 7 at 25°C.

Solution: In the preceding example, ΔG^0 for this reaction on the molar concentration scale was evaluated as +0.88 kcal/mol or 880 cal/mol. From Equation (4-39),

$$\Delta G_c^0 = -RT \ln K_c$$
$$880 = -(1.987)(2.303)(298) \log K_c$$
$$\log K_c = \overset{-}{-}0.645 = \overset{+}{1}.355$$
$$K_c = 2.26 \times 10^{-1}$$

It is instructive to combine Equations (4-37) and (4-39):

$$\Delta G = -RT \ln K + RT \ln Q = RT \ln \frac{Q}{K} \qquad (4\text{-}40)$$

In words, this relation states that the free-energy change for some arbitrary combination of product and reactant concentrations depends upon the logarithm of the ratio of the appropriately formulated quotient of these concentrations to the corresponding equilibrium quotient of the concentrations. If Q is less than K, then the tendency is for the reaction to yield more of the "products," whereas if Q is greater than K, only the production of "reactants" can occur spontaneously.

Example: Standard free energies of formation of ethyl n-butyrate and n-butyric acid at 25°C are –76,000, and –91,500 cal, respectively. How much ester is formed when 0.5 mol of n-butyric acid is mixed with 2 mol of ethanol and allowed to come to equilibrium at 25°C?

Solution: The reaction is

$$CH_3CH_2CH_2COOH(l) + C_2H_5OH(l) \rightleftharpoons CH_3CH_2CH_2COOC_2H_5(l) + H_2O(l)$$

From the free energies of formation given and those in Table 4-2,

$$\Delta G^0 = -76,000 - 56,690 - (-41,630 - 91,500)$$
$$= +440 \text{ cal}$$

Since this is a liquid-phase reaction, the standard states of the materials are the pure liquids. Since there is no change in number of molecules, the equilibrium ratio can be expressed equally well as a ratio of mole fractions or as a ratio of numbers of moles. From the free energy change,

$$\log K = -\frac{440}{(2.303)(1.987)(298)} = -0.323 = \overset{-\;+}{1.677}$$

$$K = 0.48 = \frac{y^2}{(0.5 - y)(2 - y)}$$

where y is the number of moles of ester formed and also the number of moles of water formed. This equation rearranges to

$$0.52y^2 + 1.20y - 0.48 = 0$$

of which the positive root is $y = 0.35$, so that 0.35 mol is the amount of ester formed.

4-9
EFFECT OF TEMPERATURE ON FREE ENERGY CHANGE AND EQUILIBRIUM CONSTANT

For a reversible reaction, the qualitative *principle of Le Chatelier* predicts that an increase in temperature changes the equilibrium constant in a direction such that the amount of material corresponding to the product of the endothermic reaction is increased and the amount of material from the exothermic reaction is decreased. Thermodynamics yields a quantitative relation, which we shall now develop, between the enthalpy of reaction, the temperature, and the equilibrium constant or standard free energy change.

As a starting point, we take the basic equation defining free energy:

$$G = H - TS \qquad (4\text{-}18)$$

For the enthalpy, its equivalent in terms of internal energy is substituted and differentials are taken of each term in the resulting equation:

$$dG = dE + P\,dV + V\,dP - T\,dS - S\,dT \qquad (4\text{-}41)$$

By the first law, $dE = \delta q_{rev} + \delta w_{max}$, but δq_{rev} is equal to $T\,dS$ so that these quantities cancel one another. Furthermore, for a reaction involving only work of expansion, δw_{max} equals $-P\,dV$, so that these quantities also cancel. There remains

$$dG = V\,dP - S\,dT \qquad (4\text{-}42)$$

Each term of this equation can be differentiated with respect to temperature at constant pressure. Under these conditions $dP = 0$, and only two terms remain:

$$\left(\frac{\partial G}{\partial T}\right)_P = -S \qquad (4\text{-}43)$$

Equation (4-43) applies individually to each reactant and product in a process, and subtracting the terms for reactants from those for products results in a parallel equation:

$$\left(\frac{\partial \Delta G}{\partial T}\right)_P = -\Delta S \tag{4-44}$$

If the reaction occurs from reactants in their standard states to products in their standard states, the applicable form of this equation is

$$\left(\frac{\partial \Delta G^0}{\partial T}\right)_P = -\Delta S^0 \tag{4-45}$$

The preceding equation can now be used to obtain the dependence of the equilibrium constant on temperature. It was shown above that the equilibrium constant is related to the standard free energy change by the equation

$$\Delta G_P^0 = -RT \ln K_P \tag{4-39}$$

where the standard states are those of unit pressure and the equilibrium constant is written in terms of pressures. Differentiation of Equation (4-39) with respect to temperature requires treatment of the right-hand side as a product of two functions of the variable T:

$$\left(\frac{\partial (\Delta G_P^0)}{\partial T}\right)_P = -RT \frac{d \ln K_P}{dT} - R \ln K_P \tag{4-46}$$

Multiplying through by T and substituting from Equations (4-45) and (4-39), we obtain

$$-T \Delta S_P^0 = -RT^2 \frac{d \ln K_P}{dT} + \Delta G_P^0 \tag{4-47}$$

The standard enthalpy change now replaces the sum of ΔG_P^0 and $T \Delta S_P^0$, and the equation thus becomes

$$\frac{d \ln K_P}{dT} = \frac{\Delta H^0}{RT^2} \tag{4-48}$$

Since the enthalpy change does not vary much with pressure, the superscript zero in ΔH^0 is not very important and is often omitted.

Equation (4-48) is often termed the van't Hoff equation. If the enthalpy change is independent of temperature, the equation can be integrated between two temperatures to yield

$$\ln \frac{K_2}{K_1} = \frac{\Delta H}{R} \frac{T_2 - T_1}{T_2 T_1} \tag{4-49}$$

In Section 1-10, we applied a particular form of this equation to the equilibrium between a liquid and a vapor, with the equilibrium constant represented by the pressure of the vapor.

In using Equation (4-49), it is necessary to be certain that the enthalpy change is independent of temperature, the equation can be the same reaction. For example, doubling the coefficients of all reactants and products in the equation doubles the value of the enthalpy change; at the same time, the value of the equilibrium constant is squared, thus doubling its logarithm. Accordingly, so long as ΔH and K belong to the identical chemical reaction, their values will match appropriately.

Equation (4-48) is valid only when the equilibrium constants are expressed in terms of pressures. If constants in terms of concentrations are to be used, there is a parallel equation involving the energy of reaction ΔE:

$$\frac{d \ln K_c}{dT} = \frac{\Delta E^0}{RT^2} \tag{4-50}$$

Example: The ΔG^0 value for the reaction

$$2NO_2(g) \rightleftharpoons N_2O_4(g)$$

is -1.23 kcal at $25°C$. Estimate the equilibrium constant K_p at $100°C$ if the ΔH value is -11.20 kcal.

Solution: Use the equation

$$\log K_2 - \log K_1 = \frac{\Delta H}{(2.303)(1.987)} \frac{T_2 - T_1}{T_2 T_1}$$

with the index 1 referring to $25°C$ and the index 2 to $100°C$. (It is convenient to substitute 4.576 for the product 2.303×1.987.) From the known $\Delta G^0_{298\,K}$,

$$\log K_{298\,K} = \frac{1230}{(4.576)(298)} = 0.902$$

From this $K_{298\,K}$ is equal to 8.0. Then

$$\log K_{373\,K} = 0.902 + \frac{-11{,}200}{4.576} \frac{373 - 298}{(373)(298)} = 0.902 - 1.651 = -0.749$$

Log $K_{373\,K}$ is expressible as $\overset{-\;+}{1.251}$ and $K_{373\,K}$ is 1.78×10^{-1}. Note the decrease in K from lower to higher temperature for the exothermic process.

4-10
ENERGY RELATIONS IN LIVING SYSTEMS

The living organism is enabled to carry on life processes only by obtaining a supply of energy which can be converted into work. The source of most of this energy is the process of photosynthesis, in which, with the aid of chlorophyll, the energy of sunlight is utilized to convert carbon dioxide and water into carbohydrates and other compounds that can be oxidized to produce energy. Plants bring about this conversion themselves, and animals then consume either plants or other animals that have derived their food from plants.

In order to avoid burning up the organism itself, and in order to make the energy produced from oxidation available in controlled form to do work, a whole series of stepwise processes is typically required. Thus the conditions under which glucose is oxidized to carbon dioxide and water in a living cell are much milder than those that would be required to "burn" the glucose in air, and nature achieves such reactions by following pathways with several steps in sequence, as well as by the use of enzymes as catalysts.

COUPLED REACTIONS AND THE ROLE OF ATP

To show how one reaction can be related to another in the sequence of reactions, we consider the example of the production of sucrose from glucose and fructose; for the reaction in aqueous solution we write the equilibrium equation

$$\text{glucose (aq)} + \text{fructose (aq)} \rightleftharpoons \text{sucrose (aq)} + H_2O \qquad (4\text{-}51)$$

The equilibrium constant for this reaction is about 10^{-4}. Only if the glucose and fructose concentrations are quite large and the sucrose concentration very small will the reaction proceed at all in the direction of sucrose synthesis as a spontaneous process. Let us, however, relate this reaction to another potential reaction, the hydrolysis of adenosine triphosphate (ATP) to adenosine diphosphate (ADP):

$$\text{ATP} + H_2O \rightleftharpoons \text{ADP} + \text{inorganic phosphate} \qquad (4\text{-}52)$$

The phosphate components in this reaction are ionized in aqueous solution to an extent that depends upon the pH. The formula for ATP is given on page 142. In a neutral solution, the principal ionic forms present are represented in an alternative form of Equation (4-52):

$$\text{ATP}^{4-} + H_2O \rightleftharpoons \text{ADP}^{3-} + H_2PO_4^{-} \qquad (4\text{-}53)$$

For this reaction, the equilibrium constant at pH 7 is about 4×10^5.

Adding Equations (4-51) and (4-53) yields

$$\text{ATP}^{4-} + \text{glucose} + \text{fructose} \rightleftharpoons \text{sucrose} + H_2PO_4^{-} + \text{ADP}^{3-} \quad (4\text{-}54)$$

The equilibrium constant for this overall reaction is the product of the constants for the two individual reactions, $10^{-4} \times 4 \times 10^5$, or approximately 40.

For convenience, the equilibrium constants have been written to apply to the total amount of all ionization states of each substance present in the solution. Thus, although $H_2PO_4^{-}$ is indicated in the equation, both HPO_4^{2-} and $H_2PO_4^{-}$ are included in the numerical value of K. Furthermore, the values of K are modified under physiological conditions by circumstances such as pH, the presence of metal ions, total ion concentration, and temperature, details we are ignoring for the present.

The main point of our discussion is that, if the ATP hydrolysis were linked to the sucrose synthesis, the latter could proceed. The term *coupled reactions* is applied to cases of this sort in which a process for which the equilibrium constant is low is driven by another process for which the constant is large. A more convenient scheme for describing the relations between systems that can be coupled together is by citing the standard free energy change for each reaction and adding these changes to obtain ΔG^0 for the combined process. Not only does the treatment in terms of free energy change replace multiplication by addition, but it permits easy extension to calculation of the chemical driving force for nonequilibrium conditions. For the ATP hydrolysis, the ΔG^0 value is -7.6 kcal, and for the sucrose synthesis about 5.5 kcal, leaving a net driving force for the coupled system of -2.1 kcal. Since the living organism does not violate the laws of thermodynamics, the chemical processes that occur are necessarily those leading toward equilibrium, and they are therefore those that are accompanied by a decrease in free energy. The same comments apply to the free energy changes discussed for biological systems as were made in the preceding paragraph for equilibrium constants. In addition, there is the further convention in the use of the symbol $\Delta G^{0\prime}$ for pH 7, as described on page 133.

The next question is: How are the two reactions discussed above "coupled" so that the occurrence of the spontaneous one leads to the simultaneous occurrence of the driven process? In fact, in the present example, the two reactions do not really occur as we have written them. Serving as a necessary *common intermediate* to link two steps together under physiological conditions to yield the overall process is the substance glucose 1-phosphate, and what actually happens is the sequence

$$\text{ATP}^{4-} + \text{glucose} \rightleftharpoons \text{glucose 1-phosphate}^{2-} + \text{ADP}^{3-} + \text{H}^+$$
$$\Delta G^{0\prime} = -2.6 \text{ kcal} \quad (4\text{-}55)$$
$$\text{glucose 1-phosphate}^{2-} + \text{fructose} + \text{H}^+ \rightleftharpoons \text{sucrose} + \text{H}_2\text{PO}_4{}^-$$
$$\Delta G^{0\prime} = +0.5 \text{ kcal} \quad (4\text{-}56)$$

The sum of these two equations is equivalent to Equation (4-54), and the standard free energy changes must obviously sum to the same value, in whatever manner the steps making up the overall process are represented.

The combination of Equations (4-51) and (4-53) illustrates a common convention used by biochemists for the discussion of reactions involving phosphate-containing compounds: As the basis for a scale of the tendency of these compounds to transfer phosphate groups to other compounds, the free energy of the hydrolysis reaction is used. Two hydrolysis reactions can then be combined in a way so that water cancels out in the resulting equation. Thus we *added* Equation (4-51), the reverse of a hydrolysis, to Equation (4-53), a hydrolysis reaction; if both equations were written with water as a reactant, one would be subtracted from the other to obtain the equilibrium expression.

We can compare as two further examples of representative hydrolysis reactions those of glyceryl phosphate and of 1,3-diphosphoglycerate. The first is the reaction of an "ordinary" organic phosphate ester:

$$
\begin{array}{l}
\text{CH}_2\text{—O—}\overset{\overset{\displaystyle \text{O}^-}{|}}{\underset{\parallel}{\text{P}}}\text{—O}^- \\
\text{CHOH} \quad\ \ \text{O} \\
\text{CH}_2\text{OH}
\end{array}
+ \text{H}_2\text{O} + \text{H}^+ \rightleftharpoons
\begin{array}{l}
\text{CH}_2\text{OH} \\
\text{CHOH} \\
\text{CH}_2\text{OH}
\end{array}
+ \text{H}_2\text{PO}_4{}^- \qquad (4\text{-}57)
$$

For this hydrolysis, $\Delta G^{0\prime}$ is about -2.3 kcal, a modest value compared to the $\Delta G^{0\prime}$ of about -12 kcal for the following reaction, involving an acyl phosphate:

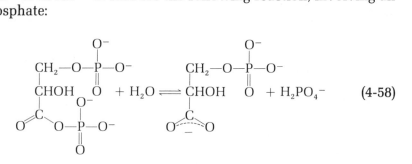

$$(4\text{-}58)$$

Since the free energy change for the hydrolysis of a phosphate compound is a measure of the relative tendency of that compound to transfer phosphate to another substance, the negative of the free energy of hydrolysis is often termed the *phosphate-transfer potential;* some values are listed in Table 4-3. Such molecules as ATP, acetyl phosphate, 1,3-diphosphoglycerate, and creatine phosphate are seen to have relatively high tendencies to transfer a phosphate group to a suitable acceptor. Of course, the actual transfer is usually that of a phosphoryl group, $-PO_3^{2-}$.

A key position in biological metabolism is occupied by the molecule adenosine triphosphate, which has the structure

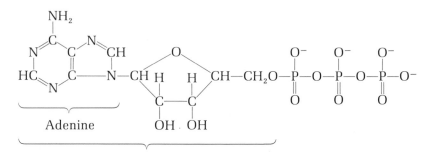

Reactions in which nutrients are consumed by an organism are coupled to drive the formation of ATP from ADP by the addition of a phosphoryl group. Thus the free energy from these reactions is neither lost nor utilized directly but is converted into chemical free energy in the form of ATP, from which it can be released as needed. The conversion of ATP back to ADP plus phosphate is coupled by what are often

Table 4-3
Phosphate-transfer potentials[a]

Compound	Potential	Compound	Potential
1,3-Diphosphoglycerate	11.8	Pyrophosphate	7.6
2-Phosphoenolpyruvate	12.8	Arginine phosphate	7.0
Creatine phosphate	10.3	Glucose 1-phosphate	5.0
Acetyl phosphate	10.1	Glucose 6-phosphate	3.3
Adenosine triphosphate	7.6	Glycerol 1-phosphate	2.3

[a]Values are the magnitudes of the free energy of hydrolysis at pH 7 in kilocalories per mole.

very complex mechanisms to the performance of muscular work, to active transport of dissolved substances against an unfavorable concentration gradient, and to the synthesis of needed chemical substances. In some reactions, incidentally, ATP loses a pyrophosphate group in a single step, becoming adenosine monophosphate (AMP) plus $P_2O_7^{4-}$, and the pyrophosphate is subsequently hydrolyzed to phosphate. Of course, each of the many reactions involved in the production and utilization of the free energy of ATP requires a specific enzyme able to catalyze that particular reaction.

In the process of muscular contraction, ATP participates along with phosphocreatine, another compound of high phosphate-transfer potential, which serves as a storage reservoir of energy in muscle:

$$HN{=}C \begin{cases} NH{-}PO_3^{2-} \\ \\ N{-}CH_2COO^- \\ | \\ CH_3 \end{cases}$$

The ADP–ATP and creatine–phosphocreatine pairs enter into an equilibrium involving the interchange of a phosphate group and catalyzed by the enzyme creatine phosphokinase. Since the phosphate-transfer potentials of creatine phosphate and ATP are not very different, this equilibrium does not lie far to one side. The system can be thought of as balanced and easily tipped to one side or the other as the requirements of the organism dictate. By one of the pathways described below, ATP is formed from ADP and then the ATP transfers phosphate to creatine and is reconverted to ADP, which is again phosphorylated in a cyclic process. When energy must be supplied rapidly, the phosphocreatine quickly returns phosphate to ADP, converting large amounts of it to the effective ATP, which participates in the muscle contraction process by phosphorylating the protein myosin which is part of the actomyosin system of filaments in the muscle.

ATP also functions in the synthesis of various chemical substances required by the living system. In the cell there are several other nucleoside 5′-triphosphates in addition to ATP. The members of one group of these, the ribonucleoside phosphates, are identical with ATP except that the adenosine group is replaced by a guanine, uracil, or cytosine group, respectively, in guanosine triphosphate (GTP), uridine triphosphate (UTP), and cytidine triphosphate (CTP). There is also a series of four deoxyribonucleoside 5′-triphosphates in which the sugar unit is 2-deoxyribose instead of ribose. *Because ADP is the only diphosphate that can accept phosphate from the fueling scheme of the cell metabolic processes,* it is the channel through which phosphate groups reach the diphosphate compounds such as GDP and UDP to convert them to GTP, UTP, and so on. The phosphate-transfer potentials of the nucleoside triphosphates are all about the same, but many synthetic processes require a specific triphosphate to carry them on. Thus CTP and ATP are both required for the synthesis of lipids, GDP participates specifically in the synthesis of cellulose, and UTP is involved in the synthesis of polysaccharides such as glycogen.

Glycogen is a high-molecular-weight material found in animal cells where it serves as a storage form of glucose. Indeed, it consists of a chain of many glucose molecules linked together by glycosidic linkages. The glycogen molecule "grows" by the addition of glucose units, and this reaction affords an illustration of synthesis in which nucleoside triphosphates participate. The molecule of glucose is first phosphorylated by ATP to glucose 6-phosphate, which is then isomerized to glucose 1-phosphate. The glucose 1-phosphate next reacts with UTP to form a pyrophosphate unit, $P_2O_7^{4-}$, plus an intermediate called uridine diphosphate glucose:

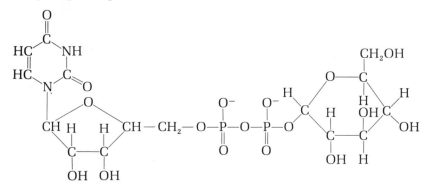

The uridine diphosphate glucose then reacts with a glycogen molecule and adds the glucose unit to the end of the polymeric chain, setting free uridine diphosphate.

USE OF FREE ENERGY
TO FORM ATP

Molecules of ATP are formed from ADP by several fairly well-defined pathways. An example is the process of *glycolysis* which occurs without the aid of oxygen and which corresponds to the breakdown of a molecule of glucose to two molecules of lactic acid, a series of reactions which is coupled to the phosphorylation of two molecules of ADP to ATP. In this sequence, the glucose molecule is first phosphorylated by a molecule of ATP, which the cell invests in the process, to produce glucose 6-phosphate. This is isomerized, in the presence of the enzyme phosphoglucomutase, to fructose 6-phosphate, which is phosphorylated in turn by a second molecule of ATP to form fructose diphosphate:

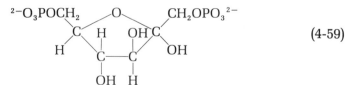

(4-59)

This molecule is then split, with the aid of the enzyme adolase, into two three-carbon units, 3-phosphoglyceraldehyde and dihydroxyacetone phosphate. The latter is isomerized in a reaction catalyzed by triose phosphate isomerase to form a second molecule of 3-phosphoglyceraldehyde:

$$CH_2OH—CO—CH_2OPO_3^{2-} \rightleftharpoons CH_2OPO_3^{2-}—CHOH—CHO \quad (4\text{-}60)$$

Each of the two molecules of 3-phosphoglyceraldehyde then combines with inorganic phosphate and is simultaneously oxidized to produce 1,3-diphosphoglycerate:

$$
\begin{array}{c}
CH_2OPO_3{}^{2-} \\
| \\
CHOH + H_2PO_4{}^- + NAD \rightleftharpoons \\
| \\
H{-}C{\diagdown}O
\end{array}
\begin{array}{c}
CH_2OPO_3{}^{2-} \\
| \\
CHOH + NADH_2 + H^+ \qquad (4\text{-}61) \\
| \\
C{=}O \\
| \\
OPO_3{}^{2-}
\end{array}
$$

This oxidation has been achieved with the participation of nicotinamide adenine dinucleotide (NAD) in its oxidized form, which accepts two hydrogen atoms in the course of being reduced. The structure and function of NAD, which serves as an electron carrier in many biological systems, will be further discussed in Chapter 7.

In Equation (4-58) we saw the hydrolysis reaction of 1,3-diphosphoglycerate, and we already know that it is a molecule with a high phosphate-transfer potential for the acyl phosphate group. This group is now transferred to ADP, forming ATP. The 3-phosphoglycerate remaining from this process is isomerized to 2-phosphoglycerate, which in turn is dehydrated to phosphoenolpyruvate, another molecule with high phosphate-transfer potential:

$$
\begin{array}{c}
CH_2OPO_3{}^{2-} \\
| \\
CHOH \\
| \\
O{-}C{-}O \\
-
\end{array}
\rightleftharpoons
\begin{array}{c}
CH_2OH \\
| \\
CHOPO_3{}^{2-} \xrightarrow{-H_2O} \\
| \\
O{-}C{-}O \\
-
\end{array}
\begin{array}{c}
CH_2 \\
\| \\
COPO_3{}^{2-} \qquad (4\text{-}62) \\
| \\
O{-}C{-}O \\
-
\end{array}
$$

Each molecule of phosphoenolpyruvate, in a reaction catalyzed by pyruvate phosphokinase, produces a molecule of ATP, as well as a molecule of pyruvate:

$$
\begin{array}{c}
CH_2 \\
\| \\
COPO_3{}^{2-} + ADP^{3-} + H^+ \rightleftharpoons \\
| \\
O{-}C{-}O \\
-
\end{array}
\begin{array}{c}
CH_3 \\
| \\
C{=}O + ATP^{4-} \qquad (4\text{-}63) \\
| \\
O{-}C{-}O \\
-
\end{array}
$$

In the absence of oxygen, the pyruvate is hydrogenated to lactate, consuming the two hydrogens produced earlier in the oxidative phosphorylation of 3-phosphoglyceraldehyde.

The scheme of glycolysis which has just been described can be summarized in the overall equation

$$
C_6H_{12}O_6 + 2HPO_4{}^{2-} + 2ADP^{3-} \longrightarrow 2CH_3CHOHCOO^- + 2ATP^{4-} + 2H_2O \tag{4-64}
$$

Two ATP molecules are used in the initial breakdown of the glucose molecule to two three-carbon fragments, but each of these fragments generates two ATP molecules from ADP, so that there is a net gain of two ATP molecules.

In cells that use molecular oxygen for respiration, the pyruvate from glycolysis is not reduced to lactate but enters into an additional sequence of reactions which leads into a series of compounds termed the *respiratory chain,* in which the effect of the molecular oxygen is utilized to produce more molecules of ATP by *oxidative phosphorylation.* The pyruvate is prepared for the reaction sequence by oxidation in the presence of pyruvic dehydrogenase with loss of a carbon atom in the form of carbon dioxide. The oxidizing agent is NAD, which again serves as an electron carrier. The acetyl group remaining from this reaction becomes attached to a molecule called *coenzyme A* (CoA):

$$NAD + CH_3COCOOH + CoA—SH \rightleftharpoons CH_3CO—S—CoA + CO_2 + NADH_2$$
$$(4\text{-}65)$$

Two-carbon fragments from other fuel sources, such as fatty acids and amino acids, also enter the oxidation scheme at this point, becoming attached by parallel reactions to CoA, which is represented as CoA—SH to indicate the active sulfhydryl group in its structure:

$$
\begin{array}{c}
\text{Adenine—ribose—P—O—P—OCH}_2\text{—C——C——C} \\
\end{array}
$$

The free energy of hydrolysis of acetyl CoA, formed in reaction (4-65), is of the order of 9 kcal/mol, and thus this molecule has a high *acetyl-transfer potential* and serves as a source of acetyl groups, such as in the reaction with oxalacetic acid to form citric acid:

$$CH_3CO—S—CoA + COOH—CO—CH_2—COOH + H_2O$$

$$
\rightleftharpoons \overset{\displaystyle OH}{COOH—\underset{\displaystyle CH_2COOH}{C}—CH_2COOH} + HS—CoA \quad (4\text{-}66)
$$

The reaction represented in Equation (4-66) is part of what is known as the *Krebs cycle.* It is termed a cycle because, at the end of a series of rearrangement, decarboxylation, and oxidation–reduction processes, the molecule of oxalacetic acid is regenerated, ready to accept another acetyl group and carry out the process again. The other products of the cycle are (a) two molecules of carbon dioxide, containing carbon atoms equivalent to those in the input acetyl group, (b) three molecules of NAD which have been reduced, and (c) one molecule of another electron carrier, flavin adenine dinucleotide (FAD), which has been reduced.

The free energy at this stage of the respiratory process resides in the reduced NAD and FAD molecules, which are then able to contribute electrons to the compounds of the respiratory chain, mentioned above and discussed further in Chapter 7, which deals with electron transfer processes of the type characteristic of that chain. As the electrons flow along the respiratory chain to their final meeting with oxygen, which they reduce to water, they cause the production of ATP from ADP— three molecules of ATP for each pair of electrons. When all processes of glycolysis and oxidation leading from 1 molecule of glucose are combined, it is found that a net of 36 molecules of ADP have been converted to ATP:

$$C_6H_{12}O_6 + 6O_2 + 2ATP + 36ADP \longrightarrow 6CO_2 + 6H_2O + 38ATP \qquad (4\text{-}67)$$

If the free energy stored in each of the 36 molecules of ATP corresponds to 7.6 kcal/mol, the total free energy that has been stored up is 273 kcal. The total free energy of the combustion reaction of a mole of glucose with oxygen to form water and carbon dioxide is -686 kcal, and the physiological processes are thus able to utilize something less than half this total.

HIGH-ENERGY
PHOSPHATE COMPOUNDS

Compounds containing a phosphate group with an unusually large negative free energy of hydrolysis are often termed "high-energy" compounds, and the bond broken on hydrolysis is called a high-energy bond. Often this bond has been represented in the structural formulas of the compounds by a wavy line, implying something unusual in its nature. These labels are somewhat misleading. To begin with, the bond does not have a high energy in the usual sense of a bond energy, but rather has a normal or smaller than normal bond energy; it is the free energy that is the important function.

The ease of hydrolysis may result from the weakness of the bond being broken. This is probably the situation for ATP, in which the row of neighboring negative charges in the ionized form of the molecule repel one another. However, there is often nothing at all unusual about the bond being broken, but instead the driving force for the reaction may be provided by the stability of the products being formed. Thus in Equation (4-58) the product of the reaction contains a carboxylate group. This has a particularly high degree of stability because electrons can be delocalized over the whole O—C—O unit, an effect sometimes called resonance, which will be discussed more fully in Chapter 9. In the hydrolysis of phosphoenolpyruvate, Equation (4-63), the driving force is evidently the greater stability of the keto form of the product compared to the enol form of the reactant.

4-11
THERMODYNAMICS
OF MIXTURES

Suppose we wish to deal with a system of which the composition may vary. The free energy of the system depends upon the pressure, the temperature, and the composition, as well as upon the amount of material comprising the system. The dependence on composition and amount of material is most conveniently expressed by writing the free energy as a function of the number of moles of each component:

$$G = f(P, T, n_1, n_2, n_3, \ldots) \tag{4-68}$$

The total differential of free energy is then

$$dG = V \, dP - S \, dT + \left(\frac{\partial G}{\partial n_1}\right)_{T,P,n_i(i \neq 1)} dn_1 + \cdots + \left(\frac{\partial G}{\partial n_j}\right)_{T,P,n_i(i \neq j)} dn_j \tag{4-69}$$

If the temperature and pressure are constant,

$$dG_{T,P} = \sum_{i=1}^{j} \left(\frac{\partial G}{\partial n_i}\right)_{T,P,\text{other } n_i\text{'s}} dn_i \tag{4-70}$$

The derivative $(\partial G/\partial n_i)_{T,P,\text{other } n_i\text{'s}}$ is the partial molar free energy, often referred to as the *chemical potential* of the ith component in the given phase and often represented by the symbol μ_i. For phase equilibria involving multicomponent phases it is, strictly speaking, μ_i that is equal in every phase for any component distributed between phases, rather than merely the molar free energy that is equal. Of course, for an ideal solution the chemical potential and the molar free energy are identical, and the chemical potential can be calculated from the concentration by rewriting Equation (4-31) as

$$\mu_i = \mu_i^0 + RT \ln c_i \tag{4-71}$$

For a real solution, the concentration c_i must be replaced by an effective or apparent concentration in order that the free energy be correctly calculated. We call the value of the concentration of material that *appears* to be present in a solution when a physical measurement related to the free energy is made upon the solution, the *activity a*. The activity thus is the number that makes the following equation valid for any particular solution:

$$\mu_i = \mu_i^0 + RT \ln a_i \tag{4-72}$$

The activity is often expressed as a coefficient, the *activity coefficient* γ_i, multiplied by the stoichiometric concentration in the solution:

$$a_i = \gamma_i c_i \tag{4-73}$$

The chemical potential is then

$$\mu_i = \mu_i^0 + RT \ln \gamma_i c_i \tag{4-74}$$

It may appear that nothing has been gained by replacing one quantity that must be determined experimentally for each solution—the chemical potential—by a second quantity that also must be found experimentally for each individual solution. To a degree, this is true. However, there are practical advantages in the use of activities and activity coefficients, and we will return to their consideration later in two important areas: first, the study of ionic equilibria in Chapters 5 and 6, and second, the interpretation of reaction rates in Chapter 10. The activity coefficient is especially helpful because the amount by which it deviates from unity for a solution gives an immediate measure of the extent to which the solution is nonideal. Furthermore, values of the activity coefficient can be conveniently estimated for dilute ionic solutions. Finally, when one deals with any sort of equilibrium, it is the "thermodynamic" equilibrium constant, which is the constant expressed in terms of activities of the several species concerned, that is truly constant, whereas the value of K in terms of stoichiometric concentrations may vary considerably as the concentration changes.

In accord with these considerations, the proper equation to write for the standard free-energy change is

$$\Delta G^0 = -RT \ln K_a \qquad (4\text{-}75)$$

The superscript zero now denotes standard states of the reactants and products in which the activity is unity. Often such a state is not physically attainable, or knowledge of the values of the activity coefficients is insufficient to establish just what the stoichiometric concentration of this state is. However, even under these circumstances, we can view the value of the standard-state free energy as a definite quantity, even if we cannot precisely state the composition of the phase to which it refers. In a sense, the value of G^0 corresponds to that of a constant of integration.

Example: At 50°C, the vapor pressure of water from a sucrose solution in which the mole fraction of sucrose is 0.1000 is 78.6 torr. Calculate the activity and activity coefficient of the water.

Solution: The vapor pressure of pure water at this temperature is 92.5 torr, so that the vapor pressure is 0.850 times that of pure water, and therefore 0.850 is the activity of the water in the solution. Since the mole fraction of water is 0.900, the activity coefficient is 0.850/0.900 or 0.944.

EXERCISES

1. Assuming that benzene forms an ideal solution in cyclohexane, calculate the changes in free energy and entropy for benzene when an amount of 0.25 M benzene solution containing exactly 0.10 mol of benzene is diluted by the addition of cyclohexane to a final concentration of 0.015 M.

2. For the reaction $N_2O_4(g) \rightleftharpoons 2NO_2(g)$, K_P at 298 K is 0.120. Calculate the density of the vapor of the equilibrium mixture formed from N_2O_4 at this temperature under a total pressure of 0.100 atm and under a total pressure of 0.010 atm.

3. At 1 atm total pressure, the fractional dissociation of hydrogen sulfide into gaseous H_2 and gaseous S_2 is 0.055 at 750°C, and 0.087 at 830°C. Calculate the enthalpy change for the dissociation reaction.

4. From the data in Exercise 3, calculate the fractional dissociation of hydrogen sulfide at 750°C and a total pressure of 0.010 atm.

5. From the ΔG^0 value for the reaction

$$2NO_2(g) \longrightarrow N_2O_4(g)$$

given on page 139, calculate the value of ΔG when NO_2 at 0.010 atm reacts to form N_2O_4 at 0.50 atm. Specify what quantity of material your value refers to.

6. Calculate the equilibrium constants for the reactions given by Equations (4-55) and (4-56).

7. One mole of cyclohexane is vaporized at 1 atm pressure and its normal boiling point 354 K. From the value of the enthalpy of vaporization of 7.19 kcal/mol, calculate q, w, ΔG, and ΔS for the process.

8. From data in the table of absolute entropies, calculate the standard entropy change of each of the following processes at 25°C:

Acetic acid(l) + ethanol(l) $\longrightarrow$
$$\text{ethyl acetate}(l) + H_2O(l)$$
$$3H_2(g) + \text{benzene}(l) \longrightarrow \text{cyclohexane}(l)$$
$$N_2(g) + 3H_2(g) \longrightarrow 2NH_3(g)$$

9. A quantity of 5.00 g of dimethyl ether is compressed isothermally at 50°C from a volume of 10 liters to a volume of 2 liters. Calculate the change in free energy and in entropy, assuming the material behaves as an ideal gas.

10. From heats of formation and standard entropies in the tables calculate the standard free energy changes for each of the following reactions at 25°C:

$$NH_3(g) + CH_3COOH(l) \longrightarrow$$
$$NH_2CH_2COOH(s) + H_2(g)$$

$$CO_2(g) + C_6H_6(l) \longrightarrow C_6H_5COOH(l)$$
$$(CH_3)_2C{=}O(l) \longrightarrow C_2H_6(g) + CO(g)$$

11. At $-10°C$, ice has a vapor pressure of 1.950 torr and liquid water has a vapor pressure of 2.149 torr. Calculate the free energy change when 1 mol of ice is transformed into liquid at this temperature and state whether or not this is a spontaneous process.

12. From the table of phosphate transfer potentials, calculate the equilibrium constants for the following reactions in a medium of pH 7:

$$ATP + \text{arginine} \rightleftharpoons$$
$$\text{arginine phosphate} + ADP$$
$$\text{glucose 1-phosphate} \rightleftharpoons$$
$$\text{glucose 6-phosphate}$$

REFERENCES

Books

David E. Green and Robert F. Goldberger, *Molecular Insights into the Living Process,* Academic Press, New York, 1967. Chapters 6, 7, and 8 deal with the role of ATP, glycolysis, and processes of biological synthesis.

Tsoo E. King and Martin Klingenberg, Eds., *Electron and Coupled Energy Transfer in Biological Systems,* Vol. 1, Parts A and B, Dekker, New York, 1972. Oxidative phosphorylation and the role of cytochromes.

Irving M. Klotz, *Energy Changes in Biochemical Reactions,* Academic Press, New York, 1967. Emphasizes applications of free energy and discusses the nature of the high-energy bond.

Irving M. Klotz, *Introduction to Chemical Thermodynamics,* W. A. Benjamin, Menlo Park, Calif., 1964. Chapters 7 through 12 discuss free energy and entropy with applications to various systems.

A. L. Lehninger, *Bioenergetics,* 2nd ed., W. A. Benjamin, Menlo Park, Calif., 1971. An excellent introductory account with specific applications to biological processes.

H. J. Morowitz, *Energy Flow in Biology,* Academic Press, New York, 1968. A more advanced treatment of biological applications.

Thomas P. Singer, Ed., *Biological Oxidations,* Interscience, New York, 1968. Good description of the processes of oxidative phosphorylation.

Jürg Waser, *Basic Chemical Thermodynamics,* W. A. Benjamin, Menlo Park, Calif., 1966. Chapters 3 to 6 cover entropy, free energy, and physical and chemical equilibria.

Journal Articles

R. A. Alberty, "Maxwell Relations for Thermodynamic Quantities of Biochemical Reactions," *J. Am. Chem. Soc.* **91**, 3899 (1969).

A. F. M. Barton, "Internal Pressure—A Fundamental Liquid Property," *J. Chem. Educ.* **48**, 156 (1971).

E. Hamori, "Illustration of Free Energy Changes in Chemical Reactions," *J. Chem. Educ.* **52**, 370 (1975).

Joel Kirschbaum, "Biological Oxidations and Energy Conservation," *J. Chem. Educ.* **45**, 28 (1968).

Martin J. Klein, "Maxwell, His Demon, and the Second Law of Thermodynamics," *Am. Sci.* **58**, 84 (1970).

Rodolfo Margaria, "The Sources of Muscular Energy," *Sci. Am.* **226**, 84 (March 1972).

John M. Murray and Annemarie Weber, "The Cooperative Action of Muscle Proteins," Sci. Am. **230,** 59 (February 1974).

Efraim Racker, "Bioenergetics and the Problem of Tumor Growth," *Am. Sci.* **60,** 56 (1972).

Efraim Racker, "The Membrane of the Mitochondrion," *Sci. Am.* **218,** 32 (February, 1968).

Daniel E. Stull, "The Thermodynamic Transformation of Organic Chemistry," *Am. Sci.* **59,** 734 (1971).

Jeffrey S. Wicken, "The Chemically Organizing Effects of Entropy Maximization," *J. Chem. Educ.* **53,** 623 (1976).

Jack M. Williams, "Combining Residual Entropy and Diffraction Results to Understand Crystal Structure," *J. Chem. Educ.* **52,** 210 (1975).

Five
Solutions
of Electrolytes

Aqueous solutions of materials such as potassium chloride, sodium hydroxide, hydrochloric acid, magnesium sulfate, sodium acetate, tetramethylammonium bromide, or acetic acid are capable of conducting an electric current to a degree far surpassing the ability of pure water or of an aqueous solution of ethyl alcohol, sucrose, or acetone. Those materials that dissolve to yield a conducting solution—chiefly acids, bases, and salts—are called *electrolytes*. Since an electric current can be carried only by the motion of electric charges, charged particles must be present in the solution of an electrolyte. These are the ions, positive and negative, comprising the solute before it dissolves, or formed from the interaction of the solvent and the solute.

In this chapter, several methods of studying solutions of electrolytes are described, and the models that have been developed to represent the behavior of ions in solution are discussed, with particular emphasis on the effects of the relatively strong electrostatic forces between ions.

5-1
STRONG AND WEAK
ELECTROLYTES

COLLIGATIVE PROPERTIES
Some insight into the nature of solutions of electrolytes can be obtained from measurements of their colligative properties. For example, the freezing point depression values for 0.1, 0.01, and 0.001 m solutions of sodium chloride in water are 0.346, 0.0361, and 0.00366°, respectively. Results of this sort are most informative if expressed as the ratio of the colligative property observed to the "normal" value expected for a nonelectrolyte of the same molal concentration, a ratio termed the *van't Hoff factor* and represented by the symbol i. For the three solutions, taken in order, the normal values of the freezing point depression are 0.186, 0.0186, and 0.00186°, so that i is 1.86, 1.94, and 1.97,

respectively. Other examples of values of i are 3.82 for 0.001 m $K_3Fe(CN)_6$, 2.84 for 0.001 m K_2SO_4, and 2.85 for 0.001 m $BaCl_2$.

Values of i may be calculated from any of the colligative properties: from the boiling point elevation, from the vapor pressure lowering, or from the osmotic pressure, as well as from the freezing point depression. The same value of i is obtained from any of the four properties, and the value obtained from measurements of one colligative property can be employed to predict accurately the value of any of the other three properties.

If we examine the numerical results for i cited above, it appears that, for each electrolyte, i is slightly smaller than the total number of ions that correspond to one molecular formula of the substance. Thus for sodium chloride, as the concentration of the solution decreases, i approaches 2, corresponding to one sodium and one chloride ion; for $K_3Fe(CN)_6$, i approaches 4 as the solution is made more dilute, corresponding to three potassium ions and one ferricyanide ion; and so on. For dilute solutions, each of the ions of the solute seems to have the same effect upon the colligative properties as does an ordinary molecule, and the dissociation of salts in these solutions is very nearly, if not entirely, complete.

Other compounds may have values of i which are somewhat greater than unity but which do not approach, in the region of concentration accessible to measurement, the number of ions expected from complete dissociation of the molecules. These materials are mostly weak acids and weak bases, such as benzoic acid or ammonium hydroxide. For these materials, which are *weak electrolytes* in contrast to salts like potassium bromide which are *strong electrolytes*, the value of a colligative property can be used as a measure of the extent of dissociation. To establish the quantitative relation, suppose that each molecule that dissociates produces n ions, and that α is the fraction of the total number of molecules that are dissociated. The fraction of molecules remaining as single particles is $1 - \alpha$. For each original molecule there are then at equilibrium $1 - \alpha + n\alpha$ particles. This quantity is equal to the van't Hoff factor i. If the equation in which it is set equal to i is solved for α, we obtain

$$\alpha = \frac{i - 1}{n - 1} \tag{5-1}$$

For an electrolyte that produces only two ions, n is 2 and Equation (5-1) reduces to $\alpha = i - 1$. This applies to acetic acid which has the ionization equilibrium

$$CH_3COOH + H_2O \rightleftharpoons CH_3COO^- + H_3O^+ \tag{5-2}$$

For example, the freezing point of an aqueous solution 0.00301 m in acetic acid is $-0.00606°C$, which corresponds to an i value of $0.00606/(0.00301)(1.86)$ or 1.08. The degree of dissociation is therefore 0.08, which means that the acid is 8 percent dissociated, or 8 molecules out of every 100 are ionized in solution, leaving 92 in intact molecular form.

THE ARRHENIUS THEORY

It should be profitable at this point to recall a bit of the historical development of our ideas of solutions of electrolytes. Early workers knew of the ability of these solutions to conduct an electric current, but they generally supposed that the application of an external electric potential was required to pull the molecules apart into ions. It remained for a graduate student in the 1880's, Svante Arrhenius, to assemble evidence clearly supporting the idea that charged particles exist in all solutions of electrolytes. Arrhenius compared the results from measurements of colligative properties and those from electric conductivity, which are described below, and showed that both lead to similar conclusions about the degree of dissociation of electrolytes. He reasoned that, since colligative properties are measured in the absence of an applied potential, dissociation occurs when the solution is initially formed. At first there was reluctance to accept the theory of Arrhenius, because the production of charged particles under other circumstances requires high energy and drastic conditions. What was not realized was the role of the solvent, water, in the process of dissociation. As stated in Chapter 1, the force of electrostatic attraction drawing together two charges, q_1 and q_2, at distance r, is equal to $q_1 q_2 / D r^2$. The Coulomb force described by this expression is quite large in air, but the dielectric constant D is large for water, 78.6 at 25°C, compared to a value for air of little more than unity. Thus the presence of water reduces tremendously the work required to separate two ions of opposite sign. Furthermore, solvation of the ions by a sheath of water molecules, each with one end of the dipole oriented toward the charged particle as shown in Figure 1-11, tends to keep the ions apart from one another.

Arrhenius supposed that there exists an equilibrium between the ions and molecules of any electrolyte in solution, and stated that the distinction between strong and weak electrolytes is simply that of a different degree of ionization. In this view, as the solution is diluted the degree of ionization of a given electrolyte increases; this is necessarily true if an equilibrium exists in which the ionization proceeds in such a way as to increase the number of particles.

For the general case of an electrolyte yielding two ions, if c is the stoichiometric concentration of the electrolyte and α is the fraction dissociated, the equilibrium-constant expression is

$$K = \frac{(\alpha c)(\alpha c)}{(1 - \alpha)c} = \frac{\alpha^2 c}{1 - \alpha} \tag{5-3}$$

which is in a form that again illustrates the point that, if c is decreased, there must be a compensating increase in α.

The Arrhenius theory applies quite well to solutions of weak electrolytes. It is supposed that weak electrolytes are in molecular form until put into solution, at which time the covalent bonds in some of the molecules are broken with the aid of the solvent and partial ionization occurs. For weak electrolytes, the process of ionization, which is the formation of the ions from uncharged particles, and the process of dissociation, which is the separation of the positive ions from the negative ions, both occur in a single common step. Many weak electrolytes are acids or bases, and these are treated in Chapter 6; in the ionization of a weak acid or base, the solvent functions by participating in the transfer of a proton from or to the solute, as well as by preventing the recombination of ions by providing a medium of high dielectric constant.

Quantitatively, however, the equilibrium treatment of Arrhenius does not fit the experimental results for strong electrolytes. In particular, the "equilibrium constants" calculated for strong electrolytes turn out to be far from constant when the electrolyte concentration is varied. Furthermore, these "constants" can also be changed by the addition of other electrolytes that do not have even one ion in common with the electrolyte for which the dissociation constant is calculated.

5-2
THE DEBYE–HÜCKEL THEORY; ACTIVITY COEFFICIENTS OF IONS

It is reasonable to suppose, on the basis of the facts just recounted, as well as of much other evidence, that *strong electrolytes are completely dissociated in moderately dilute solutions*. This theory requires, however, that there be substituted for the Arrhenius hypothesis of incomplete dissociation some other explanation, founded on a reasonably quantitative basis, of the deviation of the van't Hoff factor from whole numbers. In 1923, by considering the electrostatic interactions between ions in solution, P. Debye and E. Hückel succeeded in developing a treatment of these deviations which satisfactorily describes the experimental results for dilute solutions.

THE THEORY OF COMPLETE DISSOCIATION OF STRONG ELECTROLYTES

The theory of Debye and Hückel assumes that strong electrolytes are present in solution entirely in the form of ions. If a single ion in the solution is considered, it is evident that ions of opposite sign, drawn by electrostatic attraction, have a slightly greater probability of being in the neighborhood of the given ion than do ions of the same sign. The fleeting cloud of charge about an ion, containing an excess of charge of the opposite sign in an amount equal to the charge on the ion, is termed an *ionic atmosphere*. Because of the presence of the atmosphere, the ion is held more tightly in position than it would otherwise be, and its motion is somewhat restricted. The results of the theory are usually couched in the form of an equation for the activity coefficient of an ion, the quantity that gives the ratio of the effective concentration, or activity, to the stoichiometric concentration, and which was defined in Equation (4-73).

Even before this theory was developed, G. N. Lewis proposed an empirical generalization that the activity coefficient of an ion depends primarily upon the magnitude of its charge, rather than upon any specific aspect of its chemical nature, and upon a quantity character-istic of the solution termed its *ionic strength*. The ionic strength μ is computed in the following manner: The concentration of each ionic species in the solution is multiplied by the square of the charge on that ion type, and the results are added together and divided by 2:

$$\mu = \tfrac{1}{2} \sum c_j Z_j^2 \tag{5-4}$$

According to this definition of the ionic strength, ions with multiple charges have a much greater effect upon the behavior of a given ion than do ions with single charges. The validity of the use of the ionic strength as a parameter for characterizing deviations from ideality is confirmed by the Debye–Hückel theory as well as by experimental tests on dilute solutions.

Example: Calculate the ionic strength of a solution containing 0.010 M $Al(NO_3)_3$ and 0.025 M $MgSO_4$.

Solution: The contribution of each type of ion is first listed separately:

$$
\begin{array}{lll}
Al^{3+} & 0.010\ M \times 3^2 = 0.090 \\
NO_3^{-} & 0.030\ M \times 1^2 = 0.030 \\
Mg^{2+} & 0.025\ M \times 2^2 = 0.100 \\
SO_4^{2-} & 0.025\ M \times 2^2 = \underline{0.100} \\
& \qquad\qquad\text{Total}\quad 0.320
\end{array}
$$

The total is now divided by 2 to obtain the ionic strength, 0.160.

MATHEMATICAL BASIS OF THE DEBYE–HÜCKEL THEORY

We begin by focusing attention on some one ion, the reference ion for which some physical properties are to be calculated, which will be labeled with the index i. This ion has a charge of $Z_i \epsilon$, where ϵ is the charge of one electron and Z_i is the "valence" of the ion. If this ion is viewed apart from its environment, the total charge on the surround-ings must be equal to $-Z_i \epsilon$, equal in magnitude but opposite in sign to that of the ion itself. This charge is pictured as being distributed throughout a spherical region with the ion itself at the center of the sphere, and this region comprises the ionic atmosphere, the compo-nents of which are in very rapid motion. It is now supposed that the radial variation of ion concentration in the atmosphere with distance from the central ion is given by a Boltzmann distribution as described in Equation (1-19). If an ion in the atmosphere has a charge of Z_j units and is present in an average concentration c_j^0 in the bulk of the solu-tion, its concentration at a point is

$$c_j = c_j^0 e^{-\epsilon Z_j \psi / kT} \tag{5-5}$$

In this equation, ψ is the electric potential at the point in excess of the average potential in the solution. The potential difference between two points is a measure of the work required to transport unit positive electric charge from one point to the other. Given the chance, positive charge tends to move from a place of more positive potential to one of less positive potential. The electric field at a point, describing the force on a unit positive charge placed there, is the gradient or derivative of the potential at that point.

The Debye–Hückel theory is limited to solutions for which the ionic concentration is very small and, under this stipulation, the value of the potential difference ψ is never very large. The exponential in Equation (5-5) can be expanded in a power series in $\epsilon Z_j \psi / kT$:

$$e^{-\epsilon Z_j \psi / kT} = 1 - \frac{\epsilon Z_j \psi}{kT} + \frac{1}{2}\left(\frac{\epsilon Z_j \psi}{kT}\right)^2 + \cdots \tag{5-6}$$

and, so long as ψ is sufficiently small so that $\epsilon Z_j \psi$ is smaller than kT, higher-order terms can be neglected:

$$e^{-\epsilon Z_j \psi / kT} = 1 - \frac{\epsilon Z_j \psi}{kT} \tag{5-7}$$

From the potential we can thus calculate the charge density ρ at any point in the atmosphere by substituting Equation (5-7) in the expression for the concentration of the jth ion, Equation (5-5), multiplying the molar concentration by Avogadro's number N and dividing by 1000 to obtain the number of ions per cubic centimeter, multiplying by the charge on an ion $Z_j \epsilon$, and then summing over all the ionic species present:

$$\rho = \frac{N\epsilon}{1000} \sum_j c_j^0 Z_j \left(1 - \frac{\epsilon Z_j \psi}{kT}\right) = \frac{N\epsilon}{1000} \sum_j c_j^0 Z_j - \frac{N\epsilon^2 \psi}{1000kT} \sum_j c_j^0 Z_j^{\,2} \tag{5-8}$$

The condition that the sum of the positive charges in the solution equal the sum of the negative charges—the condition that the solution be electrically neutral—requires that the first term in the last member of this equation be equal to zero. When the definition of the ionic strength μ from Equation (5-4) is inserted into the second term, the equation becomes

$$\rho = -\frac{2N\epsilon^2 \psi}{1000kT}\mu \tag{5-9}$$

Equation (5-9) can be regarded as showing how the charge density, and therefore the distribution of ions in the atmosphere, is determined by the potential. However, the potential is, at the same time, determined by the distribution of ionic charge, and a second relation between ρ and ψ must be obtained from the principles of electrostatics. The relation between these quantities is known as Poisson's equation, and has the following form for a spherically symmetrical situation:

$$\frac{1}{r^2}\frac{d}{dr}\left(r^2 \frac{d\psi}{dr}\right) = -\frac{4\pi\rho}{D} \tag{5-10}$$

The radius r is the distance from the center of the reference ion being

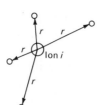

Figure 5-1
The length r is the distance to any point in the surrounding
solution from the ion for which properties are being calculated.

considered, as shown in Figure 5-1, and D is the dielectric constant
of the medium. Substitution of the charge density from Equation (5-9)
into Equation (5-10) leads to the result

$$\frac{1}{r^2}\frac{d}{dr}\left(r^2\frac{d\psi}{dr}\right) = \frac{8\pi N\epsilon^2\psi}{1000DkT}\mu \tag{5-11}$$

For convenience, we symbolize the coefficient of ψ on the right-hand
side of this equation by κ^2, so that

$$\kappa = \sqrt{\frac{8\pi N\epsilon^2\mu}{1000DkT}} \tag{5-12}$$

and Equation (5-11) becomes

$$\frac{1}{r^2}\frac{d}{dr}\left(r^2\frac{d\psi}{dr}\right) = \kappa^2\psi \tag{5-13}$$

This is a differential equation, and the problem now is to find a solution that gives the indicated equality when it is substituted in the equation and which, in addition, meets the physical requirements of the
problem that has been set up. From experience with equations of this
type, one is led to try a solution of the form

$$\psi = A\frac{e^{-\kappa r}}{r} + B\frac{e^{\kappa r}}{r} \tag{5-14}$$

where A and B are constants. Carrying out the operations indicated by
Equation (5-13) upon this expression indicates that it is indeed a solution. However, for the present problem, B must be zero; otherwise the
potential would increase without limit at very large distances from the
reference ion, and this is a physical impossibility. To evaluate A, we
consider what happens when the ionic strength approaches zero.
Since this implies that κ approaches zero, the first term in ψ approaches
A/r. At the same time the potential must approach that determined by
the reference ion itself, $Z_j\epsilon/Dr$. Therefore A must be equal to $Z_j\epsilon/D$.

The final equation for the potential around the reference ion is thus

$$\psi = \frac{Z_i\epsilon}{Dr}e^{-\kappa r} \tag{5-15}$$

The aim of obtaining this equation is to permit the calculation of
some measurable property of the ion, such as the thermodynamic
activity coefficient γ_i. For this calculation, it is simplest to work in

terms of the molar free energy, which is related to the activity by

$$G_i = G_i^0 + RT \ln a_i$$
$$= G_i^0 + RT \ln c_i + RT \ln \gamma_i \qquad (5\text{-}16)$$

If there were no ionic atmosphere, it is assumed, only the first two terms on the right-hand side would appear in the equation. The third term represents the "nonideal" contribution arising from the ionic atmosphere present around the ion.

Since the free energy change is equal to the work done by the system during a reversible process, the term $RT \ln \gamma_i$ should be equal to the difference in the work required to charge the ion in the presence of the atmosphere and that required in the absence of the atmosphere. Various investigators have devised models to calculate the excess work of charging an ion in the presence of the atmosphere. One of these models, which is relatively easy to visualize, imagines the un-charged ion to be initially located at its position in the solution, and then a magic "charge transferer" to bring the charge in from a long distance away, bit by bit, until the full amount $Z_i \epsilon$ has been placed upon the ion. The work required to transfer an increment of this charge is equal to its magnitude times the potential difference through which the charge is transferred:

$$w = \psi \, dq \qquad (5\text{-}17)$$

The potential at the point where the ion is located is the limit when r is allowed to approach zero in Equation (5-15), and this result is obtained by expanding the exponential factor in a series:

$$\psi = \frac{Z_i \epsilon}{Dr} e^{-\kappa r} = \frac{Z_i \epsilon}{Dr}\left(1 - \kappa r + \frac{\kappa^2 r^2}{2} - \cdots\right) \qquad (5\text{-}18)$$

The result obtained on neglecting all but the first two terms in the series is

$$\psi = \frac{Z_i \epsilon}{Dr} - \frac{Z_i \epsilon \kappa}{D} \qquad (5\text{-}19)$$

The first term results from the ion itself and is identical whether or not the atmosphere is present. The second term is the contribution to the potential by the atmosphere. Before the charging process is complete, the potential is given by a similar equation, but with the charge $Z_i \epsilon$ replaced by a varying smaller charge, say q_i. Integrating over the entire charging process, the extra work resulting from the second term is given by

$$w = \int_0^{Z_i \epsilon} -\frac{q_i \kappa}{D} \, dq_i = \left.\frac{\kappa q_i^2}{2D}\right|_0^{Z_i \epsilon} = -\frac{Z_i^2 \epsilon^2 \kappa}{2D} \qquad (5\text{-}20)$$

This result is multiplied by Avogadro's number to obtain the value for one mole of ions, and then the result is equated to the "nonideal" term in Equation (5-16):

$$-\frac{Z_i^2 \epsilon^2 \kappa N}{2D} = RT \ln \gamma_i \qquad (5\text{-}21)$$

The definition of κ in terms of physical quantities is now inserted into this expression, similar factors are collected, and the equation is solved for $\ln \gamma_i$:

$$\ln \gamma_i = -\left(\frac{\epsilon^2}{DkT}\right)^{3/2}\left(\frac{2\pi N}{1000}\right)^{1/2} Z_i^2\mu^{1/2} \tag{5-22}$$

USE OF THE DEBYE–HÜCKEL LIMITING EQUATION

For convenience in applying Equation (5-22) to aqueous solutions at room temperature, appropriate numerical values of the properties of water and of the other constants can be substituted in the equation and the factor 2.303 employed to convert to logarithms to the base 10. The result is

$$\log \gamma_i = -0.51\, Z_i^2\mu^{1/2} \tag{5-23}$$

Equations (5-22) and (5-23) are accurate only for dilute solutions and therefore are referred to as the *Debye–Hückel limiting equations*. The restriction is a consequence of the approximations made in the derivation, which assumed a relatively small numerical value for the ionic strength and for the potential due to the atmosphere. The upper limit of applicability is in the vicinity of an ionic strength of 0.1 m when the ion for which the activity coefficient is being calculated has a unit charge, and at a much lower ionic strength for polyvalent ions. It is very important to realize that the ionic strength factor carries in it the influence of all the ions in the solution. Not only is the concentration of the ion type for which the activity is being calculated taken into account in calculating the ionic strength, but also all species actually present in the solution as free ions are counted, including for weak electrolytes the portion of the electrolyte that is dissociated. From

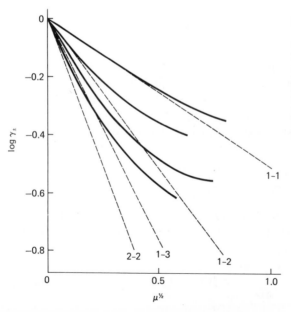

Figure 5-2
The logarithm of the mean activity coefficient as a function of the square root of the ionic strength. Dotted lines are predictions of the Debye–Hückel equation for various charge types. Solid lines are typical experimental results.

Figure 5-2, the reader can obtain some idea of how well the limiting form of the Debye–Hückel equation reproduces the experimental results.

Although this theory purports to predict the activity coefficient of an individual ion type, it is not possible to measure experimentally the activity coefficient of any single type of ion, for an ion of one charge must always be accompanied by ions of the opposite sign in order that the solution be electrically neutral. However, it is possible to determine by experiment the *mean* activity coefficient of the ions of an electrolyte, a quantity defined for sodium chloride as

$$\gamma_{\pm} = (\gamma_{Na^+}\gamma_{Cl^-})^{1/2} \tag{5-24}$$

The results of this measurement can be compared with the values calculated by a relation derived from Equation (5-23):

$$\log \gamma_{\pm} = -0.51 |Z_+| |Z_-| \mu^{1/2} \tag{5-25}$$

Here $\gamma_{\pm}$ is again the geometric mean of the activity coefficients of the two ions, and $|Z_+|$ and $|Z_-|$ are the numbers of electronic charges, without their signs, on the positive and negative ions of the electrolyte, respectively.

Example: Estimate the activity coefficients of the potassium ions and of the sulfate ions, and the mean activity coefficient for a 0.010 M aqueous solution of potassium sulfate at room temperature.

Solution: The concentration of potassium ion is 0.020, and that of sulfate ion is 0.010 M. The ionic strength is $[(1^2 \times 0.020) + (2^2 \times 0.010)]/2$ or 0.030. By the limiting equation,

$$\log \gamma_{K^+} = -0.51(1)^2(0.030)^{1/2} = -0.088 = \bar{1}.912$$
$$\log \gamma_{SO_4^{2-}} = -0.51(2)^2(0.030)^{1/2} = -0.353 = \bar{1}.647$$

From the antilogarithms of these numbers, the activity coefficients are 0.82 for the potassium ion and 0.44 for the sulfate ion. The mean activity coefficient is the nth root of the product of the activity coefficients of the n ions derived from one molecular formula of the electrolyte:

$$\gamma_{\pm} = (\gamma_{K^+}^2 \gamma_{SO_4^{2-}})^{1/3} = [(0.82)^2(0.44)]^{1/3} = 0.67$$

The mean activity coefficient of the ions of an electrolyte can be determined experimentally by the measurement of solubilities as described in the following pages, or from the electromotive force of a galvanic cell as shown in Chapter 7. In principle, activity coefficients could also be calculated from the measured colligative properties of solutions, but this is a rather complicated calculation, because the colligative properties depend upon the activity of the solvent whereas it is desired to know the activity of the solute.

A significant feature of the Debye–Hückel approach is the conceptual view of the ionic atmosphere surrounding a small ion. When the theory is examined in detail to find a quantitative description of the nature of the charge distribution in the atmosphere about an ion, one result obtained is that the average distance away from the ion for the charge in the atmosphere of that ion is equal to $1/\kappa$. Since κ is proportional to the square root of the ionic strength, an increase in the ionic strength causes the atmosphere to shrink in toward the central ion.

As we will see in Chapter 12, the effects of ionic strength are also quite important in determining the behavior of colloidal particles and macromolecules in solution. These relatively large particles have charged groups located on their surface, with counterions somewhere nearby in the solution. The application of the Debye–Hückel concept to these systems is very helpful in visualizing the influence of increase of ionic strength, which causes small ions to move closer to the surface of the macromolecule.

SOLUBILITY
OF SOLID ELECTROLYTES

The quantity of a solid electrolyte that dissolves in a given amount of water is limited, and at saturation of the solution there exists an equilibrium between the solid, undissociated—but not necessarily un-ionized—electrolyte and the ions in solution. This is one type of equilibrium in which ionic activity coefficients are usually quite significant. For the simple case of an electrolyte dissociating to form two singly charged ions, the equilibrium can be represented as

$$AB(s) \rightleftharpoons A^+(aq) + B^-(aq) \qquad (5\text{-}26)$$

For this equilibrium, the product of the ion concentrations at saturation is defined as the *solubility product* of the electrolyte:

$$[A^+][B^-] = K[AB] = K_{sp} \qquad (5\text{-}27)$$

If the electrolyte ionizes to produce more than one ion of a given type per molecule, the solubility product includes the concentration of each ion raised to a power equal to the number of that type formed in the dissociation process. For calcium phosphate,

$$Ca_3(PO_4)_2 \rightleftharpoons 3Ca^{2+} + 2PO_4^{3-} \qquad (5\text{-}28)$$
$$K_{sp} = [Ca^{2+}]^3[PO_4^{3-}]^2 \qquad (5\text{-}29)$$

This particular relationship has very important physiological applications. Thus the deposition of calcium phosphate and calcium carbonate in bone and teeth is controlled by the solubility product relationship. The blood plasma normally contains calcium and phosphate ions in amounts such that the product of the concentrations is in the vicinity of the solubility product. Bone contains an enzyme, bone phosphatase, which decomposes phosphoric acid esters to liberate phosphate ion, and the solubility product of calcium phosphate is locally exceeded, causing precipitation in the bone matrix. The function of vitamin D in ensuring proper bone structure is thought to be due to its regulatory effect upon the calcium–phosphate balance in the blood.

Returning now to a more general consideration of solubility product, we find that the expressions given above work satisfactorily for very dilute solutions. When a solute is difficultly soluble and no other electrolyte is present in the solution, this condition is met. The equation does not work, however, for moderately soluble electrolytes nor for the situation in which a difficultly soluble electrolyte is present along with a large concentration of other ions in solution. These foreign ions need not be ions in common with the ions of the precipitate in order for them to make appreciable changes in solubility. The explanation for these observations is of course that, when the ionic strength of the solution is increased, the activity coefficients of the ions are altered, and the activity of the ions must be used in the solubility product constant in place of the stoichiometric concentration. For a slightly soluble uni-univalent electrolyte, dissolving as in Equation (5-26), we can write in place of Equation (5-27) the following:

$$K_{sp,a} = a_A a_B = \gamma_A c_A \gamma_B c_B = (\gamma_\pm)^2 K_{sp} \qquad (5\text{-}30)$$

The quantity $K_{sp,a}$, in which the activities of the ions appear, is sometimes referred to as the *true*, or *thermodynamic*, solubility product.

It is possible to employ measurements of solubility as a means of evaluating the activity coefficient for an electrolyte and of ascertaining how the activity coefficient varies with ionic strength. *As the activity coefficients of the ions decrease, more of the solid dissolves to maintain the product of the ionic activities constant.* For a uni-univalent electrolyte,

$$K_{sp,a} = (\gamma_{\pm,0})^2 S_0^2 = (\gamma_\pm)^2 S^2 \qquad (5\text{-}31)$$

so that

$$\frac{\gamma_{\pm,0}}{\gamma_\pm} = \frac{S}{S_0} \qquad (5\text{-}32)$$

S_0 is the solubility of the electrolyte in water and $\gamma_{\pm,0}$ is the mean activity coefficient of its ions in water. S and $\gamma_\pm$ are corresponding values of solubility and activity coefficient in some medium containing added ions.

Example: The solubility of thallous chloride, TlCl, in water at 25°C is 0.01607 mol/liter. In the presence of 0.30 mol KNO_3 per liter the solubility is increased to 0.02312 mol/liter. Calculate the mean activity coefficient of TlCl in the latter solution.

Solution: The ionic strengths of the two solutions are 0.01607 and 0.323, respectively. The first falls in the Debye–Hückel region:

$$\log \gamma_\pm = -0.51 \sqrt{0.01607} = -0.065$$
$$\gamma_\pm = 0.86$$

The activity solubility product is

$$K_{sp,a} = (0.86 \times 0.01607)^2 = 1.91 \times 10^{-4}$$

For the second solution,

$$K_{sp,a} = (\gamma_\pm)^2 (0.02313)^2 = 1.91 \times 10^{-4}$$
$$\gamma_\pm = (0.01607)(0.86)/(0.02312) = 0.60$$

EQUILIBRIA OF IONS WITH COMPLEXES

As will be described in Chapter 9, complex ions may be formed by the donation of electrons by an electron-rich group to a metal ion, which is then covalently bonded to the group. Commonly, several different groups become attached to the same central metal ion, as in hemoglobin, where iron is held to four nitrogen atoms in the porphyrin heme and to another nitrogen atom from the protein globin. Calcium and magnesium ion are complexed by proteins in the blood and can be complexed by citrate. Metal ions, such as copper or iron, are necessary for the activity of many enzymes and are combined with portions of the enzymes as complexes.

All complexes are dissociated to a greater or less degree in solution, and this dissociation follows the laws of chemical equilibrium, so that a *dissociation equilibrium constant* characterizes the behavior of the material. The dissociation constant, or "instability constant," of the diammine–silver complex, $Ag(NH_3)_2{}^+$ for example, is 6×10^{-8}. Often the equilibrium is described by the *stability constant* of the complex, which is the reciprocal of the dissociation constant. For the copper–glycine complex, the equilibrium is

$$Cu^{2+} + 2NH_2CH_2COO^- \rightleftharpoons Cu(NH_2CH_2COO)_2 \qquad (5\text{-}33)$$

and the stability constant is

$$K = \frac{[Cu(NH_2CH_2COO)_2]}{[Cu^{2+}][NH_2CH_2COO^-]^2} = 1 \times 10^{16} \qquad (5\text{-}34)$$

Suppose that the ionic strength of a solution of this complex is increased by a moderate amount. How is the equilibrium affected? The change does not alter the activity coefficient of the neutral complex very much, but it does reduce the activity coefficients of the two ionic components. To maintain equilibrium, then, the concentration of the ions is increased at the expense of the concentration of complex.

In some systems, charged species occur on both sides of the equilibrium. As an example consider the association between ferric ions and fluoride ions:

$$Fe^{3+} + 6F^- \rightleftharpoons FeF_6{}^{3-} \qquad (5\text{-}35)$$

The thermodynamic equilibrium constant is

$$K_a = \frac{[FeF_6{}^{3-}]}{[Fe^{3+}][F^-]^6} \frac{\gamma_{FeF_6{}^{3-}}}{\gamma_{Fe^{3+}}\,\gamma_{F^-}{}^6} \qquad (5\text{-}36)$$

One can now take logarithms of both sides of this equation, rearrange, and substitute the limiting Debye–Hückel expressions from Equation (5-25) for the activity coefficients:

$$\log\frac{[FeF_6{}^{3-}]}{[Fe^{3+}][F^-]^6} - \log K_a = \log\frac{\gamma_{Fe^{3+}}\gamma_{F^-}{}^6}{\gamma_{FeF_6{}^{3-}}}$$

$$= -0.51(3)^2\mu^{1/2} - 6[0.51(1)^2\mu^{1/2}] + 0.51(3)^2\mu^{1/2} = -3.1\mu^{1/2} \qquad (5\text{-}37)$$

From this equation we see that increasing the ionic strength reduces the concentration of the complex relative to the concentrations of the free ions. This example illustrates the general rule that the effect of increasing the ionic strength is a shift in the direction that maximizes the sum of the squares of the charges on the ions, a principle which will again be encountered in Chapter 10 in the discussion of rates of reactions between ions.

One of the ways in which much information can be gained about the ions present in solutions of electrolytes is by studying the ability of these solutions to carry an electric current. The conducting ability of a solution depends upon the ease with which positive and negative particles can move through the solution.

The *ionic mobility*—the velocity of the ion under a potential drop of 1 V/cm—depends upon the size of the ion, the number of charges on the ion, the solvation of the ion, the viscosity of the solvent, and the number and nature of the other ions present in the solution. For a weak electrolyte, a knowledge of the conducting ability per ion of the electrolyte and of the conductivity of the solution permits the calculation of the number of free ions present in the solution.

MEASUREMENT OF CONDUCTANCE

In describing and measuring the conducting properties of a solution, it is necessary to apply some simple principles of electricity. For well-behaved conductors, Ohm's law is obeyed:

$$I = \frac{E}{R} \tag{5-38}$$

The current flowing is I, the potential difference between the ends of the conductor is E, and R is the resistance of the conductor. In practical electrical units, the current is expressed in amperes or coulombs per second, the potential in volts, and the resistance in ohms. The conductance L of a particular conductor is the ratio of the current that flows to the applied potential, and thus it is simply the reciprocal of the resistance:

$$L = \frac{I}{E} = \frac{1}{R} \tag{5-39}$$

Example: A current of 250 milliamperes is found to pass through a sample of material placed between two electrodes across which a potential of 15.0 V is applied. What is the resistance of the sample of material? What is its conductance?

Solution: The resistance is equal to the ratio of voltage required to current passing, or 15.0 V divided by 0.250 amperes, which is 60.0 ohms. The conductance is the reciprocal of this, or 0.01667 ohm^{-1}.

Both the resistance and the conductance are functions of the size and shape of the conductor, as well as of the intrinsic nature of the material of which it is composed. The specific conductance or conductivity κ and the specific resistance or resistivity ρ are defined as the conductance and resistance of a cube of material 1 cm on the edge. In the SI system, the meter replaces the centimeter as the unit of length. The conductivity is of course the reciprocal of the resistivity. Suppose there is a rectangular bar of conducting material a square centimeters in cross section, placed between electrodes d centimeters apart. The longer the bar, the higher its resistance; the larger the cross section perpendicular to the direction of flow, the smaller its resistance. Thus the overall resistance between the electrodes is given by

$$R = \rho \frac{d}{a} \tag{5-40}$$

The same relation can be written in terms of the conductance by taking the reciprocals of both sides of this equation:

$$L = \kappa \frac{a}{d} \tag{5-41}$$

The conductance of a solution of an electrolyte can be measured by placing the liquid solution in a glass container between two electrodes, usually made of platinum, and measuring the apparent resistance across the leads to the electrodes. Figure 5-3 shows two types of cells commonly used. In a given conductivity cell, the geometry of the conducting region of the solution and of the electrodes remains constant during various measurements; thus the dimensions of the conductor do not need to be explicitly defined but can be incorporated in a constant characteristic of the cell. This constant k is usually defined as the factor by which the observed conductance for a solution in the cell is to be multiplied to find the specific conductance of that solution:

$$\kappa = kL \tag{5-42}$$

The constant for a cell is established by measuring the conductance of the cell when it contains a solution of known conductivity, such as a standard potassium chloride solution.

In Table 5-1 are given values of the specific conductance of potassium chloride solutions over a convenient concentration range at several temperatures.

Example: A cell containing 0.0100 N KCl at 25°C has a resistance of 128.02 ohms. Calculate the cell constant.

Solution: The specific conductance of 0.0100 M KCl (normality is the same as molarity for KCl) is found from Table 5-1 to be 0.001413 at 25°C. The conductance of the cell is the reciprocal of the measured resistance, or 1/128.02. Then

$$0.001413 = k(1/128.02)$$

from which $k = 0.1809$ cm^{-1}.

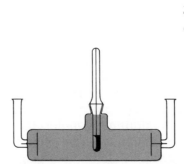

Figure 5-3
Two typical conductivity cells.

Table 5-1
Specific conductance, in ohm^{-1} cm^{-1}, of potassium chloride solutions

Molarity	Temperature		
	20°C	25°C	30°C
1.0000	0.1021	0.1119	—
0.1000	0.01167	0.01290	0.01412
0.0200	—	0.002767	0.003036
0.0100	0.001278	0.001413	—

The electric circuit used for measuring the cell resistance is a Wheatstone bridge, as shown in Figure 5-4. This can be thought of as a symmetrical square network of four resistances, R_1, R_2, R_3, and R_4. Across one pair of diagonally opposite terminals, or vertices, of the square is connected a source of potential, and across the other two terminals is connected a detector of current. R_1 is a variable resistance, which serves as a reference and has a resistance of the same order of magnitude as the cell R_2. R_3 and R_4 may be two sections of a slide-wire, with the movable tap at point c. When the bridge is balanced, a minimum of current flowing through the detector is observed. Then the values of the four resistances satisfy the equation

$$\frac{R_1}{R_2} = \frac{R_3}{R_4} \qquad (5\text{-}43)$$

Because the flow of direct current through a conductivity cell decomposes the solution, as well as setting up potential differences at the electrode–solution interface, alternating current is commonly used in a conductivity bridge. The frequent reversal of current flow tends to reverse any electrolytic process that may occur, and consequently decomposition and polarization are minimized. Polarization is a po-

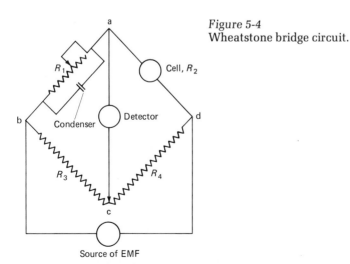

Figure 5-4
Wheatstone bridge circuit.

tential gradient produced at the surface of an electrode in order to drive onward some slow process such as the diffusion of ions up to the electrode to replace those that have reacted, or the desorption and diffusion away from the electrode of the products of the reaction.

The conductance of a solution increases relatively rapidly with temperature, usually about 2 percent per degree. It is therefore necessary in order to make measurements of reasonably good precision to maintain the temperature of the conductivity cell constant by mounting it in a constant-temperature water or oil bath.

If it is desired to make measurements of properties of solutes, then it is necessary to prepare the solvent in very pure form, so that impurities will neither contribute very much to the total conductance of the solution nor react with the solute, especially if weak electrolytes are to be studied. Aqueous solutions for conductivity measurements are prepared from "conductivity water," made by deionization of ordinary distilled water. Unless special precautions are taken to exclude air, the best water that can be obtained has a conductivity of about 10^{-6} ohm^{-1}, or a specific resistance of 10^6 ohms. This limit is set by the solubility of carbon dioxide from the air. Of course even perfectly pure water, if it could be obtained, would have a slight residual conductivity, since it ionizes to produce hydrogen and hydroxyl ions.

EQUIVALENT CONDUCTANCE

It is convenient to compare solutions of different electrolytes at concentrations at which the solute would give a standard number of posi-

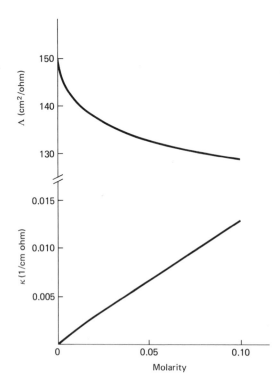

Figure 5-5
Comparison of concentration dependence of specific and equivalent conductances of potassium chloride solutions at 25°C.

tive and negative charges if it were completely ionized. For this purpose there is defined the equivalent conductance:

$$\Lambda = \frac{1000\kappa}{N} \tag{5-44}$$

Here κ is the specific conductance and N is the concentration in equivalents per liter, that is, the molar concentration of either ion of the electrolyte multiplied by the number of charges on the ion. An alternative way of describing the equivalent conductance is to state that it is equal to the conductance of a volume of solution large enough to contain exactly 1 equivalent of dissolved electrolyte when measured between parallel electrodes 1 cm apart. Thus, for a 0.01 N KCl solution, the amount containing 1 equivalent is 100,000 cm³, and the equivalent conductance is 100,000 times the conductivity; for 0.05 M Na₂SO₄, the factor is 10,000.

As a series of solutions of decreasing concentration of an electrolyte is examined, the specific conductance is found to decrease, but the equivalent conductance increases. This effect is illustrated in Figure 5-5. As the concentration of the electrolyte approaches zero, Λ approaches a limiting value Λ_0 (sometimes represented as Λ_∞), the equivalent concentration at infinite dilution. For strong electrolytes, the variation of equivalent conductance with concentration is gradual, and a plot of Λ against concentration permits easy extrapolation to the limit at zero concentration. A plot of Λ against the square root of the concentration is even more convenient, since it is linear at low concentrations. The equivalent conductance of potassium chloride at 25°C varies from 128.96 cm²/ohm at 0.10 M, through 141.27 at 0.01 M, to 146.95 at 0.001 M, and approaches the value 149.86 as a limit. For weak electrolytes, values in the experimentally attainable range of dilution do not come close enough to the limit to make extrapolation possible. Ammonium hydroxide, with a limiting value for Λ of 271, reaches only 34 at 0.001 M and 93 at 0.0001 M. Curves of Λ for typical strong and weak electrolytes are shown in Figure 5-6.

The reasons for the variation of equivalent conductance with concentration are twofold. In the first place, the mobilities of the ions are restricted in solutions of high ionic strength by interionic electrostatic forces, as well as by the existence of a counterflow of solvent attached to the ions of opposite charge which are moving in the opposite direction. The electrostatic forces drop off as the ions are farther removed from one another, and the effect of counterflow of the solvent becomes less important. These are the principal effects leading to the increase of equivalent conductance as strong electrolytes are diluted. Lars Onsager has in fact applied the theory of Debye and Hückel to this situation, obtaining an equation which, for 1:1 electrolytes in water at 25°, reduces to

$$\Lambda = \Lambda_0 - (0.2273\,\Lambda_0 + 59.78)\sqrt{N} \tag{5-45}$$

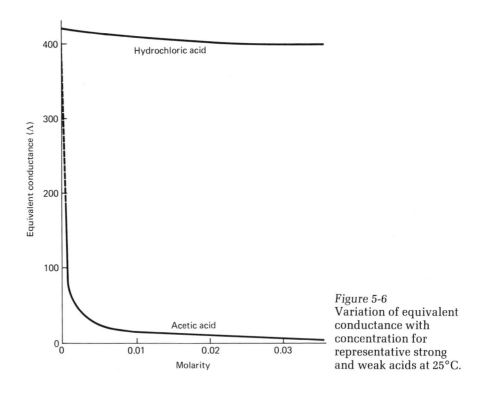

Figure 5-6
Variation of equivalent
conductance with
concentration for
representative strong
and weak acids at 25°C.

In the second place, there is for weak electrolytes the effect of increasing dissociation with increasing dilution. Because the fraction of the total solute present in the form of free ions increases, the conducting power calculated on the basis of total solute concentration, as Λ is calculated, increases. For most weak electrolytes, this change is far greater in magnitude than the change due to interionic forces, since the ionic strength of the solution of a weak electrolyte cannot become very large.

If the limiting values of equivalent conductance for various strong electrolytes are examined, it is found that they are the sums of numbers characteristic of the ions that compose each electrolyte. As infinite dilution is approached, the ions become capable of independent motion, unaffected by electrostatic forces of interionic attraction. These statements are a formulation of *Kohlrausch's law of independent migration of ions*. The limiting equivalent conductance of the electrolyte, then, is the sum of limiting equivalent conductances of the constituent ions:

$$\Lambda_0 = \lambda_{+,0} + \lambda_{-,0} \qquad (5\text{-}46)$$

The principle of independent migration of ions permits calculation of the limiting conductance for weak electrolytes when it cannot be directly determined. Specifically, it permits the combination of results of measurements of limiting conductance of strong electrolytes. For propionic acid, the limiting conductance can be obtained by using the conductances of a propionic acid salt, of some strong acid, and of a salt containing the same cation as that in the propionate employed and the same anion as that in the strong acid:

$$\Lambda_0(CH_3CH_2COOH) = \Lambda_0(CH_3CH_2COONa) + \Lambda_0(HCl) - \Lambda_0(NaCl) \quad (5\text{-}47)$$

IONIC MOBILITY
AND IONIC CONDUCTANCE

When current is passed through a solution containing several ions, each ion contributes a share to the total current. The movement of positive ions in one direction achieves the same effect as the movement of negative ions in the opposite direction. For a binary electrolyte the fraction of the current carried by each ion is termed the transport number or the transference number of that ion, and is equal to the ratio of the conductivity of that ion to the conductivity of the electrolyte. Since the conductivity of the electrolyte is the sum of the conductivities of the two ions, we have the following relations between the transference numbers, t_+ and t_-, and the ionic conductivities:

$$t_+ = \frac{\lambda_+}{\Lambda} = \frac{\lambda_+}{\lambda_+ + \lambda_-} \qquad (5\text{-}48)$$

$$t_- = \frac{\lambda_-}{\Lambda} = \frac{\lambda_-}{\lambda_+ + \lambda_-} \qquad (5\text{-}49)$$

By analyzing the transport of material from the vicinity of one electrode to the vicinity of the other electrode during electrolysis, it is possible to evaluate the transference numbers of the ions in an electrolyte, and the values of λ_+ and λ_- can thus be obtained for the solution that is electrolyzed, according to a method devised by Hittorf.

However, the equivalent conductances of the two ions in a solution are also proportional to the relative velocities at which the two ions migrate through the solution, and these velocities can be predicted from a knowledge of the ionic mobilities. Under other than unit potential gradient, the velocity is equal to the mobility multiplied by the voltage drop per unit length. In any single experiment, the potential gradient is the same for both positive and negative ions, and their relative velocities, and therefore ionic conductivities, are proportional to their mobilities.

To find a quantitative relation between mobility and conductivity, we suppose that a current is flowing through an electrolyte solution in a tube 1 cm² in cross section, with a potential drop of 1 V/cm. By the definition of the specific conductance, the magnitude of the current under these conditions is equal to κ. The current can also be defined as the charge passing in 1 sec through any plane perpendicular to the direction of ion flow. In 1 sec, all those ions within a distance from the plane equal to the velocity of the ions—in this case, their mobility—will migrate to and pass through the plane. The number of ionic charges in 1 cm³ is equal to the number of equivalents per liter divided by 1000, or $N/1000$. The total number of charges within a 1-sec range of the plane is then equal to $Nu_i/1000$, where u_i is the mobility of the ith ion. The total *amount* of charge is this number multiplied by the value of the *faraday*, or 96,494 coulombs, which is the charge on 1 equivalent of ions and is represented by the symbol $\mathscr{F}$. Combining all these relations results in

$$I = \kappa = \frac{Nu_+\mathscr{F} + Nu_-\mathscr{F}}{1000} = \frac{\Lambda N}{1000} = \frac{(\lambda_+ + \lambda_-)N}{1000} \qquad (5\text{-}50)$$

So long as the two ions contribute independently, as is the situation for sufficiently dilute solutions, the two separate equations will hold:

$$\begin{aligned} u_+\mathscr{F} &= \lambda_{+,0} \\ u_-\mathscr{F} &= \lambda_{-,0} \end{aligned} \qquad (5\text{-}51)$$

Table 5-2 presents values, for some commonly encountered ions, of conductivities and mobilities for the limiting condition of very low concentration. It is emphasized that the entries in this table are for one equivalent of ions, and therefore the number of charges per ion has been allowed for and need not be further taken into account.

Example: Calculate the limiting equivalent conductance of $Mg_3[Fe(CN)_6]_2$ at 25°C.

Solution: The equivalent conductance of Mg^{2+} is, from Table 5-2, 53.1, and that of $Fe(CN)_6^{3-}$ is 101.0, so that the value for the salt is simply the sum, 154.1.

One method of obtaining the ionic mobility is by an experiment in which the actual velocity of the ion under a known potential gradient is directly measured. This is called the *moving-boundary* method, for a boundary is formed between a portion of solution containing the ion in question and another portion of liquid not containing the ion, and the velocity of motion of the boundary is measured. If the ion is colored, the motion of the boundary can be followed by visual observation. If the ion is not colored, there is nevertheless a change in refractive index at the boundary, and this optical property can be utilized with suitable instrumentation to follow the movement of the interface.

The moving-boundary method of determining mobilities of simple ions is a form of *electrophoresis*, or study of motion in an electric field. Electrophoresis has been extensively applied to the investigation

Table 5-2
Limiting ionic conductivities and ionic mobilities[a]

Ion	λ_0	$u_0 \times 10^5$	Ion	λ_0	$u_0 \times 10^5$
H^+	349.8	362.4	OH^-	197.6	204.7
Li^+	38.7	40.1	Cl^-	76.3	79.0
Na^+	50.1	51.9	Br^-	78.3	81.1
K^+	73.5	76.1	I^-	76.8	79.6
Ag^+	61.9	64.1	HCO_3^-	44.5	46.1
NH_4^+	73.4	76.0	NO_3^-	71.4	74.0
$\frac{1}{2}Mg^{2+}$	53.1	55.0	CH_3COO^-	40.9	42.4
$\frac{1}{2}Ca^{2+}$	59.5	61.6	$C_2H_5COO^-$	35.8	37.1
$\frac{1}{2}Ba^{2+}$	63.6	65.9	$CH_3(CH_2)_2COO^-$	32.6	33.8
$\frac{1}{2}Cu^{2+}$	54	56	$C_6H_5COO^-$	32.3	33.5
$\frac{1}{2}Zn^{2+}$	53	55	$\frac{1}{2}SO_4^{2-}$	79.8	82.6
			$\frac{1}{3}Fe(CN)_6^{3-}$	101.0	104.6

[a]Values are for aqueous solutions at 25°C. Equivalent conductances at infinite dilution are given in cm²/ohm, and mobilities at infinite dilution in (cm/sec)/(V/cm).

of large molecules such as proteins, and this application, as well as the much more complex apparatus that has become available for it, will be described in Chapter 12.

APPLICATIONS OF CONDUCTANCE MEASUREMENTS

For weak electrolytes, conductance measurements afford a means of determining the degree of dissociation. Complete dissociation corresponds, except for a slight correction for interionic effects, to a condition in which the electrolyte shows its limiting conductivity. The measured equivalent conductance then bears the same ratio to the limiting value of the equivalent conductance as the ionic concentration in the solution does to the ionic concentration corresponding to complete dissociation. Thus the conductance ratio Λ/Λ_0 is equal to the fractional dissociation:

$$\alpha = \frac{\Lambda}{\Lambda_0} \tag{5-52}$$

Example: The specific conductance of a propionic acid solution of concentration 0.01 N is 1.411×10^{-4} ohm^{-1} cm^{-1} at 25°C. Calculate the fractional dissociation.

Solution: The equivalent conductance is 1000/0.01 or 10^5 times the specific conductance, or 14.11. The limiting equivalent conductance from Table 5-2 is 349.8 + 35.8 or 385.6. From these values, the conductance ratio, and therefore the fractional dissociation, is

$$\alpha = \frac{14.11}{385.6} = 0.0366$$

Conductance measurements have various other applications. In the study of protein suspensions, they can be used to check for the absence of electrolytes, which produce misleading results if present during osmotic measurements. Conductance measurements provide a rapid means of determining total electrolytes in blood serum, although the pH of the blood must also be known in order to interpret the conductivity results. The permeability of walls of living cells to ions has been investigated by following the change in conductance of the cell contents. To some extent, a living cell acts as a combined resistor and condenser in an electric circuit and, by studying the variation of electrical properties with varying frequency of alternating current, information can be gained about the cell contents without inserting electrodes into the cell.

5-4 ENTHALPIES OF SOLUTION AND OF REACTION OF IONS

Consider the process in which potassium chloride, a typical salt, is dissolved in water. This is found by experiment to be an exothermic reaction and, if the resulting solution is so dilute that no appreciable further heat change occurs on further dilution, the enthalpy of solution is found to be -4.12 kcal for each mole of potassium chloride. What is the source of this energy? The only reasonable explanation is that it results from the energy given off when each of the ions is hydrated, or surrounded by an intimately associated shell of water. Indeed, the enthalpy change observed must be much less than the full hydration energy, because a rather large amount of energy is required to pull apart the ions of opposite charge from their close juxtaposition in the crystal lattice.

As we have seen in previous sections, nonideality in solutions of electrolytes arises largely from interactions between ions or from binding of a substantial amount of the solvent to the ions. The simplest way to avoid these effects in discussing heats of solution is to compare the limiting values approached at very low concentrations. Under these circumstances, it might be expected that the heat of solution of an ionic material is the sum of contributions from the two kinds of ions of which the material is composed, and that each ionic species makes about the same contribution to the enthalpy of solution of each salt in which it occurs.

If additivity of enthalpies of solution as just described prevails, one could, for example, predict the enthalpy of solution of potassium nitrate by adding a value for the characteristic enthalpy of solution of the potassium ion to one for the nitrate ion. Unfortunately, it is not possible to obtain absolute values of individual ion contributions to the heat of solution. This is a consequence of the fact that the solution must always be electrically neutral, so that no experiment can be designed in which one ion is dissolved to the exclusion of ions of opposite charge.

Fortunately, we can pretty well circumvent this limitation by making comparisons between the enthalpies of solution of several related electrolytes. For example, it is possible to test for additivity of solution enthalpies by establishing whether the effect on the enthalpy change of substituting one particular ion for some other particular ion is always the same. If the difference is constant, regardless of the nature of the counterion, then there is reasonable justification for assuming additivity. When heats of solution of various electrolytes are compared it is found, for example, that replacement of a sodium ion by a hydrogen ion leads to a less negative, or more positive, heat of solution by about 57 kcal/mol, and replacement of a sodium ion by a potassium ion gives rise to a more negative heat of solution by about 3 kcal/mol.

Not only is the heat of solution of an electrolyte additive in the contributions of the constituent ions, but so too is the enthalpy of formation of the electrolyte in solution. This quantity is defined in the same way as is the enthalpy of formation of a substance in any other state, referring to the formation of the material from the elements. For solutions, it is necessary to specify the concentration, and the convention is used here that the solution is sufficiently dilute that no further heat change occurs upon the addition of more water. The enthalpy of for-

Table 5-3
Enthalpies of formation of electrolyte solutions at infinite dilution, in kcal/mol[a]

Cation	Anion				
	Nitrate	Chloride	Bromide	Iodide	Sulfate
Hydrogen	−49.56	−39.95	−29.05	−13.19	−217.32
Sodium	−106.65	−97.80	−86.18	−70.65	−331.46
Potassium	−109.41	−100.06	−88.94	−73.14	−336.98
Ammonium	−81.23	−71.62	−60.72	−44.86	−280.66
Magnesium	−209.15	−190.46	−168.21	−137.2	−327.3
Zinc	−135.90	−116.68	−94.88	−63.16	−254.10

[a]Values from "Selected Values of Chemical Thermodynamic Properties," Circular 500 and Technical Note 270-3, National Bureau of Standards.

mation of an electrolyte in solution can be computed by adding the enthalpy of formation of the compound in the pure state to the enthalpy change occurring on solution in water. Values for the enthalpies of formation of a few common electrolytes are given in Table 5-3, and the numbers there are seen to follow very closely the principle of additivity.

From values such as those in Table 5-3, individual ionic contributions to the heat of formation of electrolytes in solution can be evaluated. Those listed in Table 5-4 are based on the convention that the heat of formation of hydrogen ion is zero. To illustrate how other ionic contributions are evaluated, we can begin by considering the heat of formation of HCl in solution: This value, −39.95 kcal/mol, is assigned entirely to the chloride ion. Subtraction of the same value from −100.06 kcal/mol, the heat of formation of KCl solution, leaves −60.11 kcal/mol for the potassium ion. Subtracting twice the value for the potassium ion from the value for potassium sulfate, −336.98, yields −216.76 for the sulfate ion, and so on.

To show how data such as those in Table 5-4 can be applied, let us consider the problem of predicting the enthalpy for the reaction of a solution of silver nitrate with a solution of potassium chloride. The reaction that occurs when the solutions are mixed is the combination of Ag^+ with Cl^- to produce solid silver chloride. The enthalpy change

Table 5-4
Enthalpies of formation of ions in solution at infinite dilution, in kcal/mol

Hydrogen	0.000	Copper(II)	+15.39	Iodide	−13.19
Lithium	−66.55	Zinc	−36.78	Acetate	−116.16
Sodium	−57.28	Mercury(II)	+41.59	Carbonate	−161.84
Potassium	−60.04	Iron(III)	−11.4	Sulfate	−217.32
Silver	+25.31	Aluminum(III)	−127.	Bicarbonate	−165.39
Ammonium	−31.67	Hydroxide	−54.97	Cyanide	+36.0
Magnesium	−110.41	Nitrate	−49.56	Monohydrogen	
Calcium	−129.77	Fluoride	−79.50	phosphate	−308.83
Barium	−128.67	Chloride	−39.95	Dihydrogen	
Copper(I)	+12.4	Bromide	−29.05	phosphate	−309.82

for this reaction is equal to the negative of the enthalpy change for the *solution* of silver chloride, but the solubility of silver chloride is too low to permit satisfactory measurement of ΔH by direct experiment. However, the sum of the values of the enthalpies of formation of the two ions can be subtracted from the heat of formation of solid silver chloride, -30.36 kcal, to give the enthalpy change for the silver nitrate–potassium chloride reaction:

$$Ag^+(aq) + Cl^-(aq) \longrightarrow AgCl(s)$$
$$\Delta H = -30.36 - (25.31 - 39.95) = -15.72 \text{ kcal/mol} \qquad (5\text{-}53)$$

The additivity principle for heats of formation of electrolytes in solution is valid only if the substances involved are completely ionized and no reaction of association or dissociation occurs. Suppose one mixes together a solution of a strong acid with a solution containing an equivalent amount of a strong base. The reactants initially are completely ionized, but the process occurring on mixing is

$$H^+(aq) + OH^-(aq) \longrightarrow H_2O \qquad (5\text{-}54)$$

From the table of ion solution enthalpies and from the data in Table 3-3, we predict that the enthalpy change will be -68.32, the heat of formation of liquid water, less -54.97, the heat of formation of hydroxide ion, or -13.35 kcal/mol, a quantity termed the *heat of neutralization*. If the reactant acid is a weak acid, such as acetic acid or hydrocyanic acid, some energy is consumed in the ionization of the acid and the heat evolved in the neutralization process is less than 13.35 kcal by the amount of that energy.

5-5
IONIC HYDRATION
AND
THE LYOTROPIC SERIES

In this chapter, several kinds of evidence relating to the hydration of ions in aqueous solution have been pointed out. The evolution of heat on solution of many electrolytes, and indeed the very solubility that requires destruction of the crystal lattice, have been ascribed to solvation by water molecules. The extended form of the Debye–Hückel equation required to describe more concentrated solutions includes a term, linear in the ionic strength, which is attributed to binding of the solvent by ions. The coefficient of this term is often called the "salting-out" constant, since the binding of water by dissolved salts reduces the solubility of less polar molecules, an effect readily observed with gaseous solutes.

Further information about the relative extent of hydration of different ions is given by conductance and transference measurements. Examination of the relative values of ionic mobilities listed in Table 5-2 leads to the rather surprising conclusion that potassium ion moves faster than sodium ion, which in turn moves faster than lithium ion.

Why is it that the ion that would appear to be lightest and smallest has the lowest mobility of this series? The answer must lie in the extent of ionic hydration: Li^+, the smallest ion of the group when measured in the crystal or in the gas phase, has the most intense electric field in its vicinity because of its very smallness, and evidently this leads to the most extensive and strongest hydration of all the alkali metal ions. The unit $Li(H_2O)_x$ that moves through the solution gives every evidence of being larger than the hydrated sodium or hydrated potassium ion.

It is interesting that many ions were arranged in several series correlating their salting-out effectiveness by Franz Hofmeister in 1888 and 1890. The order in such an ionic series reflects the charges on the ions, their sizes, and particularly the relative extent of their hydration, and the term *lyotropic series* is frequently applied to this ordering. The mechanism of salting out of proteins from solution is similar to that described above for nonpolar solutes: The ions added tie up so much water that not enough is available to serve as solvent for the protein. Various proteins, of course, have different intrinsic tendencies to dissolve, depending in part on the number of polar groups exposed to the solvent, and are therefore salted out at different electrolyte concentrations. Thus globulins are less polar and are salted out by half-saturated ammonium sulfate, whereas albumins require saturated ammonium sulfate to be precipitated from solution. This result is the basis for the standard method of classifying or separating the two kinds of water-soluble proteins.

Examples of the lyotropic series for cations and for anions are

$$Li^+ > Na^+ > K^+ > NH_4^+ > Rb^+ > Cs^+$$
$$Mg^{2+} > Ca^{2+} > Sr^{2+} > Ba^{2+}$$
$$\text{citrate} > \text{tartrate} > SO_4^{2-} > \text{acetate} > Cl^-$$
$$> Br^- > NO_3^- > ClO_3^- > I^- > CNS^-$$

The orders given are in decreasing salting-out effect and decreasing extent of hydration. An arrangement of similar type is found to give the influence of different ions on several other phenomena, such as the effects of ions on the surface tension of water, the viscosity of salt solutions, the absorption of water by gels, the permeability of natural and artificial membranes, and the coagulation of colloidal dispersions. The ionic series arrangement is not always exactly the same for different effects and is sometimes modified by concentration. It may even be directly opposite to that expected, but this is an indication of some distinctive difference in the phenomenon itself rather than a failure of the lyotropic series to be applicable.

Finally, it should be mentioned that application of various modern instrumental techniques, such as infrared and nuclear magnetic resonance spectroscopy, has confirmed in some detail the important role ionic hydration plays in determining properties of solutions of electrolytes and has given much additional information about the number of molecules of water bound by various particular ionic species and the average length of time a given water molecule spends in association with one ion.

EXERCISES

1. Using the Debye–Hückel equation, estimate the activity coefficient of each ion in an aqueous 0.0050 M solution of zinc chloride at 25°C.

2. Carry out the algebraic operations leading to Equation (5-1).

3. The value of the freezing point of 0.0354 M acetic acid solution is -0.0684°C. Calculate the fractional dissociation of the acid in the solution.

4. A conductivity cell containing 0.1000 M KCl solution has a resistance of 762 ohms at 25°C. What is the value of the cell constant?

5. A constant current flows for 36.7 min through a coulometer, depositing 0.0683 g of metallic silver on the electrode. Calculate the magnitude of the current.

6. Calculate the ionic strength of each of the following solutions: (a) 0.050 M $K_4Fe(CN)_6$ (b) 0.0100 M $Al_2(SO_4)_3$ (c) 0.200 M Na_3PO_4.

7. Using the data in Table 5-2, calculate the limiting transference numbers at infinite dilution for solutions of KOH, NH_4NO_3, and $AgNO_3$.

8. The difficultly soluble salt $[Co(NH_3)_4(NO_2)(CNS)]^+$ $[Co(NH_3)_2(NO_2)_2(C_2O_4)]^-$ has a solubility of 3.355×10^{-4} equivalent/liter at 25°C in water. In the presence of 0.10 m KNO_3, the solubility is increased to 3.669×10^{-4} equivalent/liter. Estimate the mean activity coefficient of the ions in the presence of this concentration of KNO_3, assuming that the saturated solution of the salt alone is dilute enough to obey the Debye–Hückel limiting equation.

9. Using the data in Table 5-2, estimate the specific conductance of each of the following solutions: (a) 0.005 M $NaHCO_3$ (b) 0.001 M $Zn(CH_3COO)_2$ (c) 0.0001 M $K_3Fe(CN)_6$

10. Estimate the activity coefficients of each ion in an aqueous solution at 25°C containing 0.020 M $MgSO_4$ together with 0.050 M NaCl.

11. The value of the van't Hoff factor i for a 0.114 m solution of barium chloride is 2.52. Estimate the osmotic pressure of this solution at 27°C.

12. The term *osmolarity* is used to describe the concentration of a solution in terms of the number of moles of particles rather than in terms of the formula weight of the solute. What is the osmolarity of (a) 0.1 M $MgSO_4$ (b) 0.5 M Na_2CO_3 (c) 0.25 M $(NH_4)_2SO_4$?

13. The solubility product of $Mg(OH)_2$ is 1.2×10^{-11}. Calculate the maximum concentration of Mg^{2+} that can remain in solution in the presence of a 1.0 M concentration of potassium hydroxide. What is the ratio of this concentration to that in a saturated solution of magnesium hydroxide alone?

14. The value of i for 0.01 M $MgSO_4$ is 1.53. Calculate the freezing point, boiling point, vapor pressure at 35°C, and osmotic pressure at 35°C.

15. The equivalent conductance of 0.0200 M acetic acid is 0.001156 ohm^{-1} m^2. Calculate the fractional dissociation and the dissociation constant.

16. Calculate the enthalpy of formation in kilojoules per mole of the salt for a dilute solution of lithium sulfate.

17. Calculate the enthalpy change for the reaction of dilute solutions of calcium chloride and sodium carbonate, using the data in Table 5-4 and the fact that the enthalpy of formation of solid $CaCO_3$ is -288.45 kcal/mol.

18. A solution 0.020 M in sodium benzoate is placed between electrodes 5.00 mm apart and 2.00 cm^2 in area. A voltage of 15 V is applied between the electrodes. From the data in Table 5-2, calculate the velocity with which sodium ions move and that with which benzoate ions move in the solution, neglecting any interionic effects. What current flows through the solution?

19. What correction to the equivalent conductance of the solution in Exercise 18 is predicted by the Onsager equation?

20. Calculate the solubility product of calcium phosphate from the fact that 0.0020 g dissolves in 100 cm^3 of water.

21. What is the effect of the addition of a 0.01 M concentration of sodium chloride upon the solubility of calcium phosphate?

22. The equivalent conductance of a 0.0100 M NH_4OH solution at 25°C is 0.00113 m^2/ohm.

What is the fractional dissociation into NH_4^+ and OH^- ions?

23. Calculate the equilibrium concentration of free NH_3 in a 0.10 M solution of $Ag(NH_3)_2NO_3$. Neglect the reaction of NH_3 with water.

24. Estimate the enthalpy change for each of the following reactions:

$$Zn + 2H^+(aq) \longrightarrow Zn^{2+}(aq) + H_2(g)$$
$$CO_2(g) + H_2O(l) \longrightarrow HCO_3^-(aq) + H^+(aq)$$

25. What concentration of cupric ions can remain free in solution in the presence of a 0.001 M concentration of glycine?

26. The specific conductance contribution of AgCl in a saturated aqueous solution at 25°C is 1.75×10^{-6} ohm^{-1} cm^{-1}. What is the concentration of silver chloride in the solution?

27. Certain mathematical approximations implicit in the Debye–Hückel limiting equation were mentioned in the text. In addition, objections can be raised to the Debye–Hückel theory on the basis of oversimplifications in the mathematical model. List as many of these as you can.

28. The solubility product for calcium oxalate is 2.4×10^{-9} mol^2/liter^2. Calculate the specific conductance of a saturated solution, using a value of 74 cm^2/ohm for the ionic conductivity of the oxalate ion.

REFERENCES

Books

Henry B. Bull, *An Introduction to Physical Biochemistry,* 2nd ed., Davis, Philadelphia, 1964. Chapter 3 discusses the Debye–Hückel theory as well as ionic hydration.

Dale Driesbach, *Liquids and Solutions,* Houghton-Mifflin, Boston, 1966. Chapter 13 contains an interesting account of the history of the development of electrolytic solution theory.

Felix Franks, Ed., *Physico-Chemical Processes in Mixed Aqueous Solvents,* American Elsevier, New York, 1969. Intermediate-level reviews, including accounts of liquid structure, ionic solvation, and ionic transport mechanisms.

Bruce Martin, *Introduction to Biophysical Chemistry,* McGraw-Hill, New York, 1964. Chapter 3 describes use of the Debye–Hückel theory and effects of salts on protein solubility.

T. Shedlovsky and L. Shedlovsky, "Conductometry," in *Techniques of Chemistry,* A. Weissberger and B. W. Rossiter, Eds., Vol. 1, Part IIA, Chapter 3, Wiley-Interscience, New York, 1971.

F. Vaslow, "Thermodynamics of Solutions of Electrolytes," in *Water and Aqueous Solutions,* R. A. Horne, Ed., Wiley-Interscience, New York, 1972. A very good advanced account.

Journal Articles

B. G. Cox and A. J. Parker, "Entropies of Solution of Ions in Water," *J. Am. Chem. Soc.* **95,** 6879 (1973).

Cecil M. Criss and Mark Salomon, "Thermodynamics of Ion Solvation and Its Significance in Various Systems," *J. Chem. Educ.* **53,** 763 (1976).

J. H. Hildebrand, "A View of Aqueous Electrolytes through a Watery Eye," *J. Chem. Educ.* **48,** 224 (1971).

R. I. Holliday, "Electrolyte Theory and SI Units," *J. Chem. Educ.* **53,** 21 (1976).

Robert J. Hunter, "Calculation of Activity Coefficient from Debye–Hückel Theory," *J. Chem. Educ.* **43,** 550 (1966).

D. A. Johnson, "The Standard Free Energies of Solution of Anhydrous Salts in Water," *J. Chem. Educ.* **45,** 236 (1968).

Irving M. Klotz, "Protein Interactions with Small Molecules," *Acc. Chem. Res.* **7,** 162 (1974).

A. S. Levine and R. H. Wood, "Enthalpies of Dilution of Tetra-*n*-alkylammonium Bromides in Water and Heavy Water," *J. Phys. Chem.* **77,** 2390 (1973).

Ronald F. Probstein, "Desalination," *Am. Sci.* **61,** 280 (1973).

Michael R. Rosenthal, "The Myth of Non-Coordinating Anion," *J. Chem. Educ.* **50,** 331 (1973).

Arthur K. Solomon, "The State of Water in Red Cells," *Sci. Am.,* **224,** 88 (February 1971).

W. M. J. Strachan, A. Dolenko, and E. Buncel, "Diprotonation Equilibria Involving 4-Hydroxyazobenzene and 4-Hydroxyazobenzene-4′-sulfonic Acid," *Can. J. Chem.* **47,** 3631 (1969).

C. A. Vincent, "The Motion of Ions in Solution under the Influence of an Electric Field," *J. Chem. Educ.* **53,** 490 (1976).

Six

Acid-Base Equilibria

Many substances that are weak electrolytes are either weak acids or weak bases. In the process of ionization, a weak acid donates a proton—the nucleus of a hydrogen atom—to a solvent molecule, to form hydronium ion if the solvent is water, whereas a weak base such as ammonia produces hydroxide ions in aqueous solution by removing hydrogen ions from water molecules. These reactions have a feature in common: They are *proton-transfer* or *protolytic* processes.

   Many important reactions occurring in chemical systems, including many taking place in living organisms, are protolytic reactions. Common examples include the reaction of hydronium ion with hydroxide ion to form water, and the reaction of an amine with hydrochloric acid to form an amine hydrochloride. Proton transfers can also occur in the gas phase, as in the reaction of gaseous ammonia and gaseous hydrogen chloride to form ammonium chloride. The common occurrence of protolytic reactions is in part a consequence of the fact that the proton is a constituent of the most common solvent, water, but it is probably principally a result of the fact that the hydrogen atom has no extranuclear electrons other than the valence electron. Thus, while the hydrogen is bonded to one atom, a second atom can approach very closely to it, and the hydrogen atom can then easily transfer its attachment from the first atom to the second.

6-1
BRÖNSTED–LOWRY CONCEPT OF ACIDS AND BASES

The scope of the idea of acid and base function has been extended—as proposed originally by Brönsted and by Lowry—to include all proton-transfer processes as acid–base reactions, with the proton donor participating as the acid and the proton acceptor participating as the base. Any species capable of being a proton donor, whether it be a ca-

Table 6-1
Dissociation constants of conjugate acids and bases at 25°C

Acid	K_a	pK_a	Base	K_b	pK_b
H_3O^+	55.4	−1.74	H_2O	1.8×10^{-16}	15.74
$NH_2CONH_3{}^+$	0.67	0.18	NH_2CONH_2	1.5×10^{-14}	13.82
$COOHCOOH$	3.8×10^{-2}	1.42	$COOHCOO^-$	2.6×10^{-13}	12.58
H_3PO_4	7.5×10^{-3}	2.12	$H_2PO_4{}^-$	1.33×10^{-12}	11.88
$CH_2(COOH)_2$	1.49×10^{-3}	2.83	$CH_2(COOH)COO^-$	6.7×10^{-12}	11.17
$ClCH_2COOH$	1.4×10^{-3}	2.85	$ClCH_2COO^-$	7.1×10^{-12}	11.15
$C_6H_4(COOH)_2$	1.3×10^{-3}	2.88	$C_6H_4(COOH)COO^-$	7.7×10^{-12}	11.12
HNO_2	4.6×10^{-4}	3.34	$NO_2{}^-$	2.2×10^{-11}	10.66
$HCOOH$	1.8×10^{-4}	3.74	$HCOO^-$	5.6×10^{-11}	10.25
$CH_3CHOHCOOH$	1.37×10^{-4}	3.86	$CH_3CHOHCOO^-$	7.3×10^{-11}	10.14
C_6H_5COOH	6.6×10^{-5}	4.18	$C_6H_5COO^-$	1.5×10^{-10}	9.82
$COOHCOO^-$	4.9×10^{-5}	4.31	COO^--COO^-	2.0×10^{-10}	9.69
$C_6H_5NH_3{}^+$	2.6×10^{-5}	4.58	$C_6H_5NH_2$	3.8×10^{-10}	9.42
CH_3COOH	1.75×10^{-5}	4.76	CH_3COO^-	5.7×10^{-10}	9.24
CH_3CH_2COOH	1.34×10^{-5}	4.87	$CH_3CH_2COO^-$	7.5×10^{-10}	9.13
$C_5H_5NH^+$	6.2×10^{-6}	5.21	C_5H_5N	1.6×10^{-9}	8.79
$C_6H_4(COOH)COO^-$	3.9×10^{-6}	5.41	$C_6H_4(COO^-)_2$	2.6×10^{-9}	8.59
$CH_2(COOH)COO^-$	2.0×10^{-6}	5.70	$CH_2(COO^-)_2$	5.0×10^{-9}	8.30
H_2CO_3	4.52×10^{-7}	6.34	$HCO_3{}^-$	2.2×10^{-8}	7.66
H_2S	9.1×10^{-8}	7.04	HS^-	1.10×10^{-7}	6.96
$H_2PO_4{}^-$	6.23×10^{-8}	7.19	$HPO_4{}^{2-}$	1.6×10^{-7}	6.81
$NH_2NH_3{}^+$	3.3×10^{-9}	8.48	NH_2NH_2	3.0×10^{-6}	5.52
H_3BO_3	5.8×10^{-10}	9.23	$H_2BO_3{}^-$	1.7×10^{-5}	4.77
$NH_4{}^+$	5.7×10^{-10}	9.24	NH_3	1.75×10^{-5}	4.76
C_6H_5OH	1.02×10^{-10}	9.99	$C_6H_5O^-$	9.8×10^{-5}	4.01
$Ag(H_2O)_2{}^+$	9.1×10^{-11}	10.04	$AgOH$	1.1×10^{-4}	3.96
$HCO_3{}^-$	5.59×10^{-11}	10.25	$CO_3{}^{2-}$	1.8×10^{-4}	3.75
$CH_3NH_3{}^+$	2.5×10^{-11}	10.60	CH_3NH_2	4×10^{-4}	3.40
$HPO_4{}^{2-}$	1.7×10^{-12}	11.77	$PO_4{}^{3-}$	5.9×10^{-3}	2.23
HS^-	1.0×10^{-15}	15.0	S^{2-}	10	−1.0
H_2O	1.8×10^{-16}	15.74	OH^-	55.4	−1.74

tion, a neutral molecule, or an anion, is a potential acid. Thus, given a suitable acceptor of protons, such species as water, ammonium ion, a hydrogen acetate molecule, or dihydrogen phosphate ion can each exhibit acidic function. Any molecular or ionic species that offers a site for ready attachment of a proton is a potential base. Examples are a carbonate ion, water molecule, or complex species such as $Al(H_2O)_5OH^{2+}$.

Whenever a molecule or ion acting as an acid loses a proton, it forms a second species which can necessarily, because of the very manner of its formation, take back a proton and is therefore a potential base. This resulting base is termed the base *conjugate* to the original acid. Thus acetate ion is the base conjugate to hydrogen acetate. Likewise, for every base there is a conjugate acid formed by the addition of a proton to the base. The two members of such a couple are said to constitute a conjugate acid–base pair. Many examples of conjugate pairs can be seen in Table 6-1. The charges on the molecular or ionic species that are components of an acid–base conjugate pair are not of direct importance, except that the base is always one unit more negative than the acid to which it is conjugate.

6-2
AQUEOUS SOLUTION AND THE pH SCALE

If consideration is restricted to aqueous solutions, one must realize that no free protons can exist at equilibrium in the presence of water, but that any formed are immediately taken up by the solvent to form hydronium ion, H_3O^+. Any acid stronger than hydronium ion loses its protons to the solvent to form hydrated hydrogen ion, as does hydrogen chloride:

$$HCl + H_2O \longrightarrow H_3O^+ + Cl^- \qquad (6\text{-}1)$$

In view of these facts, it is customary and convenient to express the proton-donating tendency of an aqueous solution in terms of the concentration of hydronium ion, or "hydrogen ion." Since this concentration can vary over many orders of magnitude, and since the proton-donating tendency or potential is proportional to the logarithm of the hydrogen ion concentration, the quantity called the pH is often defined by the equation

$$pH = -\log[H_3O^+] \qquad (6\text{-}2)$$

The lowercase "p" stands for exponent or "power." A pH of 2 corresponds to a hydrogen ion concentration of 10^{-2} mol/liter; a pH of 10 indicates a concentration of 10^{-10} mol/liter.

Most methods of experimental measurement of pH, such as those described in Chapter 7, involve determination of the effective concentration, or activity, of hydrogen ions, rather than direct determination of the stoichiometric concentration. Therefore pH is properly defined by the equation

$$pH = -\log a_{H_3O^+} \qquad (6\text{-}3)$$

We sometimes use pH with this meaning, and sometimes with the meaning given by Equation (6-2), with the understanding that only in extremely dilute solutions is the pH a measure of the hydrogen ion stoichiometric concentration, and in concentrated or moderately dilute solutions it refers to the hydrogen ion activity.

6-3
WEAK ELECTROLYTE EQUILIBRIA

WEAK ACIDS AND BASES

When acids that are weak electrolytes in aqueous medium are put into solution in water, they enter into equilibrium with the solvent:

$$AH + H_2O \rightleftharpoons A^- + H_3O^+ \qquad (6\text{-}4)$$

The equilibrium constant for this reaction is given by

$$K = \frac{[A^-][H_3O^+]}{[AH][H_2O]} \qquad (6\text{-}5)$$

In reasonably dilute aqueous solutions, the concentration of water is constant at 55.51 mol/liter. This number can be multiplied by both sides of the equation to give the expression for the conventional *acid dissociation constant*:

$$K_a = 55.51K = \frac{[A^-][H_3O^+]}{[AH]} \tag{6-6}$$

An example is formic acid, HCOOH, for which the dissociation constant is 1.8×10^{-4}. Suppose it is desired to know the concentration of hydrogen ion in a 0.1 M solution of this acid. The equilibrium is

$$HCOOH + H_2O \rightleftharpoons HCOO^- + H_3O^+ \tag{6-7}$$

If the hydrogen ion concentration is set equal to y, the formate ion concentration is also equal to y and the concentration of undissociated HCOOH is $0.1 - y$. Then

$$1.8 \times 10^{-4} = \frac{(y)(y)}{0.1 - y} \tag{6-8}$$

Cross-multiplying and rearranging yields

$$y^2 + 1.8 \times 10^{-4}y - 1.8 \times 10^{-5} = 0$$

This quadratic equation can be solved by means of the usual formula:

$$y = \frac{-1.8 \times 10^{-4} \pm \sqrt{(1.8 \times 10^{-4})^2 + 4(1.8 \times 10^{-5})}}{2}$$

$$= 4.2 \times 10^{-3} \text{ mol/liter}$$

Inspection of this result shows that the concentration of undissociated molecules, which is the denominator of Equation (6-8), is only 4 percent less than the total formic acid concentration. The equation can therefore be rewritten, using in the denominator the approximation [HCOOH] = 0.1:

$$1.8 \times 10^{-4} = \frac{y^2}{0.1} \tag{6-9}$$

From this, y equals $\sqrt{18 \times 10^{-6}}$ or 4.2×10^{-3} mol/liter. This result agrees to two significant figures with that found by the solution of the quadratic equation. In view of the fact that dissociation constants are not usually reliably known to more than two significant figures and, furthermore, vary somewhat with concentration because ionic activities should be used instead of concentrations in the equation, it is often satisfactory to use the approximation embodied in Equation (6-9) in calculating ionic concentrations produced by the dissociation of weak electrolytes.

The values of the negative logarithm of K_a for some common acids are given in Table 6-1; these are designated pK_a values, by analogy with pH values.

Materials that accept protons from water dissociate in aqueous solution as weak bases:

$$C_6H_5NH_2 + H_2O \rightleftharpoons C_6H_5NH_3^+ + OH^- \tag{6-10}$$

$$H_2PO_4^- + H_2O \rightleftharpoons H_3PO_4 + OH^- \tag{6-11}$$

Although there are metal hydroxides that dissociate reversibly to form hydroxide ions, apparently without the necessity for reaction with solvent, it is probable that these, too, are reactions involving water:

$$AgOH + H_2O \rightleftharpoons Ag(H_2O)^+ + OH^- \tag{6-12}$$

The equation for the equilibrium constant of a basic ionization can be generalized:

$$K_b = \frac{[\text{conjugate acid}][OH^-]}{[\text{base}]} \tag{6-13}$$

Just as for a weak acid, the water concentration that appears in the expression for the full equilibrium constant is incorporated in the conventional *basic dissociation constant* K_b. If the base is the only solute present, then $[OH^-] = [\text{conjugate acid}]$ and the approximate equation for a small degree of dissociation is

$$K_b = \frac{[OH^-]^2}{[\text{base}]} \tag{6-14}$$

Water is a weak electrolyte, ionizing to give hydronium and hydroxide ions. The conventional ion product constant for the ionization is

$$K_w = [H_3O^+][OH^-] \tag{6-15}$$

At temperatures near room temperature the value of this product is very close to 1×10^{-14}. An equivalent statement is that, for water or any aqueous solution,

$$pH + pOH = 14 \tag{6-16}$$

The acid dissociation constant of a weak acid is obviously related to the basic dissociation constant of the base conjugate to the acid, for both are dependent upon the tendency of the base to hold a proton. For the acid HA,

$$K_a = \frac{[H^+][A^-]}{[HA]} \tag{6-17}$$

or the conjugate base A^-,

$$K_b = \frac{[HA][OH^-]}{[A^-]} \tag{6-18}$$

Multiplying through this expression in both numerator and denominator by $[H^+]$ yields

$$K_b = \frac{[HA][OH^-][H^+]}{[A^-][H^+]} = \frac{K_w}{K_a} \tag{6-19}$$

where K_a is the constant for the conjugate acid. In general, then, the product of the acid dissociation constant of an acid and the basic dissociation constant of the base conjugate to that acid is equal to the ion product of water.

POLYPROTIC
AND AMPHOTERIC SPECIES

Molecules like H_2CO_3, H_3PO_4, and $COOH-COOH$ contain more than one acidic hydrogen each, and the ionization of such species occurs in stepwise fashion. The ionization constants for successive proton transfers from a polyprotic acid are sometimes designated K_1, K_2, K_3, and so on:

$$CH_2(COOH)_2 + H_2O \rightleftharpoons H_3O^+ + CH_2COOHCOO^- \qquad K_1 = 1.49 \times 10^{-3}$$
(6-20)

$$CH_2COOHCOO^- + H_2O \rightleftharpoons H_3O^+ + CH_2(COO^-)_2 \qquad K_2 = 2.0 \times 10^{-6}$$
(6-21)

Obviously K_2 is the same quantity as the conventional acid dissociation constant of $CH_2COOHCOO^-$. Successive dissociation constants for a polyprotic species are increasingly smaller, since the ionic charge becomes increasingly unfavorable for the transfer of another proton.

If the constants of a polyprotic acid differ by several powers of 10, as they frequently do, the first ionization is the only one that contributes materially to the hydrogen ion concentration of the solution. For example, in a 0.01 M solution of carbon dioxide, the first ionization is described by the equation

$$K_1 = 4.5 \times 10^{-7} = \frac{[H_3O^+][HCO_3^-]}{[CO_2]} \qquad (6\text{-}22)$$

and the approximate hydrogen ion concentration can be calculated from this as

$$[H_3O^+] = \sqrt{4.5 \times 10^{-7} \times 0.01} = 6.7 \times 10^{-5} \text{ mol/liter}$$

Because of the presence of this amount of hydrogen ion from the first ionization step, and the fact that the second ionization constant is so much smaller than the first, the second ionization, that of bicarbonate to carbonate, may be assumed to produce a negligible increase in the hydrogen ion concentration and a negligible decrease in the bicarbonate concentration. By the stoichiometry of the first step, equal amounts of hydrogen and bicarbonate ion are formed, and the equation

$$K_2 = \frac{[H_3O^+][CO_3^{2-}]}{[HCO_3^-]} = 5.6 \times 10^{-11} \qquad (6\text{-}23)$$

consequently reduces to $[CO_3^{2-}] = 5.6 \times 10^{-11}$ mol/liter. This concentration is equal to the loss in bicarbonate and the gain in hydrogen ion in the second step, and the result we have just seen that it is only about 10^{-6} of the concentration of these ions well justifies the approximation made in obtaining it.

Similarly, in calculating the hydroxide ion concentration in a solution of trisodium phosphate, one need only consider the equilibrium

$$PO_4^{3-} + H_2O \rightleftharpoons HPO_4^{2-} + OH^- \qquad (6\text{-}24)$$

The further reaction of HPO_4^{2-} with H_2O, in the presence of the hydroxide ion formed by the phosphate ionization, produces an amount of hydroxide ion that is negligible in comparison with that from the first ionization. We shall find later some examples of systems for which the successive ionization constants do not differ greatly, because the sites of ionization are farther removed from one another, and for which the procedure of calculating the equilibrium ion concentrations is therefore somewhat more complex.

A species such as $H_2PO_4^-$, HCO_3^-, or $Fe(H_2O)_5OH^+$ is amphoteric, in the sense that a proton can be either lost or gained upon reaction with water. To calculate the hydrogen ion concentration of a solution of one of these materials, one can write the two equilibria in a general form as follows, although the particular charges on the ions vary with the charge on the parent species:

$$RH + H_2O \rightleftharpoons R^- + H_3O^+ \tag{6-25}$$

$$RH + H_2O \rightleftharpoons RH_2^+ + OH^- \tag{6-26}$$

The equilibrium constant for the first process is the acid dissociation constant for species RH:

$$K_a = \frac{[R^-][H_3O^+]}{[RH]} \tag{6-27}$$

For the second process, the equilibrium constant is the basic dissociation constant of RH, equal to the ion product of water divided by the *acid* dissociation constant of RH_2^+, the acid conjugate to RH:

$$K_b = \frac{[RH_2^+][OH^-]}{[RH]} = \frac{[RH_2^+][OH^-][H_3O^+]}{[RH][H_3O^+]} = \frac{[RH_2^+]K_w}{[RH][H_3O^+]} \tag{6-28}$$

Using the principle that every solution must remain electrically neutral, one can write

$$[H_3O^+] + [RH_2^+] = [OH^-] + [R^-] \tag{6-29}$$

Three of the variable concentrations in this equation can be expressed in terms of the fourth by substituting for $[RH_2^+]$ from Equation (6-28), for $[OH^-]$ from Equation (6-15), and for $[R^-]$ from Equation (6-27):

$$[H_3O^+] + K_b \frac{[H_3O^+][RH]}{K_w} = \frac{K_w}{[H_3O^+]} + K_a \frac{[RH]}{[H_3O^+]} \tag{6-30}$$

Solving for the hydrogen ion concentration results in the general equation

$$[H_3O^+] = \sqrt{\frac{K_w + K_a[RH]}{1 + K_b[RH]/K_w}} \tag{6-31}$$

Under the special, although commonly met, conditions that K_w is much smaller than either $K_a[RH]$ or $K_b[RH]$, this reduces to

$$[H_3O^+] = \sqrt{\frac{K_a[RH]}{K_b[RH]/K_w}} = \sqrt{\frac{K_a K_w}{K_b}} \tag{6-32}$$

This equation indicates that the hydrogen ion concentration is independent of the concentration of the solute, and this is, in many cases, found to be true.

Example: Estimate the pH of a 0.1 M solution of sodium dihydrogen phosphate.

Solution: The ion that reacts with water is $H_2PO_4^-$. The acid dissociation constant of this species, the second dissociation constant for phosphoric acid, is 6.23×10^{-8}. The dissociation constant of the acid conjugate to $H_2PO_4^-$ is that of H_3PO_4, or 7.5×10^{-3}. The hydrogen ion concentration is therefore equal to $\sqrt{6.23 \times 10^{-8} \times 7.5 \times 10^{-3}}$ or 2.2×10^{-5}, and the corresponding pH is 4.66.

ACTIVITY VERSUS CONCENTRATION

In the preceding sections, equations for equilibrium relations were shown in terms of the concentrations of ions and molecules. Only when all materials behave ideally are such equations fully accurate. The effective concentration, the activity as described in Section 5-2, must be used if the equilibrium constant is to be truly constant. Of course, if activity coefficients are not known, one can make reasonable estimates of concentrations by assuming the coefficients to be unity, and these estimates may be sufficiently accurate for many purposes, especially if the ionic strengths are kept relatively low.

6-4 EQUILIBRIA INVOLVING SEVERAL SOLUTES

All types of equilibrium systems of interest here have a common feature in that they contain both members of a conjugate acid–base pair, along with hydronium ion, and these three species are *not all derived from the dissociation of a single electrolyte.* There exists therefore the equilibrium

$$acid + H_2O \rightleftharpoons base + H_3O^+ \qquad (6\text{-}33)$$

From the equilibrium constant expression, we can write

$$[H_3O^+] = K_a \frac{[acid]}{[base]} \qquad (6\text{-}34)$$

Taking logarithms of both sides to base 10 and changing signs, one obtains

$$pH = pK_a + \log \frac{[base]}{[acid]} \qquad (6\text{-}35)$$

In applying this equation, it is necessary to remember that the concentrations of base and acid included refer to the two members of a conjugate pair. A plot of pH against the fraction α, equal to the quantity $[base]/([base] + [acid])$, is given in Figure 6-1.

COMMON ION EFFECT

In the first of the cases to which we apply Equation (6-35) and the ideas upon which it is based, there are present in solution two electrolytes, one of which is weak and the other strong, producing upon dissociation a common ion. The solutes may be a weak acid and a salt of that

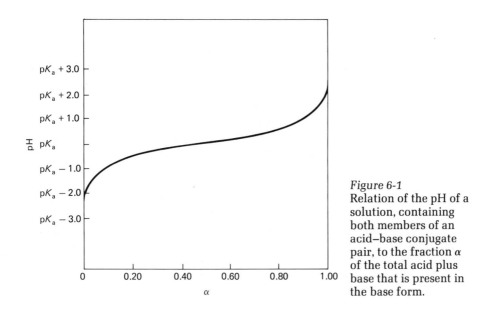

Figure 6-1
Relation of the pH of a
solution, containing
both members of an
acid–base conjugate
pair, to the fraction α
of the total acid plus
base that is present in
the base form.

acid, or they may be a weak base and a salt of that base such as ammonia and ammonium chloride.

Consider a mixture of acetic acid and potassium acetate. Since K_a
for the acid is small, provided that the acid and salt are present in concentrations of the same order of magnitude, practically all of the base,
the acetate ion, comes from dissociation of the salt, and its concentration may be set equal to the salt concentration. Further, practically no
acid molecules are lost by dissociation, and the concentration of undissociated hydrogen acetate may be set equal to the concentration of
acetic acid put into the solution. As an example, if the solution is 0.2 M
in acetic acid and 0.1 M in potassium acetate,

$$pH = 4.76 + \log \frac{0.1}{0.2} = 4.46$$

and the hydrogen ion concentration is 3.5×10^{-5} mol/liter.

BUFFER SOLUTIONS

A solution capable of resisting to a considerable extent changes in pH
that usually result upon the addition of an acid or a base is termed a
buffer solution. Such a solution can be employed simply as a standard
for reference in measurements of the pH of unknown solutions by
colorimetric or electrometric methods to be described later; another
way it can be used is as a medium for carrying out a chemical or biochemical process under conditions of constant pH.

A moderately concentrated solution of a strong acid or a strong base
is a fairly good buffer for high or low pH values.

If, however, a buffer at pH 5 is desired, a hydrochloric acid solution
would not be very satisfactory; it would contain 0.00001 mol of acid
per liter, and the addition of 0.01 mol of hydrogen ion per liter would
decrease the pH three units, to pH 2. Thus relatively small amounts of
acid or base can completely overwhelm any buffer effect the solution

possesses. It therefore becomes necessary to resort, for intermediate pH values, to a different type of buffer, one that contains both members of a conjugate acid–base pair.

Solutions containing either a weak acid along with a salt of that acid, or a weak base along with a salt of that base, are effective as buffers. If to an acetic acid–acetate buffer there is added some sodium hydroxide, the hydroxide ion is neutralized by protons furnished by the ionization of acetic acid without appreciable change in the concentration of hydronium ion; so long as much un-ionized acetic acid remains, there is a reserve of protons available. If hydrochloric acid is added to the same buffer mixture, it is neutralized by the acetate ion which acts as a sink to take up added protons. In a mixture of ammonium chloride and ammonia, the ammonium ions function as the proton reservoir and the ammonia molecules as a proton sink.

The pH of a buffer mixture obeys Equation (6-35); as applied to buffers, this equation is termed the Henderson–Hasselbalch equation. Plots of pH against the fraction of the total system that is in the base form are shown for several examples in Figure 6-2; the curves are identical to the general curve in Figure 6-1. A fraction of one-half in the base form, corresponding to a 1:1 base/acid ratio, yields a pH equal to

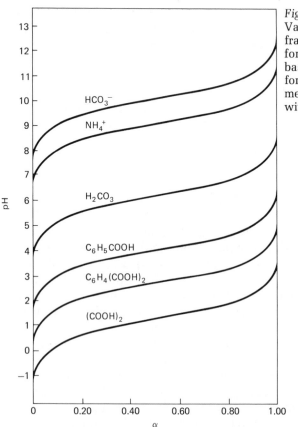

Figure 6-2
Variation of pH with fraction in the base form for several specific acid–base conjugate pairs. The formula of the acid member of the pair is given with each curve.

the pK_a of the acid member of the conjugate pair, and, for a given change in relative amounts of the two components, the change in pH is smallest in the vicinity of this point.

In choosing a buffer, one selects an acid–base system for which the acid pK_a is close to the pH at which the buffer is expected to function. Acetate buffers can be used from about pH 3.5 to pH 6 with good effectiveness. Some other useful buffers and the pH ranges covered are:

Glycine–glycine hydrochloride	1–3.7
Phthalic acid–potassium acid phthalate	2–4
Citric acid–sodium hydroxide	2.5–6
Sodium dihydrogen phosphate–disodium hydrogen phosphate	6–8
Boric acid–borax	7–9
Borax–sodium hydroxide	9–11
Disodium hydrogen phosphate–trisodium phosphate	11–12

In practical use of a buffer, several other factors must be considered. If the buffer is to be used as a reaction medium, the components must have no effect on the reaction under study; for example, phosphate buffers may influence physiological processes and therefore are not suitable for use in studying these processes. The temperature may have an effect upon the pH that a buffer of a given composition maintains. Further, the presence of a large concentration of inert ions may influence the activity coefficients of the ions of the buffer and thus change the equilibrium somewhat. Either of these effects can sometimes be minimized by the selection of an appropriate buffer system.

A favorite buffer in biochemical research has been tris(hydroxymethyl)aminomethane, $H_2NC(CH_2OH)_3$ (tris), for which the pK_a is 8.3. However, it is ineffective as a buffer in the lower range of physiological pH values, and the primary amine group has considerable reactivity. N. E. Good and co-workers [*Biochemistry* **5**, 467 (1966)] have described a series of buffers including several that have advantages over tris in biological research. Most of those included are zwitterionic amino acids (see Section 6-5), N-substituted derivatives of either glycine $(H_3N^+CH_2COO^-)$ or taurine $(H_3N^+CH_2CH_2SO_3^-)$. Zwitterions were chosen because their high polarity confers a high ratio of solubility in water to solubility in the organic media within cell structures, a correspondingly reduced rate and extent of penetration through membranes, minimum effects of the ionic strength of the medium on the pH established, and ease of purification by recrystallization. The buffer materials were tested for their effects on several biological reactions in systems already well buffered to ascertain their tendency to inhibit or uncouple the reactions, independent of their buffering capacity, and some of them were found to be substantially better than buffer systems traditionally in use. Table 6-2 lists formulas and suggested short designations, as well as pK_a values, for some of the Good buffers.

For use as basic standards, several readily reproducible buffers have been developed and calibrated at the National Bureau of Standards. Their use, as well as considerations involved in the application of buffers in general, are described by R. G. Bates in *Determination of pH: Theory and Practice*.

Table 6-2
Buffers useful for biological research

Structure	Designation[a]	pK_a at 20°C
(MES structure)	MES	6.15
(PIPES structure)	PIPES	6.8
$(CH_3)_3\overset{+}{N}CH_2CH_2NH_2Cl^-$	Cholamine chloride	7.1
$(HOCH_2)_3\overset{+}{C}NH_2CH_2CH_2SO_3^-$	TES	7.5
(HEPES structure)	HEPES	7.55
$(HOCH_2CH_2)_2\overset{+}{N}HCH_2COO^-$	Bicine	8.35
$\overset{+}{N}H_3CH_2CONHCH_2COO^-$	Glycylglycine	8.4

Example: Prepare a buffer of pH 7.00 from sodium dihydrogen phosphate and sodium hydroxide.

Solution: The pK_a of $H_2PO_4^-$ is 7.19. This is the appropriate acid component, and the conjugate base is then HPO_4^{2-}. The required ratio of the two is found from

$$7.00 = 7.19 + \log\frac{[HPO_4^{2-}]}{[H_2PO_4^-]}$$

$$\log\frac{[HPO_4^{2-}]}{[H_2PO_4^-]} = -0.19 = \overset{-}{1}.\overset{+}{8}1$$

$$\frac{[HPO_4^{2-}]}{[H_2PO_4^-]} = 0.65$$

The amount of $H_2PO_4^-$ originally present must provide the phosphate for both buffer constituents; thus if 1.65 mol of NaH_2PO_4 and 0.65 mol of NaOH are mixed, there would be produced 0.65 parts Na_2HPO_4, and 1.00 parts NaH_2PO_4 would be left. Any mixture in the molar ratio of 1.65:0.65 or 2.54:1 would give the desired pH.

INDICATORS AND COLORIMETRIC DETERMINATION OF pH

When a relatively small amount of a weak electrolyte is added to a solution containing relatively large amounts of other electrolytes, the ionization of the weak electrolyte is essentially controlled by the concentration of any common ion already present in the solution. An example of this situation is the use, for determination of the pH of a solution, of an indicator, or conjugate acid–base pair of which the members have different colors.

The ionization of the indicator takes place according to a reversible process such as

$$HIn + H_2O \rightleftharpoons H_3O^+ + In^- \tag{6-36}$$

When the indicator is placed in a solution, it does not appreciably affect the hydrogen ion concentration, but its ionization equilibrium is constrained to conform to the hydrogen ion concentration of the solution, according to the equation

$$pH = pK_{In} + \log \frac{[\text{base form of indicator}]}{[\text{acid form of indicator}]} \qquad (6\text{-}37)$$

Here K_{In} is the acid ionization constant of the acid form of the indicator, known usually as the *indicator constant*, and pK_{In} is its negative logarithm.

When the pH is plotted against the fraction of the indicator in the base form, the curve shown in Figure 6-3 is obtained. The eye cannot detect color changes if more than 90 percent of the total indicator is in one form, so that the region of visual comparison of color shades is limited to about one pH unit on either side of the half-change point, or point at which the pH equals the pK_{In}. In many cases more accurate results can be obtained by the use of optical instruments which can quantitatively measure the absorption spectrum of the solution and determine the amount of absorption of light in spectral bands characteristic of each of the indicator forms.

The color change in an indicator system is the result of a substantial change in electronic distribution in the molecule accompanying the loss or gain of a proton. For example, methyl orange in the base form contains two benzene rings and, by addition of a proton with simultaneous rearrangement of some of the double bonds, is converted into a quinoid form, a type of structure often responsible for the appearance of visible color or the shift in color from the red toward the blue end of the spectrum:

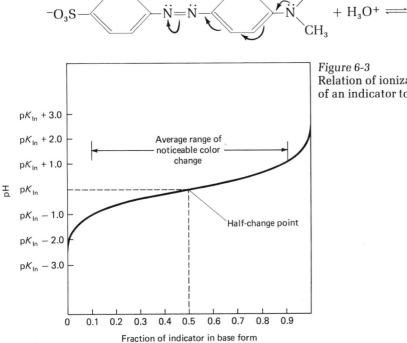

Figure 6-3
Relation of ionization of an indicator to pH.

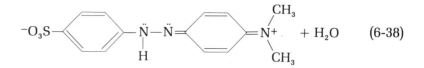

$$-O_3S\underset{H}{\overset{..}{N}}-\overset{..}{N}=\underset{CH_3}{\overset{CH_3}{N^+}} + H_2O \qquad (6\text{-}38)$$

We shall discuss electron delocalization effects further in Chapter 9.

6-5
CHARACTERIZATION OF ACID–BASE FUNCTIONAL GROUPS

TITRATION CURVES

When a neutralization reaction is carried out with measured increments of one reactant added to a certain quantity of the other reactant, the results can be plotted as a titration curve, showing the variation in pH with added volume or amount of reagent. The amount of reactant required up to the point at which there is a sudden change in pH indicates the amount of acid or base in the initial solution. It is also possible to determine the acid or basic strength of the material being titrated, by finding from the titration curve the pH at which half of the acid or base has been neutralized; this value is equal to the pK_a for the acid form of the functional group.

The pH changes for an illustrative case, the titration of 0.001 M HCl with 0.1 M NaOH, are shown in Figure 6-4. The initial pH of the acid solution is 3. After 0.1 equivalent of base has been added for each equivalent of acid, 1 in 10 of the hydronium ions has been neutralized, and the pH of the solution, now 0.0009 M in free hydronium ion if we neglect the small change in total volume brought about by the added titrant solution, is 3.05. When 0.5 equiv of base per equivalent of acid has been added, the hydronium ion concentration has been

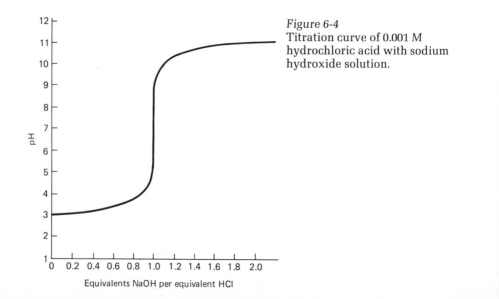

Figure 6-4
Titration curve of 0.001 M hydrochloric acid with sodium hydroxide solution.

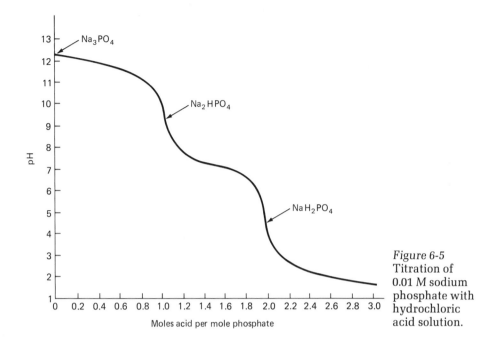

Figure 6-5
Titration of
0.01 M sodium
phosphate with
hydrochloric
acid solution.

reduced to 0.0005; the pH corresponding to this concentration is 3.30. At 0.9 equiv of base the pH is 4, and at 0.99 equiv it has risen to 5. When 1.00 equivalent of base per equivalent of acid has been added, the solution is equivalent to a solution of sodium chloride and the pH is 7. At 1.1 equivalent of base, there is an excess hydroxide concentration of 1×10^{-4}, corresponding to a pH of 10. The pH continues to increase with added base but changes somewhat more slowly. The very drastic change in pH at the end point is characteristic of the titration of a strong acid with a strong base.

In Figure 6-5 is the titration curve obtained when a 0.01 M solution of sodium phosphate, Na_3PO_4, is titrated with hydrochloric acid, the latter concentrated enough to produce only a negligible volume change. The initial pH is calculated by considering the ionization of the weak base phosphate ion by reaction with water to form monohydrogen phosphate ion. When $\frac{1}{2}$ mol of hydrogen ion has been added for each mole of phosphate ion, there is a mixture of equal parts of PO_4^{3-} and HPO_4^{2-}, and the pH is equal to the pK_a for the third dissociation of phosphoric acid. In this region, the system is a good buffer, and the pH changes slowly with the volume of acid added. When the acid/phosphate ratio is 1.0, the solution is essentially a solution of monohydrogen phosphate ions, along with sodium ions. The pH is calculated by the method applicable to amphoteric electrolytes as in Equation (6-32) and is thus equal to one-half the sum of pK_2 and pK_3 for phosphoric acid.

The titration curve for the addition of sodium hydroxide solution to the dibasic acid, phthalic acid, is sketched in Figure 6-6. The pK_a values for the acid are found to be 2.88 for the first ionization and 5.41 for the second. This is a system often employed for preparing buffers

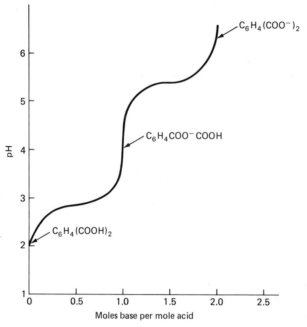

Figure 6-6
Titration of 0.01 M phthalic acid with sodium hydroxide solution.

in the pH range on the acid side of neutrality, and the reader should by now be able to relate the titration curve to the pH regions of phthalate buffer applicability.

To follow experimentally a titration of the sort we have been describing, one must measure the concentration of hydrogen ion after each addition of an increment of acid or base—this can be done electrometrically by methods described in Chapter 7. There are several options in the direction in which the titration of a system under study can be carried out. One can acidify the solution initially with strong acid, converting all ionizable groups into their acid forms, and then titrate with base until all have been converted into their basic forms. Or one can add the base initially and then titrate with acid. A third procedure is to begin with the solution of the material as it is available or as it is prepared by dissolving the solute in water, titrate one portion with acid, and then titrate another portion with base. Whatever sequence is chosen, the same titration curve can be constructed.

RESOLUTION
OF OVERLAPPING
IONIZATION CONSTANTS

Citric acid, like phosphoric acid, is tribasic, containing three replaceable hydrogens:

$$CH_2COOH$$
$$|$$
$$HOCCOOH$$
$$|$$
$$CH_2COOH$$

However, titration of sodium citrate with acid yields a curve quite unlike that of sodium phosphate, as seen in Figure 6-7, because the successive dissociation constants are about 8×10^{-4}, 4×10^{-5}, and 3×10^{-6}. These numbers are so close together that, for example, some citrate ions will have added two hydrogen ions apiece before other citrate ions have added their first hydrogen ion, and therefore the regions of the titration curve corresponding to the several successive neutralizations are not separated by discernible end points.

Resolving three close ionization constants is more of a problem than we are able to treat here, but presentation of the solution for the case of two overlapping constants will illustrate the type of approach that can be employed. In this discussion, we shall follow the procedures described in the books by Edsall and Wyman and by Martin listed at the end of the chapter.

We represent the un-ionized, bifunctional acid by the formula H_xAH_y. The equilibrium equations are written as if H_xAH_y were neutral, but the results would be similar if it happened to be a charged acid. In addition to the situation that there are two successive ionization constants lying close together, there is also the complication that there may be appreciable concentrations of each of the two possible partially ionized, or intermediate, forms, differing in which functional group has lost the proton. We symbolize these forms by AH_y^- and H_xA^- and represent the possible equilibria as

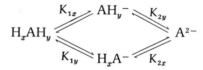

Constants for the separate ionization steps shown in the scheme above, often termed *microconstants,* cannot be determined from titration curves alone but can sometimes be evaluated by the use of spectrometric methods which can distinguish between AH_y^- and H_xA^-, or in other cases can be estimated by drawing analogies with related

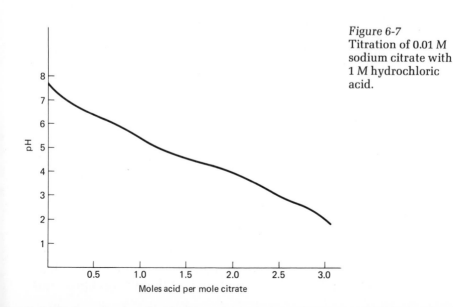

Figure 6-7
Titration of 0.01 *M* sodium citrate with 1 M hydrochloric acid.

monofunctional compounds. The conventional dissociation constants, K_1 and K_2, are *macroconstants*, in which are lumped together the concentrations of the two kinds of intermediate to give a single number.

The macroconstants can be expressed in terms of the microconstants:

$$K_1 = \frac{[H^+]([AH_y^-] + [H_xA^-])}{[H_xAH_y]} = K_{1x} + K_{1y} \qquad (6\text{-}39)$$

$$K_2 = \frac{[H^+][A^{2-}]}{[AH_y^-] + [H_xA^-]} = \frac{1}{(1/K_{2y}) + (1/K_{2x})} \qquad (6\text{-}40)$$

Multiplying these two equations together gives an equation consistent with the overall equilibrium requirements:

$$K_1K_2 = \frac{[H^+]^2[A^{2-}]}{[H_xAH_y]} \qquad (6\text{-}41)$$

One can show that the product K_1K_2 is also equal to $K_{1x}K_{2y}$ and to $K_{1y}K_{2x}$.

If proton x is lost very much more readily than proton y, one expects the microconstant K_{1x} to be much larger than K_{1y}, and K_{2x} to be larger than K_{2y}. Under these circumstances, the above equations show that K_1 is approximately equal to K_{1x} and K_2 is approximately equal to K_{2y}, and the only intermediate present then is AH_y^-. It is when the acidities of protons x and y approach one another that things become more complicated. In the limit of equal acid strength for the two groups and completely independent ionization of the two groups—the acid strength of one is not affected by whether or not the other one is ionized—the four microconstants all become equal to one another. Substitution of the equal values for this case in Equations (6-39) and (6-40) shows that K_1 is then equal to $4K_2$. There is then also no discontinuity whatever between the first and second stages of the titration curve.

Let us now consider a titration of a dibasic acid with sodium hydroxide, for example, for the case in which the first and second dissociation constants are of comparable magnitude. For this purpose there is defined the quantity $\bar{h}$, the mean number of protons removed by the base per molecule of acid originally present:

$$\bar{h} = \frac{[AH_y^-] + [H_xA^-] + 2[A^{2-}]}{[H_xAH_y] + [AH_y^-] + [H_xA^-] + [A^{2-}]} \qquad (6\text{-}42)$$

The condition of electronegativity requires

$$[Na^+] + [H^+] = [AH_y^-] + [H_xA^-] + 2[A^{2-}] + [OH^-] \qquad (6\text{-}43)$$

Substitution of this in the numerator of Equation (6-42) yields

$$\bar{h} = \frac{[Na^+] + [H^+] - [OH^-]}{C} \qquad (6\text{-}44)$$

where C denotes the total initial acid molar concentration, which is the denominator of Equation (6-42). Equation (6-44) is used with data from the titration to obtain values of the function $\bar{h}$ as the titration proceeds.

It is possible by substitution of the concentrations of various species in terms of the equilibrium constants defined in Equations (6-39) and (6-40) into Equation (6-42) to show that $\bar{h}$ is also equal to

$$\bar{h} = \frac{K_1[H^+] + 2K_1K_2}{[H^+]^2 + K_1[H^+] + K_1K_2} \tag{6-45}$$

The relation of $\bar{h}$ to pH depends only upon the value of K_1 and the ratio K_1/K_2. If $\bar{h}$ as calculated from the titration data is plotted against pH, the *shape* of the curve depends only on the ratio K_1/K_2, and this ratio can be evaluated by comparison with standard plots.

STRUCTURAL EFFECTS ON ACID–BASE EQUILIBRIA

The magnitude of the pK_a value is often a good clue to the nature of the acidic group that is ionizing. If a carboxyl group is present, for example, the pK_a is usually somewhere in the vicinity of that for acetic acid, 4.76. Substitution of an electron-withdrawing group nearly always leads to greater acidity and a smaller pK_a value. For example, chloroacetic acid has a pK_a of 2.86, benzoic acid a pK_a of 4.20, and succinic acid a pK_a of 4.21. The electronic effects of substituents are also illustrated by the pK_a values of para-substituted benzoic acids: A methyl group in the para position acts as a source of electrons and decreases the acidity slightly, leading to a pK_a value of 4.34, whereas an electron-withdrawing nitro group in the same position considerably enhances the acidity, altering the pK_a to 3.44.

Acid strengths of substituted ammonium ions can be considered in a similar qualitative way. Methylamine is quite a weak acid with a pK_a of 10.60. Benzylamine is a bit stronger with a pK_a of 9.33. Anilinium ion is a very much stronger acid with a pK_a of 4.58, and this strength results from the delocalization of electrons into the aromatic ring, as described in Chapter 9, which can occur after the proton has been removed from the nitrogen.

THERMODYNAMICS OF IONIZATION

In identifying ionizing groups, it is sometimes possible to utilize the magnitude of the enthalpy of ionization, the values of which vary according to the type of functional group. Some examples relevant to the groups most likely to be found in amino acids and proteins are given in Table 6-3. The enthalpy change on ionization is primarily a result of changes in the magnitudes of forces between atoms, usually changes

Table 6-3
Enthalpies of ionization of acid groups

Group	ΔH (kcal/mol)
Carboxyl	0–2
Imidazole	7–8
Phenol	5–8
Amino	11–14
Sulfhydryl	7
Guanidyl	12

in the nature of covalent bonding. The dissociation of water requires, as we have seen earlier, about 13.4 kcal/mol. This is the amount of energy required to break the covalent bond in the water molecule in excess of that gained by hydration of the resulting ions in solution, the largest part of which comes from hydration of the proton in the formation of hydronium ion. In much the same way, ammonium and substituted ammonium ions require 10 to 12 kcal/mol of net energy input to remove a proton.

The ionization of carboxylic acids, in contrast, requires little energy. An oxygen–hydrogen bond is broken and a new oxygen–hydrogen bond is formed in the hydronium ion, much as in the ionization of water, but the product carboxylate ion is stabilized by the possibility of electron delocalization, an energetically favorable circumstance:

$$R-C{\overset{O}{\underset{O-H}{}}} + H_2O \rightarrow R-C{\overset{O^-}{\underset{O}{}}} + H_3O^+ \tag{6-46}$$

The ionization of phenol is an intermediate case, requiring about 5 kcal/mol because there is some electron delocalization in the phenolate ion but not as much as in the carboxylate ion.

Quite different considerations apply to the values of the *entropy* of ionization. The primary effect here arises from the creation or destruction of order in solvent molecules surrounding the solute. Water molecules are oriented, restrained, and compressed by the charge on an ion, so that creation of a charge or increase in its magnitude corresponds to a loss of entropy. For example, the process, $HPO_4^{2-} + H_2O \longrightarrow H_3O^+ + PO_4^{3-}$, is accompanied by a loss in entropy of about 30 cal/(mol K), and the second ionization of carbonic acid involves an entropy decrease of approximately 20 cal/(mol K). In contrast, the acidic dissociation of a substituted ammonium ion results in no change in the number of charges, and the entropy change is small.

6-6
AMINO ACIDS
AND PROTEINS

The molecules of amino acids contain both carboxyl and amino groups. Many L-α-amino acids occur naturally, combined together with the elimination of water, in the form of polypeptides and proteins. In addition to the two ionizable groups present in every amino acid, there are also often polar groups in the side chains as well. Thus aspartic and glutamic acids have a second carboxyl group, lysine has a second amino group, histidine includes an imidazole ring, and arginine has a guanidyl group. The acid–base equilibria into which the amino acid functional groups enter are of considerable interest, as well as of great physiological significance.

Consider first a fairly simple amino acid, alanine, or α-amino-propionic acid, $NH_2CH(CH_3)COOH$. If a solution of alanine is acidi-

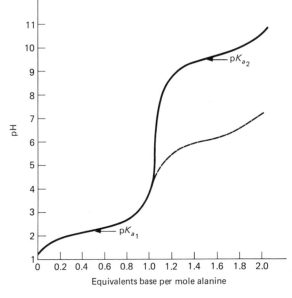

Figure 6-8
Titration of the acid form of
alanine with sodium
hydroxide solution. The
dashed line shows the result
of the titration in the
presence of formaldehyde.

fied and then titrated with base, a titration curve such as is shown by
the solid line in Figure 6.8 is obtained. The species present in the acid
solution is clearly $NH_3{}^+CH(CH_3)COOH$. The curve shows breaks
at pH 2.34 and 9.69, corresponding to the pK_a values of two acid
groups. The species at the basic end of the curve is certainly
$NH_2CH(CH_3)COO^-$. However, there are two possible intermediate
species, as represented in the scheme

$$NH_3{}^+CH(CH_3)COOH \overset{NH_3{}^+CH(CH_3)COO^-}{\underset{NH_2CH(CH_3)COOH}{\rightleftarrows}} NH_2CH(CH_3)COO^-$$

To determine which of these species of zero net charge predominates,
one can compare the observed pK_a values with those expected for the
functional groups concerned. The pK_a of the first ionization, 2.34,
matches better with that of a carboxyl group, normally somewhat over
4, and the second value, 9.69, corresponds well with the prediction of
10 for an ammonium ion. Thus there seems no doubt that the first ionization is that of the carboxyl group, and the second that of the protonated amino group. Furthermore, it seems clear that the pK_a of the
carboxyl group is below the normal value because of the positive
charge on the $-NH_3{}^+$ group which is relatively close by. The assignment of the two ionizations can be further confirmed by carrying out
the titration at different temperatures. The first ionization constant is
practically independent of temperature, corresponding to an enthalpy
change of nearly zero, behavior characteristic of the ionization of a
carboxyl group, but the second ionization constant shows a dependence on temperature yielding a ΔH value of about 11 kcal/mol, characteristic of the removal of a hydrogen ion from an ammonium group.

In view of these results, it is concluded that the amino acid alanine,
like other amino acids, which all show parallel behavior, exists be-

tween pH 4 and 8 chiefly in the dipolar or *zwitterion* form. That this form is stable compared to $NH_2CH(CH_3)COOH$ is simply a consequence of the fact that the NH_2 group is a stronger base than the COO^- group and holds on to protons more strongly. Indeed amino acids also exist in the zwitterion form in the solid state.

In further support of the conclusion that the second ionization corresponds to ionization of the hydrogen from the amino nitrogen is the result obtained when alanine is titrated in the presence of formaldehyde, shown by the dashed line in Figure 6-8. This procedure is known as a *formol* titration and is frequently employed in the study of amino acids and proteins. Formaldehyde lowers the pH at which amino groups are titrated, thus avoiding any difficulty involving carbon dioxide absorption from the atmosphere at high pH. This type of titration clearly distinguishes amino groups from other strongly basic groups which may be present. The formaldehyde reacts with the amino groups, forming N-methylol derivatives:

$$2HCHO + R-NH_2 \rightleftharpoons R-N\begin{array}{c} CH_2OH \\ \\ CH_2OH \end{array} \tag{6-47}$$

Many commonly occurring amino acids have functional groups in addition to the amino and carboxyl groups alpha to one another. Figure 6-9 shows the titration curve of glutamic acid which has two carboxyl groups. The first ionization, with a pK_a of 2.10, is assigned to the α-carboxyl group, and the second, with a pK_a of 4.07, is assigned to

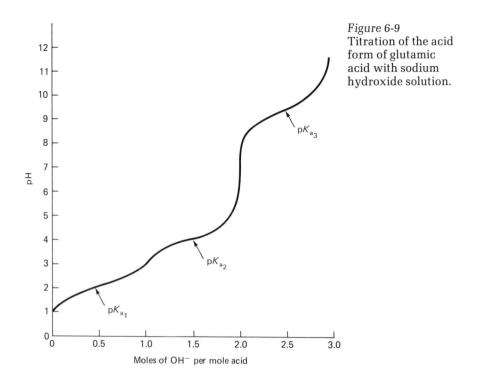

Figure 6-9
Titration of the acid form of glutamic acid with sodium hydroxide solution.

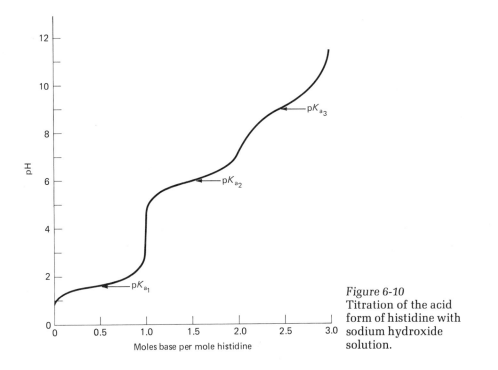

Figure 6-10
Titration of the acid
form of histidine with
sodium hydroxide
solution.

the side-chain or γ-carboxyl group, for this value is closer than the first
to the value for unsubstituted carboxylic acids such as acetic acid. The
third ionization, with a pK_a of 9.47, is that of the NH_3^+ group. According to these assignments, the ionization sequence is

$$\underset{\textstyle CH_2CH_2COOH}{NH_3^+CHCOOH} \rightleftharpoons \underset{\textstyle CH_2CH_2COOH}{NH_3^+CHCOO^-}$$

$$\rightleftharpoons \underset{\textstyle CH_2CH_2COO^-}{NH_3^+CHCOO^-} \rightleftharpoons \underset{\textstyle CH_2CH_2COO^-}{NH_2CHCOO^-}$$

If solid glutamic acid is dissolved in water, the pH of the resulting
solution is about 3.1. This is the pH predicted for the second form in
the ionization series above, the form having no net charge, by application of the general rule derived in Section 6-3 showing that the pH of a
solution of an amphoteric species is midway between the pK_a value of
the species itself and the pK_a value of its conjugate acid. This is also
referred to as the *isoelectric* pH of the amino acid, for there would be
no migration of the species with zero total charge in an applied electric
field.

Another example of a trifunctional amino acid is histidine, which
accordingly also displays three breaks in the titration curve as depicted in Figure 6-10. As with the other amino acids, the first ionization
is that of the carboxyl group, here quite acidic with a pK_a of 1.80. The
second is deprotonation of the imidazole ring, and the last is, as for the
other acids, the reaction of the substituted ammonium group. The

isoelectric point is the average of the last two pK_a values, or 7.69, and the ionization sequence is

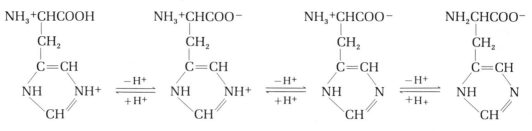

Considerable effort has been put forward in analyzing the ionization of the amino acid cysteine in terms of microconstants rather than merely the macroconstants obtained by titration. The first ionization, with a pK_a of 1.71, is clearly the deprotonation of the carboxyl group, but this is followed by hybrid, overlapping dissociations of the sulfhydryl and NH_3^+ groups with macro pK_a values of 8.33 and 10.78. The fact that the ionized thiol group, RS^-, absorbs ultraviolet radiation makes it possible to estimate the relative amounts of the species with a proton on the nitrogen and that with the proton on the sulfur atom, a ratio found to be about 2:1, although it is necessary to assume that the extent of ultraviolet absorption is independent of whether the NH_3^+ group has lost its proton or not. The negative logarithms of the microconstants are shown in the following scheme [after R. E. Benesch and R. Benesch, *J. Am. Chem. Soc.* **77**, 5877 (1955)]:

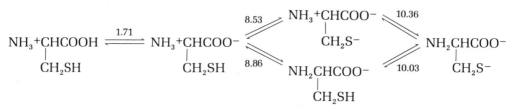

Table 6-4 lists the pK_a values and isoelectric points of several amino acids, and the reader should be able to write out schemes like those

Table 6-4
Values of pK_a and isoelectric pH for amino acids at 25°C

Amino Acid	pK_1	pK_2	pK_3	Isoelectric pH
Alanine	2.34	9.69		6.01
Arginine	1.82	8.99	12.48	10.74
Aspartic acid	1.99	3.90	9.84	2.95
Cysteine	1.71	8.33	10.78	5.02
Glutamic acid	2.10	4.07	9.47	3.09
Glycine	2.35	9.78		6.07
Histidine	1.80	6.04	9.33	7.69
Leucine	2.33	9.74		6.04
Lysine	2.16	9.18	10.79	9.99
Ornithine	1.71	8.69	10.76	9.73
Proline	1.95	10.64		6.30
Tyrosine	2.20	9.11	10.13	5.66

given above for the ionization of each of them and to justify the value given for the isoelectric point.

Polypeptides and proteins are formed of amino acid units linked together by *peptide* linkages formed by the elimination of water between the carboxyl group of one acid and the amino group of the next:

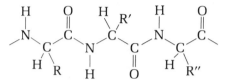

The side chains denoted by R, R', and R'' contain various functional groups such as we have seen to be present in the component amino acids, including carboxyl groups, amino groups, phenolic groups, imidazole rings, and so on.

One might expect that the nature of the polar groups in a protein could be determined by dividing the titration curve into sections corresponding to the pK_a values of various sorts of groups. Thus the carboxyl groups should titrate at pH values between 2.5 and 5, the imidazole and terminal NH_3^+ groups between 6 and 8, side-chain amino, phenolic, and sulfhydryl groups between 9 and 10.5, and the guanidyl group at 12 or above. To a certain extent such divisions can be made, although the sections of the titration curve tend to merge. It should be noted that proteins can buffer over a wide range of pH, although their buffer action is quite limited in the physiological region of pH, between about 7 and 8, where only the imidazole of histidine is effective —hemoglobin, present in red blood cells, is one of the few protein materials containing a large number of histidine units.

Detailed interpretation of a protein titration curve is complicated by several factors. One of these is the possible occurrence of a structural change or denaturation process as the pH is changed, and it is necessary to be certain that the titration is reversible before interpreting the results in terms of equilibrium constants. A second important point is that the groups in the protein do not ionize completely independently. We have seen that in amino acids neighboring groups influence acidity, and it is found further that the apparent pK_a value for a given functional group in a protein may differ from that of the same group in an isolated amino acid unit. This is usually a consequence of the change in net charge on the whole protein molecule: At low pH values, the protein is positively charged, and as the pH of the solution rises, there is an increasing negative charge. The positive charge at low pH repels hydrogen ions and facilitates their ionization, lowering the apparent pK_a of any group, whereas the negative charge at high pH attracts hydrogen ions and decreases the ease of their loss, raising the apparent pK_a value. The net effect of charge on the protein is therefore to spread the titration curve over a wider pH range. This effect tends to diminish as the ionic strength is increased, but then it is necessary to be concerned about the binding of other ions to the protein.

Semiempirical equations based on the Debye–Hückel theory— which was described in Section 5-2—have been developed to correct the observed pK_a values in a protein to the intrinsic values pK_{int} for functional groups ionizing independently in the absence of other charges on the macromolecule.

Another effect that may cause the apparent pK_a of a functional group in a protein to deviate from the value observed for independent ionization of the same group is the tendency of the protein to fold certain parts of its structure inward, creating a local environment which can be termed hydrocarbon, nonpolar, or hydrophobic. Polar groups tend to avoid this region when possible, but some of them may be forced from the aqueous medium, becoming inaccessible to ions from the solution. In addition, the effective dielectric constant of the nonpolar medium is substantially less than that of water, and ionizations that proceed in the direction of increasing numbers of charges occur with greater difficulty than in an aqueous environment.

A third influence affecting the extent of ionization of some groups is the presence of hydrogen bonds. Involvement of the proton of an acid group in a hydrogen bond makes the process of ionization more difficult than it otherwise would be, reduces the acidity, and increases the pK_a value. The involvement of a potential proton acceptor site of a base in hydrogen bonding has the opposite effect, making the addition of a proton more difficult, decreasing the basicity, and decreasing the pK_a value.

Finally, one must realize that, at any given pH, a protein solution is not really a solution of uniformly ionized molecules. The presence of many functional groups in any one protein means that there are many possible patterns of ionization corresponding to the same overall charge. The situation is indeed an extension of that described for di- and trifunctional amino acids to an n-functional molecule which can exist in any of $n + 1$ different total charge states. The numbers of different forms corresponding to the several total charge states are given by the coefficients of the binomial expansion, from unity for the most positive possible form, through a maximum at the point of zero charge, to unity for the most negative possible state.

A protein may be characterized by its isoelectric point or by its isoionic point, although the values of the two are nearly the same under usual circumstances. The isoelectric point is that pH at which the total charge on the protein is zero, so that there is no migration of the molecule when an electric field is applied. Experimental determination of the isoelectric point is usually carried out in the presence of buffers containing various ions which may be selectively bound and thus contribute to the charge along with the charges from the ionization of acid–base groups. The precise pH of the isoelectric point therefore depends upon both the composition and ionic strength of the medium. Most proteins found in animals have isoelectric points between that of casein at pH 4.6 and that of horse hemoglobin at pH 6.8; as a result they exist in animal fluids chiefly with a net negative charge. At the isoionic pH, the number of positive charges combined on basic groups is equal to the number of negative charges from acid groups from which protons have been lost. In other words, the net charge resulting from acid–base ionization is zero. Experimental determination of the isoionic point requires a solution from which all ions except the protein, hydrogen, and hydroxide have been removed by dialysis or by ion exchange.

The titration results for β-lactoglobulin provide an example of how data for a protein can be interpreted. This molecule, consisting of two polypeptide chains, has a molecular weight of 35,500. Denaturation sets in above a pH of about 9.5, so that the titration curve is reversible only to this point. Between pH 1.5 and 6.5, a total of 51 groups is titrated—these are the carboxyl groups. Eight more groups are titrated between pH 6.5 and 8.5, which should include the two α-amino groups at the peptide chain ends plus the imidazole groups in the histidine residues.

Analytical data for lactoglobulin show that there are only four rather than six histidines, and there must therefore be two carboxyl groups which are buried in the interior of the molecule and react only at a pH of about 7.5, at which point some conformational change occurs. The fact that these two carboxyl groups change in environment is probably the most significant result to come from the titration. When the denatured protein is titrated, it is found, in confirmation of this interpretation, that all 53 carboxyl groups titrate in the normal range of low pH.

In the alkaline region, the molecule is found to have a charge of -19 units at pH 8.5. All 53 carboxyl groups are negatively charged at this point, so that there should be 34 positively charged groups, which would include the side-chain amino groups in lysine and the guanidine units in arginine. A formol titration at pH 8.5 indicates the presence of 28 lysyl units, which leaves 6 arginines. As a further check, the number of cationic groups can be found from the number of protons added from the isoionic point at pH 5.4 to full protonation. The number of protons is 40, and the 6 arginines, 28 lysines, 4 histidines, and 2 α-amino groups do sum to 40.

Lactoglobulin also contains some phenolic groups in tyrosine units; these are not titrated in the available pH range before denaturation occurs. By measurement of the difference in ultraviolet absorption between the native and denatured protein, it was estimated that there are six of these, although data from analysis show eight. The probable reason for this discrepancy is that the ultraviolet absorption is also affected by some tryptophan residues which are freed in the denaturation.

The protein ribonuclease, for which the structure is well established and which consists of 124 amino acid units, provides a second example of titration results. Curves for three different ionic strengths are shown in Figure 6-11. In the region from full protonation up to pH 5, 11 carboxyl groups are titrated. In the neutral region, up to pH 8, the α-amino group at the end of the single polypeptide chain plus 4 imidazole groups are titrated. Up to the point at which denaturation begins, near pH 12, 13 more groups are titrated, 10 of which are side-chain amino groups and the other 3 of which are phenolic groups in tyrosine residues. The curve beyond this point is not reversible, but a conformational change occurs which exposes 3 more phenolic groups buried in the interior of the native protein. These groups are titratable in the

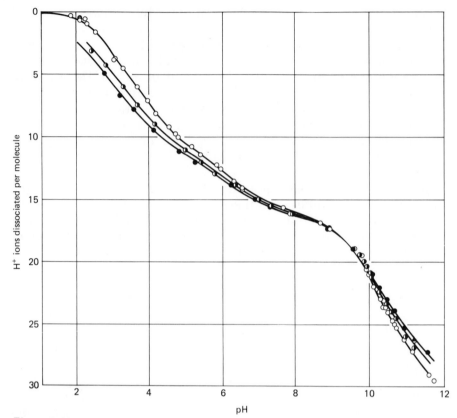

Figure 6-11
Titration of the protein ribonuclease at 25°C, at ionic strengths 0.01 (●), 0.03 (◐),
and 0.15 (○). Reprinted with permission from C. Tanford and J. D. Hauenstein,
J. Am. Chem. Soc. **78**, 5288 (1956). Copyright by the American Chemical Society.

same range as the other 3 phenolic groups when the protein is dissolved in a solution containing urea. The ribonuclease molecule contains in addition 4 guanidyl groups which would lose protons only at pH values above 12. It is not necessary to titrate these to estimate their number—one can calculate the total number of cationic groups from the number of carboxyl groups neutralized up to the isoionic point, and then obtain the number of guanidyl groups by subtracting the number of NH_3^+ and imidazole groups from this total. The most interesting feature of the titration is again demonstration of the change in accessibility of the three tyrosine units associated with the change in conformation.

6-7
IONIC EQUILIBRIA
IN THE BLOOD

Blood plasma, as well as interstitial body fluids and lymph, are well-buffered systems. The principal couples in the plasma are the carbonic acid–bicarbonate system and the plasma protein system, with a smaller amount of the dihydrogen phosphate–monohydrogen phosphate couple. The maximum buffer capacity of the carbonic acid–bicarbonate system is at a pH of 6.34, while the normal blood pH range is 7.3 to 7.5. Approximately 95 percent of the couple is in the form of bicarbonate, and only about 5 percent in the form of carbonic acid. However, most of the changes likely to disturb the blood equilibrium are in the acid direction, so that the demand for neutralization by bicarbonate, the base in the couple, is greater than the demand for carbonic acid.

In considering the acid–base balance in the blood, it has been customary in the past to refer to cations, such as sodium or potassium ion, as "bases." Of course they are not bases, but they are present along with bicarbonate ions which are the actual basic materials. Furthermore, when a base such as sodium hydroxide is added to the plasma, the hydroxyl ions are neutralized by the acid components of the buffer system, and one of the measurable results of making the solution more basic is that the sodium ion concentration of the fluid rises.

In metabolic processes, the body converts amino acids to ammonia and organic acids. The ammonia is converted by the liver to nearly neutral urea and glutamine. The organic acids are utilized in further reactions and do not contribute much in the way of acidity. However, if a sodium salt of one of these organic acids is ingested, the anion of the salt takes up protons, freeing hydroxyl ions and leading to a basic reaction, while the sodium ions remain unchanged.

Proteins also contain sulfur and phosphorus. These elements are oxidized to sulfuric and phosphoric acids, which cannot be used by the body. They are normally excreted by the body as ammonium salts. If they are produced in excess and not neutralized by ammonia, the hydrogen ion is taken up by the blood bicarbonate, increasing the acidity, and sodium salts of the acids are excreted. The kidneys also excrete such weak acids as citric acid and β-hydroxybutyric acid, which are un-ionized at the pH of 5 to 6 in the urine and thus can also carry along any excess protons.

Another buffer in the blood is the hemoglobin of the red cells. The hemoglobin molecule, with a molecular weight of about 67,000, contains four iron atoms, each of which is associated with the imidazole ring of a histidine group of the protein globin. Each iron is capable of adding a molecule of oxygen, and the greater the degree of oxygenation the stronger the acidity of the imidazole group. The pK_a for hemoglobin is 8.18, while the pK_a for oxyhemoglobin is 6.62. Thus increased oxygen in the blood lowers the pH as more of the stronger acid is formed. This occurs as blood depleted in oxygen circulates through the lungs. At the same time, the increase in acidity releases carbon dioxide from bicarbonate in the blood, and this is lost to the air in the lungs. The reverse change takes place in body tissues where the carbon dioxide produced by oxidation processes enters the blood and the oxygen attached to hemoglobin leaves the blood, with an accompanying increase in pH.

An unbalance in the blood condition may result from various causes that prevent proper functioning of the complex system of balances. Acidosis, or a decrease in pH below 7.3, may result from diabetes, improper functioning of the kidneys, lack of oxygen due to an obstruction in breathing, excess carbon dioxide formation after prolonged exertion, or loss of bicarbonate through loss of the alkaline digestive fluids as in diarrhea. Alkalosis, or an increase in pH above 7.5, may result from overventilation at high altitudes, ingestion of excess alkali as sodium bicarbonate, or excessive loss of hydrochloric acid from the stomach by vomiting.

EXERCISES

1. Estimate the pH and pOH of each of the following solutions:
   (a) 1.0×10^{-5} M HNO_3 (b) 0.0065 M KOH (c) 10 M HCl (d) 2.8×10^{-3} M NaOH

2. Write the formulas of the bases conjugate to each of the following acids:

   $H_2PO_4^-$, H_2SO_3, $Cr(H_2O)_6^{3+}$, $(CH_3)_3NH^+$, $H_3N^+CH_2CH_2SO_3^-$

3. Calculate the concentration of hydrogen ions in a 0.200 M solution of acetic acid. Compare the result with the concentration of hydrogen ions when 0.100 M potassium acetate is present along with the acetic acid.

4. Find the equilibrium constant for the reaction

   $$H_2O + H_2CO_3 \rightleftharpoons H_3O^+ + HCO_3^-$$

   from the fact that the equivalent conductance is 1.977 cm²/ohm for a solution of 0.01374 M concentration of H_2CO_3.

5. Assuming that the only acid–base functional groups are amino and carboxyl groups (or those derived from them by ionization), use the data in Table 6-4 to estimate the number of each type of functional group in the following amino acids: aspartic acid, lysine, glutamic acid, ornithine.

6. Look up the formulas of the amino acids in Exercise 5 and assign the pK_a values as specifically as possible to the various functional groups.

7. What would be the specific resistance at 25°C of water containing only hydroxyl and hydrogen ions from its own ionization and no other solutes?

8. Calculate the hydrogen and hydroxyl ion concentrations having each of the following pH values: 12.50, 2.84, 7.95.

9. Calculate the dissociation constant of NH_3 from the data in Exercise 22 of Chapter 5.

10. An acid functional group in a protein is titrated and found to have a pK_a value of 6.5 and an enthalpy of ionization of 8 kcal/mol. What is the probable nature of the group?

11. What fraction of the imidazole groups of oxyhemoglobin exists in the protonated form when the pH of the blood in 7.38?

12. Draw formulas for the amino acid tyrosine, indicating the various possible structures present at the several stages of the titration with base of the acid form, $HO-C_6H_4-CH_2CH(NH_3^+)COOH$, as was done in the text for cysteine.

13. The concentration of H_3O^+ at 25°C in pure water is 1.00×10^{-7} M. In a 0.1 M sodium chloride solution, the concentration is 1.32×10^{-7}. Suggest an explanation.

14. A determination using a spectrophotometer shows that in a particular solution to which bromthymol blue has been added the indicator, which has a pK_{In} of 7.00, exists to the extent of 36 percent in the blue or base form and 64 percent in the yellow or acid form. What is the pH of the solution?

15. Using a table of indicators in a handbook, select those that might be suitable for the titration of
    (a) Sodium bicarbonate with formic acid
    (b) Pyridinium chloride with sodium hydroxide
    (c) Sodium chloroacetate with hydrochloric acid

16. Estimate the CO_3^{2-} concentrations in 0.005 M solutions of carbonic acid if the pH is adjusted to 4, 8, and 12, respectively, by the addition of suitable amounts of acid or base.

17. Calculate the concentrations of the various ionic species present in a 0.025 M solution of H_3PO_4. For the first ionization, use the exact method.

18. For each of the following buffer solutions, calculate the ratio of the two forms present and write out the corresponding structural formulas:
    (a) Glycylglycine at pH = 8.0
    (b) Cholamine chloride at pH = 7.5

19. Calculate the pH of each of the following solutions (consider the concentrations specified to be the exact concentrations):
    (a) 0.01 M propionic acid
    (b) 0.5 M sodium nitrite
    (c) 0.1 M potassium hydrogen phthalate
    (d) 0.025 M lithium formate
    (e) 0.0003 M hydrazinium chloride
    (f) 0.1 M sodium malonate
    (g) 0.075 M anilinium nitrate
    (h) 0.5 M sodium dihydrogen phosphate
    (i) 0.5 M silver nitrate
    (j) 0.01 M potassium hydrosulfide
    (k) 0.2 M $CH_3CHOHCOONa$

20. Describe how to prepare the following buffers from the indicated solutions:
    (a) pH 3.25, from a 0.250 M solution of phthalic acid and a 0.001 M solution of sodium hydroxide
    (b) pH 5.00, from 0.1 M sodium acetate and 0.1 M acetic acid
    (c) pH 5.00, from 0.1 M acetic acid and 0.1 M sodium hydroxide

21. Calculate the concentration of H_3O^+ in a 0.2 M solution of monochloracetic acid. What would be the effect on the H_3O^+ concentration of adding $MgCl_2$ to the same solution until the salt concentration is 0.5 M? What would be the effect of adding 0.5 M sodium monochloracetate to the original acid solution?

22. What is the pH of solutions made by mixing
    (a) 100 cm^3 of 0.01 M oxalic acid with 150 cm^3 0.01 M KOH?
    (b) 100 cm^3 of 0.05 M sodium oxalate with 100 cm^3 0.02 M oxalic acid?
    (c) 100 cm^3 of 0.05 M monosodium oxalate with 50 cm^3 of 0.05 M hydrochloric acid?
    (d) 100 cm^3 of 0.05 M monosodium oxalate with 10 cm^3 of 0.2 M sodium hydroxide?

23. Work out the important points and then sketch titration curves of each of the following pairs, assuming initial concentration of the first reagent to be 0.05 M and neglecting any dilution effects:
    (a) The acid form of histidine with sodium hydroxide
    (b) The acid form of aspartic acid with sodium hydroxide
    (c) Disodium malonate with hydrochloric acid

REFERENCES

Books

A. Albert and E. P. Sergeant, *Determination of Ionization Constants,* 2nd ed., Halsted Press, New York, 1971.

R. G. Bates, *Determination of pH: Theory and Practice,* 2nd ed., Wiley, New York, 1973. A comprehensive and authoritative account.

R. P. Bell, *Acids and Bases,* 2nd ed., Halsted Press, New York, 1972. An excellent introduction.

R. P. Bell, *The Proton in Chemistry,* 2nd ed., Cornell University Press, Ithaca, N.Y., 1974. A more advanced and extensive treatment.

Henry B. Bull, *An Introduction to Physical Biochemistry,* Davis, Philadelphia, 1964. Chapter 5 describes acid–base equilibria and buffer solutions.

H. N. Christensen, *Body Fluids and the Acid–Base Balance,* Saunders, Philadelphia, 1964. A programmed-learning book, covering principles from elementary to those of advanced applications.

R. M. C. Dawson, D. C. Elliott, W. H. Elliott, and K. M. Jones, *Data for Biochemical Research,* 2nd ed., Oxford University Press, New York, 1969.

John T. Edsall and Jeffries Wyman, *Biophysical Chemistry,* Vol. 1, Academic Press, New York, 1958. Chapters 8 and 9 include extensive accounts of biochemical aspects of acid–base equilibria.

E. J. King, *Acid–Base Equilibria,* Pergamon Press, Elmsford, N.Y., 1965. A comprehensive review at an intermediate level.

R. Bruce Martin, *Introduction to Biophysical Chemistry,* McGraw-Hill, New York, 1964. Chapter 4 includes a good account of multiple acid–base equilibria and methods for analyzing them, and Chapter 5 describes protein titrations.

Charles Tanford, *Physical Chemistry of Macromolecules,* Wiley, New York, 1961. Chapter 8 presents a comprehensive, fairly advanced account of multiple equilibria, such as those between macromolecules and small ions.

Charles Tanford, "The Interpretation of Hydrogen Ion Titration Curves of Proteins," in *Advances in Protein Chemistry,* Vol. 17, Academic Press, New York, 1962.

C. A. VanderWerf, *Acids, Bases and the Chemistry of the Covalent Bond,* Reinhold, New York, 1961. A very good elementary introduction.

A. White, P. Handler, and E. L. Smith, *Principles of Biochemistry,* 5th ed., McGraw-Hill, New York, 1973. Acid–base equilibrium and its biochemical applications are well covered.

Journal Articles

C. R. Allen and P. G. Wright, "Entropy and Equilibrium," *J. Chem. Educ.* **41,** 251 (1964). Interpretation of ionization data for organic acids.

R. G. Bates, R. A. Robinson, and A. K. Covington, "pK Values for D_2O and H_2O," *J. Chem. Educ.* **44,** 635 (1967).

J. D. Burke, "On Calculating [H+]," *J. Chem. Educ.* **53,** 79 (1976).

G. E. Clement and T. P. Hartz, "Determination of the Microscopic Ionization Constants of Cysteine," *J. Chem. Educ.* **48,** 395 (1971).

H. L. Clever, "The Ion Product Constant of Water," *J. Chem. Educ.* **45,** 231 (1968).

H.-L. Fung and L. Cheng, "Linear Plots in the Determination of Microscopic Dissociation Constants," *J. Chem. Educ.* **51,** 106 (1974).

D. M. Goldish, "Component Concentrations in Solutions of Weak Acids," *J. Chem. Educ.* **47,** 65 (1970).

J. A. Goldman, "Le Chatelier's Principle and Rigorous Ionic Equilibria Equations," *J. Chem. Educ.* **44,** 658 (1967).

N. E. Good et al., "Hydrogen Ion Buffers for Biological Research," *Biochemistry* **5,** 467 (1966).

S. L. Hein, "Physiochemical Properties of Antacids," *J. Chem. Educ.* **52,** 383 (1975).

D. A. Jenkins and J. L. Latham, "Estimation of Some K_i and K_{sp} by Potentiometric Titration," *J. Chem. Educ.* **43,** 82 (1966).

William L. Jolly, "The Intrinsic Basicity of the Hydroxide Ion," *J. Chem. Educ.* **44,** 304 (1967).

P. Jones, M. L. Haggett, and J. L. Longridge, "The Hydration of Carbon Dioxide," *J. Chem. Educ.* **41,** 610 (1964).

C. Minnier, "Cystinuria: The Relationship of pH to the Origin and Treatment of a Disease," *J. Chem. Educ.* **50,** 427 (1973).

F. S. Nakayama, "Hydrolysis of Sodium Carbonate," *J. Chem. Educ.* **47,** 67 (1970).

R. W. Ramette, "Equilibrium Constants from Spectrophotometric Data," *J. Chem. Educ.* **44,** 647 (1967).

S. J. Rogers, "Composite pK's of Cysteine," *J. Chem. Educ.* **46,** 239 (1969).

D. I. Stock, "Dissociation of Weak Acids and Bases at Infinite Dilution," *J. Chem. Educ.* **44,** 764 (1967).

Jurg Waser, "Acid–Base Titration and Distribution Curves," *J. Chem. Educ.* **44,** 274 (1967).

Seven
Oxidation-Reduction Equilibria

A group of chemical reactions that can be placed beside the acid–base reactions in a position of outstanding prominence in chemistry in general and in the chemistry of living systems in particular is the class of electron-transfer or oxidation–reduction reactions. Both acid–base and oxidation–reduction reactions are intimately involved in the metabolic processes of cells, in the changes occurring in the soil, water, and air about us, and in the methods chemists employ in the analysis of solutions.

7-1
REACTION POTENTIALS FOR OXIDATION–REDUCTION

The tendency for a reaction to proceed is measured by the free energy change accompanying that reaction. This statement is true as well for electron-transfer reactions as for other types. However, it is convenient to express the driving force of an electron-transfer reaction in terms of an electric voltage, that is, an electric potential difference, instead of citing the appropriate thermochemical quantities. One reason for this situation, as we shall shortly see, is the ease of determining the driving voltage of an oxidation–reduction reaction by direct electrical measurements.

To establish the relation between free energy change and electric voltage, we recall the basic principle: The work required to transfer an electric charge from one level of potential to another equals the magnitude of the charge transferred multiplied by the potential difference through which it is transferred, and the maximum work associated with the process is equal to the free energy change. If the amount of charge transferred is $n\mathscr{F}$, where $\mathscr{F}$ is the value of the faraday, that is, the number of coulombs of charge corresponding to Avogadro's number of electrons, then

$$-w_{max} = -\Delta G = n\mathscr{F}\mathscr{E}_{rev} \tag{7-1}$$

where $\mathscr{E}_{rev}$ is the electric potential measured when the process is carried out reversibly.

Just as the symbol ΔG^0 is applied to the free energy change when all reactants and products in a reaction are in their standard states, so also the superscript zero is attached to the symbol for the potential to denote the *standard potential* applicable when *each material concerned in the reaction is in its standard state*:

$$-\Delta G^0 = n\mathscr{F}\mathscr{E}_{rev}^0 \tag{7-2}$$

Combining this equation with the expression for the equilibrium constant, Equation (4-39), and solving for the standard potential leads to

$$\mathscr{E}_{rev}^0 = \frac{RT}{n\mathscr{F}} \ln K \tag{7-3}$$

If the reactants or products or both are in conditions other than their standard states, it is necessary to use the expression for the more general free energy change, Equation (4-40):

$$\Delta G = -RT \ln K + RT \ln Q$$

When the free energy change is replaced by $-n\mathscr{F}\mathscr{E}_{rev}$ and each term is divided by $-n\mathscr{F}$, the result is

$$\mathscr{E}_{rev} = \frac{RT}{n\mathscr{F}} \ln K - \frac{RT}{n\mathscr{F}} \ln Q \tag{7-4}$$

This is equivalent to:

$$\mathscr{E}_{rev} = \mathscr{E}_{rev}^0 - \frac{RT}{n\mathscr{F}} \ln \frac{\Pi[\text{products}]}{\Pi[\text{reactants}]} \tag{7-5}$$

Here the capital Greek letter pi indicates a mathematical product is to be taken of all the product concentrations or all the reactant concentrations, respectively; each species concentration appears in the product to a power equal to its coefficient in the stoichiometric equation.

From this point on, we drop the subscript "rev" on potentials, assuming that all potential values stated are for reversible conditions. Further, stoichiometric concentrations are used in place of the values of activities required to make the equations for potentials exact.

If the value of the faraday $\mathscr{F}$ in the equations is taken as 96,490 coulombs/equivalent, the value of R to be used is 8.314 J/(mol K). When numerical values for R and $\mathscr{F}$ and a temperature of 298 K are substituted in Equation (7-5), and the logarithm scale is changed to the base 10, the equation becomes

$$\mathscr{E} = \mathscr{E}^0 - \frac{0.0592}{n} \log \frac{\Pi[\text{products}]}{\Pi[\text{reactants}]} \tag{7-6}$$

Example: The standard potential for the following equilibrium is +0.50 V at 25°C:

$$Fe^{3+} + \text{reduced cytochrome c} \rightleftharpoons Fe^{2+} + \text{oxidized cytochrome c}$$

Calculate (a) the equilibrium constant and (b) the potential when the ratio of ferrous ion concentration to ferric ion concentration is 1000:1 and the two forms of cytochrome are each present at 0.0001 M concentration.

Solution: (a) The equilibrium constant is obtained from Equation (7-3),
$2.303\, RT \log K = n\mathscr{F}\mathscr{E}^0$. Substituting the appropriate numbers for this problem:

$$\log K = \frac{1(96,490)(0.50)}{2.303(8.314)(298)} = 8.45$$

$$K = 2.8 \times 10^8$$

For part (b), Equation (7-6) is applicable:

$$\mathscr{E} = \mathscr{E}^0 - \frac{RT}{n\mathscr{F}} \ln \frac{(1000)(0.0001)}{(1)(0.0001)} = 0.50 - \frac{0.0592}{1}(+3) = 0.50 - 0.18 = 0.32 \text{ V}$$

GALVANIC CELLS AND ELECTRODE POTENTIALS

How can the driving voltage associated with a given oxidation–reduction reaction be measured? The answer to this question is quite important, and the key point in the answer is the requirement that the reacting materials be isolated from one another, connected only by a metallic conductor through which electrons can flow or by a salt bridge through which nonreacting ions can migrate.

For example, one might be interested in the reaction

$$Cd + CuSO_4(aq) \rightleftharpoons CdSO_4(aq) + Cu \qquad (7\text{-}7)$$

If a piece of metallic cadmium is dipped into a solution of copper sulfate, this reaction proceeds, but there is no way of measuring the driving force, and indeed the driving force diminishes as time goes on and cupric ions in solution are replaced by cadmium ions. As a means of carrying out the reaction under controlled conditions, a galvanic cell composed of two half-cells or electrodes can be set up as shown schematically in Figure 7-1. One of these half-cells is a vessel containing a known concentration of copper sulfate in solution, with a piece of metallic copper dipping into the solution. The other is a vessel in which a piece of cadmium dips into a cadmium sulfate solution of

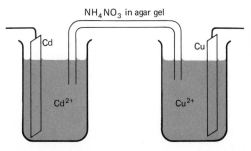

NH$_4$NO$_3$ in agar gel

Cd

Cu

Cd^{2+}

Cu^{2+}

Figure 7-1
Galvanic cell made up of a cadmium–cadmium ion couple linked by a salt bridge to a copper–cupric ion couple.

known concentration. The two vessels are connected by a bridge containing a solution of a conducting salt, such as potassium sulfate, potassium chloride, or ammonium nitrate, perhaps gelled in agar to prevent the electrode solutions from flowing into one another. If the cadmium metal and the copper metal are connected by a wire, the cadmium proceeds to dissolve, freeing electrons which pass through the wire to the copper metal. At the interface between the solid copper and the copper sulfate solution, these electrons meet copper ions which they reduce to elemental copper. In the solution, cations migrate from the vicinity of the cadmium electrode, where cadmium ions are formed, toward the copper electrode, where copper ions are consumed, and anions migrate in the reverse direction.

Under the conditions described, the concentrations within the cell are continuously changing, and the driving force, or voltage, is at any instant that characteristic of the instantaneous concentrations. To measure a voltage corresponding to the initial concentrations of the electrode solutions, we interpose in the external electric circuit a source of potential which can be adjusted to provide a voltage exactly equal and opposite to the voltage of the cell. Thus no current can flow and no reaction occurs, yet from a knowledge of the applied potential, which has been set at the value necessary to prevent current flow, the voltage of the cell is known.

An arrangement of two oxidation–reduction couples, or two electrodes, of this sort constitutes a *galvanic cell*. In addition to being employed to measure electrically the driving force of the oxidation–reduction reaction that can take place within it, this type of cell affords a means for converting the energy of a chemical reaction into electric energy which can then be made to do work. Applications to produce electric current are exemplified by dry cells, storage batteries, and fuel cells.

The potential of a cell is the resultant of the potentials of the two electrodes of which it consists; only this resultant is experimentally obtainable, and absolute single electrode potentials have not been measured. To see why this is so, suppose that it is desired to measure the potential of the ferrous–ferric ion couple. A piece of platinum wire is dipped into the solution to make electric contact. Somehow the circuit must be completed back to the solution. This cannot be done through a platinum wire or other inert conductor leading to the same solution, or there will be nothing to measure. It can only be done through a different type of return, and the only way electrons can enter or leave the solution in any other way than by interaction with the ferrous–ferric couple is by addition to or removal from the components of some other couple, which then makes its own contribution to the measured potential.

Because absolute electrode potentials cannot be determined, it is necessary, in order to give a numerical value for the potential of any cell as the algebraic sum of two tabulated electrode potentials, to set up an arbitrary scale. The origin of the scale is taken to be the value of zero for the potential of the standard hydrogen electrode—a couple of hydrogen gas at 1 atm pressure and hydrogen ion in solution at unit activity. The potential of any other electrode can then be obtained by direct comparison with the hydrogen electrode, or indirectly by measurement against another electrode which has been measured against the hydrogen electrode.

As we have already seen, the oxidized and reduced forms of an oxidation–reduction couple can be in various physical states. The hydrogen, hydrogen ion electrode is an example of a couple in which one component is a gas and the other is an ion in solution. Of course this electrode must be connected to the external circuitry by a conductor, usually a piece of platinum wire in contact both with the gas and with the solution to ensure true equilibrium. The hydrogen electrode and its reaction can be represented:

$$H^+(aq), H_2(g), Pt \qquad H^+ + e^- \longrightarrow \tfrac{1}{2}H_2 \qquad (7\text{-}8)$$

Examples of electrodes in which a metallic solid is in contact with an ion of the metal in solution are

$$Sn^{2+}(aq), Sn(s) \qquad Sn^{2+} + 2e^- \longrightarrow Sn \qquad (7\text{-}9)$$

$$Ag^+(aq), Ag(s) \qquad Ag^+ + e^- \longrightarrow Ag \qquad (7\text{-}10)$$

Two or more ions or molecules, all in solution, may constitute the members of an oxidation–reduction system, with contact to the external circuit through a platinum wire:

$$Ce^{4+}, Ce^{3+}, Pt \qquad Ce^{4+} + e^- \longrightarrow Ce^{3+} \qquad (7\text{-}11)$$

$$Mn^{2+}, H^+, MnO_4{}^-, Pt \qquad MnO_4{}^- + 8H^+ + 5e^- \longrightarrow 4H_2O + Mn^{2+} \qquad (7\text{-}12)$$

Succinic acid, maleic acid, H_3O^+, Pt

$$\begin{array}{c} CHCOOH \\ \| \\ CHCOOH \end{array} + 2H^+ + 2e^- \longrightarrow \begin{array}{c} CH_2COOH \\ | \\ CH_2COOH \end{array} \qquad (7\text{-}13)$$

In dealing with the signs of potentials of single electrodes and galvanic cells, a system of conventions is necessary in order to avoid confusion. Several versions of this system are in use, and it is therefore necessary to use care in interpreting any literature values of oxidation–reduction potentials. The feature of any system should be of course that *a more positive potential is associated with a process that has a more negative free energy change* and therefore the greater tendency to proceed.

To begin with, when the potential of a galvanic cell is measured, there is obtained a numerical value and information about which electrode is positive and which is negative when viewed from outside the cell. Although there is no particular justification for calling the voltage positive or negative, the aim is to associate the voltage with the driving force for some chemical reaction, and it is for this purpose that conventions are required.

We start therefore by writing down a schematic representation of the cell, such as

$$Pt, Fe^{2+}, Fe^{3+} \parallel H^+, H_2, Pt$$

This cell consists of a ferrous–ferric ion couple on the left and a hydrogen electrode on the right. The double vertical line indicates a salt bridge between the solutions in the two half-cells. The concentrations of ions in the solutions have not been specified, but they also could have been included in the cell representation.

The left-to-right order in which one writes the two electrodes establishes a direction for the reaction conventionally associated with the cell:

The conventional cell reaction is taken to be that in which oxidation occurs at the electrode written on the left side and reduction occurs at the electrode written on the right side.

It is to be emphasized that this implies nothing about whether the cell reaction as written is spontaneous or not. In fact, in designing or discussing the cell, we may not know in advance of making a potential measurement in which direction the spontaneous reaction will proceed. Also, it should be noted that the order in which components in the same phase, such as the ferrous and ferric ions, are written with respect to one another is immaterial.

Applying the convention to the cell representation shown above, the electrode reactions are $Fe^{2+} \longrightarrow Fe^{3+} + e^-$ and $H^+ + e^- \longrightarrow \frac{1}{2}H_2$. The cell reaction is the sum of these two reactions:

$$Fe^{2+} + H^+ \longrightarrow Fe^{3+} + \tfrac{1}{2}H_2 \tag{7-14}$$

A diagram of this cell is given in Figure 7-2. The arrows indicate the flow of electrons that would occur *if the conventional cell reaction were the spontaneous reaction.*

It is desired to assign a positive potential to a spontaneous reaction. From Figure 7-2, one can see that, when the conventional reaction is the spontaneous one, the right-hand electrode is positive when viewed from outside the cell. This leads to the second convention:

The cell potential is designated positive or negative according to whether the electrode written on the right side in the cell representation is positive or negative as viewed from outside the cell.

The description of the electrode as having a positive potential when viewed from outside the cell means that electrons tend to be con-

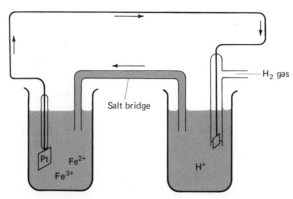

Figure 7-2
The arrows illustrate the clockwise direction of the flow of negative charge associated with the conventional cell reaction in a galvanic cell representation.

Table 7-1
Standard reduction potentials at 25°C

Reaction	Potential (V)
$Li^+ + e^- \longrightarrow Li$	-3.045
$K^+ + e^- \longrightarrow K$	-2.925
$Ca^{2+} + 2e^- \longrightarrow Ca$	-2.87
$Na^+ + e^- \longrightarrow Na$	-2.714
$Al^{3+} + 3e^- \longrightarrow Al$	-1.66
$Zn^{2+} + 2e^- \longrightarrow Zn$	-0.763
$Cr^{3+} + 3e^- \longrightarrow Cr$	-0.74
$Fe^{2+} + 2e^- \longrightarrow Fe$	-0.440
$Cd^{2+} + 2e^- \longrightarrow Cd$	-0.403
$PbSO_4 + 2e^- \longrightarrow SO_4^{2-} + Pb$	-0.356
$V^{3+} + e^- \longrightarrow V^{2+}$	-0.255
$Ni^{2+} + 2e^- \longrightarrow Ni$	-0.250
$Sn^{2+} + 2e^- \longrightarrow Sn$	-0.136
$Pb^{2+} + 2e^- \longrightarrow Pb$	-0.126
$Fe^{3+} + 3e^- \longrightarrow Fe$	-0.036
$2D^+ + 2e^- \longrightarrow D_2$	-0.003
$2H^+ + 2e^- \longrightarrow H_2$	0.000
$Sn^{4+} + 2e^- \longrightarrow Sn^{2+}$	$+0.15$
$Cu^{2+} + e^- \longrightarrow Cu^+$	$+0.153$
$AgCl + e^- \longrightarrow Ag + Cl^-$	$+0.222$
$Cu^{2+} + 2e^- \longrightarrow Cu$	$+0.337$
$Fe(CN)_6^{3-} + e^- \longrightarrow Fe(CN)_6^{4-}$	$+0.36$
$O_2 + 2H_2O + 4e^- \longrightarrow 4OH^-$	$+0.401$
$Cu^+ + e^- \longrightarrow Cu$	$+0.521$
$I_2 + 2e^- \longrightarrow 2I^-$	$+0.536$
$O_2 + 2H^+ + 2e^- \longrightarrow H_2O_2$	$+0.682$
$Fe^{3+} + e^- \longrightarrow Fe^{2+}$	$+0.771$
$Hg_2^{2+} + 2e^- \longrightarrow 2Hg$	$+0.792$
$Ag^+ + e^- \longrightarrow Ag$	$+0.799$
$2Hg^{2+} + 2e^- \longrightarrow Hg_2^{2+}$	$+0.920$
$Br_2(l) + 2e^- \longrightarrow 2Br^-$	$+1.065$
$Cr_2O_7^{2-} + 14H^+ + 6e^- \longrightarrow 2Cr^{3+} + 7H_2O$	$+1.33$
$Cl_2 + 2e^- \longrightarrow 2Cl^-$	$+1.360$
$MnO_4^- + 8H^+ + 5e^- \longrightarrow Mn^{2+} + 4H_2O$	$+1.51$
$Ce^{4+} + e^- \longrightarrow Ce^{3+}$	$+1.61$
$F_2 + 2e^- \longrightarrow 2F^-$	$+2.65$

sumed in the cell and therefore extracted from the external wire at this electrode.

For electrode potentials, or potentials of any system in which the two oxidation states of a couple are present, we wish to associate a positive or negative sign with each direction of the electrode reaction.

If the electrode reaction is a reduction reaction, then the more positive the electrode potential, the more negative the free energy for, and the more spontaneous is, the reduction process. The more positive the potential of an electrode oxidation reaction, the greater the tendency for the system to be spontaneously oxidized. A change in the direction in which the reaction is written is always accompanied by a corresponding change in the sign of the potential.

In Table 7-1 are given standard electrode potentials for various systems. The voltages cited in the table are reduction potentials, and the corresponding reduction reaction is shown for each couple. Another way of describing one of these potentials is to point out that it is the measured voltage when the electrode with all its components at unit activity is used as the right-hand half-cell connected to a standard hydrogen electrode as the left-hand half cell.

To use electrode voltages in predicting the potential of a cell, one simply matches the direction of the electrode reaction with the direction of the change indicated in the conventional cell reaction. If the two directions are the same, the electrode voltage is used with the sign as given; otherwise the sign of the electrode potential is changed. Thus it makes no difference whether data are available in the form of tabulations of oxidation potentials or of reduction potentials. Just as the cell reaction is the sum of two electrode reactions, so the cell potential is the sum of the two electrode potentials.

In the definitions of a standard potential and in calculations involving the effect of concentration or activity on potential, certain customs are observed. For example, a solid is said at all times to be in its standard state with unit activity, since its activity cannot be varied, and there is no point in trying to include an explicit numerical value. For gases, the activity is taken as equal to the pressure in atmospheres; a correction must be made for the partial pressure of water, since the gases are water-saturated when in contact with aqueous solutions. For ions, the molal rather than the molar scale has been used in most precise determinations of potentials, although use of the molar scale does not introduce a serious error for reasonably dilute aqueous solutions, and we use the molar scale in this book. The ionic strength of the solutions is usually sufficiently large so that equations in terms of stoichiometric concentration instead of activity give only a rough first approximation, but for simplicity we ignore this point in some of the illustrative calculations.

The examples following are typical of calculations involving potentials of galvanic cells.

Example: Calculate the potential of the following cell at 25°C:

$$Cu, Cu^{2+}(a = 0.1) \parallel H^+(a = 0.01), H_2(0.9 \text{ atm}), Pt$$

Solution: The conventional cell reaction is the sum of the electrode reactions, with oxidation at the left-hand electrode and reduction at the right-hand electrode:

$$Cu \longrightarrow Cu^{2+}(a = 0.1) + 2e^-$$
$$\underline{2e^- + 2H^+(a = 0.01) \longrightarrow H_2(0.9 \text{ atm})}$$
$$Cu + 2H^+(a = 0.01) \longrightarrow Cu^{2+}(a = 0.1) + H_2(0.9 \text{ atm})$$

The equation for the cell potential is:

$$\mathscr{E} = \mathscr{E}^0 - \frac{0.0592}{2} \log \frac{[Cu^{2+}]P_{H_2}}{[H^+]^2}$$

$$= -0.337 + 0 - \frac{0.0592}{2} \log \frac{(0.1)(0.9)}{(0.01)^2}$$

$$= -0.337 - 0.0874 = -0.424 \text{ V}$$

Example: Calculate the activity of chloride ions in the following cell, for which the measured potential at 25°C is +0.435 V:

$$Ag(s), AgCl(s), Cl^-(a = x) \parallel Fe^{2+}(a = 0.1), Fe^{3+}(a = 0.05), Pt$$

Solution: The cell reaction is obtained by adding the appropriate electrode oxidation and reduction reactions:

$$Ag + Cl^-(a = x) \longrightarrow AgCl(s) + e^-$$
$$\frac{Fe^{3+}(a = 0.05) + e^- \longrightarrow Fe^{2+}(a = 0.1)}{Ag + Cl^-(a = x) + Fe^{3+}(a = 0.05) \longrightarrow AgCl(s) + Fe^{2+}(a = 0.1)}$$

The equation for the cell potential is

$$\mathscr{E} = -0.222 + 0.771 - 0.0592 \log \frac{(0.1)}{(0.05)x} = +0.435 \text{ V}$$

$$\log \frac{2}{x} = \frac{-0.114}{-0.0592} = 1.93$$

$$\frac{2}{x} = 85 \quad \text{or} \quad x = 0.024$$

Example: Calculate the equilibrium constant at 25°C for the combination of hydrogen and chlorine to form aqueous hydrochloric acid:

$$H_2(g) + Cl_2(g) \rightleftharpoons 2HCl \text{ (aq)}$$

Solution: Set up a cell for which this reaction is the cell reaction. The left-hand electrode involves the oxidation of hydrogen gas to hydrogen ion:

$$Pt, H_2(g), H^+ \qquad H_2(g) \longrightarrow 2H^+ + 2e^-$$

The right-hand electrode involves the reduction of chlorine to chloride ion:

$$Cl^-, Cl_2(g), Pt \qquad Cl_2(g) + 2e^- \longrightarrow 2Cl^-$$

The standard potential of the resulting cell is

$$0 + (+1.360) = +1.360 \text{ V}$$

Setting the cell potential equal to zero to correspond to the equilibrium condition,

$$0 = 1.360 - \frac{0.0592}{2} \log K$$

$$\log K = \frac{1.360}{0.0296} = 45.9$$

$$K = 8 \times 10^{45}$$

A potential can be produced across two otherwise identical electrodes because of a difference in concentration of a dissolved ionic or molecular species involved in the electrode reaction. A galvanic cell of this sort is called a concentration cell. Consider, for example, the cell

$$Ag, Ag^+(a = 0.01) \parallel Ag^+(a = 0.05), Ag$$

The electrode reactions are

$$Ag \longrightarrow Ag^+(a = 0.01) + e^- \tag{7-15}$$

$$Ag^+(a = 0.05) + e^- \longrightarrow Ag \tag{7-16}$$

The net cell reaction is

$$Ag^+(a = 0.05) \longrightarrow Ag^+(a = 0.01) \tag{7-17}$$

Since the two electrodes are similar, the standard cell potential is zero, and the equation for the cell potential is

$$\mathscr{E} = -0.0592 \log \frac{0.01}{0.05} = +0.0414 \text{ V} \tag{7-18}$$

Example: The potential of the following cell is +0.1110 V at 25°C. Calculate the activity coefficient of silver ion in the 0.1 m solution.

$$Ag(s), Ag^+(a = 0.001\ m) \parallel Ag^+(c = 0.1\ m, a = x), Ag(s)$$

Solution: The cell reaction is: $Ag^+(a = x) \longrightarrow Ag^+(a = 0.001\ m)$. The equation for the potential, assuming the dilute solution is ideal is

$$\mathscr{E} = 0 - 0.0592 \log \frac{0.001}{x}$$

From this,

$$0.1110 = -0.0592 \log \frac{0.001}{x}$$

$$\log 0.001 - \log x = -(0.1110)/(0.0592) = -1.875$$

$$\log x = -1.125 = \bar{2}.875 \quad \text{and} \quad x = 0.075$$

$$\gamma = \frac{a}{c} = 0.75$$

7-3
TECHNIQUES OF POTENTIAL MEASUREMENT

If the result of potential measurement of a galvanic cell is to have meaning in terms of a driving force for a reaction or a measure of species concentration, the determination must be carried out under reversible conditions. The criterion for reversibility of an electrode pair is simply that, if the cell potential is opposed by an equal and opposite external potential, an infinitesimal decrease in the external potential allows the cell to discharge, whereas an infinitesimal increase causes current to flow through the cell in the opposite direction, with a consequent reversal of the chemical reaction in the cell. Some electrode systems that would be useful cannot be employed because reversible versions have not been devised. For example, no oxygen gas electrode of satisfactory performance has been devised.

Valid potential values also require measurement under conditions in which negligible current is drawn from the cell, in order to avoid a voltage drop through the cell solution from its resistance, as well as concentration gradients in the vicinity of the electrodes and polarization at the solid–solution interface, both of which contribute to the voltage.

The arrangement shown in Figure 7-3 is an example of a potentiometer circuit designed for measuring electric potentials while drawing only a minimum amount of current. Battery B supplies a voltage which is adjustable by means of the resistance R in series with it. There results a continuous and linear drop in potential from point a at one end of the slide-wire to point c at the other, and a fraction of this potential is tapped off by means of a sliding contact and opposed to the unknown X in the other branch of the circuit. When this opposing voltage is just equal to the voltage of X, no current flows through the potentiometer G. In this condition of balance, the potential drop from a to b via the slide-wire is exactly the same as the potential drop across X, with the result that both terminals of the galvanometer are at the same potential.

In measurements with this circuit, resistance R is first adjusted to null the galvanometer with the standard cell SC in the circuit and with the slide-wire set so that the potential dial attached to it reads the known voltage of the standard cell. The unknown is then switched into the circuit, and, without changing R, the galvanometer is brought to zero by adjusting the slide-wire setting. Key K_1 is in series with a protective resistance P and, when the bridge is far from balance, is closed momentarily to determine the direction in which adjustments should be made. After preliminary balancing, key K_2 is closed for the final null adjustment with maximum sensitivity.

A disadvantage of a potentiometer circuit of this sort is that a small current must be drawn from the unknown in order to provide an indication of unbalance on the galvanometer. Vacuum tube or transistorized electrometer circuits have been devised in which the input resistance is very high, and currents in the microampere range are more than adequate. Another way of understanding this problem is to realize that a device in series with the unknown, such as the galvanometer in the potentiometer circuit, if of low resistance, will have insuffi-

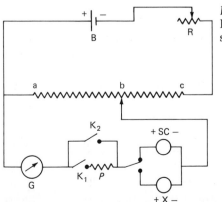

Figure 7-3
Potentiometer circuit, using a slide-wire, ac.

cient voltage across its terminals to drive it, and if of high resistance will have an appreciable voltage drop even for a small flow of current.

When potential measurements are directed toward determination of the concentrations or oxidation–reduction tendency of the components of one electrode, the circuit must be completed through a *reference* electrode of known characteristics. Such an electrode should be readily prepared in reproducible and reversible form, and that most commonly used is the calomel electrode, in which metallic mercury is in contact with solid mercurous chloride, which in turn is in contact with a solution containing chloride ion. The electrode reaction is

$$Hg_2Cl_2 + 2e^- \longrightarrow 2Hg + 2Cl^- \tag{7-19}$$

The only variable concentration is that of chloride ion, and reduction potential values at 25°C for various *stoichiometric* concentrations of potassium chloride are: saturated KCl, +0.2444 V; 1 M KCl, +0.2800 V; 0.1 M KCl, +0.3356 V.

Another frequently used electrode, sensitive also to the chloride ion concentration in solution, is the silver–silver chloride electrode. It has the advantage of ready fabrication in miniature form for insertion into small systems.

As a standard cell for voltage measurements, the Weston cell is almost invariably employed. The positive electrode is a layer of metallic mercury covered by a paste of mercurous sulfate and mercury, and the negative electrode contains cadmium amalgam and solid $CdSO_4 \frac{8}{3}H_2O$. The cell reaction is

$$Cd(s) + Hg_2SO_4(s) + \tfrac{8}{3}H_2O(l) \longrightarrow CdSO_4 \cdot \tfrac{8}{3}H_2O(s) + 2Hg(l) \tag{7-20}$$

and the voltage at 25°C is 1.01807 V. This equation and potential apply to the situation in which the electrolyte solution is saturated with respect to mercuric sulfate and cadmium sulfate; since the solubility changes with temperature, the cell potential has a high temperature coefficient. This is not true of the unsaturated Weston cell, which has a potential of about 1.0186 V at room temperature but which is not quite so reproducible.

For some oxidation–reduction reactions it is possible to set up a cell in which the two electrolyte solutions have the same composition. An example is the cell

$$H_2(g),\ HCl(aq),\ AgCl(s),\ Ag$$

in which there are electrodes reversible to each of the two ions of the solute. However, this would not be feasible if the anion were sodium ion, for instance. When the two electrodes involve different solutions, a potential difference is set up across the boundary between the solutions—the *liquid junction*.

To understand the source of the liquid junction potential, visualize the case in which the solution on the right-hand side of the junction contains a cation with higher mobility than the anion, such as 0.1 M hydrochloric acid. Suppose that the solution on the left side of the junction is 0.1 M sodium chloride, in which the anion has the higher mobility. In a conventional cell reaction, anions migrate to the left and cations migrate to the right. As the hydrogen ions move rapidly to the right, they tend to leave the right side of the junction negatively charged and, since the chloride ions move more rapidly in the left compartment than they do in the right, the left side of the junction acquires an excess of positive charge. No appreciable separation of charge can in fact occur, but the tendency to such a separation leads to a potential difference which reaches a value of over 30 millivolts for the case described. Generally, the most satisfactory procedure to minimize the effect of the liquid junction potential is to connect the two solutions by a bridge containing a salt, such as potassium chloride or ammonium nitrate, in which the transference numbers are very close to 0.5 for each of the ions.

7-4 OXIDATION–REDUCTION TITRATIONS AND INDICATORS

The value of the reduction potential in a particular solution is often quoted as a measure of the tendency of the solution to provide electrons to a reducible material. Equation (7-6) in the form

$$\mathscr{E} = \mathscr{E}^0 - \frac{0.0592}{n} \log \frac{[\text{reduced form}]}{[\text{oxidized form}]}$$

can be applied to the two components of an oxidation–reduction pair, that is, to a single electrode, and is parallel to the equation for the pH of a conjugate acid–base pair:

$$pH = pK_a + \log \frac{[\text{base}]}{[\text{acid}]}$$

These equations can be utilized from either of two viewpoints: (1) The reduction potential or the pH, as controlled by some outside agency, determines the ratio of the two forms present; or (2) the ratio of the two forms of the particular couple determines the reduction potential or the pH of the solution. If a curve of the reduction potential against the fraction of the total material of a couple in the reduced form is plotted, as in Figure 7-4, the slope varies with n, the number of electrons transferred each time the oxidation process occurs, in contrast to acid–base equilibria in which only a single proton is transferred in each process.

The sign convention used in this text determines how the electrode potential is related to the relative electron-donating tendency of the oxidation–reduction couple in the following way:

(1) The best oxidizing agent, which is the material most readily reduced, has the most positive potential.
(2) The best reducing agent, which is the material most readily oxidized, has the most negative potential.
(3) As the potential of a solution becomes more negative, the availability of electrons increases, the reducing ability increases, and more of any given couple is present as the reduced form.

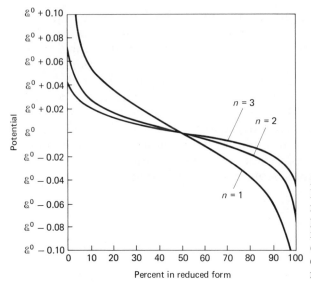

Figure 7-4
Relation between reduction potential and percent of the total material constituting an oxidation–reduction couple that is in the reduced form.

Titration of an oxidizing agent with a reducing agent can be carried out in a fashion analogous to acid–base titrations. The variation in potential throughout the titration can be followed by direct potentiometric measurement, using for example, a platinum wire and calomel reference electrode along with a potential mediator as described in Section 7-5 if needed, or the end point can be located by the change in color of an indicator that is sensitive to potential. During a titration, the potential variation curve approximates the superposition of two single-couple curves, one for each of the oxidation–reduction systems involved in the reaction.

If the two couples of which the reactants are members have standard potentials more than about 0.5 V apart, there is some value of the potential intermediate between the two standard potentials at which practically all of the more positive couple is in the reduced form and nearly all of the more negative couple is in the oxidized form. The titration involving these couples then has a sharp end point, as shown in Figure 7-5 for the addition of ferrous ion to a solution containing ceric ion.

An organic substance forming a reversible oxidation–reduction couple in which the two forms have different colors may be suitable for use to determine the reduction potential of a solution or to indicate the end point of an oxidation–reduction reaction. Just as for acid–base indicators, the amount of the *oxidation–reduction indicator* must be small enough not to disturb by its presence the potential to be measured. The requirements of stability and reversibility are more difficult to satisfy in oxidation–reduction systems than in acid–base systems and limit somewhat the availability of oxidation–reduction indicators suitable for titrations.

The iron complex of *o*-phenanthroline is one example of a useful indicator. The pale-blue ferric complex is reduced to a bright-red complex at a standard potential of $+1.14$ V, a value making it suitable for use with strong oxidizing agents such as ceric ion:

$$Fe(C_{12}H_8N_2)_3{}^{3+} + e^- \rightleftharpoons Fe(C_{12}H_8N_2)_3{}^{2+} \qquad (7\text{-}21)$$

Another example of an indicator system is diphenylbenzidine, which is violet in its oxidized form and colorless in its reduced form:

$$C_6H_5N{=}C_6H_4{=}C_6H_4{=}NC_6H_5 + 2e^- + 2H^+ \rightleftharpoons$$
$$C_6H_5NHC_6H_4C_6H_4NHC_6H_5 \quad (7\text{-}22)$$

In an oxidation–reduction reaction such as this, involving hydrogen ions, the observed potential depends on the pH of the solution as well as upon the ratio of the oxidized and reduced forms of the organic couple. A true standard potential would refer to unit hydrogen ion concentration or a pH of zero. However, this is a hydrogen ion concentration much greater than that usually present in systems where oxidation–reduction is being considered, and therefore standard potentials are often given for other specific pH values; these potentials are designated $\mathscr{E}^{0\prime}$. This amounts to incorporating the hydrogen ion concentration in $\mathscr{E}^0$, which is then a function of pH. It has become customary to associate the symbol $\mathscr{E}^{0\prime}$ with a pH of 7 whenever it is not

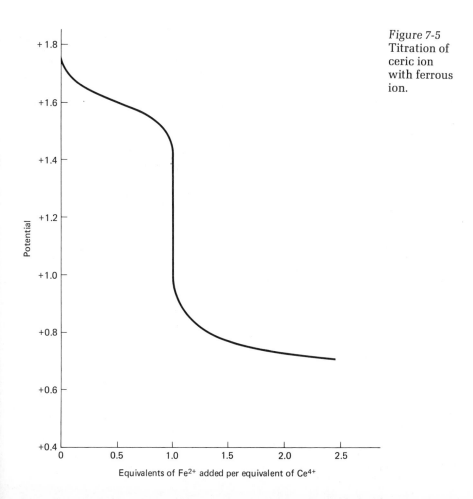

Figure 7-5
Titration of
ceric ion
with ferrous
ion.

Potential

Equivalents of Fe²⁺ added per equivalent of Ce⁴⁺

otherwise specified, and such an $\mathscr{E}^{0\prime}$ value for the system in Equation (7-22) is $+0.76$ V.

Indicators can be used to establish the potential of systems of biochemical interest, such as physiological fluids, living cells, and the respiratory enzymes described below. A very common example is methylene blue, which has a standard reduction potential of $+0.011$ V at pH 7. As a result of bacterial action, the reduction potential of milk decreases with time from $+0.25$ to -0.20 V, and the ability of a sample of milk to convert methylene blue from its blue oxidized form to its colorless reduced form is an indication that the sample has been allowed to age. The equilibrium for methylene blue can be written

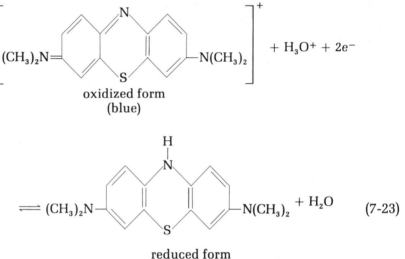

oxidized form
(blue)

reduced form
(colorless)

$$ (7\text{-}23) $$

However, the reduced form is in acid–base equilibrium with forms having one or two protons attached to the dimethylamino nitrogens.

Table 7-2
Some biochemical oxidation–reduction indicators

Substance	$\mathscr{E}^{0\prime}$ at pH 7 (V)
Methyl viologen	-0.45
Sulfonated rosindone	-0.380
Neutral red	-0.325
Safranine T	-0.289
Phenosafranine	-0.252
Cresyl violet	-0.173
Indigo trisulfonate	-0.081
Indigo tetrasulfonate	-0.046
Methylene blue	$+0.011$
Cresyl blue	$+0.047$
1-Naphthol-2-sulfonate- indophenol	$+0.123$
2,6-Dichlorophenol-indo- o-cresol	$+0.181$
Phenol-m-sulfonate-indo- 2,6-dibromophenol	$+0.273$

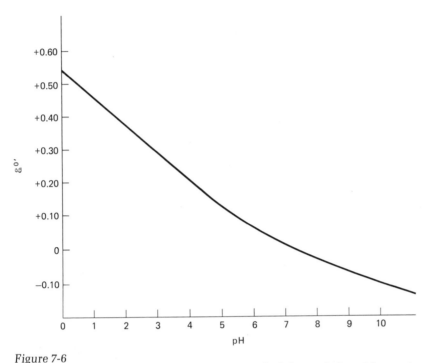

Figure 7-6
Dependence of the standard reduction potential of the methylene blue system on pH.

The more acid the solution, the larger the fraction of the reduced form that is protonated, and the smaller the concentration of the oxidizable base form. Consequently the reduction potential becomes less negative as the solution becomes more acidic. Standard potentials for the methylene blue system are shown as a function of pH in Figure 7-6.

Frequently in studying biological systems, it is more satisfactory to employ a series of indicators with fairly closely spaced potentials, each on a basis of the potential being higher or lower than $\mathscr{E}^{0\prime}$ of the indicator, rather than to attempt to match shades of color within the range of a single indicator. The standard reduction potentials in neutral solution are given for several indicators in Table 7-2.

7-5
CHARACTERISTICS OF ORGANIC OXIDATION– REDUCTION SYSTEMS

Some organic oxidation–reduction reactions take place in a substantially irreversible fashion. The products in combustion reactions, for example, cannot be readily reconverted to the reactants. Thus electrical measurements of the driving force of the reaction cannot be made, for equilibrium cannot be established. Of course, a driving force or reaction potential exists independently of whether the equilibrium point can be measured, and often it can be calculated from values of the thermodynamic functions for the materials involved.

Other organic reactions in which oxidation occurs can be made reversible under suitable conditions but are innately rather sluggish in coming to equilibrium. The reason seems to be that they involve the formation or rupture of a covalent bond, corresponding to the transfer of two electrons, and are often also accompanied by the transfer of one or two protons. Under biochemical conditions, an enzyme can serve as an agent to promote the electron-transfer reaction. Sometimes a more readily reversible oxidation–reduction couple, known as a *potential mediator,* is added to a sluggish system. In Table 7-3 are listed standard reduction potentials for a variety of organic systems, many of biological interest. In many of these couples, hydrogen ions are involved, so that the potential depends upon the pH of the solution as well as upon the concentration ratio of the organic components of the system.

To illustrate the points of the last paragraph, consider the system fumaric acid–succinic acid:

$$\begin{matrix} \text{CHCOOH} \\ \| \\ \text{CHCOOH} \end{matrix} + 2e^- + 2H^+ \rightleftharpoons \begin{matrix} \text{CH}_2\text{COOH} \\ | \\ \text{CH}_2\text{COOH} \end{matrix} \qquad (7\text{-}24)$$

For this system, the standard potential at pH 7 is $+0.026$ V. At pH 7 of course the two organic species are not present as neutral molecules, but instead as an equilibrium mixture of molecules and carboxylate ions. Accordingly, it is probably better to refer to this system as the

Table 7-3

Standard reduction potentials of some organic and biochemical oxidation–reduction systems

System oxidized, reduced	$\mathscr{E}_{25}^{0\prime}$ at pH 7.0 (V)
Acetate + CO_2, pyruvate	-0.70
Succinate + CO_2, α-oxoglutarate	-0.67
Acetate, acetaldehyde	-0.60
Plant ferredoxin (ox), plant ferredoxin (red)	-0.43
Acetyl CoA, acetaldehyde + CoA	-0.41
Pyruvate + CO_2, malate	-0.33
Acetone, propanol-2	-0.30
1,3-Diphosphoglycerate, glyceraldehyde 3-phosphate + phosphate ion	-0.29
Riboflavin (ox), riboflavin (red)	-0.21
Acetaldehyde, ethanol	-0.20
Pyruvate, lactate	-0.19
Oxalacetate, malate	-0.17
α-Oxoglutarate + ammonium ion, glutamate	-0.14
Rubredoxin (ox), rubredoxin (red)	-0.06
Fumarate, succinate	$+0.03$
Methemoglobin, hemoglobin	$+0.17$
Formaldehyde, methanol	$+0.19$
Oxygen, hydrogen peroxide	$+0.30$
High-potential ion protein (from *Chromatium*) (ox), HPIP (red)	$+0.33$

fumarate–succinate couple, rather than as an acid pair, but both terms are used. If a platinum electrode is dipped into a solution containing both components of the couple as well as the enzyme, a stable potential reading is not obtained. Addition of a potential mediator permits a reversible potential to be measured, and methylene blue serves very well as a potential mediator for this range of reducing power. As a mediator, methylene blue comes to equilibrium with the components of the organic oxidation–reduction system, so that the ratio of oxidized form to reduced form for the mediator is determined by the fumarate–succinate system; then the methylene blue in turn interacts with the metallic electrode to establish its potential.

There are a few reactions in which the addition of two electrons to the oxidized form has been shown to occur in two distinguishable steps. Most of these cases can be described as the reduction of a quinone-type molecule to a hydroquinone. Although an intermediate cannot be isolated in the reduction of unsubstituted quinone itself, we can formulate the general reaction sequence:

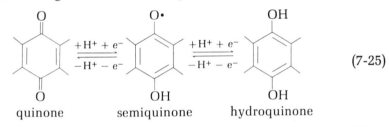

$$(7\text{-}25)$$

quinone   semiquinone   hydroquinone

The intermediate stage in stepwise oxidation or reduction differs from the terminal stages by the presence or absence of one electron. This stage, known as a semiquinone, may possibly be of rather general occurrence in organic oxidations, but, if it is present, its lifetime is not sufficiently long and therefore its concentration is not sufficiently great for it to be detected. The stability of the semiquinone for molecules in which there are two hydroxyl groups, or two amino groups, or one hydroxyl group and one amino group, substituted ortho or para to one another in an aromatic ring, seems to be the result of the fact that the odd electron is delocalized over various positions in the molecule.

Substances in the reduction of which semiquinones have been detected include various sulfonated anthraquinones and phenanthraquinones, α-oxyphenazine, tetramethyl-p-phenylenediamine, and the dye pyocyanine, for which the reduction steps are

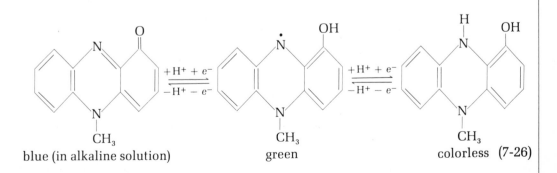

blue (in alkaline solution)   green   colorless   (7-26)

In view of the involvement of acid–base ionizable forms in the various stages of this reaction, a dependence of potential on pH is encountered here as for many other oxidation–reduction reactions; in this case, the oxidized form is protonated in acid solution, giving a pink color.

A potentiometric titration curve may indicate the presence of a semiquinone by a distinct break, or end point, when an amount of reagent equal to one-half that required for compete reduction has been added. This break is evident only if the particular semiquinone is relatively stable, and whether or not it is seen is sometimes determined by the pH of the solution. In certain systems, the appearance of a distinct color at an intermediate stage in the reaction is evidence for the presence of a semiquinone. Finally, a clear-cut proof for the existence of a semiquinone can be provided by measurement of the magnetic susceptibility of the solution during the course of the reaction, for the unpaired electron in the semiquinone makes the solution paramagnetic (see Chapter 13 for further discussion).

The structures of some of the oxidative enzymes mentioned in Section 7-6 are such as to indicate the strong possibility of formation of semiquinone intermediates in the course of their biological function. For example, a phosphopyridine nucleotide may lose two electrons in successive steps, with a semiquinone intervening, to two different enzyme–ferric ion complexes, reducing each to a ferrous complex; then the resulting quinone form of the phosphopyridine nucleotide may oxidize in a single step a species such as succinate which loses two electrons simultaneously. This mediation by the semiquinone, if it occurs, obviates the necessity for a three-body collision between two ferric complexes and the succinate ion and thus materially accelerates the reaction.

7-6
BIOCHEMICAL
OXIDATION

In living organisms, a major portion of the metabolic process consists of utilization of the free energy derived from the combination of oxygen with organic compounds to synthesize needed materials, to transport ions and molecules from place to place, to provide the means of muscular contraction, and so on. The end result of the "combustion" of organic materials with oxygen is the production of carbon dioxide and water, just as in the laboratory combustion of organic substances. However, the metabolism of the organism is not directed toward the production of heat, but toward the utilization of available energy for the functions mentioned, and thus the oxidation is carried out in a complicated sequence of intermediate steps, rather than in a single complete reaction. In Section 4-10, we mentioned the *respiratory chain,* also referred to as the *electron-transport* chain, a series of reversible oxidation–reduction systems, arranged in the order of their reduction potentials.

In higher organisms, the electron carriers of the respiratory chain are located in structural units of the cell called mitochondria. One cell

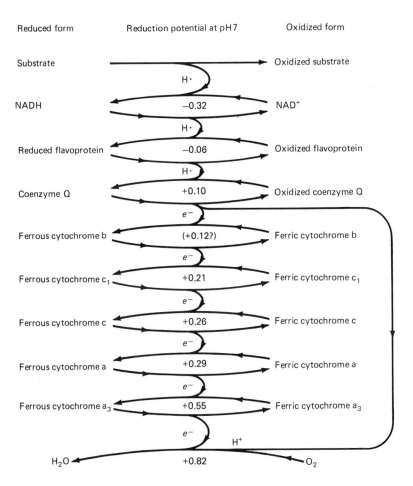

Reduced form	Reduction potential at pH7	Oxidized form

Substrate → Oxidized substrate

H·

NADH −0.32 NAD⁺

H·

Reduced flavoprotein −0.06 Oxidized flavoprotein

H·

Coenzyme Q +0.10 Oxidized coenzyme Q

e⁻

Ferrous cytochrome b (+0.12?) Ferric cytochrome b

e⁻

Ferrous cytochrome c₁ +0.21 Ferric cytochrome c₁

e⁻

Ferrous cytochrome c +0.26 Ferric cytochrome c

e⁻

Ferrous cytochrome a +0.29 Ferric cytochrome a

e⁻

Ferrous cytochrome a₃ +0.55 Ferric cytochrome a₃

e⁻   H⁺

H₂O ← +0.82 ← O₂

Figure 7-7
Typical sequence in the electron-transport chain of a mitochondrion.

can contain from 10 to several hundred of these ellipsoidal units, about a micrometer (μm) in size. Each mitochondrion includes a complicated membrane system comprising a smooth outer surface and a highly folded inner surface projecting like a series of shelves into the interior soluble matrix and bearing several thousand knoblike protuberances. The large surface area provided by this arrangement is evidently directed toward allowing accessibility of the membrane to various chemical reactants, for attached to the membrane in a definite pattern are the components of the respiratory chain. The space within the mitochondrion also contains several other enzyme systems related to energy production and utilization, including those of the Krebs citric acid cycle.

By no means all of the details of the electron-transport chain have been worked out, but the general scheme is reasonably well understood. A sequence that may represent the oxidation reactions occurring in a typical mitochondrial oxidation is shown in Figure 7-7, together with values for the standard reduction potentials for the couples involved. The reader should recognize that these values are for pH 7 and a temperature of 25°C, and that the actual potential in the cell depends upon the ratio of oxidized and reduced forms present. Accordingly the potential values under physiological conditions may differ somewhat from those given. However, the respiratory chain must represent a progression from strongest reducing system to strongest oxidizing system. For any stage represented in the diagram, electrons (sometimes in the form of hydrogen atoms) are transferred from the reduced form of one system to the oxidized form of the next couple in sequence down the chain. In this transfer, the species losing electrons is oxidized, only to gain electrons later on from the system one step more negative in potential along the chain, while the species gaining electrons is reduced and will later transfer the electrons down the chain to the next level of more positive potential.

Let us now consider individually some of the elements in the respiratory chain. The substrates may be species such as pyruvate, isocitrate, α-ketoglutarate, or malate from the Krebs cycle described in Section 4-10, or other materials such as lactate, glutamate, or glucose 6-phosphate. Hydrogen atoms are transferred from one of these to nicotinamide adenine dinucleotide (NAD, formerly known as diphosphopyridine nucleotide, DPN, or coenzyme I), a molecule in which the component parts are linked together in the following sequence:

Nicotinamide—ribose—phosphate—phosphate—ribose—adenine

In the reduced form of NAD, the nicotinamide ring has been converted into a dihydropyridine ring.

In the next step of the series, the reduced form of NAD transfers hydrogen to one of several flavoproteins, substances originally called the yellow enzymes. Each consists of a protein portion and an accompanying *prosthetic* group, the group active in the electron-transfer reaction, which contains riboflavin phosphate. In most of the flavoproteins this active group is alloxazine adenine dinucleotide, also called flavin adenine dinucleotide (FAD):

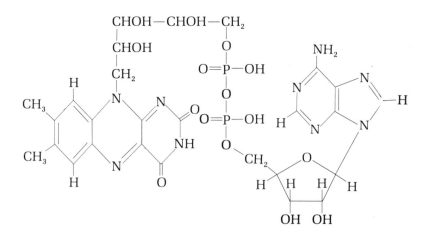

When this ring system is reduced, two hydrogen atoms are added across the conjugated $N=C—C=N$ unit in the alloxazine ring.

Following the flavoprotein is a species called coenzyme Q or ubiquinone, typically with the following formula, although the length of the side chain may vary with the source of the quinone:

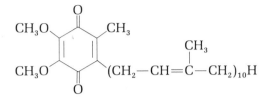

From coenzyme Q, the electrons probably flow to the first of a series of *cytochromes*, cytochrome b, although the oxidation potentials are so close together that some workers feel that cytochrome b precedes coenzyme Q. The cytochromes are a series of enzymes each containing as its active group a heme group—a substituted porphyrin ring coordinated to a central iron atom which may be in the ferrous or ferric oxidation state—attached to a protein. Each of the cytochromes has a characteristic optical absorption spectrum. The electrons are next passed along to cytochrome c_1, then to cytochrome c, then to cytochrome a, and finally to cytochrome a_3. The last of these, originally called the "respiratory enzyme" or cytochrome oxidase, is the only system that can react directly with oxygen. It is interesting that the inhibition of respiration by cyanide ion or carbon monoxide has been associated with their ability to change the reduction potential of cytochrome a_3 so that it cannot function normally.

Cytochrome iron atoms accept only electrons, not the accompanying hydrogen nuclei. At the coenzyme-Q stage, these are set free as hydrogen ions, H_3O^+, which are then finally consumed in the reduction of oxygen in the final step.

At this point, it may be wise to reemphasize the rule that the direction of electron flow is from systems of more negative reduction potential to those of higher positive potential. Many systems other than those described here are involved in oxidation–reduction processes; as they are discovered, however, each must be fitted into the overall scheme at a position appropriate to its potential.

How is the energy in the respiratory chain utilized? Per pair of electrons traversing the chain, three molecules of ADP are phosphorylated to ATP. The overall potential difference from NAD to oxygen is 1.14 V, corresponding to an energy of 2(23,060)(1.14) or 52,000 cal. The addition of each phosphate group to ADP requires about 7600 cal, so that in the process of oxidative phosphorylation about 23,000 out of 52,000 cal is utilized.

Efforts to break down the respiratory chain into smaller segments have led to separation of the systems described above into three complexes, each of which is able to carry on electron-transfer activity independently and each of which evidently exists as a unit in the mitochondrion. Apparently each complex is associated with the phosphorylation of one of the three molecules of ADP. There is, in addition, a fourth complex, which accepts electrons from succinate and delivers them to the cytochromes, independently of the NAD route. Finally, the molecules cytochrome c and coenzyme Q are fairly easily extracted from the mitochondrial material independent of the complexes and

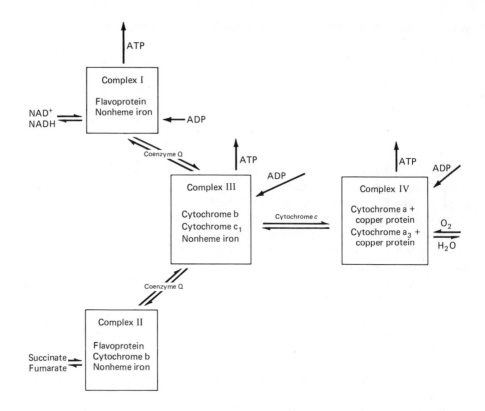

Figure 7-8
Component units of the respiratory chain in the mitochondrion.

thus can be viewed as mobile carriers which transport electrons from one complex to another.

Figure 7-8 is a schematic representation of the respiratory chain, based on the description developed by D. E. Green and co-workers at the University of Wisconsin. Much of the research on which this scheme is based was done on mitochondria from beef heart muscle, in which there have been found three complexes of type IV along with one each of type I, II, and III.

In addition to the features described above, complexes I, II, and III seem to include iron-containing proteins in which each iron atom is surrounded by four sulfur atoms from the amino acid cysteine. This structure is often referred to as *nonheme iron* in obvious distinction to what was earlier considered the principal form in which iron was found in organisms. The existence of such iron–sulfur proteins has been recognized only since about 1960, chiefly because of their high lability, and they occur frequently in the form of bacterial-type and plant-type ferredoxins as well as in animal mitochondria; their role in electron transport, however, is not yet well understood.

In complex IV, the cytochromes seem to be associated with copper-containing proteins; since copper can be in either cuprous or cupric form, it is possible that these proteins may in some way be involved in the oxidation–reduction processes. Finally, an important constituent of each type of complex is phospholipid, the material from which are built the structures of various kinds of membranes (Chapter 12). It appears that here the phospholipid has some important part in the oxidation–reduction scheme aside from simply providing support for the electron–transport systems, but it remains to be established what its full function is.

7-7
POTENTIOMETRIC DETERMINATION OF ION CONCENTRATION

Extensive effort has gone into the development of accurate and convenient methods for the measurement of pH in solutions. From this has come the glass electrode, a barrier across which the potential depends primarily on hydrogen ions. As a recent outgrowth of the investigation of barrier electrodes, a variety of electrodes selective to particular ions as a consequence of a particular structural design has been devised.

MEASUREMENT OF pH

Any electrode whose potential depends upon the hydrogen ion concentration can be considered for use in the determination of pH. The electrode, along with a reference electrode, need merely be dipped into the solution under study, and the potential measured. Choice of a hydrogen-sensitive electrode is limited by the nature of the reactions that can occur between the electrode components and constituents of the solution, by the effects of concentrations other than that of hydrogen ion on the electrode potential, and by the mechanical requirements involved in setting up the electrode.

The ultimate standard for all potential measurements is the hydrogen electrode, in which a stream of hydrogen gas is bubbled over a piece of platinum foil partially immersed in the solution, as shown in Figure 7-9.

The potential of this electrode is given by the equation

$$\mathscr{E} = -2.303RT \log \frac{P_{H_2}^{1/2}}{[H_3O^+]}$$

$$= -\frac{2.303RT}{2} \log P_{H_2} - 2.303RT \, pH \qquad (7\text{-}27)$$

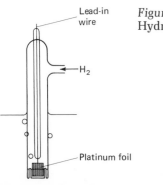

Lead-in wire

H_2

Platinum foil

Figure 7-9
Hydrogen electrode.

Aside from the dependence on hydrogen pressure, the potential is linear in pH. In fact, this equation can be taken as the defining equation for pH, since it relates the latter to a measurable quantity, the potential of a galvanic cell.

The hydrogen electrode cannot be used if the solution contains easily reducible substances or volatile materials which would be swept out by the stream of hydrogen gas.

Although other electrodes, such as one involving quinone and hydroquinone, have been utilized under certain conditions for pH measurement, by far the most convenient and versatile pH electrode is the *glass electrode*. A thin glass membrane is placed between the solution to be measured and a reference solution. Reference electrodes are then placed in each of the solutions. A potential is developed across the membrane, the magnitude of which is dependent upon the difference in hydrogen ion concentration in the two solutions bathing the membrane. At 25°C, the potential is

$$\mathscr{E} = \mathscr{E}^0 - 0.0592 \text{ pH} \tag{7-28}$$

where $\mathscr{E}^0$ is constant for a given glass membrane and pair of reference electrodes.

Usually the glass electrode is set up in the form of a glass bulb, sealed to the end of a glass tube. Inside the tube is placed one of the reference electrodes, often a silver–silver chloride electrode dipping into a solution of fixed pH, usually 0.1 N hydrochloric acid. The glass bulb and tube are then dipped into the unknown solution along with the other reference electrode, almost invariably a calomel electrode. The glass electrode has the advantage that it can be used without regard to the presence of oxidizing agents or poisons. Furthermore, it does not contaminate the sample or disturb poorly buffered solutions. It can be made in very small sizes suitable for use with microscale samples. A diagram is shown in Figure 7-10.

The composition of the glass of which the electrode is made is critical for the existence of a suitable pH response, which depends upon the exchange of hydrogen ions between the solution and a gel layer at the surface of the glass. The mechanical condition of the glass has some effect upon the potential developed and, since the membrane is fabricated in the form of a bulb, the condition on the two sides of the glass is not the same. For each electrode there is found to be an *asymmetry potential* which is compensated for by standardizing the

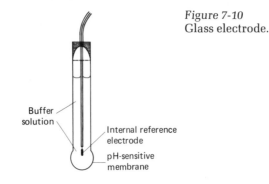

Figure 7-10
Glass electrode.

Buffer
solution

Internal reference
electrode

pH-sensitive
membrane

electrode against a buffer solution of known pH before use. New electrodes require soaking for a period of several hours in water or dilute acid before their response becomes steady.

Because the resistance of the glass electrode is very high, possibly 10^6 to 10^9 ohms, an ordinary potentiometer is of no value for determination of the potential. However, compact and dependable circuits have been developed which have a very high input resistance.

In very strongly acidic or basic solutions deviations from the linear relation of potential to pH are sometimes observed and must be corrected for, although special glasses for these extreme conditions which minimize the need for corrections can be obtained. Attempts to apply the glass electrode to measurements of acidity in nonaqueous solutions have been made. However, the results are often erratic, and their meaning may be difficult to establish because the proton may not be present in hydrated form.

It is often desirable to be able to determine the acid concentration of a solution of D_3O^+ in D_2O, in other words, to measure the pD instead of the pH. The primary standard here would be an electrode of D_2 gas with a piece of platinum making contact with the solution. However, it has been found by experiment that a satisfactory value for the pD of a solution can be obtained by taking a reading on an ordinary pH meter with a glass electrode containing a reference solution of H_3O^+ and then adding 0.40 to that pH reading.

ION-SELECTIVE ELECTRODES

As we have seen, the most convenient method of determining the activity of hydrogen ions is not by an electrode involving the element hydrogen—as hydrogen gas—and the hydrogen ion in solution, but by means of a concentration cell set up across a barrier formed by a glass membrane selective for the exchange of hydrogen ions. There are other systems for which the use of a simple element—ion electrode is difficult or impossible: An alkali metal is not suitable in the pure state, for it would react with water, although sometimes such a metal is employed as an amalgam, in mercury solution. Many anions present difficulties with respect to the availability of a suitable oxidation state to form a couple, or to the reversibility of an electrode system if one is available. In recent years, considerable effort has been expended, and some success has been achieved, in the development of barrier electrodes suitable for determination of the concentration of specific ions other than hydrogen.

Experience with early glass electrodes showed that they often developed potential difference contributions from singly charged cations other than the hydrogen ion, such as potassium and sodium ions, particularly in strongly basic solutions, and corrections had to be made for these errors. Experimentation with glass composition aimed at understanding these errors and minimizing them led to the discovery that it is possible to prepare materials selective for any one of the alkali metal ions. The degree of selectivity is not complete; that is, for most ions it is not possible to find an electrode for which the potential is completely independent of pH or of the concentrations of ions other than the one being measured, but many electrodes can be made adequate for various practical analytical requirements.

The surface of the membrane is considered to act as an ion exchanger, with the potential difference depending upon the selectivity of the surface sites for the binding of particular ions, modified by the relative mobility of the several ions through the glass. One can anticipate that a glass will have a strong tendency to bind one ionic species, only to have this advantage of favorable selectivity offset to some degree by the low mobility of that ion through the solid, simply because of a large affinity of the ion for the solid. A glass is made by fusing together an oxide of a metal in a valence state of $+3$ or $+4$, such as Al_2O_3, B_2O_3, or SiO_2, with another oxide of lower valence, such as Na_2O, K_2O, or Li_2O. Of these components, it is the low-valence cation that can migrate through the glass.

Other types of solid-state membranes that produce selective potential differences have been prepared from materials that exhibit ionic conductivity. One example is lanthanum fluoride, in which charge transport occurs entirely by the migration of fluoride ions. Thus, so long as lanthanum ions are absent from the solution, the potential difference across the electrode is determined by the fluoride ion concentration. Perhaps more surprisingly, in the absence of fluoride ions, the electrode can be used to measure the concentration of lanthanum ions, for the fluoride concentration at the surface is related to the lanthanum ion concentration in the solution by the solubility product of lanthanum fluoride.

Yet another type of barrier electrode contains a liquid phase constrained in some way to form a membrane. Possible arrangements include supporting the liquid in the interstices of a porous disk of inert material, confining it by a cellulose acetate film, or immobilizing it in a gelatinous matrix of agar or collodion. The components of the liquid phase are a water-immiscible organic solvent, and a solute that is insoluble in water but contains an ionizable functional group which can serve as the ion exchanger for ions of the opposite charge type. Examples are aliphatic amines, which are selective for certain anions, and long-chain phosphorus esters, which are selective for calcium. Nickel and ferrous complexes of phenanthroline have been found applicable as anion exchangers. Suitable solutes also include neutral species which selectively complex certain ions. For instance, the antibiotics nonactin and valinomycin have been used in electrodes specific for potassium ion, the former in solution in a mixture of Nujol and 2-octanol, and the latter dissolved in diphenyl ether. It is significant that in liquid membranes there is no effect of differential mobility, and only the complexing or ion exchange ability of the medium determines the relative transport capability for ions.

A particularly interesting electrode utilizes the specificity of an enzyme, urease, which is contained within a layer of polyacrylamide surrounding a glass electrode. When the electrode is dipped into a solution containing urea, the enzyme produces ammonium ions from urea, and these diffuse to the glass surface where they determine the potential.

EXERCISES

1. Write the cell reaction for each of the following cells, predict the potential at 25°C, and indicate whether or not the cell reaction is spontaneous:
   (a) Hg, Hg^{2+}(0.5 m) || Ag^+(0.001 m), Ag
   (b) Ag, AgCl, KCl(0.01 m), Cl_2(0.5 atm), Pt
   (c) Pt, Fe^{3+}(0.05 m), Fe^{2+}(0.01 m) || I^-(0.02 m), I_2(0.001 m), Pt
   (d) Sn, Sn^{2+}(0.03 m) || H^+(pH = 3), H_2(g, 1 atm), Pt

2. Determine the activity of hydrogen ions in the following cell, for which the measured potential at 25°C is −0.3956 V:

   Hg, Hg_2Cl_2, KCl(0.1 M) || HCl(a = ?), H_2(1 atm), Pt

3. State in words why the sign of the potential in Equation (7-18) is consistent with the expected direction of the spontaneous process in the cell.

4. Given the following reactions, set up an electrochemical cell corresponding to each:
   (a) $Fe + Cd^{2+} \rightleftharpoons Fe^{2+} + Cd$
   (b) $Cu^{2+} + \frac{1}{2}H_2 \rightleftharpoons H^+ + Cu^+$
   (c) $2MnO_4^- + 10Cl^- + 16H^+ \rightleftharpoons 5Cl_2 + 2Mn^{2+} + 8H_2O$
   (d) $2Ag + Cu^{2+} \rightleftharpoons Cu + 2Ag^+$
   (e) $2Ce^{4+} + 2Br^- \rightleftharpoons Br_2 + 2Ce^{3+}$

5. Calculate the equilibrium constant at 25°C for each reaction in Exercise 4.

6. At 25°C, a hydrogen electrode with gas bubbling through the aqueous medium out to an atmosphere of 730 torr pressure is found to be 0.5150 V negative with respect to a 1 M calomel electrode. Calculate the pH of the solution into which the electrodes are dipping.

7. From the data in Table 7-3, calculate the equilibrium constant for each of the following reactions at pH 7:
   (a) $O_2 + CH_3OH \rightleftharpoons HCHO + H_2O_2$
   (b) Pyruvate + succinate $\rightleftharpoons$ lactate + fumarate
   (c) Acetyl coenzyme A + malate $\rightleftharpoons$ acetaldehyde + oxalacetate + coenzyme A

8. Predict the effect of changing pH over the range from 1 to 4 on the potential of the fumaric acid−succinic acid couple.

9. From the standard potentials of the silver−silver chloride electrode and the silver−silver ion electrode, calculate the solubility product of silver chloride.

10. From the standard potentials of the hydrogen and oxygen electrodes and the standard free energy of formation of liquid water at 25°C, calculate the ion product of water.

11. Write out an equation for the equilibrium between oxaloacetate and lactate on the one hand, and malate and pyruvate on the other, showing the formulas of the species involved. From the data in Table 7-3, predict the extent of reaction occurring when 0.2 mol each of malate and pyruvate is mixed with 0.05 mol each of oxaloacetate and lactate at pH 7, assuming the presence of a suitable catalyst or mediator to permit attainment of equilibrium.

12. Choose an indicator that might be suitable for
   (a) Determination of the equilibrium between flavoprotein and cytochrome c
   (b) Determination of the potential in a system containing plant ferredoxin and riboflavin

13. Equation (4-48) relates the enthalpy change of a process to the temperature dependence of the equilibrium constant. Derive from this equation a relation that can be used to calculate the enthalpy change for a cell reaction from the temperature coefficient of the potential.

14. The potential of the following cell is 0.4970 V at 20°C, 0.4984 V at 25°C, and 0.5000 V at 30°C:

   Pt, H_2(1 atm), HCl(0.005 m), AgCl(s), Ag

   Calculate the enthalpy change for the cell reaction using the result from Exercise 13.

15. Calculate ΔG^0 and ΔS^0 for the reaction in Exercise 14 from the data given there.

16. Calculate the value of ΔG^0 at pH 7 for the oxidation by nicotinamide adenine dinucleotide of glutamate to α-oxoglutarate plus ammonium ion. Each of the half-reactions is a two-electron process.

17. What ratio of concentrations of oxidized and reduced cytochrome c is found at equilibrium in a solution containing equal concentrations of oxidized and reduced cytochrome a?

REFERENCES

Books

R. G. Bates, *Determination of pH: Theory and Practice,* 2nd ed., Wiley, New York, 1973. Includes extensive discussion of electrometric methods.

J. O'M. Bockris and Z. Nagy, *Electrochemistry for Ecologists,* Plenum Press, New York, 1974.

R. P. Buck, "Potentiometry: pH Measurements and Ion-Selective Electrodes," in *Techniques of Chemistry,* A. Weissberger and B. W. Rossiter, Eds., Vol. 1, Part IIA, Chapter 2, Wiley-Interscience, New York, 1971.

Henry B. Bull, *An Introduction to Physical Biochemistry,* 2nd ed., Davis, Philadelphia, 1964. Chapter 4 deals with oxidation–reduction.

W. Mansfield Clark, *Oxidation–Reduction Potentials of Organic Systems,* Williams and Wilkins, Baltimore, 1960. An introduction to both principles and techniques, with an extensive compilation of data, especially for physiologically important substances.

Richard A. Durst, Ed., *Ion-Selective Electrodes,* Special Publication 314, U.S. Government Printing Office, Washington D.C., 1969.

George Eisenman, Ed., *Glass Electrodes for Hydrogen and Other Cations,* Dekker, New York, 1967.

David E. Green and Robert F. Goldberger, *Molecular Insights into the Living Process,* Academic Press, New York, 1967. Chapter 9 describes processes in the mitochondrion and the units of the electron-transfer chain.

H. A. Lardy and S. M. Ferguson, "Oxidative Phosphorylation in Mitochondria," in *Annual Review of Biochemistry,* Vol. 38, Annual Reviews, Palo Alto, Calif., 1969. A critical summary of results in the literature.

W. M. Latimer, *Oxidation Potentials,* 2nd ed., Prentice-Hall, Englewood Cliffs, N.J., 1952. Extensive data on inorganic systems.

A. L. Lehninger, *Bioenergetics,* W. A. Benjamin, Menlo Park, Calif., 1965. A very readable account of energy transport and electron-transfer processes in living organisms.

Jürg Waser, *Basic Chemical Thermodynamics,* W. A. Benjamin, Menlo Park, Calif., 1966. Chapter 7 gives a detailed introduction to oxidation–reduction potentials and includes problems, some with answers.

S. Wawzonek, "Potentiometry: Oxidation–Reduction Potentials," in *Techniques of Chemistry,* A. Weissberger and B. W. Rossiter, Eds., Vol. 1, Part IIA, Chapter 1, Wiley-Interscience, New York, 1971.

Journal Articles

D. N. Bailey, Owen A. Moe, and J. N. Spencer, "The Relationship between Cell Potential and Half-Cell Reactions," *J. Chem. Educ.* **53**, 77 (1976).

J. O'M. Bockris, "Overpotential, a Lacuna in Scientific Knowledge," *J. Chem. Educ.* **48**, 352 (1971).

T. P. Chirpich, "Electrochemistry in Organisms," *J. Chem. Educ.* **52**, 99 (1975).

Richard E. Dickerson, "The Structure and History of an Ancient Protein—Cytochrome," *Sci. Am.* **226**, 58 (April 1972).

Richard A. Durst, "Ion-Selective Electrodes in Science, Medicine, and Technology," *Am. Sci.* **59**, 353 (1971).

Richard A. Durst, "Mechanism of the Glass Electrode Response," *J. Chem. Educ.* **44**, 175 (1967).

Robert B. Fischer, "Ion-Selective Electrodes," *J. Chem. Educ.* **51**, 387 (1974).

T. Førlund, L. U. Thulin, and T. Østvold, "Concentration Cells with Liquid Junction," *J. Chem. Educ.* **48**, 741 (1971).

Earl Frieden, "The Biochemistry of Copper," *Sci. Am.* **218**, 102 (May 1968).

William L. Jolly, "The Use of Oxidation Potentials in Inorganic Chemistry," *J. Chem. Educ.* **43**, 198 (1966).

J. Kirschbaum, "Biological Oxidations and Energy Conversion," *J. Chem. Educ.* **45**, 28 (1968).

Richard M. Lawrence and William H. Bowman, "Electrochemical Cells for Space Power," *J. Chem. Educ.* **48**, 359 (1971).

S. J. Lippard, "Iron–Sulfur Coordination Compounds and Proteins," *Acc. Chem. Res.* **6**, 282 (1973).

L. Michaelis, "Occurrence and Significance of Semiquinone Radicals," *Ann. N.Y. Acad. Sci.* **40**, 39 (1940).

Richard A. Pacer, "Conjugate Acid–Base and Redox Theory," *J. Chem. Educ.* **50**, 178 (1973).

G. A Rechnitz, "New Directions for Ion-Selective Electrodes," *Anal. Chem.* **41,** 109A (1969).

G. A. Rechnitz, "Membrane Electrode Probes for Biological Systems," *Science* **190,** 234 (1975).

John H. Riseman, "Electrode Techniques for Measuring Cyanide in Waste Waters," *Am. Lab.* p. 63 (December 1972).

Peter A. Rock, "The Design of Electrochemical Cells without Liquid Junction," *J. Chem. Educ.* **47,** 683 (1970).

W. H. Slabaugh, "Corrosion," *J. Chem. Educ.* **51,** 218 (1974).

N. Sutin, "Electron Transfer in Chemical and Biological Systems," *Chem. Brit.* **8,** 148 (1972).

Ashok K. Vijh, "Electrochemical Principles Involved in a Fuel Cell," *J. Chem. Educ.* **47,** 680 (1970).

Colin A. Vincent, "Thermodynamic Parameters from an Electrochemical Cell," *J. Chem. Educ.* **47,** 365 (1970).

David F. Wilson, P. Leslie Dutton, Maria Erecinska, J. Gordon Landsay, and Nobuhiro Sato, "Mitochondrial Electron Transport and Energy Conservation," *Acc. Chem. Res.* **5,** 234 (1972).

Eight
Electromagnetic Radiation and the Structure of Atoms

In most of the preceding material in this book, we have been concerned with matter in bulk. The principles of thermodynamics developed in earlier chapters apply to aggregates containing many molecules. Furthermore, we have been content to apply these principles to systems at equilibrium. Without very much regard to the detailed structure of individual molecules, we have treated phase equilibria, oxidation–reduction equilibria, and acid–base equilibria. At this point, our emphasis is somewhat altered as we turn to a consideration of chemical phenomena on an atomic and molecular scale, indeed on an electronic scale, from a viewpoint in which we examine details of atomic and molecular arrangement, bonding and valence forces between atoms, the tools by which molecular structure can be probed, and dynamic processes involving changes in bonds between atoms. The principles presented apply as well to the structure and behavior of the chemical constituents of living cells.

A powerful means for investigating the structure of matter, as well as a source of energy for physical and chemical changes, is provided by electromagnetic radiation. The most familiar example of this radiation, and the one longest known to humanity, is visible light, but electromagnetic radiation also includes many other forms, such as radio waves, microwaves, infrared radiation, ultraviolet radiation, and x rays. In the first part of this chapter, the nature of electromagnetic radiation is examined, and ways of visualizing and representing its behavior are described. We shall find that under certain circumstances radiation displays a particle or quantum character, while under other conditions it exhibits the characteristics of waves. This dual nature of radiation is paralleled by a duality in the behavior of material particles —electrons, protons, and neutrons—which may sometimes profitably be regarded as wavelike. With these principles as a basis, the structure of atoms is considered, laying the foundation for the discussion of bonding in molecules and of some of the kinds of interaction of radiation with molecules that follows in Chapter 9. Many recent developments in molecular biology are based on the application of specific knowledge of the structure and geometry of molecules growing out of these methods.

8-1
WAVE CHARACTER OF ELECTROMAGNETIC RADIATION

Visible light was the first form of electromagnetic radiation to be recognized and investigated. Speculation concerning the nature of light came from such scientists as Isaac Newton and Robert Hooke in England, Rene Descartes and Pierre Laplace in France, and Christiaan Huyghens in Holland during a period beginning in the seventeenth century and extending through the early part of the nineteenth century. Differences in views between those who preferred a model of light depicting it as a stream of particles and those who preferred a model representing it as a continuous series of waves resulted in vigorous, even bitter, controversies. Later developments demonstrated that each of these descriptions of light is partially correct.

Beginning in the latter portion of the nineteenth century, physicists discovered other forms of radiation, all traveling with the same velocity in a vacuum as light. This velocity, usually denoted by the symbol c, is approximately 2.998×10^8 m/sec, a value that can be rounded off for most purposes to 3×10^8 m/sec. In addition to the common velocity, other types of behavior shared by all forms of radiation include the ability to be refracted or diffracted under suitable circumstances. These characteristics, which are described in the following pages, can be explained readily on the basis of the wave model. Another way in which the wave nature of radiation is manifested is the cyclic fluctuation in electrostatic force experienced by an electron or proton as the radiation passes the place where the charged particle is located. Now the *electric field* at a point in space is by definition equal in magnitude and direction to the force felt by a unit positive charge when it is placed at that point. Thus the wave character of electromagnetic radiation is associated with the fact that it includes an electric field which fluctuates with time at any point and, at any instant, fluctuates with location in space.

The electric field of a beam of radiation, and therefore the resulting force felt by a charged particle, is transverse, that is, perpendicular to the direction in which the ray is moving. To visualize the situation better, imagine a series of ripples going along the surface of a pond. An instantaneous cross section of the surface would resemble the curve shown in Figure 8-1. As the ripples travel along, the elevation of the surface at any particular location on the pond fluctuates periodically. Thus a cork floating on the surface bobs up and down, executing motion *perpendicular* to the direction in which the ripples are moving. The periodic rise and fall of the cork can be represented as in Figure 8-2 by a plot of the vertical displacement from the normal or average position as a function of time. Transverse waves are to be contrasted with longitudinal waves, such as those that carry sound through a gas or a

Figure 8-1
Cross section of a water surface along which a train of waves is traveling. The broken line shows the normal or "equilibrium" level of the water surface. The diagram is a plot of displacement as ordinate against location in space as abscissa, at some particular value of time.

liquid: In the latter type of wave, molecules vibrate in the direction *parallel* to the direction of propagation of the sound, and they convey the energy of the sound wave by collisions with molecules ahead of and behind them along the path.

The fluctuating electric field of light is accompanied by a fluctuating magnetic field, perpendicular both to the direction of travel of the ray and to the direction of the electric field, as shown in Figure 8-3. In a beam of light emitted by a typical source, the directions of the electric field vibrations of the various components of the beam are tilted at a great many different angles about the axis of the beam. Rather than attempting to depict the behavior of the entire ray, it is simpler to describe only the component of the beam oscillating in one particular plane, and the equations and diagrams given in most of this chapter are to be understood as representing one such selected component.

We now present several alternative quantities by which the various forms or "colors" of electromagnetic radiation can be distinguished and quantitatively described. The first of these is the frequency ν which is equal to the number of waves passing a fixed point per second, as well as to the number of waves in the distance traveled by the ray in 1 sec. If a train of waves is generated by the vibrations of an electric charge, such as that on an electron, about an equilibrium position, the frequency is then equal to the number of oscillations executed by the source in 1 sec.

The frequency is usually quite a large number—the distance traveled by the wave in 1 sec is of the order of 3×10^{10} cm—and the same information is conveyed by stating the value of the *wave number* $\tilde{\nu}$ which is defined as the frequency divided by the velocity in a vacuum:

$$\tilde{\nu} = \frac{\nu}{c} \tag{8-1}$$

The physical significance of the wave number is that it is equal to the number of cycles or wavelengths that would be found in a length of 1 cm of the ray in a vacuum. For infrared and visible radiation, the

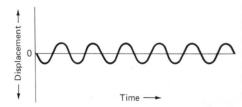

Figure 8-2
Plot of the vertical displacement of a cork floating on the surface of water as a traveling wave passes by the location of the cork. The equilibrium position of the cork is taken to be zero displacement.

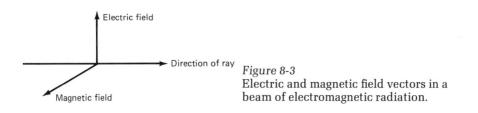

Figure 8-3
Electric and magnetic field vectors in a
beam of electromagnetic radiation.

wave number is especially convenient, for its numerical value is within the general region of 10 to 10,000 cm^{-1}.

The *wavelength* is another parameter frequently used to describe radiation and is often measured by experiment more directly than the frequency. Represented by the symbol λ, the wavelength is the spatial extent, in the direction of the ray, of a single cycle or wave of radiation, as shown in Figure 8-4. Obviously the wavelength in centimeters in a vacuum is the reciprocal of the wave number:

$$\lambda = \frac{1}{\tilde{\nu}} \tag{8-2}$$

Consistent with Equations (8-1) and (8-2) is the relation between wavelength and frequency

$$\lambda \nu = c \tag{8-3}$$

where λ and c are expressed in the same units of length. This equation can also be written down directly from the fact that the velocity of a wave train is equal to the number of waves passing a given point per second multiplied by the length of each wave, just as the rate at which a column of marching people moves forward is equal to the number of rows of people passing an observer in unit time multiplied by the distance between two successive rows.

If a radiation source oscillates at a frequency ν in simple harmonic motion—the type of motion required to generate a sinusoidal wave—the displacement of the source from its equilibrium position can be described by the equation

$$\phi = A \sin 2\pi \nu t \tag{8-4}$$

where ϕ is the displacement at time t and A is the amplitude or maximum displacement attained in the cycle. The electric field of the radiation at a given distance from the source undergoes the same variation with time as that observed for the source, except that it lags behind

Figure 8-4
Some of the characteristic quantities associated
with a wave train.

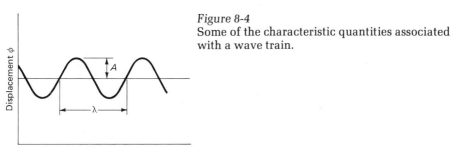

Table 8-1
Regions of the electromagnetic spectrum

Nature	λ (m)	λ (μm)	λ (nm)	λ (Å)	$\bar{\nu}$ (cm^{-1})	ν (sec^{-1})	Source
	10^3	10^9	10^{12}	10^{13}	10^{-5}	3×10^5	
Radio							Alternating electric currents, nuclear spin transitions
	10^{-1}	10^5	10^8	10^9	10^{-1}	3×10^9	
Microwave							Electron beams, molecular rotations, electron spin transitions
	10^{-3}	10^3	10^6	10^7	10^1	3×10^{11}	
Far infrared							Molecular rotations or vibrations
	2×10^{-5}	2×10^1	2×10^4	2×10^5	5×10^2	1.5×10^{13}	
Near infrared							Molecular vibrations
	7.5×10^{-7}	7.5×10^{-1}	7.5×10^2	7.5×10^3	1.3×10^4	3.9×10^{14}	
Visible							Valence-shell electron transitions
	4×10^{-7}	4×10^{-1}	4×10^2	4×10^3	2.5×10^4	7.5×10^{14}	
Near ultraviolet							Valence-shell electron transitions
	2×10^{-7}	2×10^{-1}	2×10^2	2×10^3	5×10^4	1.5×10^{15}	
Vacuum ultraviolet							Valence-shell electron transitions
	2×10^{-9}	2×10^{-3}	2×10^0	2×10^1	5×10^6	1.5×10^{17}	
X rays							Inner-shell electron transitions
	10^{-11}	10^{-5}	10^{-2}	10^{-1}	10^9	3×10^{19}	
Gamma rays							Nuclear transformations
	10^{-13}	10^{-7}	10^{-4}	10^{-3}	10^{11}	3×10^{21}	

in time by the interval it takes for the wave to travel the distance from the source to the point. Thus we can write for the fluctuating electric field at a point x distance from the source an equation in which x/λ, the number of cycles by which the motion at the point lags behind the motion of the source, is subtracted from νt, the number of cycles executed by the source since time zero:

$$\phi = A \sin 2\pi\left(\nu t - \frac{x}{\lambda}\right) \tag{8-5}$$

In this equation, we have extended the meaning of the function ϕ to represent the displacement of the wave—the magnitude of the electric field—at any point along its course and at any time. The function repeats its value at intervals of distance x equal to λ, so that λ is indeed the length of a complete cycle from one point on the wave to the next corresponding point.

In Table 8-1 are given approximate ranges for several kinds of radiation in the electromagnetic spectrum. The categories of radiation vary in the nature of the sources that generate them, as well as in the effects they produce when they impinge on matter. The parameters cited are those measured as the radiation traverses a vacuum, values that are very close to those when the radiation passes through a gas at atmospheric pressure. However, as we shall find in Section 8-2, the passage of radiation through matter reduces its velocity below that in a vacuum. The frequency is a fixed characteristic of the radiation, independent of the medium, for the wave must be continuous. The wavelength, however, is decreased in direct proportion to the reduction in velocity, and the wave number is proportionately increased.

8-2
REFRACTION

When a beam of light passes, in a direction forming an oblique angle to the interface, from a region containing a low concentration of atoms, such as the air, into a region containing more dense matter, such as a liquid or solid, the path of the ray is bent toward the normal to the surface. On passage from a more dense phase into a less dense one, the effect is reversed and the beam is bent away from the normal. These phenomena are examples of *refraction,* and the ratio of the sine of the angle of incidence (the angle between the path of the incident ray and the normal to the surface) to the sine of the angle of refraction (the angle between the ray in the second phase and the normal) is found to be a constant for differing angles. This ratio is by definition equal to the ratio of the *index of refraction* of the second phase to the index of refraction of the first phase, with the index of refraction of empty space—of a vacuum—defined as unity. Incidentally, the interface between two phases of the same refractive index is invisible, because light passes through it without being bent.

What is the explanation for the phenomenon of refraction? For the answer, we must look at the way in which the light ray is affected by the transparent matter through which it is passing: Its velocity is reduced. The index of refraction of a substance is simply equal to the ratio of the velocity of light in a vacuum to its velocity in the substance. Because the velocity in matter is always less than it is in a vacuum, the index of refraction of any substance is greater than unity.

Figure 8-5 illustrates the relation between the refraction of a light ray and the velocity of light for the case of light passing from air into glass, the more dense phase. In this diagram, lines drawn parallel to the wave front represent successive maxima in the traveling wave. Since

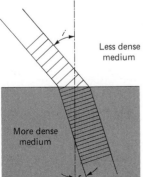

Figure 8-5
Refraction of light at the interface between two phases. Lines perpendicular to the direction of the light ray show the successive positions of the advancing wave front at equal time intervals. The angle of incidence is i, and the angle of refraction is r.

the velocity is less in the glass, the wavelength is shorter and the lines are closer together. If, as in the situation in this diagram, one side of the wave front reaches the glass first, it is slowed down sooner than the opposite side, which travels farther in air. The result is that the wave is bent toward the perpendicular to the surface. Of course, when the same wave emerges from the glass into air, the reverse effect occurs and the ray is bent away from the perpendicular to the surface.

Frequently the index of refraction of a particular material for light depends upon the wavelength—and therefore the frequency—of the light. A piece of flint glass may have a refractive index varying from 1.643 for red light to 1.685 for violet light. If white light is allowed to enter and leave a glass prism by a pathway such that the deviation due to refraction at the entering surface is in the same direction as the deviation resulting from refraction at the leaving surface, as shown in Figure 8-6, the prism can be used to separate the light into its component wavelengths. Since the prism serves to *disperse* the radiation, any variation of refractive index with wavelength has come to be called *dispersion*. For ultraviolet radiation, quartz is sufficiently transparent to be used as the material for a prism.

By what mechanism does matter slow radiation? The process is in essence one in which the fluctuating electric field of the radiation sets into vibration the positively and negatively charged particles of matter with respect to one another. The electric field pulls the positive charges one way, and the negative charges in the opposite direction; as the electric field fluctuates, nuclei and electrons move in synchronization with it. The oscillating dipoles created serve as centers which reemit the radiation, unchanged except for a time lag which results in a decrease in the velocity of propagation of the beam.

The extent to which a sample of matter slows light depends upon the number of electrons per unit volume of the sample; this is the reason why the refractive index of a solid or liquid is larger than that of a gas. The other important factor contributing to the magnitude of the refractive index is the electrical polarizability of the molecule. Substances containing electrons less tightly bound, such as the electrons in carbon–carbon double bonds in olefins or aromatic compounds, have

Figure 8-6
Action of a prism in separating a beam of white light into its component wavelengths to produce a spectrum.

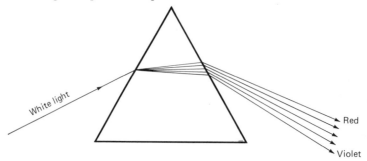

Table 8-2
Molar refraction equivalents for the sodium D line

Carbon	2.418	Oxygen in —C—O—C—	1.643
Hydrogen	1.100	Double bond, C=C	1.733
Chlorine	5.967	Triple bond, C≡C	2.398
Bromine	8.865	Four-membered ring	0.48
Oxygen in —O—H	1.525	N in a primary amine	2.322
Oxygen in —C=O	2.211		

higher refractive indices than organic molecules containing only carbon–carbon and carbon–hydrogen single bonds. An iodine atom, with its valence electrons far out from the nucleus and thus more easily influenced by outside forces, contributes much more to the refractive index than does a fluorine atom.

From the index of refraction of a compound, it is possible to calculate the molar refraction R_M, a quantity related to the molecular structure of the compound. The relation between R_M and the refractive index n is

$$R_M = \left(\frac{n^2 - 1}{n^2 + 2}\right)\frac{M}{\rho} \tag{8-6}$$

where ρ is the density of the material at the temperature at which the index of refraction is measured and M is the molecular weight. The molar refraction is independent of the temperature or physical state of the substance, and it can be predicted to a good approximation by adding together contributions from each atom and each particular structural feature in the molecule. Values of some molar refraction equivalents are listed in Table 8-2.

The molar refractions of some molecules containing conjugated double bonds are higher than those predicted from comparisons with unconjugated molecules. The reason for this is believed to be that the electrons are more mobile in such molecules, having a larger volume over which they can move, and thus can be set in vibration more easily by the incident radiation.

8-3
THE SUPERPOSITION PRINCIPLE AND DIFFRACTION

Two or more waves of electromagnetic radiation that happen to reach the same region of space interfere with one another. In other words, when they pass one another, they become a single entity, with the electric fields of each of the component waves at a particular point adding to give a resultant electric field at that point. This leads to an *interference pattern*—some new and different waveform.

If the two waves being superimposed are of the same wavelength and are *in phase*, which means that a peak of one coincides with a peak of the other, the resultant wave has an amplitude equal to the

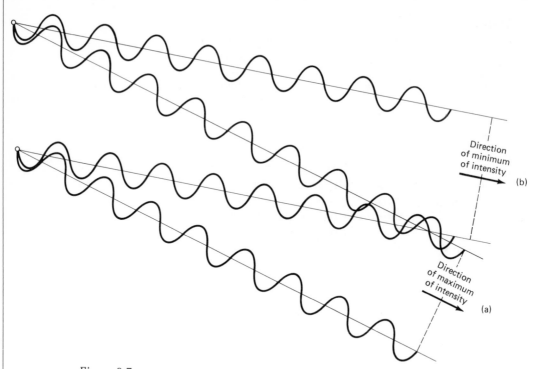

Direction
of minimum
of intensity

(b)

Direction
of maximum
of intensity

(a)

Figure 8-7
Constructive (a) and destructive (b) interference in wave trains sent out by two neighboring sources in different directions.

sum of the amplitudes of the individual waves. If the two waves of the same wavelength are *out of phase,* with a peak of one coinciding with a valley of the other, the two waves tend to cancel one another, and the amplitude of the resultant is smaller than the amplitude of either of the original waves. That this behavior is quite different from what experience teaches us to anticipate for macroscopic matter is an indication that electromagnetic radiation is not adequately described as a stream of particles.

Suppose that electromagnetic radiation is emitted by two sources at a distance apart fairly small compared to the wavelength of the radiation being produced. An observer traveling along a closed path encircling the two sources in a plane containing both of them sees alternating regions of high and low intensity. Figure 8-7 shows the situation at two different directions in which the observer could be located: In Figure 8-7a the observer sees the rays from the two sources as being in phase with one another and there is an intensity maximum, whereas in Figure 8-7b the rays are out of phase with one another and the intensity is zero.

A diffraction grating consists of a plane surface ruled with a series of evenly spaced parallel lines very close together. A cross section of a grating is shown in Figure 8-8. When light is transmitted or reflected by a piece of glass or plastic ruled as a grating, each of the grooves acts as a secondary source sending out radiation in all directions. Intensity maxima are observed at those angles which meet the requirement that the difference, Δ, in path length between parallel rays from any two neighboring sources be an integral number of wavelengths, so that the light rays from the several rulings arrive in phase with one another. The beam for which the path difference between rays from any two neighboring rulings is one wavelength is called the first-order diffracted beam, that for a difference of two wavelengths, the second-order beam, and so on. There is of course always a zero-order beam, parallel to the direction of the incident radiation. If the incident light is perpendicular to the plane of the grating, diffraction maxima are observed at angles satisfying the condition

$$\Delta = n\lambda = d \sin \phi \qquad (8\text{-}7)$$

where n is the order of the diffracted beam, λ is the wavelength of the radiation, d is the grating interval, and the angle ϕ is defined in Figure 8-8.

Since the angle of maximum intensity of the diffracted beam depends upon the wavelength of the radiation, a diffraction grating separates mixed radiation into a spectrum of its various components. Diffraction gratings are often used in spectrophotometers to select monochromatic radiation from a beam containing a mixture of wavelengths.

Diffraction gratings can in practice be ruled with parallel lines spaced several hundreds of nanometers (nm) apart, but nature also provides ordered arrays in which the repeat distances are as small as nanometers or even tenths of nanometers. These arrays are formed by regular placement of the structural units—atoms, ions, or molecules—in crystalline solids. Readily available x rays have wavelengths of a magnitude similar to interatomic distances, and the diffraction of x rays has been utilized to measure distances on the scale of molecular and atomic dimensions. The technique of determining repeat distances

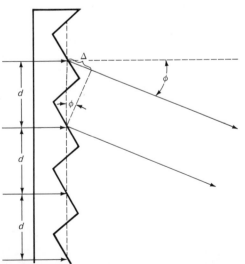

Figure 8-8
Cross section of a diffraction grating with a repeat distance between rulings equal to d.

by x-ray diffraction is also applicable to liquids, as mentioned in Chapter 1. In Chapter 12 we will describe how this method has been applied to structural determinations of macromolecules.

8-4
POLARIZED RADIATION

Radiation, including visible light, can be produced in such a way that the fluctuating electric field components appear to an observer, looking along the direction of the ray, to be oriented at a preferred angle rather than distributed randomly over all angular directions. As an example, imagine a radio transmitter with an antenna wire stretched horizontally between two towers some distance apart. A current of electrons flows in the antenna, reversing its direction several hundred thousand times per second. From the antenna there are emitted radio waves with a frequency equal to that of the alternating current in the wire. As seen by an observer with a radio receiver located several miles away from the antenna and somewhere near the surface of the earth, the radio waves have electric field fluctuations exclusively in a horizontal direction, parallel to the length of the wire; that is, the radio waves are polarized horizontally.

Light reflected at an angle from a surface is often partially polarized. Picture the plane defined by the beams of incident and reflected light in a situation in which a surface is acting as a mirror. Looking back along the reflected beam toward the light source, an observer sees in full view the effect of that component of the electric field that is at right angles to this plane, but the fluctuations parallel to this plane appear much smaller because the observer sees only their projection, that is, only a small part of their full magnitude. A sheet of Polaroid film, or a Nicol prism properly cut from a piece of quartz, can produce polarized light by transmitting only rays with a certain orientation of the electric field.

Molecules that do not have a plane of symmetry are capable of rotating the plane of polarized light and are said to be *optically active*. A plane of symmetry is a plane that can be passed through the molecule so that the part of the molecule on one side of the plane is an exact mirror image of the half on the other side of the plane. Some examples of optically active molecules are shown in Figure 8-9.

The chromium complex in the figure has six oxygen atoms around the central atom in an octahedral configuration, one in which the six

Figure 8-9
Some optically active compounds.

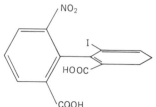

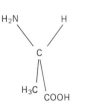
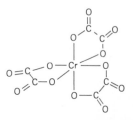

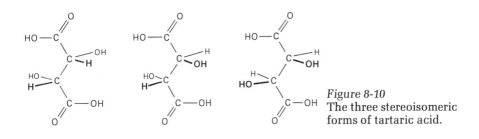

Figure 8-10
The three stereoisomeric forms of tartaric acid.

positions of the surrounding atoms correspond to the positive and negative directions of the x, y, and z axes. There are two forms of this complex, differing in the arrangement of the rings of bonded atoms in the same way that a right-handed screw thread differs from a left-handed thread. One of these forms is an exact mirror image of the other; two such isomers related to one another like an object and its mirror image, but which are not superimposable on one another, are termed *enantiomers*. One of these rotates the plane of polarized light clockwise by a certain angle; the other rotates it counterclockwise by the same angle.

The diphenyl compound shown in Figure 8-9 is prevented from assuming a planar, optically inactive structure by the steric repulsions of the groups in the ortho positions of the two rings, and here again there are two possible enantiomers. One of the forms is labeled D and the other L, standing for dextrorotatory and levorotatory, respectively. This label does not necessarily refer to the actual direction, right or left, in which the molecule rotates the plane of polarized light, but rather to its steric relationship or resemblance to a standard or reference compound for which the forms have been assigned absolute configurations.

Although an asymmetric carbon atom is not a necessary or sufficient condition for optical activity, it is convenient in many organic compounds to consider optical activity as the result of centers of asymmetry—carbon atoms to which four differing groups are attached. The three stereoisomeric forms of tartaric acid, shown in Figure 8-10, illustrate the situation in which there are two asymmetric centers. In this molecule, the two centers have identical surroundings, so that when their configurations are opposite, as in the DL and LD combinations, termed the meso form, there is no ability to rotate polarized light. The meso form, shown in the center of Figure 8-10, can more easily be seen to have a plane of symmetry if one-half of the structure as drawn is imagined to be rotated about the middle carbon–carbon bond. The other two forms, with both centers D or both centers L, respectively, are optically active.

Stereochemistry is quite important to the biochemist, for many organisms are stereochemically specific in function. Some enzymes act only on the D form of a molecule, and others are specific to the L form. During fermentation processes, some bacteria produce only one optical isomer of the two possible. Alkaloids such as strychnine, brucine, and cinchonine occur naturally in optically active form, as do many sugars.

The optical rotation of most materials is measured conveniently in solution; that of liquids can also be measured in the pure liquid state. The instrument used is termed a *polarimeter*. It consists of a light source, a polarizer, usually containing a Nicol prism, which removes all of the light beam except that polarized in one plane, a trough for the tube containing the sample, and an analyzer, also a Nicol prism or its equivalent, which is used to find the plane of polarization of the light that has passed through the liquid in the sample tube. When the instrument contains no sample or an optically inactive sample, minimum light is transmitted when the polarization planes of the polarizer and the analyzer are at right angles to one another. Insertion of a sample changes the angle at which the light entering the analyzer is polarized, and thus to find the new position of minimum intensity the analyzer must be rotated. The change in angular position of the analyzer corresponds to the angular rotation effected by the sample.

The rotational angle depends upon the concentration of the sample and the length of the tube containing the liquid, as well as upon the nature of the solute. The specific rotation of the solute $[\alpha]$ is defined by the equation

$$[\alpha]_\lambda^T = \frac{\alpha}{lc} \tag{8-8}$$

where α is the observed angle of rotation, c is the concentration of the solution in grams per cubic centimeter, and l is the length of the light path through the tube in decimeters. The superscript of the symbol indicates the temperature of the measurement, and the subscript indicates the wavelength of the light used. A subscript "D" refers to the commonly used orange line of sodium, designated the D line. If the sense of the rotation of light is clockwise as one looks through the sample toward the source, a positive sign is assigned to $[\alpha]$; if the sense is counterclockwise, $[\alpha]$ is given a negative sign. The molar rotation is sometimes defined as the specific rotation times the molecular weight divided by 100.

To visualize the mechanism by which an optically active material rotates the plane of polarization of light, it is helpful to consider plane-polarized light as the sum of two circularly polarized components.

Figure 8-11
Counter-rotating components of a ray of polarized radiation. In going from
(a) to (b) the component rotating in one direction has moved more rapidly than
that rotating in the other, and the resultant, represented by the heavy arrow,
has changed angular orientation.

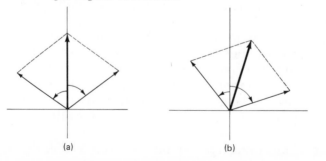

(a)   (b)

In light that is circularly polarized, the electric vector, the line representing the direction and magnitude of the electric field, rotates about the axis of the ray in a regular way, tracing out a spiral path, as the beam travels along. Superposition of such a ray having the vector rotating clockwise on another ray polarized counterclockwise but otherwise identical yields a resultant that is plane-polarized. This can be seen by reference to Figure 8-11, which is drawn from a view looking along the axis of propagation of the rays. So long as the electric vectors of the two components rotate at the same velocity but in opposite directions, the resultant remains in the same plane, denoted here as the yz plane, and has a magnitude equal to twice the projection of either vector along the y direction; this projection varies periodically with distance along the direction of travel of the ray and periodically with time at any one place. The resultant in the x direction is always zero, since the two projections on the x axis are always equal and opposite.

In an optically active material, the asymmetry of the molecules or of the crystal structure causes the index of refraction for the clockwise component to be different from the index of refraction for the counterclockwise component. This means that one component travels faster than the other and the rotations of the two are no longer at exactly the same rate; consequently, the resultant of the vectors at successive points along the ray is rotated increasingly away from the original plane of polarization.

In Chapter 9, we shall explore some of the effects resulting from the wavelength dependence of optical rotation, *optical rotatory dispersion,* and from preferential absorption by a substance of one of the two circularly polarized components of plane-polarized light, *circular dichroism.*

8-5
THE QUANTUM NATURE
OF RADIATION

When certain experiments involving electromagnetic radiation are performed, the results cannot be satisfactorily interpreted by applying the wave model of radiation together with the assumption that matter can absorb any arbitrary amount of energy from a beam of radiation. One set of data of this kind, a group of results that greatly troubled the physicists of the late nineteenth century, comprises the distribution over various wavelengths of the energy in the radiation within a heated furnace. Because it can be shown that the radiation confined in a closed chamber in such a way that it comes to equilibrium with the heated walls must have the same wavelength distribution of energy as the radiation emitted at the same temperature by a black body—one that absorbs all incident radiation and reflects none—thermal radiation as obtained in a heated furnace is referred to as *black-body radiation.* This radiation has a continuous distribution throughout the spectrum from long waves to short waves and is independent of the material of which the walls of the furnace are composed. At low temperatures, most of the energy is in the infrared spectral region, but the

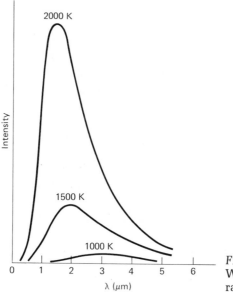

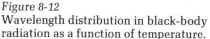

Figure 8-12
Wavelength distribution in black-body
radiation as a function of temperature.

maximum in energy shifts to shorter wavelengths with increasing temperature, as illustrated in Figure 8-12, until it reaches the visible region at high temperatures. This is in accord with our observations from experience that solid objects heated from ordinary temperatures first glow a dull red, then a bright red, then yellow, and finally reach "white heat."

Earlier attempts to describe black-body radiation quantitatively regarded the interior of the furnace as a collection of oscillations or waves, each obeying the principle of equipartition of energy. This principle, described in Section 3-4, predicts that the average energy associated with an oscillational mode will be kT—the sum of $\frac{1}{2}kT$ for potential energy and $\frac{1}{2}kT$ for kinetic energy. We saw earlier that the principle fails for heat capacities at low temperatures, and it fails as well for black-body radiation at short wavelengths. When the types of vibrations that can be fitted into a closed chamber are analyzed mathematically, the number of them that can occur in the vicinity of a given wavelength increases as the wavelength decreases. If each kind of vibration has the same amount of energy, the energy per unit wavelength interval should increase *without limit* as the wavelength becomes shorter and shorter. This predicted behavior, which was termed the "ultraviolet catastrophe," is not in accord with observations.

To explain the experimental results, Max Planck proposed in 1901 what was at that time a revolutionary way of looking at the energy levels of an oscillating system. He suggested that the energy of an oscillator is limited to one of a series of definite and discrete values, numerically equal to an integral multiple of a basic unit $h\nu$, where ν is the frequency of oscillation and h is a fundamental constant with a numerical value of 6.6×10^{-27} erg sec or 6.6×10^{-34} J sec. At any temperature, the oscillators are distributed over the energy levels according to the

Boltzmann equation, which was mentioned in Chapter 1. Equation (1-19) can be written as

$$P_i = P_0 e^{-E_i/RT} \qquad (8\text{-}9)$$

where P_i and P_0 are the populations of the ith and the zeroth levels, respectively, and E_i is the excess energy of a mole of the material at the ith level above its energy in the zeroth level. For the relative populations of two levels, designated 1 and 2, this equation becomes

$$\frac{P_2}{P_1} = e^{-(E_2 - E_1)/RT} \qquad (8\text{-}10)$$

This distribution has the characteristic behavior for a series of *equally spaced* levels that the ratio of the population in the $(n + 1)$st level to the population in the nth level is the same for any pair of levels, as illustrated in Figure 8-13. If the energy of the lowest state above the ground level is greater than that of the ground state by a quantity sufficiently large compared to RT, then nearly all the oscillators are in the ground state. Thus, for black-body radiation, the oscillations corresponding to very short wavelength, which would correspond to a high energy per oscillation, are not active at any attainable temperature. In parallel fashion, the vibrational contribution to the heat capacity of a molecular system is likely to be small at low temperatures, for practically all the molecules are in the ground vibrational state, and they stay in that state even if the temperature is increased by several degrees.

Planck's proposal is the basis of the quantum theory; the *quantum* is the term applied to a unit of energy, and the energy of an oscillator is said to be *quantized*. Other types of energy in an atomic or molecular system are likewise quantized, being restricted to certain definite values, although for translational energies quantum effects turn out to

26 *Figure 8-13*
Example of a Boltzmann distribution in which $e^{-E/RT}$ is equal to 0.40. The numbers represent relative populations of the five levels.

ΔE

64

ΔE

160

ΔE

400

ΔE

1000

be negligible. Albert Einstein in 1905 extended the idea of quantization to electromagnetic radiation, proposing that radiation occurs in distinct units, called *photons,* rather than as a continuous wave. To each photon, he assigned an energy $h\nu$, where h is again Planck's constant and ν is the frequency of the electromagnetic radiation represented by the photon.

Einstein developed the idea of photons in his successful effort to explain the photoelectric effect. This effect is the emission of electrons by a metallic surface, particularly the surface of an alkali metal, when, having been placed in an evacuated chamber and freed of surface contaminants such as oxide, it is irradiated through a quartz window with visible or ultraviolet light of sufficiently short wavelength. In the same chamber there can be placed an electrode at an adjustable positive potential; this anode collects the photoelectrons, and they are returned to the emitting metal through a meter which measures the magnitude of the electron current. A schematic diagram of the setup is shown in Figure 8-14. It is of interest to note that many radiation detectors now in practical use in the visible or ultraviolet region are photocells employing the same principle.

The results observed when the photoelectric effect is studied quantitatively are:

(1)   For a given metal, there is a threshold frequency, a critical frequency below which electrons are not ejected from the metal surface.

(2)   As the light frequency is increased above the threshold value, the maximum kinetic energy of the ejected electrons increases linearly with the frequency.

(3)   An increase in the brightness of the light incident upon the surface increases the number of electrons emitted, but the *maximum kinetic energy* of the electrons is affected only by the frequency of the light.

(4)   When a weak beam of light is incident upon the surface, the first electron is emitted so quickly after the radiation is first admitted, that the energy from the entire beam must be available to that electron rather than being distributed among many electrons.

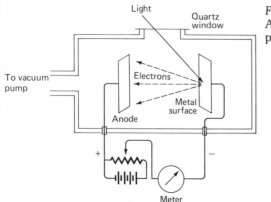

Figure 8-14
Arrangement of apparatus for studying the photoelectric effect.

The explanation proposed by Einstein is simply that *it requires one photon to eject one electron, and all the energy in that single photon is available to pull the electron out of the surface and then to supply it with kinetic energy.* If one photon does not have enough energy to eject the electron from the metal, then a large number of them will do no better. The amount of energy available for a single electron is determined only by the energy per photon and not by the number of photons. The kinetic energy per electron that corresponds to the maximum is equal to the difference between $h\nu$, with h equal to Planck's constant, and the threshold value E_0 or $h\nu_0$:

$$E_{max} = \tfrac{1}{2}mv_{max}^2 = h\nu - E_0 = h\nu - h\nu_0 \qquad (8\text{-}11)$$

This interpretation of the photoelectric effect can be extended to a variety of spectral and photochemical processes. If a transition of a single atom or molecule requires a change in energy of amount ϵ, then a photon of frequency ϵ/h carries sufficient energy to bring about the transition.

8-6
THE WAVE NATURE OF MATERIAL PARTICLES

The observation that light rays seem to have a dual character, behaving sometimes as waves and sometimes as particles, leads to the expectation that wave properties might also be observed for electrons and other atomic "particles." In the early 1920's, C. J. Davisson and L. H. Germer in fact showed that a beam of electrons could be diffracted by the regular array of atoms in a metal crystal.

In accordance with a proposal by Louis de Broglie, it appears that any particle of matter has associated with it a wave having a length λ given by the equation

$$\lambda = \frac{h}{mv} \qquad (8\text{-}12)$$

where h is Planck's constant, m is the mass of the particle, and v is its velocity. The product mv is the momentum of the particle. For any object of macroscopic size, the mass is sufficient to cause the wavelength to be negligible compared to the size of the particle, but the mass of the electron is small enough so that the wavelength associated with it can be many times larger than the "size" of the particle, and the wave nature becomes quite important.

Energy is usually supplied to electrons in a beam by accelerating them through an electric potential difference. For example, the electrons can be emitted by a heated filament kept at a negative voltage and then allowed to pass through an aperture in a barrier that is at a more positive voltage. The electric field does work on each electron, equal in magnitude to the potential difference E multiplied by the charge on the electron e, and the kinetic energy acquired by the electron is equal to this work:

$$\tfrac{1}{2}mv^2 = Ee \qquad (8\text{-}13)$$

Combining this equation with Equation (8-12) leads to

$$\lambda = \frac{h}{(2Eem)^{1/2}} \tag{8-14}$$

For example, one finds that the wavelength of an electron accelerated through a potential difference of 1000 practical volts is about 0.4 Å or 0.04 nm.

Electron diffraction has been fairly extensively used in the study of interatomic distances of gaseous molecules, somewhat in the way x rays are employed to measure repeat distances in solids.

A corollary of the wave nature of matter is the *uncertainty principle,* associated with the name of Werner Heisenberg, which states that the more closely an experimenter attempts to locate a particle, the less accurate is the information that can be obtained about the velocity of the particle, and that the more closely the velocity is evaluated, the less certain is the measurement of its position. The product of the uncertainty in position Δx and the uncertainty in momentum Δmv is of the order of magnitude of Planck's constant:

$$\Delta x \, \Delta mv \simeq h \tag{8-15}$$

An equivalent expression can be written in terms of the uncertainty in the energy of an atomic or molecular state $\Delta \epsilon$ and the lifetime of that state Δt:

$$\Delta \epsilon \, \Delta t \simeq h \tag{8-16}$$

In summary, some properties of atoms and molecules are more readily explained by the particle model and others more readily by the wave model. Each picture has its contribution to make toward our understanding of the behavior of matter, and in later sections of this book we shall apply both models.

8-7
THE NUCLEAR ATOM

Implicit in our allusions to atomic structure up to this point has been a picture of the atom as consisting of a small positively charged nucleus and a surrounding region occupied by mobile electrons. It is worthwhile at this point to examine some of the evidence from which the nuclear model of the atom and the actual existence of electrons were deduced.

If an electric potential is applied across a pair of electrodes sealed in a glass tube containing a gas at a fraction of a torr pressure, a current flows between the electrodes. The gas becomes luminous as a result of the formation of ions, which migrate under the influence of the potential difference toward the electrode of charge opposite their own. One of several processes by which ions are formed is the emission of *cathode rays* from the surface of the negative electrode. The cathode rays themselves are negatively charged, carry enough momentum to cause a paddle wheel to rotate and thus are material particles, heat an object on which they impinge, and cause fluorescence when they strike the glass walls of the tube. In 1897, by measuring their deflection in electric and magnetic fields, J. J. Thomson, a British physicist, established the identity of these particles as what we now know as electrons. He showed them to be a universal constituent of matter, for their charge/mass ratio is the same regardless of the material of the electrodes or the composition of the gas in the discharge tube. Further study of properties of the electron shows that its mass is 9.108×10^{-28} g, or about 1/1840 that of the hydrogen atom, and the charge is 4.803×10^{-10} esu or 1.602×10^{-19} coulomb.

It was not obvious to early workers in the field of atomic structure just how the electrons, present in every atom, are related to the positively charged part of the atom. In 1909 H. Geiger and E. Marsden first performed an experiment which laid the basis for the present picture of the arrangement of the atom. A beam of alpha particles, particles equivalent to doubly ionized helium atoms and emitted by some radioactive materials, was allowed to strike a metal foil. It was found that many of the particles passed through the foil almost undeflected, but some were deflected through large angles, and in fact some were scattered back toward the source. The large-angle scattering could not be explained as a result of the cumulative effect of a series of individual encounters with atoms, each giving a small deflection. It was concluded by E. Rutherford in 1911 that the material causing scattering is concentrated in small centers, or positively charged nuclei, which repel the alpha particles strongly, and that most of the atomic volume is "empty," containing only electrons, which have a mass too small to deflect an alpha particle appreciably. The number of these electrons is equal to the number of positive charges on the nucleus and to the atomic number of the element.

When a gaseous discharge tube is set up with a perforated cathode, positively charged particles are found to stream through the holes in the cathode. By study of the deflection of such gaseous ions in a combination of electric and magnetic fields, the charge/mass ratio of a particle can be determined, and the positive rays are generally found to be mixtures of various species. This kind of experiment led to development of the modern *mass spectrometer*, used for determining the charge/mass ratio of molecular ions as well as of atomic ions.

Among the earliest mass spectrometric results were the following:

(1) The lightest particle corresponds to the nucleus of the hydrogen atom, assigned the name *proton*. Protons can also be obtained in other ways, as, for example, by a reaction of an atomic nucleus.

(2) Heavier particles have masses almost equal to integral multiples of the mass of the proton.

(3) Many elements are mixtures of different nuclear species, each of which has the same atomic number but which differ in mass. Thus chlorine is a mixture of about 25 percent of the *isotope* of mass 37 and 75 percent of that of mass 35, resulting in a chemical atomic mass of 35.5. It is possible, in the disintegration of nuclei, to obtain both protons and other particles with approximately unit mass and no charge, *neutrons*. The nuclear mass, to the nearest whole

number, is equal to the total number of protons and neutrons in the nucleus. A particular nuclear species, with a certain atomic number and atomic mass, is referred to as a *nuclide* and can be represented by a symbol such as $^{14}_{6}C$, which indicates a species with an atomic number of 6 and an atomic mass of 14.

8-8
ATOMIC SPECTRA
AND THE PARTICLE
MODEL OF THE ATOM

NATURE OF ATOMIC SPECTRA

When atoms of an element are heated in a flame or an electric discharge, radiation is emitted consisting of a series of lines at definite frequencies, a spectrum characteristic of the element. If, in contrast, radiation containing a distribution of frequencies is passed through the vapor of an element at lower temperatures, radiation of certain frequencies, many of them identical with the characteristic emission frequencies, is absorbed by the atoms. Thus both sodium chloride decomposed into its elements by heating in a burner flame and sodium vapor heated in an electric arc give off strong orange light of wavelength 589 nm. This spectral line is called the D line because it is the fourth in the series of prominent dark lines observed by J. Fraunhofer in the spectrum of sunlight, absorption lines produced as the light from the hotter central portion of the sun passes through the cooler solar "atmosphere". Processes in which atoms or molecules increase in energy by absorbing photons of electromagnetic radiation yield an *absorption spectrum;* those transitions in which they lose energy in the form of photons give rise to an *emission spectrum.*

At ordinary temperatures, sodium atoms are all in their normal or ground electronic state, and their electrons are in the most stable paths or *orbitals* available to them. When an atom in this state absorbs a photon of radiation of wavelength 589 nm, we describe the process as the transition of an electron to the orbital of lowest energy above the ground state, or the change of the entire atom to its *first excited state.* When the excited atom loses energy by emitting a photon of radiation, usually after an extremely short lifetime, it drops back to the ground state; that is, the electron returns to the orbital of lowest energy.

An atomic emission spectrum differs from black-body radiation, emitted by a heated solid as described in Section 8-5, because it consists of a relatively few lines of certain definite frequencies, whereas the spectrum of the solid includes all frequencies over a wide range. Indeed, it is the very feature of the atomic spectrum that it is discontinuous that leads us to the concept of definite or *quantized* energy states for the atom.

The experimental data one obtains from a spectrum are the changes in energy occurring as the system undergoes changes from one state to another. The first problem in the interpretation of these data is that of fitting together the observed transition energies to produce the pattern of energy levels. The number of transitions is usually greater than the

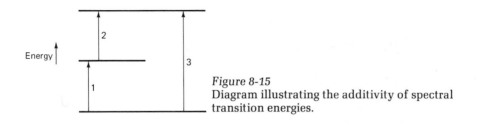

Figure 8-15
Diagram illustrating the additivity of spectral transition energies.

number of levels, so that the problem is overdetermined. For example, if there are 10 different levels available to the system, there are 45 different possible combinations of 2 levels between which transitions can occur. Usually not every one of the 45 possible changes can occur by emission or absorption of a photon, for some of them are *forbidden*, but in general more than enough are observed to allow the pattern of levels to be worked out. To illustrate the kind of reasoning employed: If the energy for an observed transition 1 when added to the energy of another transition 2 is equal to the energy of transition 3, the scheme of levels responsible is likely to be as shown in Figure 8-15.

HYDROGEN—THE BOHR MODEL

The spectrum of atomic hydrogen consists of several series of lines, some of which are indicated in Figure 8-16. In each series, the intervals between the lines become smaller as the frequency increases. Any line in the spectrum can be expressed as the difference between two *term values* of the form $\tilde{\nu} = R/n^2$, where n is an integer and R is the *Rydberg constant*. If the term energies are given in cm^{-1}, the value of R is 109,677.58 cm^{-1}. Thus for the lines of the Balmer series, named after its discoverer and lying in the visible region,

$$\tilde{\nu} = R\left(\frac{1}{2^2} - \frac{1}{n^2}\right) \tag{8-17}$$

where n is 3 for the first line, 4 for the second line, and so on. For the series discovered by Lyman, which is in the ultraviolet region, the equation is

$$\tilde{\nu} = R\left(\frac{1}{1^2} - \frac{1}{n^2}\right) \tag{8-18}$$

where n is 2 for the first line, 3 for the second line, and so on.

Figure 8-16
The spectrum of atomic hydrogen in the visible and ultraviolet regions. Wavelengths are given in nanometers. The dotted lines represent the limits approached by the two spectral series.

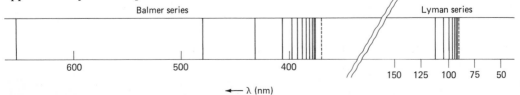

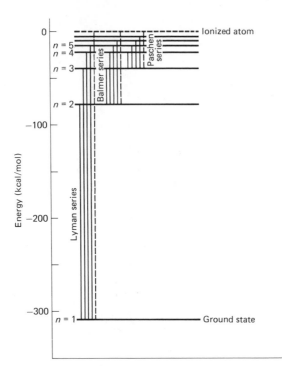

Figure 8-17
Energy level diagram for the hydrogen atom. Higher levels, too close together to be seen separately, are also possible. The integers, or quantum numbers, which label the levels, are given at the left. Vertical lines represent possible spectral transitions, with the limit of each series shown as a dotted line.

Figure 8-17 shows the pattern of energy levels for the hydrogen atom and the relation of these levels to the series of spectral transitions. The lowest level on the diagram is the ground state of hydrogen, the state in which atoms are usually found at ambient temperature. The zero level at the top of the diagram corresponds to an energy just sufficient to separate the atom into the positive nucleus and the negative electron, that is, to ionize the atom. The reason for taking this level as zero is the simple form the term values then take, as in Equations (8-17) and (8-18), and the consequence is that the term values are proportional to the binding energy of the electron to the nucleus.

How can we justify the observed energy levels on the basis of a model or theory of the hydrogen atom? One answer was given by Niels Bohr, who in 1913 combined the quantum theory with the nuclear picture of the atom described in the previous section. Viewed as particles, electrons must have some motion to keep from falling into the nucleus, like satellites moving around the earth or planets moving around the sun, and Bohr assumed that each electron travels in a circular path of radius r about the nucleus as a center with a velocity such that the centrifugal force exactly balances the Coulomb electrostatic attraction of the positive nucleus:

$$\frac{mv^2}{r} = \frac{e^2}{r^2} \tag{8-19}$$

Solving this equation for the velocity leads to

$$v = \left(\frac{e^2}{mr}\right)^{1/2} \tag{8-20}$$

The energy of an electron in a Bohr orbit is simply the sum of the kinetic energy, the usual $\frac{1}{2}mv^2$, and the potential energy associated with the presence of two charges at a distance r from one another, a quantity negative because charges of opposite sign tend to be drawn together and equal to the product of the charges divided by the distance between them:

$$E = \frac{mv^2}{2} - \frac{e^2}{r} \qquad (8\text{-}21)$$

Substitution of the value of mv^2 from Equation (8-19) yields

$$E = \frac{e^2}{2r} - \frac{e^2}{r} = -\frac{e^2}{2r} \qquad (8\text{-}22)$$

Now the line spectrum of hydrogen indicates that the energy levels of the atom are limited to certain values; to explain these limitations, Bohr introduced quantum restrictions, based on Planck's treatment of cyclic processes, according to which the angular momentum of the electronic motion must be an integral multiple n of the basic unit $h/2\pi$. For a particle of mass m moving in a circular path of radius r, the angular momentum is the product of the linear momentum mv and a "lever-arm" factor r. Therefore

$$mvr = \frac{nh}{2\pi} \qquad (8\text{-}23)$$

If the velocity from Equation (8-20) is substituted in this expression and the resulting equation solved for the radius, there is obtained

$$r = \frac{n^2h^2}{4\pi^2me^2} \qquad (8\text{-}24)$$

The energy corresponding to an allowed orbit can be calculated by combining this equation with Equation (8-22):

$$E = -\frac{2\pi^2me^4}{n^2h^2} \qquad (8\text{-}25)$$

At this point we have reached a critical test of the Bohr theory: Does the predicted energy agree with the experimental energy? If we express the energy of a photon as $hc\tilde{v}$, where h is Planck's constant, and substitute numerical values for the constants, Equation (8-25) becomes

$$\tilde{v} = -\frac{2\pi^2me^4}{n^2h^3c} = -\frac{109{,}735}{n^2} \text{ cm}^{-1} \qquad (8\text{-}26)$$

The value of the numerator is slightly larger than the observed value of the Rydberg constant, $-109{,}678$ cm^{-1}, but the discrepancy is completely removed if allowance is made for the fact that the nucleus does not remain stationary, but instead the combination of electron and nucleus moves about a common center of gravity.

Let us now summarize the Bohr model of the hydrogen atom. Each energy state of the atom corresponds to occupation by the electron of an orbit characterized by the positive integer n, called the *principal quantum number.* In the ground state, *n* has a value of unity and the radius of the electron path as calculated by Equation (8-24) is 0.53 Å. For states of higher energy, the radius of the electron orbit increases as the square of *n*, and the binding energy of the electron varies inversely with the square of *n*. In any one atomic state, the electron travels in a particular circular orbit and the energy remains constant as given by Equation (8-25); a spectroscopic transition corresponds to the transfer of an electron from one orbit to another.

8-9
POLYELECTRONIC ATOMS

ENERGY LEVELS OF ALKALI METAL ATOMS

Those elements that have one valence electron, the alkali metals, have relatively simple atomic spectra; as an example, we shall discuss the spectrum of sodium and show how information about the energy levels can be obtained from the spectrum and how it can be interpreted in terms of electronic behavior.

Various series of lines have been recognized in the sodium spectrum. These resemble the hydrogen series in converging to a limiting value at the high-frequency end, but they differ in that the term values have nonintegral denominators. Furthermore, several different spec-

Figure 8-18
Energy-level diagram for the sodium atom, with energies in wave numbers, showing some transitions from the prominent spectral series. For comparison, some of the hydrogen atom energies are shown in the right-hand column.

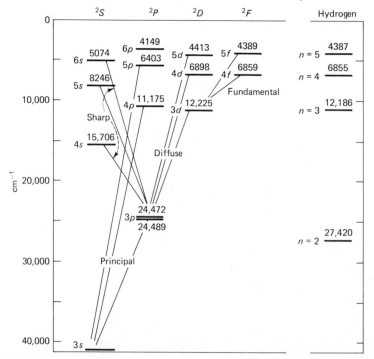

tral states are described by the same value of the principal quantum number n. To explain these results, a second quantum number is introduced, symbolized by l and termed the azimuthal quantum number. This is permitted to have any integral value from zero up to one less than n. The value of l is often represented by a code:

l	0	1	2	3	4	.	.	.
Symbol	s	p	d	f	g	.	.	.

The energy level diagram of sodium, shown in Figure 8-18, has been divided into a series of columns, each column corresponding to a particular value of l and the successive entries from the bottom to the top of each column representing increasing values of the principal quantum number n. The reader should realize that this energy level diagram has been pieced together like a jigsaw puzzle from the energy *changes* associated with the observed transitions.

To explain the variation in energy for a constant principal quantum number n found for sodium and other alkali metal atoms, A. Sommerfeld suggested that the Bohr circular orbits be modified by allowing the electron to change its distance from the nucleus, as well as to travel about the nucleus. Thus the total momentum, still determined in magnitude by the principal quantum number, is divided into two parts, an angular part for which the magnitude is measured by l, and a radial part contributing the rest of the momentum. For reasons having to do with the uncertainty principle, the magnitude of the angular momentum is found to be $\sqrt{l(l+1)}(h/2\pi)$, rather than merely $l(h/2\pi)$, although when the quantum number l becomes large, these two quantities approach one another.

For a large azimuthal quantum number, most of the motion of the electron is around the nucleus; for a smaller l, more of the electronic motion consists of approaching the nucleus and receding from it, and for an s orbital which has $l = 0$, the electron has no angular momentum and no net movement about the nucleus, and has only radial motion. For the hydrogen atom, the shape of the orbit has practically no effect on the electronic energy, which is determined by the principal quantum number alone. However, for any atom with more than one electron, there are additional energy contributions involving the repulsions of the electrons for one another, and the magnitudes of these repulsive energies depend upon the shapes of the orbitals.

Consider, as an example, the valence electron of a sodium atom, an electron with a principal quantum number of 3, which is outside the filled $n = 1$ shell of two electrons and the filled $n = 2$ shell of eight electrons. If the valence electron is in the $3d$ orbital, which is always well outside the filled inner shells, it is attracted by a charge of $+11$ on the nucleus but repelled by a charge almost equivalent to -10 for the inner-shell electrons. From Figure 8-18 one can see that a $3d$ electron is held with very nearly the same energy as an $n = 3$ electron in a hydrogen atom. In a $3p$ orbital, and even more so in a $3s$ orbital, the electron approaches close to the nucleus part of the time and, when it is between the nucleus and the inner-shell electrons, it is held by an effective positive charge much greater than one unit. Thus in the energy level scheme for sodium, electron occupation of different $n = 3$ orbitals corresponds to different binding energies for the electron, the most stable situation being occupation of the 3s orbital.

From an examination of the spectral series of sodium, as represented in Figure 8-18, it is seen that the lines of the *principal* series correspond to transitions ending on the ground state, which has the electron in the 3s orbital, and beginning on the various p states with $l = 1$ but with successively higher principal quantum numbers. The *sharp* series consists of transitions ending on the first excited state, the 3p level, and beginning on successively higher levels with $l = 0$, and the *diffuse* series includes transitions also ending on the 3p level but beginning on the levels with $l = 2$, the d levels. It is thus evident that it is from the rather arbitrary names of the spectral series that the alphabetic code for numerical values of the azimuthal quantum number arises.

One other observation from Figure 8-18 is significant: The value of l always changes by one unit, going from 0 to 1, 1 to 0, 2 to 1, or 3 to 2, but transitions in which l remains unchanged, namely, transitions between levels in the same column in the diagram, and transitions in which l changes by two units or more, are *forbidden*. These statements are generalized by a *spectroscopic selection rule*, $\Delta l = \pm 1$. Several other spectroscopic selection rules will be encountered, both for atomic and for molecular transitions. Many of these rules were first developed empirically; modern theory shows how they can be justified. The existence of selection rules greatly reduces the complexity of the spectral patterns produced by atoms, and the application of these rules assists in assigning the transitions that are found.

When the transitions of alkali metal atoms are examined closely with a spectroscope of good resolution, many of them are found to be composed of two or three components very close together. Thus the orange sodium D line is found to be a doublet with the components at wavelengths of 589.0 and 589.6 nm. To explain the fine structure of the spectral lines, the electron is assigned a magnetic moment, which in a magnetic field from another source is limited by quantum restrictions to either of two possible orientations, one orientation relative to the field termed *parallel,* and the other *antiparallel.* One might visualize the electron as being magnetized, like a compass needle. However, things on an atomic scale are somewhat different from what our observations of a compass needle lead us to expect: The electron magnetic moment can never be completely lined up with the applied magnetic field because of the uncertainty principle, so that it remains tilted at a constant angle away from the field direction; furthermore, once it is in a parallel or antiparallel state, it is locked there until a suitable external torque is applied to it to reorient it to the other possible state.

How does the magnetic moment of the electron arise? Since it is a charged body, rotation of the electron about its axis is a plausible explanation of the source of the magnetism, for motion of charge in a closed path produces a magnetic field. This explanation pictures an electron in an atom as behaving something like the earth in the solar system: spinning about its own axis while revolving in a path about the nucleus. We do not know if the electron is literally spinning, but

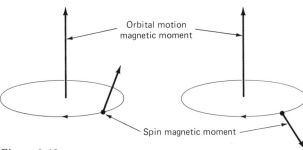

Figure 8-19
Parallel and antiparallel relationships of the orbital and spin magnetic moments.
According to the uncertainty principle, the two moments cannot be exactly
aligned with one another.

what can be observed is the small magnetic field it produces *as if it
were spinning.*

The orbital motion of the electron also produces a magnetic field,
since this motion, too, produces an electric current in a closed loop.
The orbital magnetic field provides a reference direction for the spin
magnetic field: If the two are parallel, the energy of the atom is higher,
but if they are opposite, or antiparallel, the energy of the atom is lower,
just as the more favorable situation of two bar magnets lying side by
side is that in which the north pole of one is next to the south pole of
the other. Figure 8-19 shows these two possible relationships, with
the spin and orbital magnetic moments each represented by a vector.

If the electron has a spin, it should have a corresponding angular
momentum of the spinning motion, and there is assigned to the electron a *spin quantum number s,* which measures its angular momentum, with a value *always* equal to 1/2. The observed behavior of the
electron fits nicely into the general pattern for angular momenta described in the following paragraphs, for a quantum number of this
magnitude leads to prediction of the two possible orientations in an
applied magnetic field that are found by experiment.

The ground state of the sodium atom is the state in which the electron is in a 3s orbital and therefore has no net orbital motion about the
nucleus, no angular momentum about the nucleus, and no orbital
magnetic moment. Thus the ground state, as well as any other s state,
is not split into two states by the effect of spin, because the spin magnetic moment has no other magnetic field with which to interact. The
situation here is somewhat like that of a free electron, one not in an
atom but traveling by itself in a region of high vacuum, which also
does not produce a magnetic field by orbital motion.

In Figure 8-20 is shown an energy diagram for the two components
of the sodium D line. The diagram is not to scale because the splitting
is small compared to the energy of the transition. The two sublevels
into which the 3p level is split by spin–orbit interaction are distinguished by a fourth quantum number *j,* which describes the total
angular momentum for the sodium electron and varies according to
the geometric relationship of its two ingredients, the spin angular
momentum and the orbital angular momentum. If the two magnetic
moments are parallel, which means that the angular momenta are in

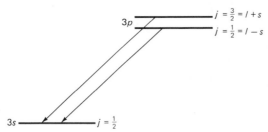

Figure 8-20
Energy levels involved in the
transition corresponding to the
sodium D line.

the same sense, j is the sum of l and s; for the upper energy sublevel
of a p state, the value is $\frac{3}{2}$. If the magnetic moments are antiparallel,
the value of j is equal to $l - s$, corresponding to $\frac{1}{2}$ for the lower energy
sublevel of a p state.

If we place an ordinary sodium lamp in a magnetic field and examine under good spectroscopic resolution the lines emitted when the
lamp is turned on, we find much structure present that is absent in
the absence of the field: The D line now has 10 components instead
of 2. What has happened is that atoms can be in one of a limited number of different orientations with respect to the applied field, and the
atomic magnetic moment interacts with the external field with an
energy that varies with this orientation. The results can be fitted into a
general scheme according to the following basic rule: The number of
possible orientations in a magnetic field and therefore the number
of energy sublevels into which the level is split by application of a
magnetic field is twice the angular momentum quantum number j
plus one more:

$$N = 2j + 1 \qquad (8\text{-}27)$$

For the $j = \frac{3}{2}$ state of the 3p electron, the external magnetic field splits
the energy into four possible levels, since $2(\frac{3}{2}) + 1 = 4$; and for the
$j = \frac{1}{2}$ state, there are two possible orientations and two possible energy
sublevels. Likewise, the 3s ground state, with $j = \frac{1}{2}$, is split into two
sublevels. The transitions observed can be rationalized in terms of the
selection rule, $\Delta m_j = \pm 1$ or 0.

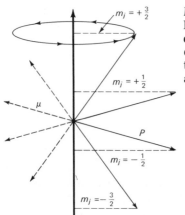

Figure 8-21
Possible orientation of a $j = \frac{3}{2}$ vector in an
external magnetic field which is in the direction
of the heavy arrow. The vector *precesses* about
the field: The tip of the vector describes a circle
as shown for the $m_j = +\frac{3}{2}$ case.

In Figure 8-21 are shown the four possible orientations of the $j = \frac{3}{2}$ state in the magnetic field. The vectors directed along the angular momentum axis, shown by the solid lines, and those directed along the axis of the magnetic dipole, shown by the dotted lines, are necessarily collinear with one another. However, their sense is opposite, according to the usual conventions, because of the fact that electronic motion corresponds to motion of *negative* charge. The length of the vector representing the angular momentum is proportional to $\sqrt{j(j+1)}(h/2\pi)$, or in this particular case $\sqrt{(\frac{3}{2})(\frac{5}{2})}(h/2\pi)$, but the largest value of the component of the angular momentum vector in the direction of the magnetic field is $\frac{3}{2}(h/2\pi)$, and the other possible values of the component differ from one another by one unit of $h/2\pi$, until the largest possible negative value is reached, which is $-j$, in this case $-\frac{3}{2}$. These components are represented by the symbol m_j, and the four values of m_j for $j = \frac{3}{2}$ are $+\frac{3}{2}$, $+\frac{1}{2}$, $-\frac{1}{2}$, and $-\frac{3}{2}$.

A different effect is observed on the spectral lines of sodium when the atoms are present in a very powerful magnetic field. Such a field interacts so strongly with the spin magnetic moment of the electron and with the orbital magnetic moment of the same electron that it causes each of these atomic magnets to be separately oriented in the external field rather than oriented with respect to one another first and *then* related to the magnetic field. When the spin–orbit coupling is thus overwhelmed, the *orientations* of the s and l vectors in the external field can be labeled by the *quantum numbers* m_s and m_l.

ELECTRONIC ORBITS
IN POLYELECTRONIC ATOMS

In order to explain the sequence of filling atomic orbits in polyelectronic atoms as the atomic number increases—that is, as one goes from element to element through the periodic scheme—Wolfgang Pauli developed a rule stating that no two electrons can have all four quantum numbers of the set n, l, m_l, and m_s identical. Since m_s can have the two values, $+\frac{1}{2}$ or $-\frac{1}{2}$, it is therefore possible for two electrons at most to occupy one spatial orbit, that is, to have n, l, and m_l identical. For a given value of n, there can be a maximum of two electrons with $l = 0$, six electrons with $l = 1$, ten electrons with $l = 2$, or fourteen electrons with $l = 3$. These numbers allow for the two m_s values and the $2l + 1$ values of m_l that are possible.

The reader should be familiar with the notation in which the number of electrons occupying each orbit in an atom is indicated by an exponent. For example, the ground state of the element vanadium is represented as

$$1s^2 2s^2 2p^6 3s^2 3p^6 4s^2 3d^3$$

For the same element, the arrangement of electrons can be represented in a qualitative energy level diagram in which the relative directions of spin of the electrons are indicated and the various subshells are divided into compartments representing various possible m_l values, as illustrated in Figure 8-22. It makes no difference in which of the five $3d$ orbits the first of the three $3d$ electrons is placed, because these orbits differ in energy only if there is an external field providing a reference

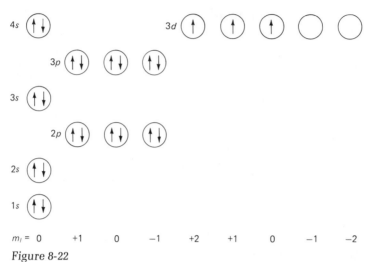

$m_l = 0$ +1   0   −1   +2   +1   0   −1   −2

Figure 8-22
Schematic orbital energy level diagram showing assignment of electrons in an atom of vanadium. The labeling by m_l values has no real meaning here; it is primarily an aid in bookkeeping the electrons. Furthermore, the difference in energy between the 2s and 2p orbitals is probably insignificant when higher orbitals are occupied by electrons.

frame. Once the first electron is present, however, it does make a difference in energy how the second and third electrons are assigned, because the electrons interact with one another by virtue of their electric and magnetic fields.

When there are several electrons in an atom, the state of the atom is characterized by quantum numbers which are the resultant of combining the quantum numbers of the individual electrons in a way permitted by quantum restrictions. Typically, the l values of the electrons are combined to give an orbital angular momentum quantum number L for the atom, and the s values combined to give a spin quantum number S for the atom. Then L and S combine to give a grand total quantum number J for the atom. To fill a shell (for example, the $n = 2$ shell in sodium) or even to fill a subshell (for example, the 3p subshell in vanadium), the electrons must pair off both with respect to m_l values and m_s values, meaning that positive m_l's cancel negative m_l's and positive m_s's cancel negative m_s's, with the result that L, S, and J are all zero. This means that we need only be concerned with electrons in partially filled subshells. In sodium, for which some of the electronic states were described above, only the single valence electron contributes anything beyond zero to the quantum numbers for the atom, and these numbers L, S, and J are simply equal to the corresponding numbers l, s, and j for that electron.

To follow all the details of atomic spectroscopy is far beyond our present needs, but there are several general ideas worth pursuing. Consider the element titanium, which has a completed subshell of two 4s electrons and two more electrons in the partially filled 3d subshell. The value of l for each of the 3d electrons is 2, and therefore L can have any of the values 4, 3, 2, 1, or 0. These numbers correspond to the sum

$l_1 + l_2$, the difference $l_1 - l_2$, and all intermediate values spaced by an integer from the extremes. The value of the spin quantum number for the atom, S, can be either 1 or 0, these numbers representing again the sum and the difference of the two electronic s values of $\frac{1}{2}$, but with no space for intermediate values. If L is 4, J may be 5, 4, or 3; if L is 3, J may be 4, 3, or 2; and so on, according to the same kind of quantum sum rule. A detailed, and complicated, analysis leads to the conclusion that some of these combinations are forbidden by the Pauli exclusion principle when both electrons are $3d$ electrons, although any of them would be allowed if one electron were in a $3d$ orbit and the other in a $4d$ orbit; however, this restriction cannot be made obvious on the basis of any simple interpretation.

One of the combinations of quantum numbers described in the preceding paragraph represents the most stable or ground state of the titanium atom, and many of the others represent possible excited states. The ground state is found to be that in which L is 3, S is 1, and J is 2. The most important point to keep in mind is that the spin quantum number S is 1, not zero, for this means that the two electrons are in a more stable state when their spins are parallel—consequently, when they are in different orbits—than when their spins are paired. This is an example of a general rule that the state of maximum spin for a group of electrons in an atom, molecule, or ion is the most stable state, provided that orbital energy does not have to be sacrificed to achieve it, a principle often known as *Hund's rule.*

Because a state in which $S = 1$ may have as many as three values of J—equal to $L + 1$, L, and $L - 1$, respectively—it is referred to as a *triplet* state. This term has come to be used generally to denote a state in which there are two electrons with parallel spins, electrons not paired with electrons of opposite spin. In contrast, when $S = 0$, there is only one possible J value for that state, because zero combined with L can yield as a resultant only a value equal to L; the term *singlet* state is synonomous with $S = 0$.

Finally, an atomic state is often represented by a term symbol in the form

$$^{2S+1}L_J$$

Thus the ground state of titanium is labeled a $^3F_2$ state, read "triplet F two." From the term symbol, the values of 3, 1, and 2 for L, S, and J, respectively, for the ground state of titanium should be immediately apparent.

An energy level diagram for mercury is shown in simplified form in Figure 8-23, with the levels designated by term symbols and with some spectroscopic transitions indicated.

For many years, chemists left the subject of atomic spectroscopy pretty much to physicists. There is at present a growing need for the chemist and biochemist to understand at least the language of the subject, for the detailed consideration of atomic energy levels and orbitals is becoming more and more important in the clarification of reaction mechanisms, and the tools of molecular spectroscopy are powerful and versatile in investigating the composition and behavior of matter,

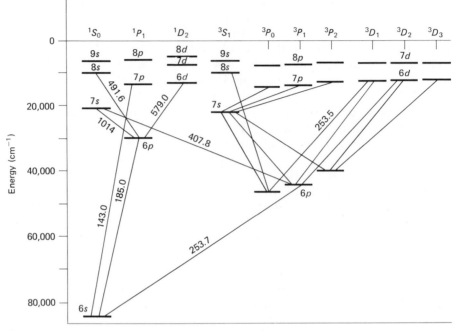

Figure 8-23
Energy level diagram of the mercury atom, which has two 6s valence shell electrons when in the ground state. Numbers on the transitions are wavelengths in nanometers.

not to mention the fact that atomic spectroscopy itself has become an analytical tool of major importance in seeking the trace amounts of metals that growing environmental awareness leads us to consider significant quantities.

8-10
THE WAVE MODEL
OF THE ATOM

In view of the experimental results demonstrating the wave nature of the electron, the Bohr theory of the atom and its extensions cannot be literally correct. The implications of the uncertainty principle are that the motion of an electron cannot be as precisely defined as the calculations of Bohr require. Scientists in the 1920's were therefore led to seek a description of electronic states in atoms based upon the mathematical description of waves.

The wave associated with an electron can be a *traveling wave* or a *standing wave*. For a free electron that is part of a beam passing through an evacuated space, the traveling wave is appropriate, but for an electron occupying an atomic *orbital,* or orbitlike pattern, altered only by absorption or emission of a photon, it is the standing wave that is applicable.

A standing wave is a wave pattern that maintains its arrangement and continues to oscillate over a period of time. To better understand what is meant by the term, let us examine some macroscopic phenomena which provide analogies. When plucked or struck, a string on a guitar or piano vibrates at a particular frequency, called the fundamental, as well as at integral multiplies of that frequency, called overtones, as shown in Figure 8-24. The allowed modes of vibration are governed by the limitation that the ends of the strings, which are fixed, must have zero displacement at all times. For the fundamental vibration, the wavelength is twice the distance between the ends of the string. For the first overtone, the wavelength is exactly equal to the length of the string, for the second overtone, the wavelength is two-thirds the length of the string, and for the nth overtone in a string of length d,

$$(n + 1)\frac{\lambda}{2} = d \qquad\qquad (8\text{-}28)$$

For the nth overtone, there are n equally spaced points along the string, called *nodes,* which, along with the fixed points at the two ends of the string, remain undisplaced during the vibration. The regions between nodes, called *loops,* are displaced in sine-shaped waves, first in one direction from the equilibrium position and then in the other. The frequency of the vibrations of a stretched string depends upon the material of which the string is composed and on the force with which it is stretched.

A standing wave is equivalent to the interference pattern of two traveling waves, of identical frequencies and amplitudes, passing each other in opposite directions, and a mathematical description of a standing wave can be obtained by adding together the equations of oppositely moving waves in the form of Equation (8-5). One can pic-

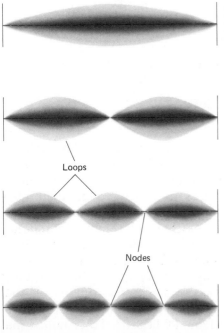

Figure 8-24
Fundamental and first three overtones in the vibration of a string with fixed ends.

Loops

Nodes

ture the standing wave in a string to be the result of a traveling wave that is reflected at the ends and meets itself coming and going.

If a standing wave is two-dimensional, as for example the resonant vibrations in a drum head, the nodes are lines instead of points, and, in a three-dimensional situation, such as the vibrations of sound waves in an organ pipe, the nodes are surfaces. It is as a three-dimensional standing wave that the pattern of an electron in an atomic orbital is described.

Physicists of the nineteenth century developed a general equation applicable to all types of waves, whether they are sound waves in air, electromagnetic radiation, or ocean waves. One form of the wave equation is

$$\frac{\partial^2 \psi}{\partial x^2} + \frac{\partial^2 \psi}{\partial y^2} + \frac{\partial^2 \psi}{\partial z^2} = -\frac{4\pi^2}{\lambda^2} \psi \tag{8-29}$$

The quantity ψ is the magnitude of the displacement of the wave from zero, a function that varies from place to place in the wave, and λ is the wavelength. It is convenient to represent the sum of the three partial derivatives on the left side of this equation by the symbol $\nabla^2 \psi$, and the equation then becomes

$$\nabla^2 \psi = -\frac{4\pi^2}{\lambda^2} \psi \tag{8-30}$$

Erwin Schrödinger developed a version of the wave equation suitable for an electron in an atom by replacing λ by the de Broglie wavelength of the electron h/mv. Further, the kinetic energy T of the electron is $mv^2/2$, so that $m^2v^2 = 2mT$, and the kinetic energy is equal to the total energy less the potential energy, $E - U$. Making these substitutions, we obtain

$$\nabla^2 \psi = -\frac{4\pi^2 m^2 v^2}{h^2} \psi = -\frac{8\pi^2 mT}{h^2} \psi = -\frac{8\pi^2 m}{h^2}(E - U)\psi \tag{8-31}$$

For a hydrogen-like atom, the potential energy is just that used by Bohr:

$$U = -\frac{Ze^2}{r} \tag{8-32}$$

where e is the electronic charge and Ze is the charge on the nucleus. Because the potential energy depends on r, which is $(x^2 + y^2 + z^2)^{1/2}$ in the system of rectangular coordinates, the latter is not a convenient coordinate frame in which to work. Instead, things are much simpler if one chooses a set of coordinates of which one is the distance r. Spherical polar coordinates meet this need, and their relation to rectangular coordinates is shown in Figure 8-25. The coordinate θ, which is defined as the angle between the z axis and the radius vector, the line from the origin to the point in question, can vary between $0°$ for the positive z direction to $180°$ for the negative z direction. The coordinate ϕ is the angle between the positive x axis and the projection of the radius vector on the xy plane and can vary from $0°$ to $360°$. To visualize the

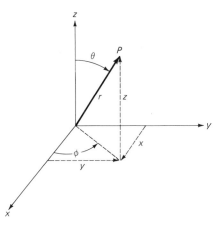

Figure 8-25
Location of a point P in the Cartesian coordinate system (x, y, z) and in the spherical polar coordinate system (r, θ, ϕ).

angular coordinates of this system, it is helpful to think of their being analogous to the navigational coordinates by which a location on the earth's surface is specified. The positive z direction corresponds to the north polar axis, the negative z direction to the south polar axis, and the xy plane to the equatorial plane. The angle ϕ is then the longitude, and the angle θ is equivalent to the complement of the latitude.

If relations between the derivatives in terms of x, y, and z and those in terms of r, θ, and ϕ are substituted in the expression for ∇^2 and the potential energy for the hydrogen atom from Equation (8-32) is used, the Schrödinger equation becomes

$$\nabla^2\psi = \frac{1}{r^2}\frac{\partial}{\partial r}\left(r^2\frac{\partial\psi}{\partial r}\right) + \frac{1}{r^2\sin^2\theta}\frac{\partial^2\psi}{\partial\phi^2} + \frac{1}{r^2\sin\theta}\frac{\partial}{\partial\theta}\left(\sin\theta\frac{\partial\psi}{\partial\theta}\right) =$$
$$-\frac{8\pi^2 m}{h^2}\left(E + \frac{e^2}{r}\right)\psi \quad (8\text{-}33)$$

This equation in spherical polar coordinates looks more complicated than the equation in rectangular coordinates, but the method of solving it, as well as the form of the solutions, turn out to be much simpler. The solutions, or wave functions, the expressions that satisfy this equation, are products of three factors, each depending on a single one of the three coordinates:

$$\psi = R(r)\Theta(\theta)\Phi(\phi) \quad (8\text{-}34)$$

This notation means that the factor R is a function only of the variable r, Θ is a function only of the variable θ, and Φ is a function only of the variable ϕ.

It is appropriate to ask what the physical counterpart of the function ψ is, but it is not possible to give a precise answer to this question. Clearly it is the displacement of some kind of wave, and one can think of it as resembling in behavior the electric field associated with electromagnetic radiation. The square of the wave function ψ^2 has a more concrete meaning: It is proportional to the probability that the electron will be in a unit volume at any point in space. Thus it is a measure of the electron concentration. Although ψ itself may be positive in various regions of space and negative in others, ψ^2 is everywhere a positive quantity. Furthermore, we want *one* wave function to represent one

Table 8-3
Wave functions for a hydrogen-like atom[a]

$$\psi \qquad 1s = \frac{1}{\sqrt{\pi}}\left(\frac{Z}{a}\right)^{3/2} e^{-Zr/a}$$

$$\psi \qquad 2s = \frac{1}{\sqrt{32\pi}}\left(\frac{Z}{a}\right)^{3/2}\left(2 - \frac{Zr}{a}\right)e^{-Zr/2a}$$

$$\psi \qquad 2p_x = \frac{1}{\sqrt{32\pi}}\left(\frac{Z}{a}\right)^{5/2} re^{-Zr/2a}\sin\theta\cos\phi$$

$$\psi \qquad 2p_y = \frac{1}{\sqrt{32\pi}}\left(\frac{Z}{a}\right)^{5/2} re^{-Zr/2a}\sin\theta\sin\phi$$

$$\psi \qquad 2p_z = \frac{1}{\sqrt{32\pi}}\left(\frac{Z}{a}\right)^{5/2} re^{-Zr/2a}\cos\theta$$

$$\psi \qquad 3s = \frac{1}{81\sqrt{3\pi}}\left(\frac{Z}{a}\right)^{3/2}\left(27 - 18\frac{Zr}{a} + 2\frac{Z^2r^2}{a^2}\right)e^{-Zr/3a}$$

$$\psi \qquad 3p_x = \frac{\sqrt{2}}{81\sqrt{\pi}}\left(\frac{Z}{a}\right)^{5/2}\left(6 - \frac{Zr}{a}\right)re^{-Zr/3a}\sin\theta\cos\phi$$

$$\psi \qquad 3p_y = \frac{\sqrt{2}}{81\sqrt{\pi}}\left(\frac{Z}{a}\right)^{5/2}\left(6 - \frac{Zr}{a}\right)re^{-Zr/3a}\sin\theta\sin\phi$$

$$\psi \qquad 3p_z = \frac{\sqrt{2}}{81\sqrt{\pi}}\left(\frac{Z}{a}\right)^{5/2}\left(6 - \frac{Zr}{a}\right)re^{-Zr/3a}\cos\theta$$

$$\psi \qquad 3d_{xy} = \frac{1}{81\sqrt{\pi}}\left(\frac{Z}{a}\right)^{3/2} r^2e^{-Zr/3a}\sin^2\theta\sin 2\phi$$

$$\psi \qquad 3d_{xz} = \frac{\sqrt{2}}{81\sqrt{\pi}}\left(\frac{Z}{a}\right)^{7/2} r^2e^{-Zr/3a}\sin\theta\cos\theta\cos\phi$$

$$\psi \qquad 3d_{yz} = \frac{\sqrt{2}}{81\sqrt{\pi}}\left(\frac{Z}{a}\right)^{7/2} r^2e^{-Zr/3a}\sin\theta\cos\theta\sin\phi$$

$$\psi \qquad 3d_{x^2-y^2} = \frac{1}{81\sqrt{\pi}}\left(\frac{Z}{a}\right)^{7/2} r^2e^{-Zr/3a}\sin^2\theta\cos 2\phi$$

$$\psi \qquad 3d_{z^2} = \frac{1}{81\sqrt{\pi}}\left(\frac{Z}{a}\right)^{7/2} r^2e^{-Zr/3a}(3\cos^2\theta - 1)$$

[a]Z is the number of charges on the nucleus; a is the unit of distance, equal to 0.53 Å.

electron, so that we require that the integral over all space—which means all of the universe—be equal to unity for the probability function:

$$\int \psi^2 \, d\tau = 1 \qquad (8\text{-}35)$$

If the wave equation has been solved in such a way that this stipulation is not met, it is possible simply to multiply the expression for ψ by a suitable numerical factor such that Equation (8-35) will be satisfied by the result. Multiplication of this sort is permissible, since the wave equation is of such a form that the product of any of its solutions and a number is also a solution. Finally, we should mention that any possible solution that does not have a real positive square so that the electron probability can be a real positive number, or for which the square does not have a finite integral over all space, is immediately excluded from consideration as not representing a physically possible description of an electron in a stationary state.

A particular wave function, one of the physically acceptable solutions of the Schrödinger equation, is characterized by a set of orbital quantum numbers such as those we encountered earlier: n, l, and m_l. Examples of the mathematical form of these functions are shown in Table 8-3. The principal quantum number n retains the physical significance of describing the distance from the nucleus at which there is maximum probability of occurrence of the electron. Indeed, the quantity a, in terms of which wave functions are often written, is simply the radius of the first Bohr orbit in hydrogen, 0.53 Å.

However n also has another meaning, for it turns out to be equal to the total *number of nodal surfaces* present in the wave function to which it applies. Since the electronic wave functions obtained as solutions of the Schrödinger equation are products of separate factors, as indicated in Equation (8-34), it is necessary only that one of the factors be zero for the product of all three to be zero, regardless of the values of the other two coordinates. Thus we can associate nodes of spherical shape with particular values of r, planar nodes with particular values of ϕ, and pairs of cones having a common vertex with particular values of θ. Since the orbital quantum numbers represent numbers of various kinds of nodal surfaces, it is easy to see why they are limited to integral values.

Consider first the ϕ nodes, each of which consists of all the points at some value ϕ_k which causes Φ, and therefore also ψ, to be zero. The surface generated is a plane containing all the points having the value ϕ_k as well as those at $\phi_k + 180°$, a plane of which the z axis is a part and which, by analogy with navigational terminology, is called a *meridian* plane. The number of such planes in a wave function is equal to the magnitude of the quantum number m_l. For a wave to be a standing wave, the wave function must close upon itself in any curved path around the z axis; furthermore, the number of units of angular momentum of motion about that axis is equal to the magnitude of m_l and to the number of nodal planes that one would pass through in following a complete circuit around that axis, as illustrated in Figure 8-26.

Figure 8-26
Diagrammatic cross section of standing waves in circular paths. (a) With one unit of angular momentum around the z axis and one nodal plane. (b) With two units of angular momentum around the z axis and two nodal planes.

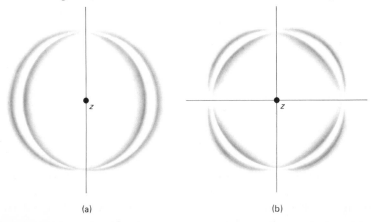

(a)   (b)

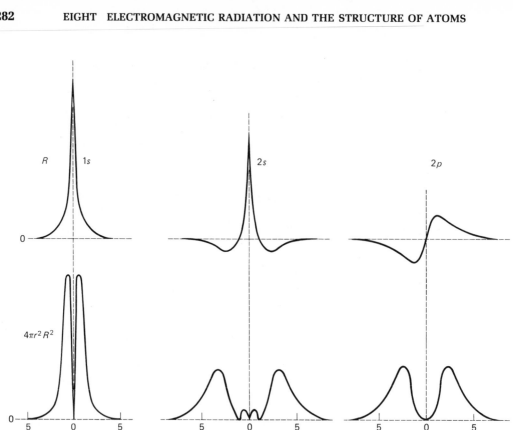

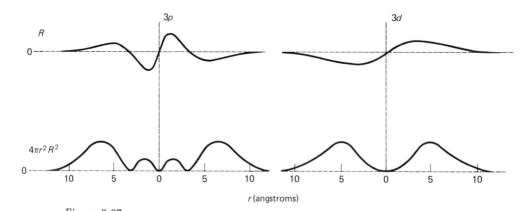

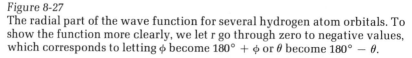

Figure 8-27
The radial part of the wave function for several hydrogen atom orbitals. To show the function more clearly, we let r go through zero to negative values, which corresponds to letting ϕ become $180° + \phi$ or θ become $180° - \theta$.

Nodes associated with values of θ come in pairs, one a cone centered about the positive z axis and the other a cone centered about the nega-tive z axis; the vertex of each cone is at the origin. The cones represent all points having some value θ_k, as well as those having θ equal to

$180° - \theta_k$. If the number of θ nodes is odd, one of them must be the xy plane, the equatorial plane, for which $\theta = 180° - \theta = 90°$; this counts only as a single node.

Both θ and ϕ nodes can be generated by moving a straight *line* about in space—a mathematician calls them ruled surfaces. The total number of such linear nodes in an orbital is equal to the value of the quantum number l for the orbital. The number of spherical nodes is equal to $n - l$, and for every orbital there must be at least one spherical node, the surface corresponding to an infinitely large value of r, so that the wave function approaches zero as r approaches infinity. Otherwise the square of the function would be an infinite number. Each spherical node corresponds to one unit of radial momentum—of "in-and-out" motion of the electron as contrasted to angular motion.

In visualizing a wave function, it is convenient to divide it into a radial part R and an angular part $\Theta\Phi$. In Figure 8-27 are plotted values of the R factor for several hydrogen orbitals, along with corresponding graphs of the function $4\pi r^2 \psi^2$ which gives the total probability that the electron will be found anywhere in a spherical shell of radius r. In Figure 8-28 are shown visual representations of the electron "cloud," depicting the regions in space in which an electron in a particular orbital has a high probability of being found.

An s-orbital pattern is spherically symmetrical about the nucleus and has no dependence on direction: The same variation of wave function and electron density with r is found, regardless of direction. In a 1s orbital, the only nodal surface is the sphere for $r = \infty$. The wave function and the electron density have maximum values at the nucleus, decrease with increasing distance from the nucleus, and approach zero asymptotically at large distances. For a 2s orbital there is, in addition, a spherical node at a finite value of r. The wave function has a large value at the nucleus, decreases to zero at the location of the finite node, changes sign at that distance, goes to a negative extreme, and then approaches zero at large distances. Thus the orbital consists of an inner core of electron density surrounded by an outer spherical shell of electron cloud, the two separated by the spherical nodal surface. For s orbitals with various values of n, the electron density pattern is simply divided into n concentric shells by the spherical nodal surfaces.

For any value of the principal quantum number equal to 2 or more, there are three different p orbitals with their axes mutually perpendicular. It is convenient to take these axes as coinciding with the x, y, and z axes of a rectangular coordinate system. Each 2p orbital has one planar node which is perpendicular to the axis of the orbital and divides the electron cloud into two portions, one along the axis in either direction from the nodal surface, so that the electron distribution is shaped something like a dumbbell. For quantum numbers higher than 2, p orbitals have the same directional characteristics, with one planar node passing through the nucleus and with the electron density concentrated along a line perpendicular to the nodal surface; however, each portion of the electron cloud is divided into successive shells by spherical nodes.

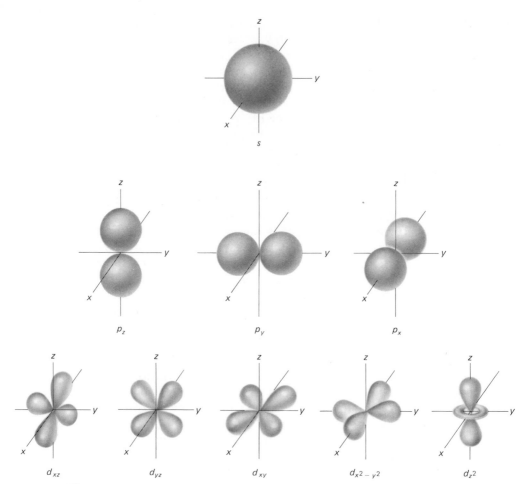

Figure 8-28
"Pictures" of the electron cloud for some hydrogen atom orbitals, representing regions of high electron density.

If one views a p_z orbital along the direction of the z axis, it is seen to be circularly symmetrical: A circuit around the z axis passes through no nodal planes. This orbital corresponds to an m_l value of zero. A section through a p_x or p_y orbital perpendicular to the z axis shows either of the orbitals divided into two parts by a nodal plane; these orbitals correspond to magnitudes of m_l of 1. Strictly speaking, values of $+1$ and -1, respectively, correspond to electrons with one unit of angular momentum of motion clockwise and counterclockwise around the z axis; since these represent traveling rather than standing waves, linear combinations of the wave functions are taken to yield the p_x and p_y orbitals conventionally pictured.

For any value of the principal quantum number n equal to 3 or more, there are five d orbitals, two corresponding to m_l of ± 2, two to ± 1, and one to 0. The last is concentrated along the z axis and has a pair of conical nodes. The regions of maximum electron density are two lobes, one along each of the cone axes, and a doughnut-shaped section in the xy plane. The other four orbitals are each shaped somewhat like a fourleaf clover, with four lobes of electron density in a generally planar arrangement. As in the case of the p orbitals, an increase in the principal quantum number does not change the directional characteristics but only introduces additional spherical nodes which divide the various lobes into inner and outer sections.

Since the solutions of the Schrödinger equation have the property that any linear combination of solutions is also a solution, there are ways other than those represented in Figure 8-28 in which one can sort out the five members of a set of d orbitals. In the absence of external electric or magnetic fields, all five have the same energy, and the energy is the same whatever combination of them is chosen. However, when atoms are attached to the atom to which the orbitals belong, the electronic energies are modified by the fields of the attached atoms. The orbitals we have described, which are represented in the figure, have been chosen because they are valuable in describing the bonding characteristics of transition metal ions, as they appear, for example, in hemoglobin, in cytochrome, and in many enzymes.

As implied in the preceding paragraph, it is convenient in discussing the electronic behavior of atoms containing more than one electron to use the electronic orbitals that have been obtained for hydrogen. As the atomic number increases, these orbitals shrink in toward the nucleus for a given principal quantum number, but an even more serious effect results from the electrostatic repulsions between electrons, for which allowance was not made in determining the hydrogen wave functions and which make the Schrödinger equation much more difficult to solve when they are introduced in the form of addition potential energy terms in that equation. In fact, it is not possible to obtain explicit algebraic expressions for the wave functions for any atom more complex than hydrogen. The wave equation can be solved numerically by successive approximations, given sufficient patience and a large enough computer, but the better the solutions become quantitatively, the more difficult it is to understand them in pictorial terms in a way chemically useful. It is therefore of considerable utility to extend to these atoms the directional characteristics of the orbitals found for the hydrogen atom.

Whatever model one chooses to visualize electronic orbitals and energy states of polyelectronic atoms, these are still characterized correctly by the quantum numbers introduced in Section 8-9. The mathematics of the wave model, indeed, provides a means of establishing rigorous links between quantum numbers, angular momenta, and energies, and justifies some of the assumptions of the simple Bohr theory.

EXERCISES

1. Calculate the energy in joules equivalent to 6×10^{23} photons of wavelength 500 nm.

2. What wavelength is associated with an electron of velocity 10^3 m/sec? What wavelength is associated with a proton of the same velocity?

3. Radiation has a wavelength of 1500 nm. Calculate the frequency, wavelength, and energy in ergs per photon. In what spectral region does the radiation fall?

4. Calculate the binding energy of the electron in a hydrogen atom and the radius of the Bohr electronic orbital for $n = 3$.

5. A wave obeys the equation
$$\phi = 6 \sin (5\pi t - 10\pi).$$
What are the amplitude, frequency, and wavelength?

6. What is the highest-order diffraction beam observable for light of wavelength 500 nm with a grating of spacing 1.2×10^{-6} cm?

7. Consider a set of vibrational energy levels uniformly spaced with an interval such that a photon of radiation of frequency 3×10^{14} sec^{-1} induces transitions from one level to the next higher level. What is the ratio of the population in the first excited level to that in the ground state at 300 K?

8. For Cs, the threshold photoelectric frequency corresponds to a 680-nm wavelength, for Zn, to a 340-nm wavelength. What are the minimum energies in ergs required to eject an electron from each of these metals? Plot diagrams showing the variation of the maximum kinetic energy of ejected photoelectrons with the frequency of the incident light for each metal.

9. Calculate the energy of a transition for a hydrogen atom in which the principal quantum number of the electron changes from 4 to 2. What is the frequency of the photon emitted in this change?

10. Carry out the algebraic steps leading to Equation (8-14).

11. Calculate the transition energy and wavelength for the third line of the Balmer series. In what spectral region would you attempt to observe this line?

12. An atomic state has an L value of 3 and an S value of $\frac{3}{2}$. What values of J are possible?

13. An atomic transition requires an energy of 10 kJ/mol. What are the frequency and wavelength of the photon required to induce this transition in one atom?

15. Sketch an energy diagram of the $^2S_{1/2}$, $^2P_{1/2}$, and $^2P_{3/2}$ levels of the sodium atom in a magnetic field in order to show the ten allowed transitions into which the D line is resolved.

16. Sketch a section in the yz plane of a $3d_{z^2}$ orbital, showing the intersections with nodal surfaces and the regions of maximum electron density.

17. List the number of various kinds of nodal surfaces in a 4f orbital with m_l equal to 2.

18. Draw diagrams like Figure 8-22 for the elements, C, Mg, Mn, Cu, Cd, La, Fe, I, and S.

19. An electron volt is defined as the amount of energy acquired by an electron as it is accelerated through a potential difference of 1 practical volt. Remembering that electrical work is equal to charge multiplied by voltage, and that there are 300 practical volts in 1 electrostatic volt, calculate the number of electron volts required to ionize a hydrogen atom.

20. What is the physical significance of the existence of three different 6p states of the mercury atom? How many different orientations can each of these states have in a magnetic field?

21. A beam of light passes from air into the plane surface of a transparent glass with a refractive index of 1.35. If the beam in air is at an angle of 30° from the perpendicular to the glass surface, what is its direction inside the glass?

22. A molecule which has the empirical formula C_3H_6O shows a molar refraction for the sodium D line of 17.08. What compound is it?

23. Radiation striking perpendicularly a diffraction grating of 200 lines per centimeter comes from the grating with the first-order diffracted beam at an angle of 15° from the initial direction. What is the wavelength of the radiation?

24. Calculate the specific rotation of a sugar if the observed angle of rotation is $+23.2°$ for a solution of 0.205 g/cm^3 concentration when measured in a 2-dm-length tube in a polarimeter.

REFERENCES

Books

Harry B. Gray, *Electrons and Chemical Bonding,* W. A. Benjamin, Menlo Park, Calif., 1964. Primarily concerned with bonding, this book has some introductory material concerning atoms and electrons.

N. N. Greenwood, *Principles of Atomic Orbitals,* Royal Institute of Chemistry, London, 1964. An excellent elementary introduction.

G. Herzberg, *Atomic Spectra and Atomic Structure,* Dover, New York, 1944. An old book, but still the classic, at an elementary to intermediate level.

R. M. Hochstrasser, *Behavior of Electrons in Atoms,* W. A. Benjamin, Menlo Park, Calif., 1964. An introduction.

G. W. King, *Spectroscopy and Molecular Structure,* Holt, Rinehart, and Winston, New York, 1964. Contains good coverage at an intermediate level.

H. G. Kuhn, *Atomic Spectra,* Academic Press, New York, 1969. A very useful detailed treatment at a fairly advanced level; much specific data is included.

W. J. Moore, *Physical Chemistry,* 4th ed., Prentice-Hall, Englewood Cliffs, N.J., 1972. Contains a helpful introduction.

Robert M. Rosenberg, *Principles of Physical Chemistry,* Oxford University Press, New York, 1977. Quite mathematical, but Chapter 4 includes a variety of helpful diagrams.

W. A. Shurcliff and S. S. Ballard, *Polarized Light,* Van Nostrand, Reinhold, New York, 1964.

H. E. White, *Introduction to Atomic Spectra,* McGraw-Hill, New York, 1934. Written from the physicists' viewpoint, this is an old but still valuable book.

Journal Articles

K. E. Banyard, "Electron Correlation in Atoms and Molecules," *J. Chem. Educ.* **47,** 668 (1970).

R. Stephen Berry, "Atomic Orbitals," *J. Chem. Educ.* **43,** 283 (1966). Very good, with an extensive bibliography.

Irwin Cohen and Thomas Bustard, "Atomic Orbitals—Limitations and Variations," *J. Chem. Educ.* **43,** 187 (1966).

Pierre Connes, "How Light is Analyzed," *Sci. Am.* **219,** 72 (September 1968).

Gerald Feinberg, "Light," *Sci. Am.* **219,** 50 (September 1968).

Jon A. Kapecki, "An Introduction to X-Ray Structure Determination," *J. Chem. Educ.* **49,** 231 (1972).

E. Lazzarini and M. M. Bettoni, "Teaching Moseley's Law," *J. Chem. Educ.* **52,** 454 (1975).

Kenneth J. Miller, "The Spectrum of Atomic Lithium," *J. Chem. Educ.* **51,** 805 (1974).

E. A. Ogryzlo and G. B. Porter, "Contour Surfaces for Atomic and Molecular Orbitals," *J. Chem. Educ.* **40,** 256 (1963).

Berta Perlmutter-Hayman, "The Graphical Representation of Hydrogen-Like Wave Functions," *J. Chem. Educ.* **46,** 428 (1969).

F. Dow Smith, "How Images Are Formed," *Sci. Am.* **219,** 97 (September 1968).

R. Marshall Wilson, Edward J. Gardner, and Richard H. Squire, "The Absorption of Light by Oriented Molecules," *J. Chem. Educ.* **50,** 94 (1973).

Nine
Bonding and
Molecular
Spectroscopy

This chapter develops further and applies to the structure of molecules the elementary principles of radiation and atomic structure introduced in Chapter 8. In the first portion, we consider how the combination of atoms to form molecules can be described and understood in terms of suitable models of electronic behavior. First, diatomic molecules are treated, and then polyatomic molecules, for which questions of interbond angles and electron delocalization are often important, are discussed. A further section deals with the formation of "coordinate bonds" between electron donor groups and the ions of the transition metals, bonds which represent a complex mixture of covalent, dipolar, and ionic character, and which are of considerable biological importance, as, for example, in the binding of metals in porphyrin molecules.

The second portion of the chapter deals with the broad subject of molecular spectroscopy—the absorption and emission of radiation by molecules as they are transformed from one energy state to another—and how we learn about molecular structure by examining the characteristics of the various discrete energy levels in which the molecules may exist. Moreover, molecular spectra, like atomic spectra, provide qualitative and quantitative means for determining the presence of various species, as well as of particular kinds of functional groups or structural units. Included in this portion of the chapter are accounts of rotational, vibrational, and electronic spectroscopy. Finally we consider optical dispersion—the behavior of the refractive index in the vicinity of an absorption band. Some applications of bonding considerations and spectroscopic information to problems of biological interest are examined: hydrogen bonding; singlet oxygen; the structure of amide units in proteins.

9-1
IONIC AND
COVALENT BONDING

The formation of certain types of compounds is based primarily upon the electrostatic attraction between positive and negative ions. For instance, sodium chloride is formed by the transfer of electrons from sodium atoms to chlorine atoms, resulting in positive sodium ions and negative chloride ions. The fairly great stability of sodium chloride in the solid state results from the circumstance that six chloride ions can surround each sodium ion and that six sodium ions in turn can surround each chloride ion. Since electrostatic forces do not saturate—the magnitude of one interaction is not reduced by the presence of other interactions—the total energy involved per ion is thus several times what it is for a single sodium ion–chloride ion pair.

Our primary concern is not with ionic bonding, however, but rather with covalent bonding, which differs in that the bonds are directional and that the number of bonds formed by an atom is limited by the number of orbitals in that atom that are of appropriate energy to overlap profitably with those of other atoms. A covalent bond can be simply described as an arrangement in which one or two electrons—usually two—occupy an orbital spread over more than one atom and thus link the atoms together. Such an orbital is appropriately termed a *molecular orbital*.

Of course a covalent bond may have some ionic character as a consequence of unequal sharing of the electrons in the bond by the atoms that it joins. If the two atoms have different tendencies to attract electrons, that is, different *electronegativities*, the one with the greater electronegativity has a larger share of the electrons, so that there is some separation of charge and some electrical polarity associated with the bond. The bond is then also more stable than it would be in the absence of ionic character. An example is hydrogen chloride, for chlorine is more electronegative than hydrogen, and the strength of the bond in this molecule is consequently greater than the geometric average of the strengths of a hydrogen–hydrogen bond and a chlorine–chlorine bond, a bond strength that might be expected if there were no ionic character. Although most covalent bonds have some ionic character or polarity, we shall not concern ourselves much with this aspect of their nature in the present discussion.

9-2
BONDING IN
DIATOMIC MOLECULES

As the simplest example of a covalently bonded unit, consider first the hydrogen molecule, H_2. Each of the two atoms has available a 1s orbital. To approximate the molecular orbital for this molecule, we imagine each atom contributing its 1s orbital to overlap with that of the other atom, so as to form a molecular orbital as represented in the lower part of Figure 9-1. To obtain a mathematical expression for the wave function of the molecular orbital, we add together the wave function expressions for the two 1s orbitals. The molecular orbital is occupied by the two available electrons, one from each atom, and in order that the two electrons can be in the same orbital, their spins must be opposite; consequently, the total spin of the molecule is zero. The electron pattern for this molecular orbital has a region of fairly high density between the nuclei, which is favorable for bonding.

The process of combining two atomic orbitals must lead to two molecular orbitals—no more, no less. If we choose for the H_2 molecule the combination described, which results in the strongest bond when the orbital is occupied by a pair of electrons, then the requirement that the two molecular orbitals be mathematically independent of each other—in more precise terms, that they be *orthogonal* to one another— necessitates that the second molecular orbital be made by subtracting one atomic 1s orbital from the other 1s orbital, as shown in the upper part of Figure 9-1. In this combined orbital, the two contributing wave functions tend to cancel one another in the region between the nuclei, and therefore occupation of the orbital is unfavorable for bonding; that is, its energy is well above that of the contributing atomic orbitals. Of course it is vacant in the H_2 molecule, but in an He_2 molecule there would be four valence electrons to be accommodated, two of which would occupy the low-energy orbital only to have their effect offset by the other two which would be forced to occupy the high-energy orbital; as a consequence, an He_2 molecule is not stable.

The orbital occupied in the H_2 molecule is termed a *bonding* orbital, since the electrons in this orbital have a more favorable energy situation than if they were in 1s orbitals in hydrogen atoms. An orbital of

Figure 9-1
Formation of molecular orbitals by two hydrogen atoms, A and B. The wave function ψ for the antibonding orbital σ^*1s has opposite signs in the two halves of the orbital.

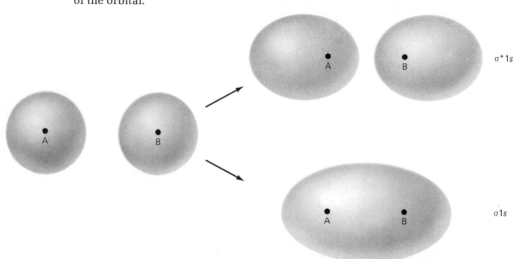

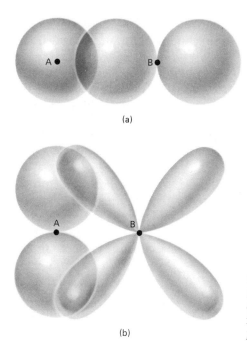

(a)

(b)

Figure 9-2
Possible geometric relations of orbitals
on two atoms, A and B, overlapping to
form a bond. (a) s and p orbitals. (b)
p and d orbitals.

higher energy than the atomic orbitals from which it is formed, an
orbital of the sort that would have to be occupied in an He_2 molecule,
is termed an *antibonding* orbital and is designated by attaching an
asterisk to its symbol.

In an imaginary cross section through either the bonding orbital
or the antibonding orbital of the hydrogen molecule, the wave function appears circularly symmetric and there is no evident division, by
nodes, of the orbital into regions of opposite sign. An electron occupying a circularly symmetric orbital has no angular momentum of motion
about the interatomic axis. By extension of the code in which an electron l value of 0 is represented by a lowercase s and an atomic L value
of 0 is represented by an uppercase S, we assign to this orbital the
Greek lowercase letter sigma (σ) to indicate that the value of the orbital
angular momentum quantum number λ is 0. The bonding orbital for H_2
is thus labeled σ1s, and the antibonding orbital σ*1s. The molecule Li_2
consists of two lithium atoms held together by the overlap of a 2s orbital from each atom to form σ2s and σ*2s molecular orbitals, of which
only the bonding orbital is occupied, much as for the H_2 molecule. In
addition, four electrons occupy the σ1s and σ*1s orbitals.

For diatomic molecules formed of two like atoms of the elements
boron and beyond in the second row of the periodic table, elements
in which p orbitals are involved in bond formation, it is necessary to
consider the requirement that a nodal plane directed from one atom
toward the atom bonded to it must be common to both the overlapping
orbitals. If the two orbitals concerned do not mesh together properly,
the result is that positive regions of one cancel the negative regions
of the other and there is no buildup of electron density between the
atoms to form a bond. To illustrate, an s orbital can overlap a p orbital

from the direction along the axis of the p orbital, as shown in Figure 9-2, but not along a direction lying in the nodal plane of the p orbital. One way in which a d orbital can overlap with a p orbital is also shown.

In a molecular orbital formed by two parallel p orbitals, an observer looking along the bond axis sees the wave function divided into two parts by the nodal plane. The electron has one unit of angular momentum about the bond axis, and the orbital is termed a π orbital, or said to form a π bond, indicating a value of 1 for the quantum number λ. Multiple bonds between atoms are usually formed by simultaneous bonding through a σ bond and one or two π bonds.

Figure 9-3 shows diagrams of molecular orbitals formed from $2p$ atomic orbitals in diatomic molecules. In describing these, the z direction is taken as the direction of the molecular axis, and the x and y directions are perpendicular to the axis.

Relative energies of the orbitals for some diatomic molecules are shown qualitatively in Figure 9-4. The energies vary with atomic size, for it is the size that determines the amount of overlap possible. One consequence of this effect is the crossing of the energies for the $\sigma 2p_z$ and the $\pi 2p_x$ and $\pi 2p_y$ levels. Electrons in any particular diatomic molecule are assigned to molecular orbitals in the order of increasing orbital energy, from the bottom to the top of the diagram, continuing

Figure 9-3
Molecular orbitals formed by the overlap of $2p$ orbitals on two like atoms, A and B. Antibonding orbitals are designated by an asterisk.

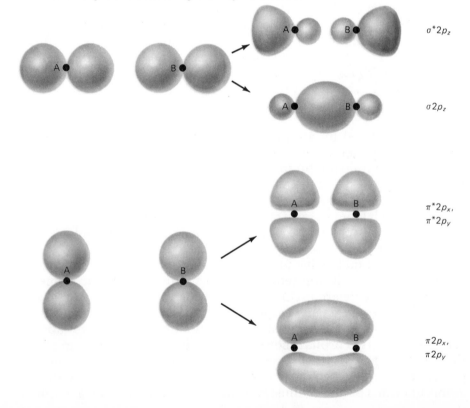

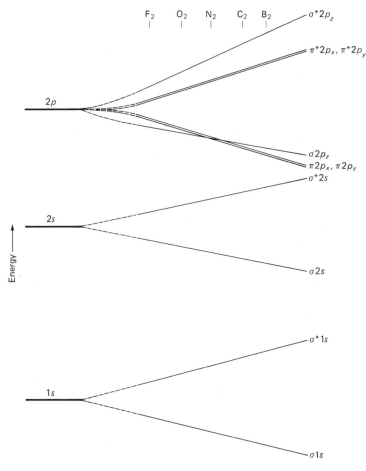

Figure 9-4
Schematic diagram of molecular orbital energies for homonuclear diatomic
molecules. Formulas are placed at positions approximately corresponding to
relative internuclear distances.

until all the electrons have been disposed of. When two electrons are
to occupy any one orbital, their spins must be opposite to one another.
In N_2, 14 electrons are assigned in pairs to the first seven molecular
orbitals from the bottom of the diagram. Since only two of these orbitals are antibonding, whereas five are bonding, there is an excess of
three pairs of bonding electrons over antibonding electrons, and the
nitrogen–nitrogen bond can be regarded as a triple bond. Indeed, it is
about three times as strong as the bond in F_2, which has an excess of
only one pair of bonding electrons, the equivalent of a single bond.

In assigning electrons to orbitals there is one choice we have not yet
provided for: how to distribute two electrons in two orbitals of equal
energy. Experience, embodied in the generalization called Hund's rule,
which was mentioned in Section 8-9, dictates that the electrons be
assigned to different orbitals of the same energy whenever possible,
rather than to the same orbital, in order that their spins may be par-

allel. One important consequence of this is that the oxygen molecule in its ground state has a total spin of 1, for the last two electrons can be assigned one each to the π^*2p_x and π^*2p_y molecular orbitals without any disadvantage in orbital energy, and thus the favored state of parallel spins can be achieved. As a result, the oxygen molecule has a permanent magnetic moment, in contrast to most stable diatomic molecules with an even number of electrons, which have all spins paired.

9-3
HYBRID ORBITALS

When an atom is linked to more than one other atom in a molecule, it is possible to determine by experiment the angle between the several bonds, and to deduce from this angle something about the orbitals used by the central atom in bonding. An atom tends to use those orbitals that (a) give the most extensive overlap with the orbitals of the bonded atoms and (b) minimize the repulsions between the several atoms attached, as well as repulsions involving unshared electron pairs. In many cases, it is not the s orbitals or p orbitals or d orbitals that are employed, but rather combinations or hybrids of several of these orbital types. Orbitals in the s, p, or d form can be used with no more than partial effectiveness, because that half of the orbital on the other side of the atom from the attached unit cannot contribute effectively to the overlap in the bond. As we shall see, mixed or *hybrid* orbitals tend to be concentrated on one side of the nucleus and therefore can overlap with orbitals of other atoms to a much greater extent. The mathematical justification for forming hybrid orbitals is the principle, mentioned in Chapter 8, that any linear combination of solutions of the wave equation is also an acceptable solution of the equation, and the wave functions for hybrid orbitals are obtained by simply taking linear combinations of s, p, and d atomic wave functions.

As an example, it is found that the molecule $BeCl_2$ is linear in the gas phase with the structure

$$Cl-Be-Cl$$

The lowest empty orbitals of the beryllium atom are 2p orbitals, and bonding by these would lead to a 90° angle between the two bonds, rather than the 180° angle observed. However, if in a beryllium atom a 2s orbital is combined with a 2p orbital, there result two hybrids which are mathematically independent of one another; the normalized wave functions for these can be written

$$\psi_{\mathrm{I}} = \frac{1}{\sqrt{2}}(\psi_{2s} + \psi_{2p_x}) \tag{9-1}$$

$$\psi_{\mathrm{II}} = \frac{1}{\sqrt{2}}(\psi_{2s} - \psi_{2p_x}) \tag{9-2}$$

The shapes of the two orbitals corresponding to these expressions are shown in Figure 9-5a, and it is obvious that they are concentrated in

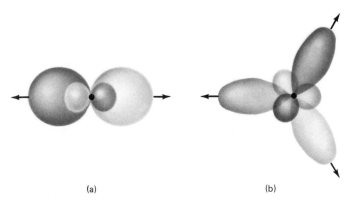

(a)   (b)

Figure 9-5
Hybrid orbitals, with bond-forming directions indicated by arrows. (a) Two sp
hybrids on one atom. (b) Three sp^2 hybrids on one atom.

directions diametrically opposed. Each orbital is circular in cross section and thus can participate in σ bonds. In $BeCl_2$ each bond to beryllium consists of a beryllium sp hybrid overlapping a valence orbital
of chlorine, with the resulting joint orbital occupied by two electrons.
Since chlorine is bonded to only one other atom, it is not easily established whether its valence orbital is a p orbital or an sp hybrid.

The molecule acetylene provides another example of sp hybridization. In the linear unit H—C—C—H, the two carbon atoms are joined
by a σ bond formed by the overlap of an sp hybrid from each carbon,
and each carbon atom also uses an sp hybrid for bonding to a hydrogen atom. There remain on each carbon atom two p orbitals not involved in the hybridization, each with its axis perpendicular to the
molecular axis. Overlap of these p orbitals leads to two π bonds between the carbon atoms, with their nodal planes at right angles to
one another.

The molecule BF_3 is planar with angles of 120° between any two
boron–fluorine bonds. Again no set of three orbitals from among the
group 2s, $2p_x$, $2p_y$, and $2p_z$ on the boron atom can provide these directional characteristics. However, it turns out that a hybrid made up of
the 2s orbital and two of the 2p orbitals does give the appropriate directions. The equations for combining the wave functions are

$$\psi_I = \frac{1}{\sqrt{3}}\psi_{2s} + \frac{\sqrt{2}}{\sqrt{3}}\psi_{2p_x} \tag{9-3}$$

$$\psi_{II} = \frac{1}{\sqrt{3}}\psi_{2s} - \frac{1}{\sqrt{6}}\psi_{2p_x} + \frac{1}{\sqrt{2}}\psi_{2p_y} \tag{9-4}$$

$$\psi_{III} = \frac{1}{\sqrt{3}}\psi_{2s} - \frac{1}{\sqrt{6}}\psi_{2p_x} - \frac{1}{\sqrt{2}}\psi_{2p_y} \tag{9-5}$$

Corresponding diagrams are given in Figure 9-5b. Note here that, in
forming hybrid orbitals, it is the wave functions themselves, complete
with sign, that are combined. Only after the linear combination has
been made does one compute ψ^2 to determine the distribution of electron density.

In the molecule ethylene, each carbon atom uses three sp^2 hybrid orbitals to form bonds, one to the other carbon and one each to two hydrogen atoms. Left over from the formation of the σ-bond structure of the molecule is a p orbital on each carbon atom, perpendicular to the plane of the CH_2 unit and the sp^2 hybrids. These two p orbitals overlap one another to form a π bond between the two carbons. Such overlap is possible only if the molecule assumes an overall planar structure. The presence of a π bond in ethylene and similar molecules restricts rotation of the two parts of the molecule about the C—C bond.

The carbon atom of a carbonyl group is bonded in a fashion quite similar to that of an olefinic carbon. The bond angles about the carbon are approximately 120°, but with one of the three coplanar bonds involving the oxygen atom. The third p orbital on the carbon overlaps a p orbital of oxygen to form a π bond, although the electrons in this bond tend to be displaced toward the oxygen rather than shared equally as in a C—C double bond. It is not possible to say with certainty what type of orbital the oxygen contributes to the bond, for there is no bond angle criterion; the oxygen may be using a p orbital or an sp hybrid or a combination having some intermediate degree of hybridization. In any event, there is one unhybridized p orbital on the oxygen available for the π bond to carbon, and there are four unshared $n = 2$ electrons on the oxygen atom.

Saturated hydrocarbons, such as methane, have a tetrahedral arrangement of four bonds surrounding each carbon atom, with an interbond angle of 109°28′. The bonding orbitals of the carbons are sp^3 hybrids, with an orientation that can be pictured by placing the atomic nucleus at the center of a cube and drawing the axes of the orbitals out toward four of the cube vertices, so that each of the vertices chosen is diagonally across a cube face from each of the other three.

In discussing bonding in molecules having more than two atoms, we have ignored the presence of the antibonding orbitals also formed for each bonding orbital. It has not been necessary to consider these because, in the examples we have given, there are just enough electrons to fill the bonding orbitals, and the antibonding orbitals remain empty.

Some simple molecules do not fit precisely into any scheme of hybridization. An example is water, which has a bond angle of 105°, intermediate between the 90° angle for a set of pure p orbitals and the 109°28′ angle of tetrahedral hybridization. One possible description of the water molecule is to take the p-orbital angle as a starting point and imagine the hydrogen atoms to be forced somewhat apart by the electrostatic repulsions of the electrons joining them to the oxygen atom. An opposite approach is to consider the molecule as derived from a tetrahedron with the two unshared electron pairs on the oxygen occupying two of the four tetrahedral directions and the hydrogen atoms occupying the other two. The unshared electron pairs repel one another more strongly than the electrons involved in the bonds to hydrogen, because the latter are farther away from the oxygen. As the angle between the sp^3 orbitals for the unshared pairs is enlarged slightly above the tetrahedral angle, that between the two bonds to hydrogen must be reduced slightly below the tetrahedral value to compensate. The second of these models is probably more satisfactory than the first, but one must remember that in formulating either one a scientist is trying to describe a structure that does not fall neatly into place in a simple scheme; indeed, every molecular structure is the result of the balance of many forces and is only approximated by models which we devise to permit visualization of the bonding met.

**9-4
ELECTRON
DELOCALIZATION**

Perhaps the classic example of the importance of electron delocalization is the series of aromatic compounds including benzene and its derivatives. Structural studies, as well as earlier chemical deductions based on the number of different products of substitution in the benzene ring, led to the conclusion that benzene has all of its atoms in a single plane, with each of the six carbon atoms at one vertex of a regular hexagon. Since the six hydrogen atoms are also hexagonally arranged, the H—C—C bond angles, as well as the C—C—C angles, are 120°. From this geometry, one concludes that each carbon atom employs sp^2 hybrids for the bonds to the neighboring carbons and to the attached hydrogen. On each carbon there is, in addition, a p orbital which is perpendicular to the plane of the hybrids—and to the plane of the molecule—and which is not involved in σ-bond formation. To be distributed in the p orbitals are six electrons, unused in the σ-bond structure, one contributed by each carbon atom.

The parallel p orbitals can be combined to form π bonds. One set of such bonds is shown in Figure 9-6a, with pairs of p orbitals on neighboring carbons allowed to overlap so as to form three π bonds. However, there are several objections to this structure for the molecule:

(1) Carbon–carbon double bonds are usually about 1.33 Å long as compared to 1.50 Å or more for single bonds, whereas all six bond distances in the benzene ring are equal, about 1.40 Å in length.
(2) The molecule does not show the chemical reactivity of olefins, as, for example, in the reaction with bromine, which does not add across two neighboring carbons but rather replaces a hydrogen atom, leaving the π system intact.
(3) When hydrogen is added to a double bond in cyclohexene to form cyclohexane, the enthalpy of the reaction is -28.6 kcal/mol, and one would expect the hydrogenation of three double bonds in benzene to be associated with an enthalpy of reaction three times as great, or -85.8 kcal/mol. However, the experimental value is only -49.80 kcal/mol, which means that benzene has a more negative energy than expected by about 36 kcal/mol.
(4) There is another, equally good, structure, represented in Figure 9-6b, which differs only in the way in which the carbon atoms are paired off to form π bonds.

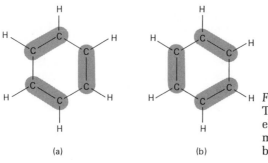

Figure 9-6
Two possible ways of assigning π electrons pairwise in a benzene molecule. Sigma bonds are indicated by lines.

A description of the electronic distribution in benzene more suitable than one of the forms in Figure 9-6 is represented by a structure obtained by superimposing an equal contribution from each. Each of the six carbon–carbon bonds is then identical and can be described as one-half single and one-half double, or as having a bond order of 1.5. This approach is often described as the *resonance* or *valence bond* method: Formulas are drawn in which each electron pair is assigned either to an orbital located on a single atom or to an orbital joining two neighboring atoms, and whenever necessary the true molecular structure is described as a superposition, or resonance hybrid, of the several contributing structures. If the contributing structures are exactly alike, as for benzene, then the molecule is said to be "stabilized by resonance" to a high degree. If the several structures differ one from the other in energy, then the actual molecule is more stable than any one of them, but the energy benefit is less than in the case where there are several equivalent forms.

The necessity of describing a molecule in terms of several contributing resonance "forms" corresponds to the situation in which electrons, usually π electrons, are not confined to a particular atom or pair of neighboring atoms but are spread out or delocalized over a larger region in the molecule. Electron delocalization corresponds to a favorable energy situation.

An alternative method of describing electron delocalization, and a method that has some advantages over the resonance description in treating electronic transitions to excited molecular states, to be discussed later in this chapter, is the *molecular orbital* method. The explanation we gave earlier of bonding in diatomic molecules was in essence an application of this method, one of the features of which is the introduction of antibonding as well as bonding orbitals. In applying the molecular orbital model to the description of electron delocalization in a molecule containing more than two atoms, one proceeds by setting up suitable combinations of atomic orbitals, usually but not necessarily of p orbitals, over as large a region as feasible, and then assigning the available electrons to these orbitals, beginning with the lowest-energy one and going in order of increasing energy until one has disposed of all the electrons.

How can we apply this to the benzene molecule? The combination to be made here is that of the six p orbitals on the six carbon atoms, and there must result six molecular orbitals—no more, no less. To combine the orbitals, we simply multiply each one by a positive or negative weighting factor indicating the extent of its contribution, and sum the results. A mathematical analysis shows that six suitable combinations for benzene are

$$\psi_{\text{I}} = \frac{1}{\sqrt{6}}(\psi_1 + \psi_2 + \psi_3 + \psi_4 + \psi_5 + \psi_6) \tag{9-6}$$

$$\psi_{\text{II}} = \frac{1}{\sqrt{12}}(2\psi_1 + \psi_2 - \psi_3 - 2\psi_4 - \psi_5 + \psi_6) \tag{9-7}$$

$$\psi_{\text{III}} = \frac{1}{2}(\psi_2 + \psi_3 - \psi_5 - \psi_6) \tag{9-8}$$

$$\psi_{\text{IV}} = \frac{1}{\sqrt{12}}(2\psi_1 - \psi_2 - \psi_3 + 2\psi_4 - \psi_5 - \psi_6) \tag{9-9}$$

$$\psi_{\text{V}} = \frac{1}{2}(\psi_2 - \psi_3 + \psi_5 - \psi_6) \tag{9-10}$$

$$\psi_{\text{VI}} = \frac{1}{\sqrt{6}}(\psi_1 - \psi_2 + \psi_3 - \psi_4 + \psi_5 - \psi_6) \tag{9-11}$$

These six molecular orbitals are represented in Figure 9-7. The orbital of lowest energy ψ_1 has no nodal surfaces perpendicular to the molecular plane, and the two regions of high electron density are doughnut-shaped, one above the plane of the nuclei and the other

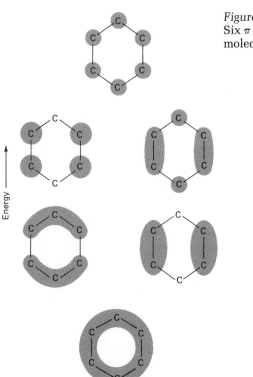

Figure 9-7
Six π molecular orbitals for the benzene molecule. Hydrogen atoms are not shown.

below. The next-higher-energy orbitals, ψ_{II} and ψ_{III}, each have one perpendicular nodal plane, dividing the electron cloud for each into four segments. These three wave functions represent bonding orbitals and contain the six available electrons. Orbitals ψ_{IV} and ψ_V each have two nodes perpendicular to the molecular plane, and orbital ψ_{VI} has three such nodes; with increasing fragmentation of the electron cloud by additional nodes, the energy increases, so that these orbitals are antibonding. The stability of the benzene ring is seen to be due to the fact that all the π electrons are in bonding orbitals and none are in antibonding orbitals, just as for the diatomic nitrogen molecule.

Consider the bonding in butadiene-1,3, with the σ structure,

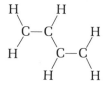

Each carbon atom uses three sp^2 hybrid orbitals to form its σ bonds, leaving a p orbital perpendicular to the plane formed by the three attached atoms, and there is one electron per carbon atom left over to form the π cloud of the molecule. One might suppose there to be, as the name of the molecule indicates, two double bonds, but the planarity of the molecule with the associated restriction on rotation about the bond between the second and third carbon atoms indicates that this bond has some partial π character. The molecular orbital approxima-

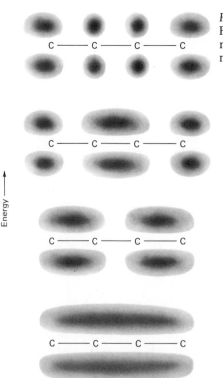

Figure 9-8
Four molecular orbitals of the butadiene molecule, viewed in the direction of the molecular plane.

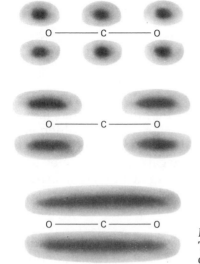

Figure 9-9
The three π molecular orbitals of carbon dioxide.

tion for butadiene combines the four carbon-atom p orbitals to give four orbitals extending over the whole carbon skeleton of the molecule. Of course, this requires that the nodal planes of the p orbitals form a single plane, which in turn establishes the rigidity of the whole molecule. As is the situation for benzene, the lowest-energy orbital, depicted in Figure 9-8, the one most effective in bonding, has no nodal surfaces aside from that inherited from the atomic p orbitals, and consequently electrons in the orbital are delocalized from one end of the carbon skeleton to the other. Energy increases with the number of nodal surfaces, the two antibonding orbitals having two and three nodes perpendicular to the molecular plane.

For another example, we turn to the three-center system in carbon dioxide. Since the molecule is linear, we know that the carbon atom is using two sp hybrid orbitals to form σ bonds to the two oxygens. We cannot determine with certainty what orbital an oxygen atom is contributing to its σ bond; it may be a p orbital, leaving the unshared pair of electrons in the s orbital, or, more likely, an sp hybrid, leaving the unshared pair of electrons in the other sp hybrid. There are then available on the carbon atom and on each oxygen atom two p orbitals perpendicular to the molecular axis. There remain eight electrons available for the π orbitals. Since the two directions perpendicular to the axis of the molecule are equivalent to one another—we arbitrarily label these the x and y directions—and are independent of one another, we assign four electrons to each direction and treat each one separately. The resulting molecular orbitals in the xz plane are shown in Figure 9-9. The molecular orbital of intermediate energy is termed *nonbonding*, for its occupation by electrons neither adds to nor subtracts from the molecular energy. A parallel set of molecular orbitals is concentrated in the yz plane and also holds four electrons.

In a carboxylate ion derived from an aliphatic carboxylic acid, there is a similar situation for the orbitals perpendicular to the plane of the ion, with one pair of electrons in the bonding orbital and one in the nonbonding orbital. In the corresponding resonance model these electron pairs are represented alternately as unshared on an oxygen atom and as located in a π bond between the carbon and an oxygen.

Similar delocalization of electrons in the undissociated molecule, RCOOH, would lead to a separation of electric charge, which is energetically unfavorable, whereas delocalization in the ion leads to the favorable result of distribution of the negative charge over a larger region. Such stabilization of the ionic conjugate base, the carboxylate ion, compared to the proton donor acidic species provides an explanation of the much greater acid strength of the OH in a carboxyl group than of an alcoholic OH.

We turn now to other examples in which partial double-bond character leads to restricted rotation. In the molecule N,N-dimethylacetamide, spectroscopic evidence indicates that the two N-methyl groups are in different environments, one cis to the oxygen and the other trans to the oxygen. Free rotation about the bond between nitrogen and carbon would lead to a time-averaged, identical environment for the two methyls. The partial double-bond character of the N—C bond can be represented by a resonance formulation:

The nitrogen and carbon atoms each use a trio of sp^2 hybrids to form σ bonds, and the third p orbital on the carbon must remain parallel to the third p orbital of the nitrogen if delocalization is to occur. The peptide linkage in proteins represents a further example of the amide arrangement, and the planarity of the unit formed by nitrogen, the carbonyl group, and the three directly attached atoms is very significant in determining the conformations assumed by protein molecules.

The N-nitroso-amines form another series of compounds with rather large barriers to rotation about what might be thought to be a single bond. For example, ethyl phenyl nitrosamine exists in two isomeric forms:

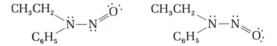

Each of these forms has a substantial contribution from the structure

$$\underset{R''}{\overset{R'}{>}}\overset{\oplus}{N}=N-\overset{..}{\underset{..}{O}}{:}^{\ominus}$$

In these molecules, as in amides, the fact that the negative charge resulting from electron delocalization falls upon an oxygen atom, an electronegative atom, enhances the stability of the structure with charge separation. Notice that we have chosen to use the valence bond symbolism to describe delocalization because it is easier to picture; the reader should attempt to visualize the corresponding molecular orbital representation.

9-5
COMPLEXES FORMED BY METAL IONS

It is quite common for metal ions to form complexes with electron-rich groups. To indicate that the latter donate electron pairs to occupy vacant orbitals on the metal, rather than sharing one electron in a pair, the bonds in these complexes are often termed *coordinate bonds*. Typical electron-donating groups are a halide ion, cyanide ion, amino group, or the carboxylate ion of an organic acid, and water or ammonia molecules. The ion or molecule *bound* to the metal is termed a *ligand*. Often one ligand molecule, such as ethylenediamine or glycine, contains two donor groups each of which forms a bond to a metal ion; the ligand is then said to be bidentate.

The species $AlCl_4^-$, which is formed when Al^{3+} accepts four electron pairs from four chloride ions, is a relatively simple example of a complex. In view of the tetrahedral structure of the $AlCl_4^-$ ion, it appears that the aluminum atom is using sp^3 hybrid orbitals to accommodate the four electron pairs.

For elements with atomic numbers greater than 20, particularly those elements referred to as transition metals, d orbitals come into play as acceptors of electrons in coordination bonds. Complexes involving d orbitals exist with a variety of coordination numbers and geometric structures, but perhaps the most common type is the octahedral complex in which six groups surround the central metal atom at locations corresponding to the positive and negative x, y, and z axial directions of a rectangular coordinate system, thus with interbond angles of 90°, as seen in Figure 9-10.

One explanation of the disposition of electrons in an octahedral complex proceeds in the following way: Each of the donor groups is an electron-rich center which repels the electrons already in the valence shell of the metal ion. Because the d_{z^2} and $d_{x^2-y^2}$ orbitals are concentrated along the x, y, and z axes, electrons occupying these orbitals are repelled by the donors more than electrons occupying the d_{xy}, d_{xz},

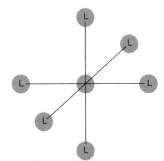

Figure 9-10
Geometry of an octahedral complex, with interbond angles of 90° and 180° for the bonds from a central metal atom to six ligand groups.

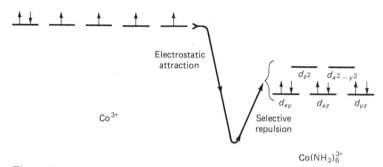

Figure 9-11
Energy levels of a free cobalt ion and a strong-field octahedral cobalt(III)
complex.

and d_{yz} orbitals, which are concentrated at directions intermediate between the axes. Thus, while the electron occupation of any of the five
orbitals is less favorable when ligands are present than in their absence, the energies of the d_{z^2} and $d_{x^2-y^2}$ orbitals are raised more than
the energies of the other three orbitals. This effect is represented schematically in Figure 9-11.

Consider the complex formed by a cobalt(III) ion and six ammonia
molecules, $Co(NH_3)_6^{3+}$. A cobalt atom has an electron configuration
equivalent to that of argon plus $3d^7 4s^2$. The formation of the tripositive
ion involves the loss of the two $4s$ electrons and one $3d$ electron, leaving six $3d$ electrons. Figure 9-11 shows the distribution of electrons
among the five $3d$ orbitals in the free Co^{3+} ion, corresponding to a total
spin of 2, and the redistribution of electrons to the three lower-energy
orbitals under the influence of the ligands, giving a complex in a singlet
state with zero spin.

In other complexes, the ligand groups have a smaller effect upon the
relative energies of the d orbitals. If the orbital energy difference is
smaller than the energy required to pair the spins, then the complex
may retain the same spin as the free ion. An example is CoF_6^{3-}, termed
a *high-spin* or *weak-field* complex, as contrasted with $Co(NH_3)_6^{3+}$
which is a *low-spin* or *strong-field* complex.

The model just described is based entirely upon electrostatic attractions between the positive metal ion and the negative charge, or the
negative end of an electric dipole, on the ligand group. Undoubtedly
there is also a major contribution to complex formation from covalent
bonding. Thus in the low-spin complex of cobalt, one can view the d_{z^2}
and $d_{x^2-y^2}$ orbitals as hybridized with $4s$ and $4p$ orbitals to give a set of
six $d^2 sp^3$ hybrids of octahedral geometry and directed so that one
hybrid orbital can be used in a bond with each of the ligand groups,
with the ligand viewed as contributing both electrons of the pair occupying the bond.

In addition to covalent σ bonds, complexes can also be stabilized by
the formation of π bonds between the metal ion and the ligand. For
example, electrons in the d_{xy}, d_{xz}, or d_{yz} orbital can be "back-donated"
by the metal ion to the ligand if the latter has available orbitals of suitable symmetry to overlap with the appropriate d orbitals. This effect
explains why unsaturated ligands, such as imidazole or o-phenanthro-

line, are much more strongly bound by metal ions than would be expected from their basicity as determined by other measures.

Complexes of iron are of particular biological importance, constituting the key functional groups of hemoglobin, of cytochromes, and of enzymes such as catalase and peroxidase, as well as being present in "nonheme" iron compounds such as ferredoxin, as described in Chapter 7. Figure 9-12 shows the distribution of electrons in octahedral ferrous and ferric complexes for both low- and high-spin cases.

Hemoglobin and myoglobin contain octahedral ferrous complexes in which four positions about the metal ion are occupied by nitrogen atoms which are part of a planar porphyrin ring, forming a heme group; the fifth position is occupied by a nitrogen atom from an imidazole group of a histidine residue in the protein globin; and the sixth position can be occupied by water, oxygen, or carbon monoxide. When the ligand is water, the complex is of the weak-field type, with a spin of 2 since there are four unpaired electrons; but when the ligand is O_2, the iron becomes low-spin and the electrons in the oxygen molecule become paired, so that the whole system has zero magnetic moment, although the iron is still in the ferrous oxidation state. Oxidizing agents other than oxygen convert the ferrous proteins into high-spin ferric forms called metmyoglobin and methemoglobin. Strong-field ligands such as CN^- can replace the water molecule in the sixth position, converting methemoglobin to a low-spin ferric state.

In addition to four polypeptide chains, two designated α and two designated β, a hemoglobin molecule contains four heme groups, each with an iron atom capable of taking up an oxygen atom. An interesting, and by no means fully explained, feature of the behavior of hemoglobin is the cooperative nature of oxygen binding. Contrary to the behavior often found for a series of equilibria occurring in steps, the binding constant for oxygen becomes greater, the more oxygen molecules there are already bound by a hemoglobin molecule. This has the physiological benefit of making oxygen less tightly bound and therefore more readily *available* when the concentration is low, but it implies transmission through a considerable distance—perhaps 25 Å— in the hemoglobin molecule of the information that an oxygen molecule is attached at one site.

There are other types of coordination complexes in addition to the octahedral structures we have discussed. Some complexes in which

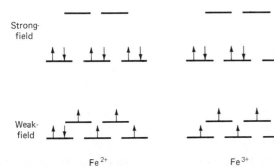

Figure 9-12
Energy levels and electron occupation of the levels in strong- and weak-field ferrous and ferric octahedral complexes.

the coordination number is 4, such as that of cupric ion and glycine, are arranged with the ligands at the vertices of a square planar pattern:

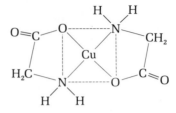

A similar square planar complex is $Ni(CN)_4^{2-}$, in which the eight $3d$ electrons occupy all the $3d$ orbitals except the $3d_{x^2-y^2}$, the lobes of which are directed toward the four donor groups and which therefore has a very high energy.

Other four-coordinate complexes, such as VCl_4 and $CoCl_4^{2-}$, are tetrahedral in shape, with donor groups situated in positions intermediate between the three coordinate axes, in each of four alternating octants about the metal ion. Thus the d_{xy}, d_{xz}, and d_{yz} orbitals are of higher energy for electrons from the metal but are in the proper orientation to be hybridized with the $4s$ orbital to give a set of sd^3 hybrids to form σ bonds containing electron pairs donated by the ligands.

9-6
VIBRATIONS
IN DIATOMIC MOLECULES

In the discussion of interatomic bonding in the previous sections, the concept *bond strength* was used qualitatively, without a specific definition. We consider here two ways of assigning a concrete meaning to this term. The first is by determining the energy required to pull apart the atoms forming the bond so that the bond is broken, and the second is by measuring the resistance of the molecule to distortion from its normal equilibrium geometry. Distortion of the molecular shape is related to the excitation of vibrations within the molecule; during these vibrations the distances between atoms fluctuate. Vibrations can be produced thermally by collisions with other molecules, or they can be spectroscopically induced by absorption of energy from a photon of radiation. The study of vibrational spectra is fruitful as a means of establishing the magnitude of the forces resisting molecular deformation.

Energy is plotted as a function of interatomic distance for a typical diatomic molecule as the solid curve in Figure 9-13. The minimum of this "potential energy" curve corresponds to the equilibrium internuclear separation r_e. From the curve, one can determine the amount of energy required to increase or decrease the interatomic distance by any desired amount. At small distances, the energy rises sharply because the electron clouds of the two atoms repel one another and, at very small distances, the two positive nuclei repel each other. Deformed in the opposite sense, to distances greater than the equilibrium value, the bond finally reaches a point at which it breaks; the plateau at the right of the diagram corresponds to the energy level at which dissociation of the molecule occurs. The energy difference between the minimum in the curve and the plateau is the energy of the bond, one measure of its strength.

A diatomic molecule behaves much like a system of two masses on the ends of a spring: If the masses are pulled apart, so as to stretch the spring, and then released, the spring pulls them back inward to the equilibrium position. However, they reach the equilibrium position traveling with some velocity, and their inertia causes them to over-shoot so that they continue to approach one another until all their kinetic energy has been converted into potential energy associated with repulsive forces and they slow to a stop; they then move apart, again overshoot the equilibrium position, and are pulled back once more after their kinetic energy has been used up in stretching the va-lence bond. If this vibrational motion is analyzed as described in the following section, it is possible to determine the ratio of the stretching force applied to the bond to the amount by which the bond stretches, a quantity known as the *force constant* for the bond, which provides a second parameter that can be used to characterize bond strength.

THE HARMONIC OSCILLATOR

As a first approximation, one can treat a vibrating molecule as a har-monic oscillator, a system in which the restoring force is proportional to the displacement from equilibrium. An example of a harmonic

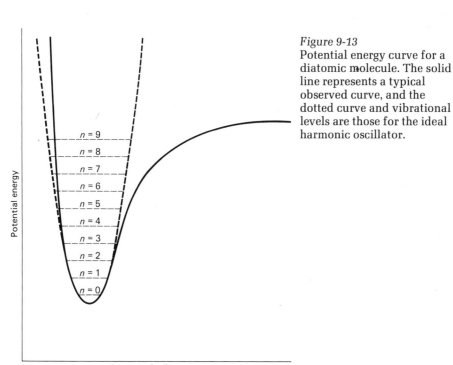

Figure 9-13
Potential energy curve for a diatomic molecule. The solid line represents a typical observed curve, and the dotted curve and vibrational levels are those for the ideal harmonic oscillator.

$n = 9$
$n = 8$
$n = 7$
$n = 6$
$n = 5$
$n = 4$
$n = 3$
$n = 2$
$n = 1$
$n = 0$

Potential energy

Interatomic distance

oscillator is a spring that obeys Hooke's law:

$$f = -kx \tag{9-12}$$

where f is the force exerted by the spring when its displacement from equilibrium length is x. The coefficient k is independent of x so long as the spring behaves ideally. Imagine a body suspended on the end of a Hooke's law spring. The acceleration the body undergoes is equal to the ratio of the force exerted upon it to its mass:

$$a = \frac{f}{m} \tag{9-13}$$

or

$$\frac{d^2x}{dt^2} = -\frac{k}{m}x \tag{9-14}$$

The solutions of this differential equation are functions showing how x varies with t, the time, and the equation implies that any such function, $x = f(t)$, has a second derivative equal to the function itself multiplied by a negative number. Either the sine or cosine function satisfies this requirement. We can use either or a linear combination of both; since the sine function has the value zero when the time is zero, we choose that function:

$$x = A \sin 2\pi\nu t \tag{9-15}$$

This relation was written earlier as Equation (8-4) to describe the motion of a source of electromagnetic radiation. Since the maximum value of the sine function is unity, the quantity A corresponds to the maximum value of the displacement, which is termed the *amplitude* of the vibration. The sine function repeats every 2π units of angle, so that if the time at the beginning of a cycle is zero and at the end of the cycle is τ, then $\nu\tau$ must be unity to make the sine again be zero; τ is termed the *period* of the vibration, and its reciprocal ν is the *frequency*.

When the function x and its second derivative with respect to t are substituted in Equation (9-14), there is obtained

$$-A(4\pi^2\nu^2) \sin 2\pi\nu t = -\frac{k}{m}A \sin 2\pi\nu t \tag{9-16}$$

which shows that our solution is indeed a satisfactory one, provided that

$$4\pi^2\nu^2 = \frac{k}{m} \tag{9-17}$$

From this we can obtain an expression for the frequency of the oscillation in terms of the force constant k:

$$\nu = \frac{1}{2\pi}\sqrt{\frac{k}{m}} \tag{9-18}$$

The same equation applies to a system consisting of two masses, m_1 and m_2, such as the two atoms in a diatomic molecule, except that there is substituted for the quantity m a *reduced mass* μ defined by

$$\mu = \frac{m_1 m_2}{m_1 + m_2} \tag{9-19}$$

With the aid of the resulting equation

$$\nu = \frac{1}{2\pi}\sqrt{\frac{k}{\mu}} \tag{9-20}$$

the force constant k of a bond can be evaluated from a measured value of the vibrational frequency.

Vibrations in molecules are limited by quantum restrictions which lead to the following rule for the allowed energies of a harmonic oscillator:

$$E = (n + \tfrac{1}{2})h\nu \tag{9-21}$$

The vibrational quantum number n is zero or a positive integer. The vibrational energy levels constitute a series of evenly spaced steps with the interval equal to $h\nu$. When n is zero, the oscillator is in its ground state, but there still remains a quantity of energy equal in magnitude to $\tfrac{1}{2}h\nu$, the *zero-point* energy. This residual energy, present even at the absolute zero of temperature, is related to the uncertainty principle, which would be violated if the atoms were exactly at their equilibrium locations and also had velocities exactly equal to zero. The potential energy curve of a harmonic oscillator is a symmetrical parabola as shown by the dotted curve in Figure 9-13; the allowed energy levels are also represented by dotted lines in this figure.

POTENTIAL ENERGY
FUNCTIONS
OF REAL MOLECULES

As discussed above for the experimental potential energy curve in Figure 9-13, the vibrations of a real molecule deviate slightly from exact harmonic behavior. The average interatomic distance in any energy level is always slightly greater than r_e, the distance corresponding to the bottom of the potential energy curve, because of the effective weakening of the bond by vibration. Anharmonicity also has the effect of causing the energy levels to deviate slightly from equal spacing, and come closer together at higher values of the vibrational quantum number, as indicated in Figure 9-14. In connection with the scheme of energies shown in Figure 9-14, it should be noted that the bond energy, defined as the difference between the minimum in the curve and the dissociation plateau and labeled D_e on the diagram, is larger by an amount equal to $\tfrac{1}{2}h\nu$ than the amount of energy D_0 required to dissociate an actual molecule in the ground state.

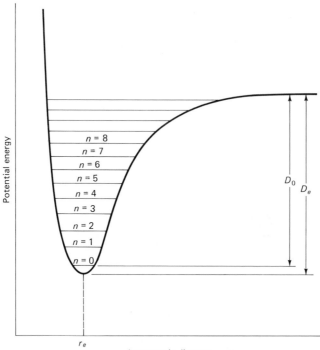

Figure 9-14
Potential energy curve and vibrational levels for a real molecule. The spacing
between successive vibrational energy levels decreases regularly with increasing
quantum number. The dissociation energy D_0 represents the experimental value,
measured from the ground state; the energy D_e would be the parameter to be
used when the curve is to be represented by a mathematical function.

In Table 9-1 are listed the equilibrium internuclear distances, the
vibrational frequencies, and the energies of dissociation of several
common diatomic molecules.

Table 9-1
Properties of some diatomic molecules

Molecule	Equilibrium interatomic distance (Å)	Energy of dissociation (kcal/mol)	Vibrational frequency (cm⁻¹)
H_2	0.74	103	4405
Li_2	2.67	26	351
B_2	1.59	83	1051
N_2	1.10	170	2360
O_2	1.21	117	1580
I_2	2.67	36	215
CO	1.13	211	2170
$H^{35}Cl$	1.28	102	2990

At this point it is desirable to place in perspective the nature and magnitudes of the various kinds of energies molecules can possess, such as the energy of vibration, and of the various types of spectroscopic transitions with which we shall be dealing. The energy of a molecule can be divided into translational energy associated with the motion of the center of gravity of the whole molecule, rotational energy associated with tumbling of the molecule as a rigid unit about certain axes, vibrational energy concerned with deformations from equilibrium geometry such as those just described for diatomic molecules, and electronic energy determined by the orbitals occupied by the electrons in the molecule.

The *translational* energy levels of a molecule are so close together that translational energy can be regarded as effectively continuous, or unquantized. Molecules gain or lose translational energy primarily through collisions and not through spectral transitions, and the spectral significance of this type of energy is that it broadens the levels associated with other kinds of energy: Infrared vibrational spectra of gases at high pressures show "collision broadening," and transitions for the liquid phase may be more or less completely smeared together into a continuous absorption by rapid intermolecular exchange of translational and rotational energy.

Next larger than translational energy in magnitude is the energy of *rotation*. For the free rotation of molecules, as occurs in the gas phase, this consists only of the kinetic energy of motion of the various atoms labeled by the index i:

$$E = \sum_i \tfrac{1}{2}m_i v_i{}^2 \tag{9-22}$$

The rotation of a diatomic molecule, as described in Section 3-4, consists of motions about two perpendicular axes. These two types of rotation are identical in characteristics because the relation of the atomic positions to one axis is the same as to the other axis. Expressed in terms of the distance r between the two atoms, the kinetic energy of rotation of a diatomic molecule is

$$E_{\text{rot}} = \tfrac{1}{2}\mu r^2 \omega^2 \tag{9-23}$$

In this equation, μ is the reduced mass of the molecule defined in Equation (9-19) and ω is the angular velocity of rotation.

Rotational motion is limited by quantum restrictions to values of the angular momentum given by the equation

$$\mu r^2 \omega = \sqrt{J(J+1)}\,\frac{h}{2\pi} \tag{9-24}$$

where J is a quantum number which can have any positive integral value from zero upward. Squaring Equation (9-24), dividing by μr^2, and substituting into Equation (9-23) leads to the energy condition

$$E_{rot} = J(J + 1)\frac{h^2}{8\pi^2 \mu r^2} \qquad (9\text{-}25)$$

Because of the way J enters into this equation, the spacing between rotational levels increases as the energy increases, and therefore, although the spectroscopic selection rule under most circumstances is that J changes by one unit, a whole series of different transition frequencies, uniformly spaced, can be observed. However, only if a molecule has a permanent dipole moment is a rotational transition spectroscopically active.

We shall not explore further the details of rotational energy levels and transitions for diatomic molecules, nor shall we delve into the complexities that arise for larger molecules, except to comment that analysis of rotational spectra obtained in the far infrared and microwave regions often allows the determination of interatomic distances in molecules.

Larger yet are the energies of vibrational transitions, which correspond to the energies of photons in the near-infrared region. Vibrational transitions are usually accompanied by simultaneous changes in rotational state, the effect of which is to increase or decrease by a small amount the energy required for the vibrational change. Thus, instead of appearing as a single line, the vibrational spectrum is a broad band, which can be resolved for gas phase samples into a series of discrete rotational lines, each corresponding to a transition between a particular rotational state in the lower vibrational level and a particular rotational state in the upper vibrational level.

The largest energy change, corresponding in magnitude to the energies of photons in the visible region, is associated with the transfer of an electron from one molecular orbital to another, a transition usually accompanied by vibrational changes as well as rotational changes. Since the energies used or released in these smaller changes add to or subtract from the energy required for the electronic change, an electronic spectrum spreads over a wide range of frequencies. In the gas phase, it can be resolved into a series of bands, termed a band system, each band corresponding to an increase or decrease in one kind of vibrational energy, and in turn composed of all the lines corresponding to the many possible upward and downward rotational changes that occur along with the electronic and vibrational transitions.

Some of the various types of molecular states are represented schematically in the energy level diagram in Figure 9-15. As a first approximation, it is convenient to consider for any one kind of molecule that the group of rotational sublevels in one vibrational level has the same pattern as the groups in other vibrational levels, and that the vibrational energies of any electronic level have the same arrangement as those of other electronic levels, so that spectra are regularly arranged and easily interpreted. In practice, however, the energy levels deviate

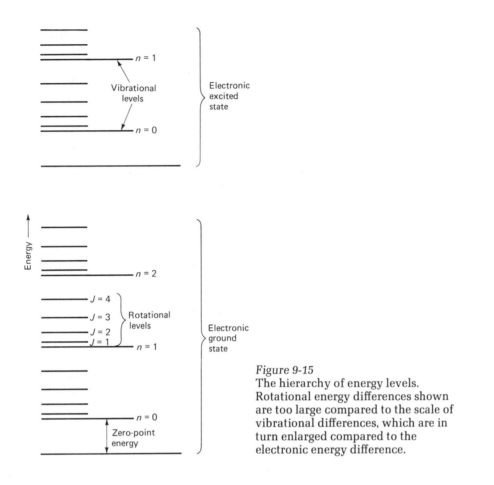

Figure 9-15
The hierarchy of energy levels.
Rotational energy differences shown
are too large compared to the scale of
vibrational differences, which are in
turn enlarged compared to the
electronic energy difference.

from this regularity. One cause is centrifugal distortion: As the molecule rotates more rapidly, the bonds are stretched, increasing the moment of inertia, which affects in turn the energies of the quantized rotational levels. Another cause is the difference in force constant from one electronic state to another, because of the difference in bonding. Consequently, it is often necessary to work out the energy level scheme from a group of complex overlapping bands or band systems; nevertheless, physicists have managed to extract a large amount of information about diatomic molecules from their electronic spectra.

In the following sections, we attempt to show some of the types of knowledge that can be obtained about the structure and behavior of a variety of molecules by investigation of their vibrational and electronic spectra, including detailed information about the geometry of several species that participate in physiological processes.

9-8
VIBRATIONAL SPECTRA

Vibrational spectra, observed in the infrared region, are used as an important tool in the detection of molecular species, in the identification of functional groups, and in the study of molecular behavior. An infrared spectrum is usually obtained by placing the sample in a cell having windows of sodium chloride, potassium bromide, or other special material that is relatively highly transparent to infrared radiation, passing a beam of radiation through the cell, and measuring the ratio of the intensity of the radiation leaving the cell to that of the incident radiation as a function of frequency. The frequency is selected by a prism or diffraction grating which can be slowly turned so as to scan through the spectral region of interest. Correction is made for absorption by the cell windows by comparing the results with those obtained for a blank cell not containing the material under study but perhaps containing a pure solvent.

DIATOMIC MOLECULES

If a molecule is to be set into vibration by absorption of a photon of radiation, the motions involved in that vibration must change the dipole moment of the molecule. Otherwise, the electric field of the radiation has no handle by which to exert a force on the molecule to influence the vibration. For a diatomic molecule, this condition requires that the two atoms be unlike, since no deformation can produce a dipole in a homonuclear molecule such as N_2.

For a harmonic oscillator, the selection rule governing a spectroscopic transition is that the quantum number n increase or decrease by one unit. Since the energy levels of the harmonic oscillator are spaced at equal intervals, one expects to see only one frequency of radiation absorbed by one species of diatomic molecule; *the frequency of that radiation is equal to the vibrational frequency of the molecule.*

Figure 9-16 shows a gas-phase, high-resolution vibrational absorption band for hydrogen chloride. The individual lines arise from the various possible changes in rotational energy accompanying the vibrational change. Each of the lines is split into two because of the

Figure 9-16
The gas-phase fundamental vibration band of HCl. The doubling of each peak is a result of the presence of both ^{35}Cl and ^{37}Cl. The peaks are labeled with the quantum numbers of the initial and final rotational states.

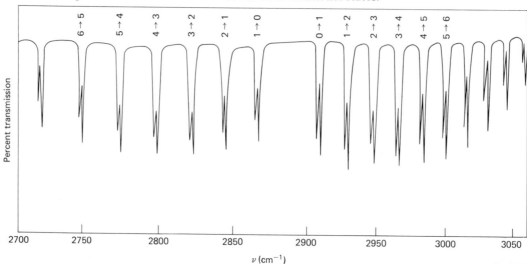

presence of two isotopes of chlorine, ^{35}Cl and ^{37}Cl, in a ratio of 3:1, for the difference in isotopic mass of the chlorine causes a slightly different reduced mass for the molecule.

Example: Calculate the force constant of the HCl molecule from the vibrational absorption frequency.

Solution: Careful evaluation of the positions of the lines in Figure 9-16 shows that the center of the band for $H^{35}Cl$ is at 2886 cm^{-1}. The corresponding frequency is obtained by multiplying by c:

$$\nu = 2886 \text{ cm}^{-1} \times 3.00 \times 10^{10} \text{ cm/sec} = 866 \times 10^9 \text{ sec}^{-1}$$

The reduced mass is

$$\mu = (35.0)(1.008)/(35.0 + 1.0)(6.02 \times 10^{23}) = 1.627 \times 10^{-24} \text{ g}$$

The force constant is calculated from a form of Equation (9-20):

$$k = 4\pi^2\nu^2\mu = 4(3.14)^2(866 \times 10^9 \text{ sec}^{-1})^2(1.627 \times 10^{-24} \text{ g})$$
$$= 4.82 \times 10^5 \text{ g/sec}^2 = 4.82 \times 10^5 \text{ dyn/cm}$$

In a real molecule, deviations from strictly harmonic behavior as discussed in Section 9-6 lead to weakening of the spectral selection rule, so that quantum number changes greater than unity become allowed, although their probability is relatively small. There are accordingly often observed *overtone* absorptions at frequencies approximately two or even three or more times the frequency of the strong fundamental.

POLYATOMIC MOLECULES

For a molecule consisting of more than two atoms, the number of independent types of vibration can easily be established by the arguments given in Section 3-4: After separating out the translational and rotational variables there remain $3N - 6$ variables for a molecule of N atoms if it is nonlinear or $3N - 5$ variables if it is linear, to be used in describing possible motions of the atoms that deform the molecule from its normal, equilibrium geometry.

For each molecule, it is possible to choose a particular set of vibrational modes such that they provide an especially simple description of the deformations of that molecule. The members of this set are called the *normal vibrations* for the molecule. If one imagines a molecule to be struck a vigorous blow, as by a collision with another molecule, the atoms within it can be set into vibration in any one of many ways, but any of these can be described by some linear combination of the normal modes for the molecule. Furthermore, the various members of the set of normal modes are independent of one another: No one of them can be generated by combining several of the others.

If a molecule is to absorb a photon of infrared radiation, the energy of the photon must correspond to that required to excite one of the normal vibrations. Spectroscopically, each of the normal modes of a complex molecule appears much like the stretching vibration of a diatomic molecule: If the vibration changes the dipole moment of the molecule, a fundamental band with rotational structure can be observed at a frequency equal to the vibrational frequency, and possibly

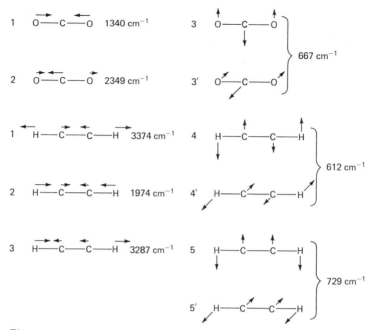

Figure 9-17
The normal vibrational modes of carbon dioxide and acetylene. Parallel
vibrations are given in the left column, and perpendicular vibrations in the
right column.

weaker overtone bands may also be observed at higher frequencies.

To determine the exact form of the normal vibrations of a polyatomic molecule requires the solution of a system of simultaneous
mathematical equations involving the parameters of molecular geometry, the atomic masses, and the bond force constants. However, we
can obtain some feeling for the nature of normal modes by examining
these modes for some simple molecules, such as are represented in
Figure 9-17. In a normal vibration, all the atoms in the molecule exe-
cute simple harmonic motion, moving in phase with one another and
passing through their extreme positions at the same instant in every
cycle. The diagrams in Figure 9-17 show the directions of motion of the
atoms during one half of the vibrational cycle; when they reach the
extreme of the vibration in the direction indicated, they reverse their
motions and move in the opposite directions.

Certain of the normal modes, those numbered 1 for carbon dioxide
and 1, 2, and 4 for acetylene, do not change the dipole moments of the
molecules, and infrared bands corresponding to their frequencies are
not found; they are said to be infrared-inactive. For linear molecules,
modes can be classified as parallel vibrations, which move atoms
parallel to the long axis of the molecule, or as perpendicular vibrations, which move atoms perpendicular to that axis. Perpendicular
vibrations always occur in pairs with the same frequency, for the
direction of bending can be in either of two planes which are orthogonal to each other. Examples are the vibrations numbered 3 and 3' for
carbon dioxide, and those numbered 4 and 4' and 5 and 5' for acetylene. For a planar molecule, such as ethylene, normal vibrations can be

classified as those that involve only atomic motions in the plane of the molecule and those in which the molecule is bent out of the plane.

There are some circumstances in which a normal mode in a complex molecule approximates a relatively simple motion, a motion confined almost entirely to the stretching or bending of one bond in the molecule. For purposes of identification of a functional group or for the evaluation of the effects of substitution in a molecule on the force constant of a bond, this is an extremely helpful situation, for the vibrational frequency is directly related to the characteristics of the particular bond. This simplification occurs if the stretching or bending of a particular bond has a frequency either much larger or much smaller than other vibrations in the molecule. When there are several motions with frequencies of the same magnitude, these couple with one another and all contribute to a very complex normal vibration.

One kind of vibrational absorption which usually appears distinct from other motions in the spectrum of an organic compound is the stretching of the bond of a hydrogen atom to some other atom. As a consequence of the very small mass of the hydrogen atom, the reduced mass of the system is only slightly different from unity, and the stretching vibration consists mostly of the motion of the hydrogen atom while the rest of the molecule is almost stationary. Since the force constants for bonds of hydrogen to other atoms differ relatively little in magnitude, the hydrogen stretching vibrations have absorptions in a fairly narrow portion of the spectrum, in the vicinity of 3000 cm^{-1}, which corresponds to about 3 μm wavelength. Some specific values are listed in Table 9-2, along with other characteristic vibrational frequencies. At the other end of the infrared frequency range from the hydrogen stretching are the low frequencies of stretching vibrations of bonds between an organic unit and a heavy atom such as chlorine, bromine, or a heavy metal.

Another kind of vibration of comparatively high frequency is the stretching vibration of a multiple bond, for which the force constant is greater than that of the corresponding single bond. For example, absorption bands characteristic of the stretching of double bonds between carbon atoms and other carbon atoms, or between carbon and oxygen, or between carbon and nitrogen appear in the vicinity of 1600 to 1800 cm^{-1}. The stretching frequencies of triply bonded units, C≡N

Table 9-2
Characteristic stretching vibrational frequencies in cm^{-1} of various structural units

C—Br	500–600	C=C=C	2000–2400
C—Cl	600–800	C≡C	2100–2300
S—S	700–900	C≡N	2200–2400
C—O (alcohols)	1050–1150	S—H	2500–2600
C—C	600–1500	C—H (alkanes)	2850–3000
C—F	1100–1400	C—H (alkenes)	2950–3100
C=C (aromatic)	1450–1700	C—H (aromatic)	3000–3150
C=C (olefinic)	1620–1750	C—H (alkynes)	3200–3300
C=N	1630–1690	O—H, N—H	3200–3700
C=O	1700–1900		

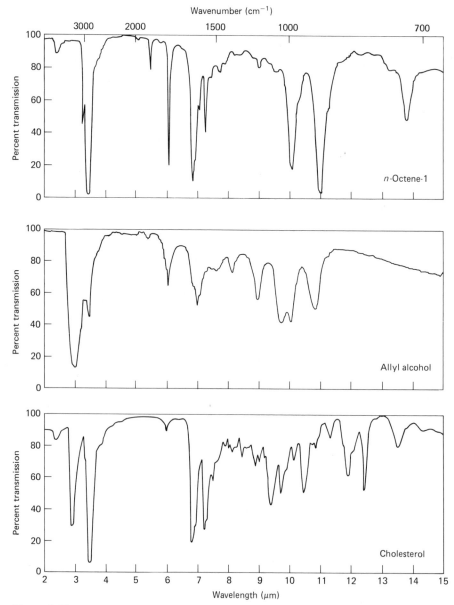

Figure 9-18
Typical infrared absorption spectra of several organic compounds. Features of special interest are the C=C, C—H, and O—H stretching bands and, in the octene spectrum, a small peak at about 3080 cm^{-1} (3.25 μm) which is characteristic of the C—H stretch for a terminal olefin. The spectrum of cholesterol has been redrawn from D. R. Johnson *et al.*, *J. Am. Chem. Soc.* **75**, 54 (1953) with permission. Copyright by the American Chemical Society.

and C≡C, lie in the range of 2100 to 2250 cm^{-1}, the increase over double-bonded frequencies corresponding to the larger force constants of the triple bonds.

Substitutional influences on stretching frequencies for multiple bonds can be illustrated by the effects observed for the carbonyl absorption frequency. The absorption of an aliphatic aldehyde is at about 1720 cm^{-1}, whereas in benzaldehyde conjugation with the aromatic ring lowers this value to 1696 cm^{-1}, a change due principally to a reduction of the double-bond character of the carbon–oxygen bond. In an α,β-unsaturated aldehyde, the carbonyl absorption appears at 1670 to 1680 cm^{-1}, the reduction again being the result of conjugation. If an electronegative atom is substituted on the carbonyl carbon, as in an acid chloride or bromide, the frequency is raised to the region of 1800 cm^{-1}, because the electron-withdrawing effect of the halogen reduces the length of the carbon–oxygen bond and increases its force constant.

Figure 9-18 shows typical infrared spectra of several organic compounds, together with assignments of some of the characteristic absorption bands.

HYDROGEN BOND EFFECTS

The participation of a hydrogen atom in a hydrogen bond (see Section 1-9) leads to a substantial change in the location of the hydrogen stretching band. The change is always toward a lower frequency, because the hydrogen bond weakens the covalent bond—to oxygen or nitrogen or other electronegative atom—of which the stretching frequency is being observed. Cholesterol, for example, contains a hydroxyl group which exhibits a single sharp band at 3620 cm^{-1} in dilute solution in carbon tetrachloride. At concentrations of about 0.05 M, a second, broader, band appears at 3470 cm^{-1} corresponding to dimers formed by intermolecular hydrogen bonding. At still higher concentrations, of 0.1 to 0.2 M, a third band appears at 3330 cm^{-1}, which probably arises from a trimer or larger aggregate. From the temperature dependence of the monomer–dimer equilibrium, as estimated from infrared measurements, the enthalpy of dimerization of cholesterol has been found to be about -2 kcal/mol.

INFRARED SPECTRA OF AMIDES

Because of their relevance to the conformational behavior of polypeptides and proteins, which will be discussed in Chapter 12, the vibrational spectra of amides are of much interest, and special attention has been devoted to their study. Furthermore, these spectra provide additional illustrations of some of the features of infrared spectra we have been describing, including the effects of conjugation and hydrogen bonding, as well as the pitfalls of interpreting the significance of vibrational bands having contributions from several types of atomic motion.

In addition to the NH stretching modes at high frequency, seven different vibrational bands have been recognized as characteristic of amides. These are designated by Roman numerals to avoid the hazard of premature conclusions about their origin.

The term amide I band refers to an absorption produced by a mode that is primarily the stretching of the carbonyl double bond but includes small contributions from stretching of the C—N bond and bending of the N—H bonds. The frequency varies from 1630 to 1680 cm^{-1} in the solid state—where the molecules are hydrogen-bonded—

up to 1715 cm^{-1} in dilute solution. Because this is a mixed mode, one cannot determine without a complete mathematical analysis whether conjugation of the unshared electrons on the nitrogen has any effect on the force constant of the carbonyl group, but the frequencies are generally below those of aliphatic aldehydes or ketones.

The amide II band appears in dilute solution at 1590 to 1620 cm^{-1} for primary amides, and at 1500 to 1550 cm^{-1} for secondary amides. In the solid, values are about 20 to 30 wave numbers *higher*. The corresponding motion consists of about 40 percent stretching of the C—N bond and 60 percent in-plane bending of the hydrogens on the nitrogen atom. This band decreases in intensity when the NH hydrogens are replaced by deuterium and was thought at one time to be entirely due to NH bending. In an NH$_2$ group, and in a CH$_2$ group as well, the pair of hydrogens may have various bending motions: out-of-plane in the same direction, called wagging; out-of-plane in opposite directions, or twisting; in-plane in the same direction, termed rocking; in-plane in opposite directions, or scissoring. It is the last that contributes most to the amide II vibration in primary amides.

The amide III band appears near 1290 cm^{-1} in secondary amides. It is associated with a mode which is about one-third C—N stretching, one-third N—H in-plane bending, and the remainder mostly C=O stretching, O—C—N bending, and C—C stretching. The amide V band represents the N—H out-of-plane bending motion and occurs at about 720 cm^{-1}. Amide bands IV and VI in the vicinity of 600 cm^{-1} arise from O—C—N bending vibrations, and the amide VII band near 200 cm^{-1} corresponds to torsional vibrations about the C—N bond, the resistance to this rotation coming from the partial double-bond character in this link.

RAMAN SPECTROSCOPY

In addition to the excitation of a molecule to a higher rotational or vibrational level by the addition of a photon carrying just the correct amount of energy for the transition, there is an alternative method of excitation called a Raman process. In this event, a photon of visible or ultraviolet light interacts with the molecule, loses some of its energy to the molecule, and comes away from the encounter as a photon of lower frequency. The change in energy of the photon is exactly equal to the amount required for the molecular excitation. Perhaps somewhat surprisingly, Raman processes can also deactivate molecules that happen to be in an excited state, the departing photon carrying away more energy than it had when it arrived and therefore appearing in the spectrum at a higher frequency than the incident light. Figure 9-19 is an example of a Raman spectrum showing vibrational lines.

Because the probability of Raman interaction is relatively low, most of the incident radiation from a light source passes through the sample. It is therefore necessary to measure the scattered radiation by viewing the sample at right angles to the incident beam. Even then, most of the scattered intensity comes from what is called Rayleigh scattering, occuring without change in frequency in a way that will be further described in Chapter 12, and the Raman spectrum appears as small

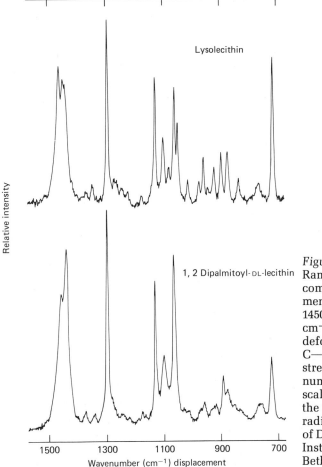

Lysolecithin

1, 2 Dipalmitoyl-DL-lecithin

Relative intensity

1500   1300   1100   900   700

Wavenumber (cm⁻¹) displacement

Figure 9-19
Raman spectra of two components of biological membranes. The peaks near 1450, 1295, 1100, and 720 cm⁻¹ represent the CH₂ deformation, CH₂ twisting, C—C stretching, and C—N stretching, respectively. The numbers on the frequency scale are differences from the frequency of the exciting radiation. Spectra courtesy of Dr. Ira W. Levin, National Institutes of Health, Bethesda, Maryland.

lines flanking the central peak of light. A typical spectrometer setup is shown in Figure 9-20.

In the last few years, the characteristics of laser sources, described in Chapter 14, which produce light of high intensity in a narrow beam with a well-defined frequency, have been applied to great advantage in

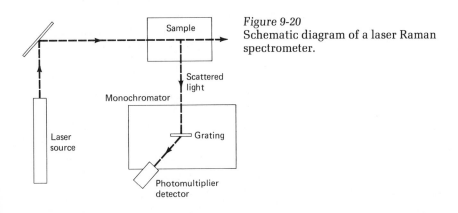

Figure 9-20
Schematic diagram of a laser Raman spectrometer.

Raman spectroscopy. It has also been found that sensitivity is greatly enhanced by using light of a frequency in a *resonance* region of the spectrum, that is, of a frequency somewhere near where the scattering molecule undergoes an electronic transition.

Despite difficulties with sensitivity, Raman spectroscopy has several advantages which make it extremely valuable. Raman excitation gives information about molecular vibrations that are inactive in the infrared because of molecular symmetry. A given vibration may be active in both types of spectrum but, if it is inactive in one, then it is certain to be active in the other, so that Raman spectroscopy complements infrared spectroscopy very well. The stretching vibrations of homonuclear diatomics, such as H_2, N_2, and O_2, are active in the Raman, as are the symmetrical stretch of CO_2 (normal mode 1 in Figure 9-17) and normal modes 1, 2, and 4 of acetylene.

Another advantage of Raman spectroscopy is that it utilizes visible or ultraviolet light rather than infrared radiation. Consequently the walls of the cell and the other units of the optical system can be made of glass or quartz rather than of special materials transparent to infrared radiation. Furthermore, it is possible to work conveniently with aqueous media, a great advantage in the study of biological systems, for water is much more transparent in the visible and ultraviolet regions than it is in the infrared. Raman spectroscopy has been applied to the study of the conformation of polypeptides and proteins in solution. For example, the amide III region in Raman spectra contains lines characteristic of the β-conformation of polypeptides as distinguished from the α-conformation (see Chapter 12). Raman spectra of polynucleotides have been shown to be very sensitive to the unstacking of the bases, the disordering of the backbone chain, and the breaking of hydrogen bonds between base pairs.

9-9
ELECTRONIC
TRANSITIONS

Absorption by a molecule of a photon of suitable frequency can excite an electron from its normal condition in the ground state of the molecule to a vacant orbital of higher energy. The energies for electronic transitions vary greatly from molecule to molecule, but we are usually concerned with those that fall in the visible and near-ultraviolet regions. The lifetime of the excited state is so short that there is very little chance that a second electron will be excited before the first has returned to the ground state.

In determining quantitatively the amount of radiation absorbed by a sample in an electronic transition, a beam of the radiation is passed through the sample in a glass or quartz cell with flat, parallel sides, and the amount of radiation leaving is compared with the amount of radiation entering. Suitable correction of course must be made for absorption of light by the solvent, if one is used, and by the cell. In order to convert the corrected experimental results to quantities characteristic of the absorbing molecules, it is customary to use the Beer–Lambert equation:

$$\log\frac{I_t}{I_0} = -\epsilon cl \qquad (9\text{-}26)$$

where I_0 is the intensity of the incident radiation, I_t is the intensity of the transmitted radiation, c is the concentration of the material responsible for absorption, and l is the length of the light path through the sample. The quantity ϵ is called the *extinction coefficient,* and it is this quantity that measures the absorbing power of a substance at a particular wavelength.

DIATOMIC MOLECULES

The first group of electronic transitions we consider includes those in diatomic molecules. Transitions observed in an absorption spectrum usually correspond to the transfer of an electron from an occupied orbital of an energy near the upper end of the energy range of these orbitals to an orbital in the lower energy range of those that are vacant. For example, in a halogen molecule such as Br_2 or Cl_2, the orbital of highest energy from which an electron can be removed is a π^*p orbital, and the lowest vacant orbital is a σ^*p orbital. These are both antibonding orbitals, lying fairly close together in energy, so that the photon required to excite the electron corresponds in frequency to the visible region of the spectrum and these molecules are colored.

For molecules such as H_2 or N_2, the highest occupied orbital in the ground state is a bonding orbital and the lowest vacant orbital is an antibonding orbital; there is an energy difference between the two sufficiently great so that absorption occurs only in the short-wavelength region of the ultraviolet.

The oxygen molecule is rather unusual in that the ground state is a triplet state, and therefore absorption of a photon induces a transition to an excited triplet state rather than a singlet state. The excited triplets of lowest energy arise from excitation of an electron from the $\pi 2p$ bonding orbital to one of the $\pi^* 2p$ antibonding orbitals, which, although they already contain two electrons, have locations available for two more. There are two of these triplet states, which differ from one another according to the geometric relation of the orbital from which the electron is removed to that to which it goes. The lower one lies about 36,000 cm^{-1} above the molecular ground state, and transitions to it would appear in the near ultraviolet. However, transitions from the ground state to this level are forbidden by a selection rule. Transitions to the upper level, which is about 50,000 cm^{-1} above the ground state, are allowed and are responsible for the onset of absorption by molecular oxygen, which begins at about 200 nm and becomes increasingly strong farther into the high-energy ultraviolet. In fact, it is this absorption that sets a lower wavelength limit to the ultraviolet spectroscopic measurements that can be performed in a conventional ultraviolet spectrometer containing air. To do spectroscopy at wavelengths much below 200 nm, it is necessary to use a nitrogen- or helium-filled spectrometer, or one that is evacuated to low pressures.

In addition to the excited triplet state, the oxygen molecule also has two levels lying much closer to the ground state in energy and differing from this state by the fact that the two electronic spins are paired rather than parallel. In one of these states, which is about 13,100 cm^{-1} above the ground state and has an average lifetime in the gas phase of about 7 sec, the electronic orbital motions of the two high-energy electrons are opposite in sense, just as they are in the ground state, leading to the designation of this as a Σ state, where the uppercase Greek letter denotes zero angular momentum about the molecular axis for the whole molecule just as a lowercase σ indicates that an individual electron has zero angular momentum about a bond axis. In the other low-lying singlet state, about 7880 cm^{-1} above the ground state, the electronic orbital motions are of the same sense and add to give an electronic orbital angular momentum quantum number value of 2, indicated by the symbol Δ. The lifetime in the gas phase of this state is about 45 min, quite long because transitions to the ground state are forbidden by both spin and symmetry selection rules; in aqueous solution, the lifetime is about 10^{-5} sec.

Singlet oxygen molecules can be formed as the product of certain chemical reactions, such as that between sodium hypochlorite and hydrogen peroxide, as well as from the decomposition of superoxide ion, O_2^-, in living cells. It has been suggested that singlet oxygen molecules are involved in a variety of biological processes, including erythrocyte disfunctions associated with a genetic deficiency of the enzyme glucose-6-phosphate dehydrogenase, bacteriocidal activities of phagocytes, metabolic hydroxylation reactions, aging, and carcinogenesis, as well as in some photochemical reactions leading to atmospheric smog.

The description of energy levels in terms of orbitals occupied by electrons is more complex for heteronuclear diatomic molecules such as CO, HCl, ClF, CaO, BH, NO, and so on, but these molecules yield electronic spectra similar to those for homonuclear diatomics, usually with absorption bands in the near ultraviolet and with complex emission patterns both in the visible and ultraviolet regions.

ORGANIC MOLECULES

Turning now to polyatomic organic molecules, one finds that saturated hydrocarbons have all their electrons in σ orbitals, and the energy required for excitation from these bonding orbitals to antibonding orbitals is quite large, so that absorption of radiation occurs only in the far ultraviolet, at wavelengths shorter than about 150 nm, the "vacuum" ultraviolet region.

Olefinic hydrocarbons absorb energy at wavelengths of about 160 to 170 nm, accompanying the promotion of an electron from a π to a π^* orbital. This type of absorption is in a range of energy still too high to be conveniently accessible without special equipment, and it is only when the double bond is conjugated with another double bond or with some other kind of unsaturation that the electronic spectrum moves into a more accessible region. This will be discussed further in Chapter 14. Acetylene shows an absorption near 180 nm.

In addition to olefinic and acetylenic units, other functional groups that permit absorption of energy in the near-ultraviolet or visible region of the spectrum, and are thus termed *chromophores,* include the nitro, sulfoxide, azo, nitroso, carbonyl, and thiocarbonyl groups, as well as aromatic rings. In some of these, the electrons undergo $\pi-\pi^*$ transitions. In other groups, the electrons that are excited move from orbitals in which they are not involved in bonds to antibonding orbitals. The orbitals containing unshared electrons are designated n orbitals, and thus this type of transition is referred to as an $n-\pi^*$ transition. In molecules with atoms having unshared electrons but no π orbitals, $n-\sigma^*$ transitions may also be observed; examples are methylamine, in which an unshared electron on the nitrogen is excited by photons with a wavelength of 213 nm, and methyl iodide, for which absorption occurs at 259 nm.

CARBONYL COMPOUNDS

The behavior of the carbonyl chromophore has been extensively investigated, both because its absorption falls in a convenient region of the spectrum and because a wide variety of compounds containing this group is readily available. There are several different kinds of electronic changes that can occur, involving the removal of electrons from either an n or a π orbital and excitation to either a π or σ antibonding orbital. Diagrams of the various orbitals involved and their approximate relative energies are given in Figure 9-21. The energies of various states of the molecules depend upon the combination of orbitals that is occupied as well as upon the relative spins of the electrons in these orbitals. For the same orbital distribution of electrons, the triplet state

Figure 9-21
Valence molecular orbitals in the carbonyl group and their relative energies.

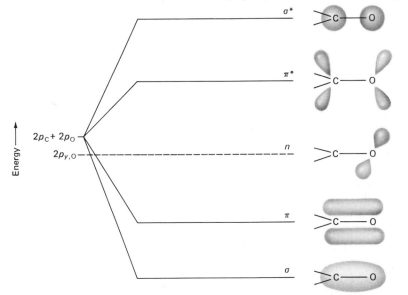

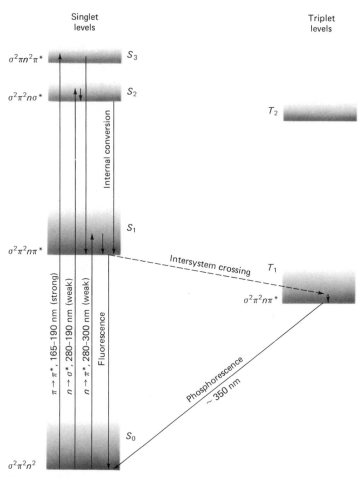

Figure 9-22
Energies of various singlet and triplet states of a carbonyl system. Electron configurations are given in terms of occupation of the orbitals depicted in Figure 9-21.

with spins parallel is, for carbonyl compounds as well as for other systems, somewhat lower in energy than the corresponding state with spins paired. Typical energies of excited carbonyl states compared to the ground state are shown in Figure 9-22.

The state marked S_1 is the lowest excited singlet state. The spectroscopic transition from the ground state requiring the least energy is excitation to this state, which is about 36,000 cm^{-1} above the ground state and requires photons of 280- to 300-nm wavelength. In this transition, the electron leaves an orbital primarily on the oxygen atom and goes to an antibonding π^* orbital; accordingly, it is designated an n–π^* transition. Because of the special selection rule for spin, the excited electron maintains its spin opposite that of the electron left behind in the oxygen orbital.

The state S_1 is usually formed in an excited vibrational condition. The vibrational energy is likely to be lost more rapidly than the electronic energy, this loss occurring by molecular collisions or by intramolecular transfer rather than by emission of radiation. A photon emitted by a vibrationally deactivated molecule as it returns to the ground state is less energetic than the photon absorbed in the transition in which it was activated. Such emission of energy differing in frequency from that absorbed is referred to as *fluorescence*. For the carbonyl group, a typical lifetime of the excited electronic state before a return to the ground electronic state is 10^{-8} sec.

It is also possible for an excited molecule to lose energy by changes referred to as *internal conversion*. The potential energy curves of excited states may cross one another in such a way that a lower vibrational level of a higher electronic state corresponds to a higher vibrational level of a lower electronic state. Thus the upper electronic state can be collisionally deactivated down to the energy level at which the curves cross, and then the molecule can change to the lower electronic level. This simply means that electronic energy is converted into vibrational energy at this point; the vibrational energy can then be dissipated by further molecular collisions.

Often, however, the excited electronic state is converted into the corresponding triplet state of somewhat lower energy, labeled T_1, before the molecule has a chance to return to the ground electronic state. A singlet–triplet transition is spectroscopically forbidden, but it can occur by a radiationless process, such as intermolecular collision or internal conversion, and the change from a singlet to a triplet state is referred to as an *intersystem crossing*. The triplet state formed has no ready pathway to lose energy, for the only lower state is the ground singlet state. However, although the selection rule predicts that a change in spin is improbable, it is not entirely impossible, and the triplet eventually does drop to the ground state with the emission of a photon, provided that it has not been deactivated first by a radiationless process such as a collision. Because of the delay that may occur in the emission of a photon, the average lifetime of a triplet state is usually relatively long, often from 10^{-4} sec up to 1 sec or more. The energy of the emitted photon is usually substantially different from that of the photon absorbed originally in exciting the molecule from the ground state—it is typically 29,000 cm^{-1} for a carbonyl compound—and this fact, combined with the time delay, causes the emission of radiation by the state T_1 to be termed *phosphorescence*.

Fluorescence and phosphorescence occur in a variety of systems other than carbonyl compounds. Their study helps to supply information about excited states in molecules, states that are important intermediates in many reactions, particularly photochemical reactions, as described in Chapter 14.

Another type of transition a ground-state carbonyl group can undergo is excitation of an unshared electron to an antibonding σ orbital, an $n–\sigma^*$ transition. This requires radiation of about 52,000 cm^{-1} for formaldehyde. A transition from a π orbital to an antibonding π^* orbital is also a possibility, usually requiring slightly more energy than the $n–\sigma^*$ transition, typically about 60,000 cm^{-1}.

The polarity of the solvent has a substantial effect on the wavelength of the $n–\pi^*$ transition, for the unshared electrons reside primar-

Table 9-3
Wavelength in nanometers of maximum absorption of carbonyl compounds

$(CH_3)_2CO$	280
$CH_3CH_2CH_2CHO$	290
$CH_3CH{=}CHCHO$	310
$(CH_3)_2C{=}CHCOCH_3$	325
$CH_2{=}CHCHO$	340
$C_6H_5COCOC_6H_5$	370
Camphor quinone	466

ily on the oxygen atom, but in the antibonding orbital they are about equally shared between the carbon and the oxygen atoms. Thus the excitation process reduces the negative charge on the oxygen atom of the carbonyl group and therefore reduces the polarity of the group. The ground state is consequently stabilized relative to the excited state by a solvent of greater polarity.

When two olefinic double bonds, or two carbonyl groups, or a carbonyl group and an olefinic group are conjugated with one another, the energy of the ultraviolet absorption is reduced, as illustrated in Table 9-3. Similarly, aromatic systems, in which conjugation is even more extensive, absorb in the near ultraviolet.

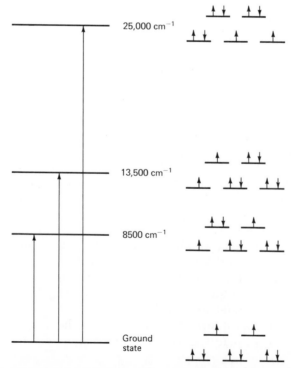

Figure 9-23
Ground state and several excited states of the d^8 electronic system in $Ni(H_2O)_6{}^{2+}$. These assignments of orbitals for the excited states permit a prediction of total spin to be made, but are not literally correct, for each of the states is actually a linear combination of several such electronic arrangements.

25,000 cm^{-1}

13,500 cm^{-1}

8500 cm^{-1}

Ground
state

TRANSITION METAL IONS

A different class of molecular system in which there is relatively facile excitation of electrons by ultraviolet or visible radiation includes a wide variety of complexes of transition metal ions. The general nature of bonding in these complexes was considered earlier in this chapter. A simple example is the hydrated titanium ion, $Ti(H_2O)_6^{3+}$, in which absorption of a photon of wave number 20,300 promotes an electron from one of the group of orbitals, $3d_{xy}$, $3d_{xz}$, $3d_{yz}$—these orbitals have one electron among them in the ground state of the complex—to one of the pair of orbitals, $3d_{z^2}$, $3d_{x^2-y^2}$, which are vacant in the ground state. Another example is hydrated ferric ion, a weak-field complex with one electron in each of the five $3d$ orbitals. It is yellow, but its color is quite pale because any change in electronic level within the system of five electrons in five $3d$ orbitals means that one electron must move into an orbital already occupied by another electron, and this necessitates a reduction in total spin, violating the selection rule which states that spectroscopic changes corresponding to changes in spin are improbable. A more complicated example is $Ni(H_2O)_6^{2+}$. There are eight $3d$ electrons, and in the ground state six are in the three lower-energy levels and two are in the two higher-energy levels. The spins of the latter two are parallel, and the ground state is a triplet state. In the spectrum, three bands are observed, all for triplet–triplet transitions. Two of these, at 8500 and 13,500 cm^{-1}, correspond to two ways in which an electron can be excited from the lower group of levels to the upper group. The third, at 25,300 cm^{-1}, corresponds to the excitation of two electrons to the upper level. The several levels are represented diagramatically in Figure 9-23.

9-10
OPTICAL DISPERSION

The absorption of energy from light by a molecule can be represented, as an alternative to the quantum picture, in terms of a mechanical model. Although this model is not an exact description of what happens, it serves as an aid in visualizing the process. As described in Section 8-2, the electric field of a beam of radiation exerts a force on the electric charges in a molecular system, which causes them to oscillate synchronously with the field. In most of the spectral region, the oscillating charges in the molecule simply reradiate the energy, and the only effect is the time lag evident in the slowing of the radiation, a slowing that corresponds to an index of refraction greater than unity. As the frequency of the impinging radiation approaches that at which a spectroscopic transition—an electronic transition if the light is in the visible or ultraviolet region—can be produced, the charges in the molecule are pulled more and more strongly in the direction of the state that would result from the transition, and a countervailing effect, a sort of frictional force, is developed which causes absorption of energy from the light beam. As rather extreme analogies of this situation then, one could propose mechanical processes such as the shattering of a piece of crystal when a singer's high note happens to match its resonant frequency, or the collapse of a bridge when wind gusts set it vibrating at a particular frequency at which it continuously absorbs energy.

The probability that a spectroscopic transition does result from incident radiation becomes larger as the frequency of the radiation moves nearer the center of an absorption band, and, at the same time, the tendency for oscillations of charge in the molecule to be distorted because of the incipient occurrence of the transition causes rather drastic changes in the refractive index. Within the absorption band, it is no longer valid to say that the interaction of the beam of radiation with the molecule is completely reversible or elastic, for some of the energy of the beam is now taken up in exciting the molecule. Figure 9-24 shows the behavior of the extinction coefficient and the index of refraction in the vicinity of the absorption band. The refractive index rises as the frequency increases, reaching a maximum, then decreases, passing through its mean value at the frequency of the maximum absorption, goes to a minimum, and recovers asymptotically toward a value smaller than the low-frequency value. On the high-frequency side, the radiation field oscillates too rapidly for the electrons in question to keep up, so that their contribution to the refractive index drops out.

The dispersion effect described in Section 8-2 results from the variation in refractive index of a substance near one of its absorption bands. Thus a glass prism disperses visible radiation because it has an absorption band in the near ultraviolet, and a prism of sodium chloride can be used effectively to disperse infrared radiation in the spectral range from about 3000 to about 650 cm^{-1}, since sodium chloride has an absorption band beyond 650 wave numbers.

The rotation of light by optically active compounds is found to depend in an analogous way on the relation of the frequency of the radiation to the frequency of an "optically active" absorption band, an effect termed *optical rotatory dispersion* (ORD). For an absorption band to be optically active, it is necessary that it be associated with a functional group that is itself asymmetric or in an asymmetric local environment. If the wavelength of light is varied through the spectral region in which the material absorbs, the specific rotation changes rapidly as the band is approached, reaches an extreme in one sense, goes back through zero at the wavelength of maximum absorption, reaches an extreme in the other direction, and then diminishes. This change in rotation, known as the *Cotton effect,* parallels the dispersion

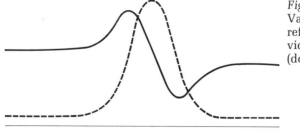

Figure 9-24
Variation in the index of refraction (solid curve) in the vicinity of an absorption band (dotted curve).

Frequency ⟶

of the refractive index in the vicinity of an absorption band, a not un-expected situation since optical rotation is related to the difference between the refractive indices of the medium for the two components of the polarized ray.

In practical cases, one deals with optical rotation in the visible and near-ultraviolet regions as affected by absorption bands which may be anywhere from the visible to the far-ultraviolet region. Even if there are no optically active absorptions in the region over which rotation is measured, there are likely to be such absorption bands in the far ul-traviolet for which one observes only the wing of the dispersion region without ever reaching a trough or a peak. A monotonic dispersion curve found in this situation is said to be a *plain* curve and can be de-scribed by the Drude equation

$$[m] = \frac{a}{\lambda^2 - \lambda_0^2} \tag{9-27}$$

where $[m]$ is the molar rotation, λ is the wavelength at which the rota-tion is measured, and a and λ_0 are constants which are adjusted to fit the experimental data. If a Cotton effect curve is observed in the vis-ible or near ultraviolet, it is often unsymmetrical in appearance, be-cause it is situated on the wing of a dispersion region in the far ultra-violet, such as causes a plain curve.

Whenever the difference in refractive index for the two circularly polarized components of plane polarized light varies with frequency, then the two components can also be expected to have different ex-tinction coefficients at any given frequency and to be absorbed by the sample to different extents. In this circumstance, the oscillating vec-tors of the two components become *unequal in amplitude* as the light passes through the sample. Since the two components no longer cancel each other in any direction, the light can be described as elliptically polarized, rather than plane-polarized, as shown in Figure 9-25. This

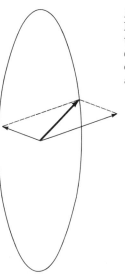

Figure 9-25
Elliptical path traced out by the resultant of two counter-rotating components of polarized light of unequal amplitude.

phenomenon is termed *circular dichroism* (CD), and it is measured for a sample by determining, after passage through the sample, the ellipticity of light that is initially plane-polarized. Circular dichroism has the same relation to optical rotary dispersion as the absorption spectrum has to ordinary optical dispersion. Figure 9-26 illustrates the relation between ultraviolet, CD, and ORD curves for D-(+)-camphor.

A group of compounds for which ORD and CD have been extensively employed are those containing the carbonyl group. These com-

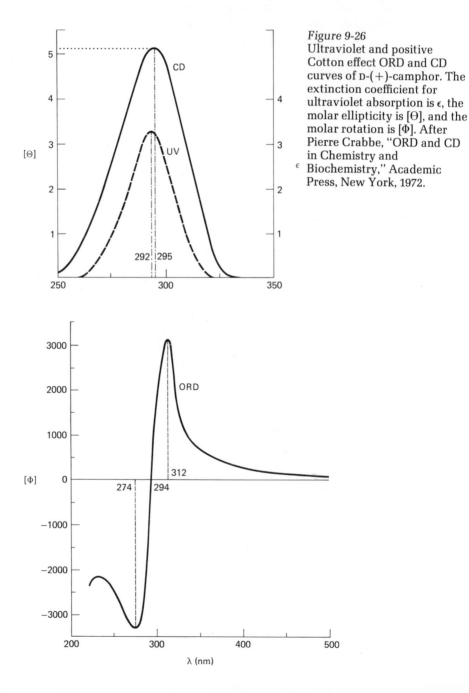

Figure 9-26
Ultraviolet and positive Cotton effect ORD and CD curves of D-(+)-camphor. The extinction coefficient for ultraviolet absorption is ϵ, the molar ellipticity is $[\Theta]$, and the molar rotation is $[\Phi]$. After Pierre Crabbe, "ORD and CD in Chemistry and Biochemistry," Academic Press, New York, 1972.

pounds have a region of absorption, as mentioned in Section 9-9, at about 290 to 300 nm in the near ultraviolet, corresponding to the $n-\pi^*$ transition. It is convenient to measure optical rotation in the vicinity of this absorption region, because the frequency is low enough so that the experimental difficulties of the far ultraviolet are not encountered, and the extinction coefficient is relatively small so that enough light is transmitted by the sample to permit adequate sensitivity.

Empirical rules have been developed for the assignment of absolute configurations of asymmetric molecules from the sign of the Cotton effect. To apply the rule for saturated carbonyl-containing compounds, the molecule is viewed along an axis from the oxygen to the carbon atom. The molecule is first divided into quadrants by two planes intersecting at this axis, one plane the nodal plane of the π cloud and the other perpendicular to the nodal plane. A third plane, perpendicular to the first two and halfway between the carbon and the oxygen, completes the division of the molecule into eight sections, or octants. The octants are classified into two sets of four, the members of one set alternating with the members of the other set. Atoms or substituents situated in one set contribute to the Cotton effect of one sign; those in the other set contribute with the opposite sign. This rule has been successfully applied to a number of keto steroids in which the carbonyl group is part of a five- or six-membered ring.

EXERCISES

1. For which of the following molecules can spectroscopic transitions be observed in which only the rotational energy level changes and the vibrational and electronic states remain unchanged: carbon monoxide, nitrogen, acetylene, carbon tetrachloride, hydrogen bromide, chloroform?

2. The molecule $^{12}C^{16}O$ has a vibrational "frequency" of 2170 cm^{-1}. Calculate the force constant for the molecule.

3. Calculate in ergs the zero-point energy of the molecule in Exercise 2. What is the equivalent in kilojoules per mole?

4. Predict the frequency and wave number for the fundamental vibrational absorption of $^2H^{35}Cl$.

5. How many normal vibrations are there for each of the following molecules: sulfur dioxide, vinyl chloride, benzene, 1,1-dibromoethane?

6. A 0.15 M solution of a material in a transparent solvent absorbs 27 percent of the incident radiation of a wavelength of 530 nm in a 1.00-cm path.

(a) Calculate the extinction coefficient for this radiation.
(b) What fraction of the incident radiation of this wavelength would be absorbed in a 2-cm path? In a 10-cm path?

7. Compare the electronic structure of N_2 with that of acetylene.

8. Draw a diagram of the projected view of three tetrahedral bond directions in methane as seen along the fourth carbon–hydrogen bond.

9. Justify in terms of the molecular orbital model the fact that a Be_2 molecule has not been found to have a stable existence.

10. Postulate a description of the bonding in an ammonia molecule, in which the H—N—H bond angles are 107°.

11. What orbitals are used for bonding in each of the following molecules: linear Cl—Hg—Hg—Cl, carbon tetrachloride, planar BF_3, cyclohexene, acetone?

12. Develop a valence bond picture of the CO_2 molecule that explains its cylindrical symmetry.

13. Based on Equation (9-25), prepare an energy level diagram for the rotation of the molecule $^1H^{35}Cl$ with an interatomic distance of 1.28 Å. Assuming that spectral transitions are governed by the selection rule, $\Delta J = \pm 1$, predict the nature of the rotational spectrum of HCl.

14. Propose an explanation of how the ligands are held to the central metal atom in a high-spin cobalt(III) complex.

15. N-phenyl-N-methylacetamide is found to exist in two isomers. Suggest an explanation.

16. The molecular ions N_2^- and N_2^+ have, respectively, one electron more and one electron less than N_2. Compare the bonding energies of the three species.

17. Explain why triphenylamine is a planar molecule, although trimethylamine has pyramidal bonds about the nitrogen. Predict the relative basic strengths of the two compounds.

18. Prepare an energy level diagram for the ground and excited states of molecular oxygen which are described in the text. Calculate the wavelength of radiation emitted in the transition from the $^1\Sigma$ excited state to the $^3\Sigma$ ground state.

19. Vinyl ethyl ether is found to exist in two different isomeric forms.
    (a) Show the geometry of the two isomers and explain in terms of an electron-delocalization bonding picture.
    (b) How many distinct isomers of divinyl ether would be expected on the same basis?

20. The hydroxyl group substituted in a benzene ring as in phenols is found to facilitate reactions at the ortho and para positions with reagents that are *electrophilic* or electron-seeking. Draw resonance diagrams explaining this effect.

21. Predict the direction in which the frequency of the NH stretching band for adenine at 3300 cm^{-1} shifts when acetone is added to a dilute solution of adenine in chloroform, and explain the basis for your answer.

22. Draw energy diagrams showing the occupied orbitals in the ground state of each of the diatomic molecules C_2, O_2, and F_2.

23. For the $^{12}C^{16}O$ molecule, the rotational constant $h/8\pi^2 \mu r^2$ is found to be 57,630 megahertz. What is the interatomic distance in this molecule?

24. Show how the bonding in butadiene could be represented by resonance structures.

25. Explain why the carbon–oxygen stretching frequency of CH_2=CH—CHO is lower than that in CH_3CH_2CHO.

26. Vibrational transitions corresponding to a change of 2 in quantum number are allowed, although weak, for a real, anharmonic, oscillating molecule. Formulate a rule relating the frequency of such a transition to the frequency of the corresponding fundamental.

27. Draw resonance structures for the CO_3^{2-} ion, which is planar and has threefold symmetry.

28. Show the distribution of electrons in the 3d orbitals of strong-field and weak-field complexes of Mn^{2+} ion, and predict the total spin of each type.

29. Two of the normal vibrations of water are represented by the diagrams

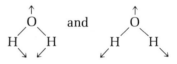

Assign the observed frequencies of 1595 and 3652 cm^{-1} to the two vibrations.

30. The allyl radical has the nominal formula CH_2=CH—$CH_2\cdot$, but in fact the two CH_2 groups are entirely equivalent because of electron delocalization, and the radical is much less reactive than the n-propyl radical. Draw a diagram showing the geometry of the radical, and give both resonance and molecular orbital formulations of the bonding that explain the properties described.

31. The following are some of the normal vibrations of the nitrate ion. Indicate which of them are active in the infrared spectrum and which are not and explain.

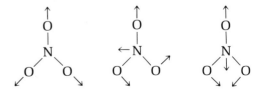

32. For the molecules COF_2 and $COBr_2$, the carbon–oxygen bond stretching force constants are found by a complete analysis of the molecular vibrations to be 12.85×10^5 and 12.83×10^5 dyn/cm, respectively. Explain the relatively large difference in frequency between the infrared bands assigned to the carbonyl stretching vibrations in the two molecules, which appear at 1928 cm^{-1} and 1828 cm^{-1}, respectively.

33. The wavelength of incident light in a Raman experiment is 480.10 nm, and the wavelength of a band in the scattered spectrum is 504.90 nm. What is the frequency in wave numbers of the energy transferred to the sample, and what type of transition has been stimulated in the molecules of the sample?

34. The two lowest excited states of the nitrogen molecule are a singlet state at about 69,000 cm^{-1} and a triplet state at about 50,000 cm^{-1} above the ground state. How could these states be represented in terms of electronic orbital occupation? What is the longest wavelength at which an absorption band of appreciable intensity would be expected for ground-state N_2 molecules?

REFERENCES

Books

C. N. Banwell, *Fundamentals of Molecular Spectroscopy,* 2nd ed., McGraw-Hill, New York, 1973. Excellent, intermediate-level overview.

Gordon M. Barrow, *The Structure of Molecules,* W. A. Benjamin, Menlo Park, Calif., 1964. A very good, elementary introduction.

J. C. D. Brand and J. C. Speakman, *Molecular Structure,* 2nd ed., Halsted Press, New York, 1975. A more advanced, comprehensive account.

R. G. Brewer and A. Mooradian, Eds., *Laser Spectroscopy,* Plenum Press, New York, 1974. Detailed accounts of applications.

R. T. Conley, *Infrared Spectroscopy,* 2nd ed., Allyn and Bacon, Boston, 1972. Intermediate level; emphasizes methods and applications.

C. A. Coulson, *The Shape and Structure of Molecules,* Oxford University Press, New York, 1973. An excellent, brief introduction.

Pierre Crabbe, *ORD and CD in Chemistry and Biochemistry,* Academic Press, New York, 1972. Intermediate level, with many examples.

J. R. Dyer, *Applications of Absorption Spectroscopy of Organic Compounds,* Prentice-Hall, Englewood Cliffs, N.J., 1965. Excellent examples and exercises in structural assignment.

Thomas M. Dunn, Donald S. McClure, and Ralph G. Pearson, *Some Aspects of Crystal Field Theory,* Harper and Row, New York, 1965. Brief but mathematical discussion of the interpretation of properties of metal–ion complexes.

R. D. B. Fraser and E. Suzuki, "Infrared Methods," in *Physical Principles and Techniques of Protein Chemistry,* Part B, S. J. Leach, Ed., Chapter 13, Academic Press, New York, 1970. Describes both methods and their applications.

T. R. Gilson and P. J. Hendra, *Laser Raman Spectroscopy,* Wiley, New York, 1970.

Harry B. Gray, *Electrons and Chemical Bonding,* W. A. Benjamin, Menlo Park, Calif., 1964. Good, intermediate-level treatment of bonding.

M. W. Hanna, *Quantum Mechanics in Chemistry,* 2nd ed., W. A. Benjamin, Menlo Park, Calif., 1969. An introduction to the mathematical aspects of bonding theory.

M. N. Hughes, *The Inorganic Chemistry of Biological Processes,* Wiley-Interscience, New York, 1973. A fairly advanced treatment of the role of metal bonding in biology.

G. W. King, *Spectroscopy and Molecular Structure,* Holt, Rinehart, and Winston, New York, 1964. Excellent, advanced treatment.

J. P. Mathieu, Ed., *Advances in Raman Spectroscopy,* Vol. 1, Heyden, London, 1973. Section V deals with applications to macromolecules and biological systems.

Kurt Mislow, *Introduction to Stereochemistry,* W. A. Benjamin, Menlo Park, Calif., 1965. A very good account of molecular geometry.

Walter J. Moore, *Physical Chemistry,* 4th ed., Prentice-Hall, Englewood Cliffs, N.J., 1972. Chapter 15 discusses chemical bonding at a more advanced level than this text; spectroscopy is covered in Chapter 17.

Linus Pauling, *The Nature of the Chemical Bond,* 3rd ed., Cornell University Press, Ithaca, N.Y., 1960. A classic.

Andrew Streitweiser, *Molecular Orbital Theory for Organic Chemists,* Wiley, New York, 1961. Application of bonding theory to organic molecules.

K. E. Van Holde, *Physical Biochemistry,* Prentice-Hall, Englewood Cliffs, N.J., 1971. Chapter 10 gives a clear description of ORD and CD.

Journal Articles

John C. Bailar, "Some Coordination Compounds in Biochemistry," *Am. Sci.* **59,** 586 (1971).

J. Bland, "Biochemical Effects of Excited State Molecular Oxygen," *J. Chem. Educ.* **53,** 274 (1976).

G. W. Rayner Canham and A. B. P. Lever, "Simple Models of Iron Sites in Some Biological Systems," *J. Chem. Educ.* **49,** 656 (1972).

Irwin Cohen and Janet Del Bene, "Hybrid Orbitals in Molecular Orbital Theory," *J. Chem. Educ.* **46,** 487 (1969).

T. R. Dyke and J. S. Muenter, "An Undergraduate Experiment for the Measurement of Phosphorescence Lifetimes," *J. Chem. Educ.* **52,** 251 (1975).

M. S. Feld and V. S. Letokhov, "Laser Spectroscopy," *Sci. Am.* **229,** 69 (December 1973).

John G. Foss, "Absorption, Dispersion, Circular Dichroism, and Rotary Dispersion," *J. Chem. Educ.* **40,** 592 (1963).

D. T. Haworth and K. M. Elsen, "Spectral Comparison of Geometrical Isomers," *J. Chem. Educ.* **50,** 300 (1973).

G. L. Henderson, "Optical Rotatory Dispersion in Transparent Media," *J. Chem. Educ.* **45,** 515 (1968).

Raold Hoffman, "Interaction of Orbitals through Space and through Bonds," *Acc. Chem. Res.* **4,** 1 (1971).

Hans H. Jaffe and Albert L. Miller, "The Fate of Electronic Excitation Energy," *J. Chem. Educ.* **43,** 469 (1966).

Takashi Kajiwara and David R. Kearns, "Direct Spectroscopic Evidence for a Deuterium Solvent Effect on the Lifetime of Singlet Oxygen in Water," *J. Am. Chem. Soc.* **95,** 5886 (1973).

Norman I. Krinsky, "Singlet Excited Oxygen as a Mediator of the Antibacterial Action of Leukocytes," *Science* **186,** 363 (1974).

Peter F. Lott and Robert J. Hurtubise, "Instrumentation for Fluorescence and Phosphorescence," *J. Chem. Educ.* **51,** A315 (1974).

Thomas H. Maugh, "Singlet Oxygen: A Unique Microbicidal Agent in Cells," *Science* **182,** 44 (1973).

D. F. Mowery, "Criteria for Optical Activity in Organic Molecules," *J. Chem. Educ.* **46,** 269 (1969).

William S. Murphy, "The Octant Rule," *J. Chem. Educ.* **52,** 774 (1975).

Richard J. Olcott, "Visualization of Molecular Orbitals—Formaldehyde," *J. Chem. Educ.* **49,** 614 (1972).

G. R. Penzer, "Applications of Absorption Spectroscopy in Biochemistry," *J. Chem. Educ.* **45,** 692 (1968).

N. F. Phelan and M. Orchin, "Cross Conjugation," *J. Chem. Educ.* **45,** 633 (1968).

K. J. Rothschild, I. M. Asher, E. Anastassakis, and H. E. Stanley, "Raman Spectroscopic Evidence for Two Conformations of Uncomplexed Valinomycin in the Solid State," *Science* **182,** 384 (1973).

W. H. Sawyer, "Demonstration of Allosteric Behavior. The Hemoglobin—Oxygen System," *J. Chem. Educ.* **49,** 777 (1972).

B. Z. Shakhashiri and L. G. Williams, "Singlet Oxygen in Aqueous Solution," *J. Chem. Educ.* **53,** 358 (1976).

Julian W. Snow and Thomas M. Hooker, "Contribution of Side Chain Chromophores to the Optical Activity of Proteins," *J. Am. Chem. Soc.* **96,** 7800 (1974).

Thomas G. Spiro, "Resonance Raman Spectroscopy: A New Structure Probe for Biological Chromophores," *Acc. Chem. Res.* **7**, 339 (1974).

R. Stuart Tobias, "Raman Spectroscopy in Inorganic Chemistry," *J. Chem. Educ.* **44**, 2, 70 (1967).

Kin-Ping Wong, "Optical Rotary Dispersion and Circular Dichroism," *J. Chem. Educ.* **51**, A573 (1974).

J. J. Worman, G. L. Pool, and W. P. Jensen, "Comparison of Electronic Spectra — Carbonyls, Thiocarbonyls, and Azomethines," *J. Chem. Educ.* **47**, 709 (1970).

Ten
Kinetics of
Chemical
Reactions

The subject of chemical kinetics is concerned with the measurement of the rates at which reactions occur, and with the interpretation and application of the results of these measurements. Knowledge about kinetics is utilized, in a very practical way, to maximize the yields of desired products in chemical reactions. For example, conditions can be chosen so as to accelerate selectively the reaction desired as compared to competing reactions. Perhaps an even more significant application is the analysis of rate behavior in order to gain information about the pathway or mechanism of a reaction. Much of the knowledge of the manner in which enzymes function, for example, has been derived from a study of the kinetics of the reactions for which they serve as catalysts.

10-1
RATES AND
THEIR MEASUREMENT

The rate of a process is defined to be the amount of that process occurring divided by the time interval in which it occurs. For example, if in a chemical reaction 3.2 mol of material reacts in 50 min, the rate is 0.064 mol/min. If a reaction is carried out in the liquid phase, so that the total volume of the reactants plus products remains essentially constant, or is carried out in the gas phase in a reactor of fixed volume, the rate can be expressed, as is usually done by chemists and as we shall do here, as the change in concentration per unit time rather than in terms of the amount of material reacted. Thus, if the reactant concentration decreases from 2.55 mol/liter to 2.14 mol/liter in 10 min, the rate of disappearance of reactant is 0.041 mol/liter min.

If the stoichiometric relation describing a reaction is such that some of the coefficients in the equation for the reaction are other than unity, it is necessary to specify which material in the equation is taken as the basis for calculating the rate. Thus, in the reaction between iodide ion and persulfate ion, for which the stoichiometric equation is

$$2I^- + S_2O_8^{2-} \longrightarrow I_2 + 2SO_4^{2-} \tag{10-1}$$

the rate at which iodide ion disappears is the same as the rate at which sulfate ion is formed, but each of these rates is equal to twice the rate at which persulfate ion disappears and at which iodine molecules are formed. One can choose to use either the larger or smaller quantity as the reaction rate, so long as the choice is clearly specified.

The experimental method by which one determines the rate of a reaction is, more often than not, an indirect one. In the usual procedure, the concentrations of as many reactants and products as possible are measured as functions of time, and then the rate is calculated from the concentration values by suitable mathematical methods which are described later.

A chemical reaction can be followed by any analytical method suitable for measuring the concentrations desired. If the reaction is carried out in the liquid phase, samples can be withdrawn periodically and analyzed by any convenient method—titration, spectrophotometric measurement, refractive index determination, electrochemical methods, or other means. When this procedure is followed, precautions must be taken to stop the reaction immediately upon withdrawal of the sample from the reaction mixture. This can sometimes be accomplished by rapid cooling of the sample to a temperature at which the reaction proceeds at a negligible rate or perhaps, if the reaction is one that takes place in an acidic or basic medium only, by neutralizing the sample by adding base or acid, respectively.

More convenient, however, is a procedure in which some property of the reaction mixture is continuously followed, without the removal of samples. If the reaction involves optically active substances, it can be carried out in a polarimeter tube, jacketed to maintain it at constant temperature, and the optical rotation can be measured as frequently as desired. If the reaction involves ions and a change in conductivity occurs as it proceeds, it can be carried out in a conductivity cell, permitting continuous measurement. If a change in volume takes place during the reaction, it may be allowed to take place in a dilatometer. This is a vessel which can be entirely sealed except for an attached calibrated capillary tube; the vessel is filled with the liquid reaction mixture, and the liquid meniscus is observed as it moves up or down the capillary. If a gas is consumed or produced in the reaction, the volume of the gas can be measured. If the infrared spectrum or the nuclear magnetic resonance spectrum of the reaction mixture contains absorptions which can be assigned to one or more of the reactants or products, the reaction can simply be carried out in the cell of the spectrometer and the spectrum obtained at suitable time intervals.

10-2
KINETICS OF THE OVERALL REACTION

The experimental results of a rate measurement are embodied in a kinetic equation in which the rate is given as a function of the concentrations of the reactants and of one or more rate constants; the latter are conventionally represented by the lower case letter k. The magnitude of a rate constant is specific for a given reaction at a given temperature, usually increasing as the temperature rises, but the value of the constant is independent of the concentration. Unless a reverse reaction, in which reactants are regenerated from products, occurs at the same time, the concentrations of the products do not enter into the rate equation.

As an example of how the kinetic behavior of a reaction is represented, the experimental result that the rate of decomposition of nitrogen pentoxide, either in the gaseous state or in solution in carbon tetrachloride, is proportional to the first power of the concentration of the nitrogen pentoxide is described by the equation

$$-\frac{d[N_2O_5]}{dt} = k[N_2O_5] \tag{10-2}$$

This reaction is therefore said to be "first-order in nitrogen pentoxide," which is equivalent to stating that the rate is proportional to the first power of the concentration of N_2O_5. The reaction of hydrogen iodide to form hydrogen molecules and iodine molecules is found to have a rate proportional to the second power of the concentration of hydrogen iodide and therefore is described as being "second-order in hydrogen iodide":

$$-\frac{d[HI]}{dt} = k[HI]^2 \tag{10-3}$$

An example of a process involving two different reactants is the reaction of hydroxyl ion with an ester:

$$CH_3COOCH_3 + OH^- \longrightarrow CH_3OH + CH_3COO^- \tag{10-4}$$

The observed kinetics for this reaction are first-order in the ester and first-order in hydroxyl ion, and therefore the overall order is second. The rate equation is

$$\text{rate} = k[CH_3COOCH_3][OH^-] \tag{10-5}$$

The order of a reaction need not be integral. For example, gaseous acetaldehyde decomposes thermally according to the stoichiometric equation

$$CH_3CHO \longrightarrow CH_4 + CO \tag{10-6}$$

at a rate given by

$$-\frac{d[CH_3CHO]}{dt} = k[CH_3CHO]^{3/2} \tag{10-7}$$

Accordingly, the order of this reaction is said to be three-halves. Obviously, three-halves of a molecule cannot undergo a decomposition reaction, and we shall find that such behavior indicates that several different steps occur in the reaction in such a way that the overall kinetic equation and order depend upon their composite effect.

Indeed, many reactions cannot be said to have a specific order. For instance, the rate of combination of gaseous hydrogen and gaseous bromine is given by

$$-\frac{d[H_2]}{dt} = -\frac{d[Br_2]}{dt} = \frac{1}{2}\frac{d[HBr]}{dt} = \frac{k'[H_2][Br_2]^{1/2}}{1 + k''([HBr]/[Br_2])} \quad (10\text{-}8)$$

This reaction can be described as first-order in hydrogen, but there is no definite order in bromine. The complexity of the kinetic equation arises because this is a chain reaction, consisting of a series of processes, as described in Section 10-5, and because some of these processes are reversible.

We can compare Equation (10-8) with the equation for the reaction of hydrogen with iodine, provided we take into account for the latter the opposing process, for which the rate expression was given in Equation (10-3). The net rate is then

$$-\frac{d[H_2]}{dt} = -\frac{d[I_2]}{dt} = \frac{1}{2}\frac{d[HI]}{dt} = k_f[H_2][I_2] - k_r[HI]^2 \quad (10\text{-}9)$$

Here we have used the subscripts "f" and "r" to distinguish the rate constants for the forward and reverse reactions.

In fact, very few reactions, even those that have simple rate equations and a definite order, occur as simple, uncomplicated, single-step processes. In Section 10-9 we consider the rates of enzyme-catalyzed biological reactions, and for these there is observed the characteristic feature that the apparent order depends, for a given concentration of the catalyzing enzyme, upon the concentration of the reactant.

To summarize, *for the great majority of chemical reactions, the stoichiometric equation as written does not portray a change that can really take place as a single step.* Accordingly, the rate equation for the overall reaction can never safely be written down from mere inspection of the stoichiometric equation.

In contrast, no matter how complex a reaction is, it occurs by a sequence of individual steps, or elementary processes, for each of which the rate is equal to a rate constant for that process multiplied by the concentrations of the reactants entering into the process. In other words, a process that is really a single-step event, truly an elementary process, does conform to the *law of mass action*.

10-3
REACTION ORDERS
AND RATE CONSTANTS

We turn now to the problem of finding the form of the kinetic equation for a reaction—that is, establishing its order—and evaluating the rate constants appearing in the equation. Consideration is confined for the most part to relatively simple examples, and we assume that the equilibrium constant is so large that the reverse reaction has a negligible effect.

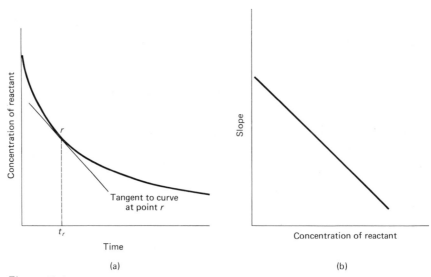

Figure 10-1
Method of establishing a rate relationship from experimental time–concentration results. (a) Rate is calculated as the slope of a concentration–time plot. (b) The slope from (a) is plotted as a function of reactant concentration; the straight line shown corresponds to a first-order reaction.

Suppose that we wish to deal with a reaction into which only one substance enters. A general procedure is to follow the change in concentration of the reactant during a run, plot the concentration against time, and evaluate the slope of the plot at each of a series of concentrations by drawing a tangent to the curve at the desired point, as shown in Figure 10-1a. The slope of the tangent at a point is equal to the rate at which the reaction is occurring at that point and therefore gives the rate for the concentration corresponding to that point. The rate can then be plotted against the concentration, and the form of the kinetic equation obtained by inspection of this plot. An example is shown in Figure 10-1. Further information can be gained by measuring the initial rates for different initial concentrations of the reactant. For reactions involving several different reactants, the method of initial rate measurements is particularly advantageous, for the concentration of each of the reactants can then be varied independently of the other concentrations.

However, the operation of drawing tangents to the curve, a procedure that is equivalent to the use of data from successive points separated by very small time intervals, is not very accurate. Alternative methods employ the variation with time, for a given run, or with initial concentration, for a series of runs, of the time interval required for a given conversion to occur. In these methods, the integrated form of the kinetic equation is applied, rather than the differential form.

DERIVATION
OF INTEGRATED
KINETIC EQUATIONS

Suppose a reaction is first-order in a single reactant, A. Let the concentration of A, which varies with time t, be symbolized by [A]. The rate equation can then be written

$$-\frac{d[A]}{dt} = k[A] \tag{10-10}$$

where k is the rate constant.

Before integrating this equation, one must separate variables, placing functions of [A] on one side of the equation and of t on the other side:

$$-\frac{d[A]}{[A]} = k\, dt \tag{10-11}$$

Integration of each side from time zero at which [A] has the value $[A]_0$ to a particular time t yields

$$-\ln [A] \Big|_{\text{time } 0}^{\text{time } t} = kt \Big|_{\text{time } 0}^{\text{time } t} \tag{10-12}$$

or

$$\ln \frac{[A]_0}{[A]} = kt \tag{10-13}$$

Another way of writing this equation is

$$\ln [A] = \ln [A]_0 - kt \tag{10-14}$$

From this, one sees that a plot of $\ln [A]$ against t should yield a straight line of slope $-k$.

Taking antilogarithms of each side of Equation (10-13) and rearranging, one obtains

$$[A] = [A]_0 e^{-kt} \tag{10-15}$$

Thus the concentration of a reactant in a first-order kinetic process decays *exponentially* with time.

An alternative way of integrating Equation (10-11) is from one time t_1 to another time t_2:

$$-\ln \frac{[A]_{t_2}}{[A]_{t_1}} = k(t_2 - t_1) \tag{10-16}$$

For some purposes, it is convenient to express the concentration in terms of a *transformation variable,* usually symbolized by x and defined as the amount by which the concentration of some one reactant has decreased from the start of the reaction up to the time at which the variable is evaluated. If a is the *initial reactant* concentration, the concentration at time t is $a - x$, and Equation (10-10) becomes

$$\frac{dx}{dt} = k(a - x) \tag{10-17}$$

The integrated form, equivalent to Equation (10-13), is

$$\ln \frac{a}{a - x} = kt \tag{10-18}$$

Very often the rate constant of a first-order reaction is described by specifying the half-life, defined as the time required for the concentration of reactant to be reduced to one-half of its initial value; after an interval equal to the half-life, the transformation variable equals $a/2$. The half-life of a first-order reaction has the distinguishing characteristic that it is independent of the concentration at the beginning of the period for which the half-life is measured. To demonstrate this, we merely substitute $x = a/2$ in Equation (10-18):

$$\ln \frac{a}{a - (a/2)} = 2.303 \log 2 = kt_{1/2} \tag{10-19}$$

Thus $t_{1/2}$, the half-life, is equal to $0.69/k$.

Example: The half-life of a first-order reaction is 35 min. If the initial concentration of reactant is 0.8 mol/liter, what length of time is required for the concentration to drop to 0.1 mol/liter?

Solution: The first period of 35 min reduces the concentration to 0.4, the second to 0.2, and the third to 0.1 mol/liter. Therefore three half-lives, or 105 min, is required.

If a reaction involves only a single reactant and is second-order with respect to that reactant, the differential equation is

$$-\frac{d[A]}{dt} = k[A]^2 \tag{10-20}$$

The integrated form of this equation is

$$kt = \frac{1}{[A]} - \frac{1}{[A]_0} \tag{10-21}$$

Thus a plot of $1/[A]$ against t is linear with a slope of k.

For a reaction of two substances, A and B, which is first-order in each and thus is second-order overall, it is convenient to use the transformation variable x with initial reactant concentrations a and b. If one molecule of B reacts with one molecule of A in the stoichiometric equation,

$$[A] = [A]_0 - x \equiv a - x \tag{10-22}$$

$$[B] = [B]_0 - x \equiv b - x \tag{10-23}$$

The rate is then given by

$$\frac{dx}{dt} = k(a - x)(b - x) \tag{10-24}$$

Integration after separation of the variables x and t leads to

$$\frac{1}{a - b} \ln \frac{b(a - x)}{a(b - x)} = kt \tag{10-25}$$

If $a = b$, these equations become indeterminate, but in this case the two initially equal concentrations change in exactly parallel fashion throughout the reaction and Equation (10-21) can be used.

For a third-order reaction between like molecules of reactant A the differential rate equation is

$$\text{rate} = -\frac{d[A]}{dt} = k[A]^3 \tag{10-26}$$

and the concentration–time equation is

$$\frac{1}{2}\left(\frac{1}{[A]^2} - \frac{1}{[A]_0^2}\right) = kt \tag{10-27}$$

For a zero-order reaction, the rate is independent of concentration and the integrated equation obtained is

$$[A] = [A]_0 - kt \tag{10-28}$$

APPLICATION OF INTEGRATED KINETIC EQUATIONS

To establish from experimental results the order of a reaction involving *a single substance,* and at the same time to calculate the rate constant, the relations derived above can be applied in any of several ways:

(1) The integrated equation for a particular order reaction can be solved for k by successively substituting pairs of corresponding values of time and concentration. If the numerical value of k is found to be constant, within experimental error, over a reasonable range of concentration and time, the equation is considered to apply to the reaction.

(2) The length of time required for some given fraction, usually one-half, of the reactant present at the beginning of the period to disappear can be determined for various initial concentrations. For zero-order reactions this time is proportional to the initial concentration, for first-order reactions it is independent of the initial concentration, and for second-order reactions it is inversely proportional to the initial concentration.

(3) A plot against time can be made of the appropriate quantity which is expected to be linear with time. The quantities to plot and the slopes of the resulting lines are given in Table 10-1 for reactions of various order. Linearity of a plot is a good indication that the reaction is of the chosen order, and the rate constant can then be calculated from the slope of the plot or by applying a least-squares treatment to the data points that have been plotted. The entry in Table 10-1 for a reaction of order n includes all the other orders—except first—as special cases and applies also to reactions of fractional order, such as one-half or three-halves.

Certain characteristics are unique to first-order kinetics: The half-life is independent of initial concentration and the numerical value of the rate constant is independent of the units used to express concentration. Quite familiar, for example, is the practice of quoting half-lives for the first-order radioactive decay processes of various isotopes as inverse measures of the rate constants. As a consequence of these

Table 10-1
Integrated kinetic equations and significant quantities for the evaluation of the rate constant for a reaction of a single starting material

Order of reaction	Integrated rate equation	Function to plot	Slope of plot	Expression for half-life	Units of k
0	$kt = x$	x	k	$\dfrac{a}{2k}$	$\dfrac{\text{mol}}{\text{liter sec}}$
1	$kt = \ln \dfrac{a}{a - x}$	$\log(a - x)$	$-\dfrac{k}{2.303}$	$\dfrac{\ln 2}{k} = \dfrac{0.693}{k}$	$\dfrac{1}{\text{sec}}$
2	$kt = \dfrac{1}{a - x} - \dfrac{1}{a}$	$\dfrac{1}{a - x}$	k	$\dfrac{1}{ka}$	$\dfrac{\text{liter}}{\text{mol sec}}$
3	$kt = \dfrac{1}{2}\dfrac{1}{(a - x)^2} - \dfrac{1}{a^2}$	$\dfrac{1}{(a - x)^2}$	$2k$	$\dfrac{3}{2ka^2}$	$\dfrac{\text{liter}^2}{\text{mol}^2\,\text{sec}}$
n	$kt = \dfrac{1}{n - 1}\dfrac{1}{(a - x)^{n-1}} - \dfrac{1}{a^{n-1}}$	$\dfrac{1}{(a - x)^{n-1}}$	$(n - 1)k$	$\dfrac{2^{n-1} - 1}{(n - 1)ka^{n-1}}$	$\dfrac{\text{liter}^{n-1}}{\text{mol}^{n-1}\,\text{sec}}$

characteristics, it is possible to determine the extent of a first-order reaction without ever bothering to evaluate the concentration itself.

Consider, as an example, the reaction in which sucrose in aqueous solution is hydrolyzed in the presence of an acid to glucose and fructose. The angle by which polarized light is rotated by the reacting mixture is additive in the concentrations of the three optically active sugars present. At any time during the course of the reaction, the fraction of sucrose that has reacted is equal to the fractional change in optical rotation that has occurred from the initial value characteristic of all sucrose toward the final value characteristic of a 1:1 mixture of the same concentration of glucose and fructose. If the initial rotation of the sucrose is represented by α_0, the product rotation by α_∞, and the rotation at time t by α_t, then the fraction of sucrose reacted up to time t is given by $(\alpha_0 - \alpha_t)/(\alpha_0 - \alpha_\infty)$ and the fraction of sucrose remaining is equal to $(\alpha_t - \alpha_\infty)/(\alpha_0 - \alpha_\infty)$. Because the kinetics of this reaction are first-order in sucrose, with no apparent effect during the course of a

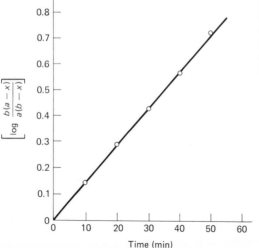

Figure 10-2
Linear plot of data for a second-order reaction. Values are for the hydroxyl ion–ethyl acetate reaction and have been taken from Table 10-2.

Table 10-2
Reaction of hydroxyl ion and ethyl acetate at 30°C

Time (min)	Ester concentration (mol/liter)	Hydroxide concentration (mol/liter)	$\dfrac{b(a-x)}{a(b-x)}$	$\log\dfrac{b(a-x)}{a(b-x)}$	Second-order rate constant (liter/mol min)
0	0.100	0.0500	1	0	—
10	0.078	0.0282	1.39	0.143	0.66
20	0.068	0.0176	1.93	0.286	0.66
30	0.062	0.0116	2.67	0.427	0.66
40	0.058	0.0079	3.67	0.565	0.65
50	0.055	0.0052	5.30	0.724	0.67

reaction run from changing concentrations of other substances, the equation for the rate constant can accordingly be written

$$k = \frac{2.303}{t} \log \frac{(\alpha_0 - \alpha_\infty)}{(\alpha_t - \alpha_\infty)} \tag{10-29}$$

Turning now to a second-order reaction involving two reactants, we present in Table 10-2 some typical results for the basic hydrolysis of ethyl acetate in an alcohol–water solution containing sodium hydroxide. The reaction proceeds according to the equation

$$CH_3COOC_2H_5 + OH^- \longrightarrow C_2H_5OH + CH_3COO^- \tag{10-30}$$

The concentration of hydroxyl ion is obtained at various times by titration of samples of the reaction mixture, and the concentration of ester is calculated by subtracting the amount of hydroxyl ion reacted from the initial ester concentration. In the last column of the table is the rate constant calculated for a second-order reaction. The values show good constancy, giving a strong indication that the reaction is second-order. A plot of $\log [b(a - x)/a(b - x)]$ against time has been made in Figure 10-2, and the slope of the straight line obtained is 0.0141. Multiplication of the slope by $2.303/(a - b)$ yields a value for the rate constant of 0.65 liter/mol min. This is an averaged value, for it has been determined by constructing the best straight line through the points representing all the available data for the run.

If the same reaction is carried out with various initial concentrations of ester and base but keeping the two initial concentrations always equal, it is possible to determine the time at which one-half of each reactant has disappeared. The results obtained are represented by the values shown in Table 10-3. The results are in keeping with the

Table 10-3
Half-times of reaction of hydroxyl ion and ethyl acetate at 30°C

Initial concentration, a (mol/liter)	Half-time, $t_{1/2}$ (min)	$k = \dfrac{1}{at_{1/2}}$
0.005	304	0.66
0.010	154	0.63
0.050	30.6	0.65
0.100	15.2	0.66

inverse proportionality of the half-life $t_{1/2}$ to the initial concentration a, which is characteristic of a second-order reaction. It is also possible, as shown in the table, to calculate a numerical value of the rate constant from the half-time and the initial concentration.

10-4
COMPLEX REACTIONS

In this section we examine the relation of the series of elementary processes—the pathway—by which reactants are converted into products to the overall kinetic equation for a complex reaction. Typically one has available an observed rate equation and wishes to test a proposed pathway against that equation to determine whether the pathway is an acceptable one. It is to be emphasized that obtaining agreement in this way between prediction and experiment by no means establishes the proposed pathway as *the* correct one, because there may be other pathways which lead to rate equations of the same form.

Our considerations in this section are limited to reactions in which the intermediates are present at any time in relatively low concentrations. There are of course many reactions in which intermediate products are formed in concentrations sufficient for them to be determined by direct analysis. This type of reaction is not further discussed here, for rather formidable mathematics is often involved in solving the applicable systems of simultaneous differential equations.

As a first example of a complex reaction, the conversion of sucrose in aqueous solution in the presence of acid to a mixture of glucose and fructose, as described in Section 10-3, is considered. A reaction run can be carried out by mixing an acid, say hydrochloric acid, with an aqueous solution of sucrose, placing the mixture in a polarimeter tube and allowing it to react, and following the optical rotation as the sucrose is hydrolyzed. The acid concentration remains unchanged, and the water is present in so large an excess that its concentration is effectively constant during the run. If the results of such a reaction run are treated in any of the ways described earlier for obtaining the form of the rate equation, it is found that the reaction is first-order in sucrose:

$$\text{rate} = k[\text{sucrose}] \tag{10-31}$$

If, now, the reaction is carried out with a different initial concentration of the catalyzing acid, the value of the rate constant, k in Equation (10-31), is found to change parallel with the acid concentration. The effect of water concentration on the apparent rate constant cannot easily be measured, but water is consumed in the reaction:

$$C_{12}H_{22}O_{11} + H_2O \longrightarrow 2C_6H_{12}O_6 \tag{10-32}$$

The following pathway affords a satisfactory explanation of the observations:

$$C_{12}H_{22}O_{11} + HA \underset{k_{-1}}{\overset{k_1}{\rightleftharpoons}} C_{12}H_{23}O_{11}^+ + A^- \tag{10-33}$$

$$C_{12}H_{23}O_{11}^+ + H_2O \overset{k_2}{\longrightarrow} 2C_6H_{12}O_6 + H^+ \tag{10-34}$$

$$H^+ + A^- \longrightarrow HA \qquad\qquad (10\text{-}35)$$

According to this scheme, in which HA represents the catalyzing acid, the reaction consists of a series of successive bimolecular processes. Equation (10-33) represents two such processes, the combination of the sucrose with the catalyzing acid, proceeding with a rate constant k_1, and the reverse transfer of a proton back to the base A^- conjugate to the catalyst, with a rate constant k_{-1}. To explain the observed kinetic equation, one may assume that the process given by Equation (10-34), with a rate constant k_2, is the slowest of all the steps indicated, constituting for the reaction a bottleneck or *rate-determining step*. The process of Equation (10-35) is required to restore the original form of the catalyst but, since it occurs subsequent to the rate-determining step, the kinetic behavior gives no information regarding it.

One method of mathematical treatment of a complex reaction assumes that equilibrium is maintained in one or more reversible processes *up to the rate-determining step*. For sucrose hydrolysis, equilibrium is supposed to exist in the reactions of Equation (10-33) with an equilibrium constant

$$K = \frac{[C_{12}H_{23}O_{11}^+][A^-]}{[HA][C_{12}H_{22}O_{11}]} = \frac{k_1}{k_{-1}} \qquad (10\text{-}36)$$

The overall rate of the reaction is now set equal to the rate of the key step—the rate-determining step—equal to the rate constant of that step times the equilibrium concentrations of the species entering into it, or $k_2[C_{12}H_{23}O_{11}^+][H_2O]$. The concentration of the intermediate cannot be experimentally determined, but it can be expressed in terms of measurable concentrations by Equation (10-36):

$$[C_{12}H_{23}O_{11}^+] = \frac{k_1[HA][C_{12}H_{22}O_{11}]}{k_{-1}[A^-]} \qquad (10\text{-}37)$$

Substitution of this in the rate expression for the key step leads to

$$\text{rate} = \frac{k_1 k_2}{k_{-1}} \frac{[HA][C_{12}H_{22}O_{11}][H_2O]}{[A^-]} \qquad (10\text{-}38)$$

Since HA and A^- are not consumed or produced in the overall reaction, their concentrations remain constant during a reaction run, and the apparent or measured value of the first-order rate constant in Equation (10-31) is really a composite, equal by this scheme to $k_1 k_2[HA][H_2O]/k_{-1}[A^-]$. Since the form of the rate expression, Equation (10-38), is compatible with experimental results, the postulated pathway is an acceptable one. However, the measured rate constant cannot by kinetic results alone be resolved into its components k_1, k_{-1}, and k_2.

The alert reader may ask: "How can you be certain that the first step, the transfer of a proton from the acid to sucrose, is not the rate-determining step?" This is a reasonable question, because such a pathway should yield the same kinetic behavior as has been described. The answer lies in the interpretation of the results obtained when the catalyzing acid is changed, when the concentration of acid is changed at constant pH in a buffer, and when the medium is changed from H_2O to D_2O. Experiments of this type are discussed in subsequent sections.

Under some circumstances it is either desirable or necessary to use an alternative approach, called the method of the *stationary state*, for the development of a rate equation for a complex reaction. This is based upon the premise that the rate of change in concentration of an intermediate present in concentrations too small to be detectable can never be more than negligible compared to other rates in the reaction. If the intermediate decreased in concentration at an appreciable rate, there would be none remaining, and if it increased in concentration, its presence would become detectable. From Figure 10-3, it is evident that the *net* rate of change in concentration of B is equal to zero.

In the sucrose hydrolysis reaction, the reactive intermediate having a stationary concentration is $C_{12}H_{23}O_{11}^+$, and the difference between the total of the rates of all steps in which it is formed and the total of the rates of the steps in which it disappears is set equal to zero:

$$\frac{d[C_{12}H_{23}O_{11}^+]}{dt} = k_1[C_{12}H_{22}O_{11}][HA] - k_{-1}[C_{12}H_{23}O_{11}^+][A^-]$$
$$- k_2[C_{12}H_{23}O_{11}^+][H_2O] = 0 \quad (10\text{-}39)$$

The solution of this equation for the stationary-state concentration of the intermediate yields

$$[C_{12}H_{23}O_{11}^+] = \frac{k_1[C_{12}H_{22}O_{11}][HA]}{k_{-1}[A^-] + k_2[H_2O]} \quad (10\text{-}40)$$

When this equation is compared with Equation (10-37), which was obtained by assuming equilibrium up to the rate-determining step, it is observed that the denominator now contains an additional term. If the rate at which the intermediate forms products, $k_2[H_2O][C_{12}H_{23}O_{11}^+]$, is much smaller than the rate at which it returns to starting materials, $k_{-1}[A^-][C_{12}H_{23}O_{11}^+]$, then Equation (10-40) reduces to Equation (10-37), which is quite consistent since the assumption of equilibrium is satisfied under these conditions.

We turn now to gaseous reactions to illustrate further the relation between the pathway and the kinetic equation for complex reactions. An interesting type of reaction is that in which two molecules of nitric

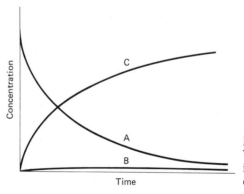

Figure 10-3
Variation in concentration with time in the reaction of A to form C by way of the reactive intermediate B.

oxide combine with one molecule of chlorine, of bromine, or of oxygen. The three equations for the overall changes are

$$2NO + Cl_2 \longrightarrow 2NOCl \tag{10-41}$$

$$2NO + Br_2 \longrightarrow 2NOBr \tag{10-42}$$

$$2NO + O_2 \longrightarrow 2NO_2 \tag{10-43}$$

Each of these reactions is found to have a third-order kinetic equation; for the combination of nitric oxide and chlorine, the rate equation is

$$\frac{d[NOCl]}{dt} = k[Cl_2][NO]^2 \tag{10-44}$$

One explanation for this is that the reaction is indeed a one-step, termolecular process, involving all three molecules in one collision with a concerted rearrangement of bonds. However, the simultaneous encounter of three molecules in the gas phase is a process with a relatively low probability of occurrence, and it thus seems relevant to explore the question of whether there is an alternative pathway leading to the same form of kinetic equation. Furthermore, all these nitric oxide reactions are rather unusual in that their rate constants have a very small dependence upon temperature. Indeed, for the combination with oxygen the apparent rate constant decreases with increasing temperature, a behavior contrary to that of almost all other rate constants.

A reasonable pathway to investigate is a sequence of two bimolecular processes. The first may well be the dimerization of nitric oxide:

$$2NO \underset{k_{-1}}{\overset{k_1}{\rightleftharpoons}} N_2O_2 \tag{10-45}$$

This would be followed by combination of the dimer with the third molecule. For example, in the chlorine reaction, there would occur

$$N_2O_2 + Cl_2 \overset{k_2}{\longrightarrow} 2NOCl \tag{10-46}$$

If the first step is rapid and reversible and the second step is rate-determining, so that k_2 is small compared to both k_1 and k_{-1}, the rate of the reaction is equal to the rate of the second step:

$$\text{rate} = k_2[N_2O_2][Cl_2] \tag{10-47}$$

However, since the concentration of the dimeric intermediate is not known from an experimental result, it must be expressed in terms of the concentrations of other species. The assumption of equilibrium in the first step yields the relation between the dimer concentration and the measurable nitric oxide concentration:

$$[N_2O_2] = \frac{k_1}{k_{-1}}[NO]^2 \tag{10-48}$$

and the rate equation is therefore

$$\frac{d[NOCl]}{dt} = \frac{k_1 k_2}{k_{-1}}[NO]^2[Cl_2] \tag{10-49}$$

which is consistent with the observed kinetic results.

If, rather than assuming that equilibrium is maintained in the first step, we apply the stationary-state assumption to the intermediate, then the predicted rate equation is

$$\frac{d[NOCl]}{dt} = \frac{k_1[NO]^2[Cl_2]}{k_{-1} + k_2[Cl_2]} \qquad (10\text{-}50)$$

Since there is no evidence that the order in chlorine is reduced below first when the chlorine concentration is high, the equilibrium approach is satisfactory.

When is it necessary to use the method of the reactive intermediate rather than the equilibrium assumption in treating a complex reaction? When conversion of the intermediate to products is more than a few percent of the total rate at which the intermediate reacts, the reactive intermediate treatment is more satisfactory; but when conversion to products occurs rarely compared to the return reaction to the reactants, the equilibrium assumption is applicable. The kineticist may or may not be able to determine from the experimental results with certainty which situation applies.

In some reactions the complications of the pathway are such that no single step can be designated rate-determining; in this case, the method of the reactive intermediate must necessarily be employed. The chain reactions described in Section 10-5 illustrate this point.

A final general reminder about a complex reaction of the sort we have been considering: Kinetic behavior reflects only the details of the pathway up to the rate-determining step and tells us nothing about the details of what occurs in more rapid processes after the bottleneck.

10-5
CHAIN REACTIONS

A chain reaction involves a cyclic sequence of at least two steps, in which a product of the first step is a reactant in the second, but a product of the second step is also a reactant in the first. Thus once the chain has started, the cycle can go around and around many times, producing products. The average number of times a chain reaction completes the cycle from its beginning until something happens to interrupt it is the *chain length*. The chain length differs from reaction to reaction, but it may be hundreds or thousands of cycles or more.

The chain reactions we deal with here are radical chains. The carriers that pass from one step to the next are free radicals, species with unpaired electrons, the initiation of the chain corresponds to the generation of radicals by breaking a covalent bond in a suitable species, and the ending of the chain occurs by loss of one or more radicals from the cyclic process.

The experimentally determined equation for the rate of reaction of hydrogen with bromine to form hydrogen bromide was given in Section 10-2. A free radical chain mechanism which reproduces the kinetics begins with an *initiation* step, the thermal decomposition of bromine molecules into bromine atoms:

$$Br_2 \xrightarrow{k_1} 2Br\cdot \qquad \text{Step 1} \qquad (10\text{-}51)$$

A bromine atom then reacts with a hydrogen molecule:

$$Br\cdot + H_2 \xrightarrow{k_2} HBr + H\cdot \qquad \text{Step 2} \qquad (10\text{-}52)$$

The hydrogen atom produced here can react with another bromine molecule:

$$H\cdot + Br_2 \xrightarrow{k_3} HBr + Br\cdot \qquad \text{Step 3} \qquad (10\text{-}53)$$

The chain is continued as this bromine atom reacts with a hydrogen molecule, in step 2. Step 2 and step 3 are both *propagation* steps, because they contribute to the route to the product but do not diminish the number of radicals capable of carrying on the chain. So long as nothing happens to deplete the stock of hydrogen or bromine atoms, the chain process goes merrily on around and around the cycle.

Occasionally the reaction may proceed along a nonproductive byway:

$$H\cdot + HBr \xrightarrow{k_4} H_2 + Br\cdot \qquad \text{Step 4} \qquad (10\text{-}54)$$

However, this represents only a temporary setback, for there is no net loss in the number of radicals, and the bromine atom formed is able to continue on in the chain.

Eventually the chain comes to an end, and the most likely *termination* step is the combination of two bromine atoms, reversing the initiation step:

$$2Br\cdot \xrightarrow{k_5} Br_2 \qquad \text{Step 5} \qquad (10\text{-}55)$$

For this scheme of reactions, the overall rate of the reaction can be expressed either as the rate of appearance of HBr or as the rate of disappearance of H_2, but we choose, only because it involves the simplest algebra, to write it as the rate of disappearance of Br_2:

$$\text{rate} = -\frac{d[Br_2]}{dt} = k_1[Br_2] + k_3[H\cdot][Br_2] - k_5[Br\cdot]^2 \quad (10\text{-}56)$$

Here the contributions of all the steps in which Br_2 is consumed or produced have been combined with the appropriate signs. However, the equation involves the concentrations of the two species, $Br\cdot$ and $H\cdot$, neither of which can be measured. To express the two concentrations, the method of the reactive intermediate can be applied. Thus from the sequence of five steps listed, all the processes in which bromine atoms are products or reactants are combined to give the overall rate of change in bromine atom concentration, which is taken to be approximately zero:

$$\frac{d[Br\cdot]}{dt} = 2k_1[Br_2] - k_2[Br\cdot][H_2] + k_3[H\cdot][Br_2]$$

$$+ k_4[H\cdot][HBr] - 2k_5[Br\cdot]^2 = 0 \quad (10\text{-}57)$$

In a similar way, the hydrogen atom concentration changes are summed to zero:

$$\frac{d[\text{H}\cdot]}{dt} = k_2[\text{Br}\cdot][\text{H}_2] - k_3[\text{H}\cdot][\text{Br}_2] - k_4[\text{H}\cdot][\text{HBr}] = 0 \tag{10-58}$$

When these equations are added together and the resultant solved for the bromine atom concentration, there results

$$[\text{Br}\cdot] = \left(\frac{k_1}{k_5}[\text{Br}_2]\right)^{1/2} \tag{10-59}$$

The concentration of hydrogen atoms can then be obtained from either Equation (10-57) or (10-58) by substituting this expression for $[\text{Br}\cdot]$. Choosing Equation (10-58),

$$k_2 k_1^{1/2} k_5^{-1/2}[\text{Br}_2]^{1/2}[\text{H}_2] - k_3[\text{H}\cdot][\text{Br}_2] - k_4[\text{H}\cdot][\text{HBr}] = 0 \tag{10-60}$$

Substitution of the bromine atom concentration from Equation (10-59) into the rate equation, Equation (10-56), simplifies the latter to

$$\text{rate} = k_3[\text{H}\cdot][\text{Br}_2] \tag{10-61}$$

Into this equation we now substitute the hydrogen atom concentration as evaluated from Equation (10-60) and then divide both numerator and denominator by $k_3[\text{Br}_2]$, to find

$$\text{rate} = \frac{k_1^{1/2} k_2[\text{Br}_2]^{1/2}[\text{H}_2]}{k_5^{1/2}(1 + k_4[\text{HBr}]/k_3[\text{Br}_2])} \tag{10-62}$$

This equation is consistent with the experimental results and shows how the apparent rate constants are related to the rate constants of the individual steps in the reaction scheme.

The objection may now be raised that some possible contributing steps have been omitted:

Initiation: $\text{H}_2 \longrightarrow 2\text{H}\cdot$ (10-63)

Propagation: $\text{Br}\cdot + \text{HBr} \longrightarrow \text{H}\cdot + \text{Br}_2$ (10-64)

Termination: $2\text{H}\cdot \longrightarrow \text{H}_2$ (10-65)

$\text{H}\cdot + \text{Br}\cdot \longrightarrow \text{HBr}$ (10-66)

Each of these may well occur, but the fact that they are not needed to explain the experimental results indicates that they occur infrequently compared to the other processes. Furthermore, the reactions in both Equations (10-63) and (10-64) are strongly endothermic and, because of the strength of the bonds involving hydrogen, the concentration of hydrogen atoms remains small compared to the concentration of bromine atoms, so that the reactions in Equations (10-65) and (10-66) are also relatively unlikely.

Another reaction that proceeds by a radical chain mechanism is the thermal decomposition of hydrocarbons. What was originally thought to be an extremely complex process because of the many products formed was shown by F. O. Rice and K. F. Herzfeld to be understandable in terms of a relatively few reaction types. Illustrative is the cracking of ethane, leading to ethylene and hydrogen as predominant products. The proposed scheme is

Initiation: $C_2H_6 \xrightarrow{k_1} 2CH_3 \cdot$ (10-67)

$CH_3 \cdot + C_2H_6 \xrightarrow{k_2} CH_4 + C_2H_5 \cdot$ (10-68)

Propagation: $C_2H_5 \cdot \xrightarrow{k_3} C_2H_4 + H \cdot$ (10-69)

$H \cdot + C_2H_6 \xrightarrow{k_4} C_2H_5 \cdot + H_2$ (10-70)

Termination: $H \cdot + C_2H_5 \cdot \xrightarrow{k_5} C_2H_6$ (10-71)

The rate of the overall reaction is equal to the rate of either step 3 or step 4, rates that are equal if the chain is reasonably long. We choose step 3 to express the rate:

$$+\frac{d[C_2H_4]}{dt} = -\frac{d[C_2H_6]}{dt} = k_3[C_2H_5 \cdot] \qquad (10\text{-}72)$$

The hydrogen atom and the ethyl radical are each treated as a reactive intermediate:

$$\frac{d[H \cdot]}{dt} = k_3[C_2H_5 \cdot] - k_4[H \cdot][C_2H_6] - k_5[H \cdot][C_2H_5 \cdot] = 0 \qquad (10\text{-}73)$$

$$\frac{d[C_2H_5 \cdot]}{dt} = 2k_1[C_2H_6] - k_3[C_2H_5 \cdot] + k_4[H \cdot][C_2H_6]$$
$$- k_5[H \cdot][C_2H_5 \cdot] = 0 \quad (10\text{-}74)$$

To obtain the last equation, it has been assumed that step 2 immediately follows every occurrence of step 1. Adding Equations (10-73) and (10-74), and solving for $[H \cdot]$, we find

$$[H \cdot] = \frac{k_1[C_2H_6]}{k_5[C_2H_5 \cdot]} \qquad (10\text{-}75)$$

Substitution of this expression in Equation (10-73) leads to

$$k_3[C_2H_5 \cdot] - k_4 \frac{k_1[C_2H_6]^2}{k_5[C_2H_5 \cdot]} - k_1[C_2H_6] = 0 \qquad (10\text{-}76)$$

The last term, equal to the rate of initiation, is relatively small for a chain of any appreciable length and is therefore neglected in comparison with the other terms. Solution of the remainder of the equation for $[C_2H_5 \cdot]$ yields

$$[C_2H_5 \cdot] = \left(\frac{k_1 k_4}{k_3 k_5}\right)^{1/2} [C_2H_6] \qquad (10\text{-}77)$$

The rate equation then simplifies to one that is first-order in ethane:

$$-\frac{d[C_2H_6]}{dt} = \left(\frac{k_1 k_3 k_4}{k_5}\right)^{1/2} [C_2H_6] \qquad (10\text{-}78)$$

As is usual for chain reactions, the choice of a different termination step leads to a different kinetic order.

A related reaction which proceeds by a radical chain is the oxidation of hydrocarbons, which results from traces of peroxides as impurities. The initiation step is the disproportionation of two peroxide molecules:

$$2ROOH \xrightarrow{k_1} R\cdot + ROO\cdot + H_2O_2 \tag{10-79}$$

The propagation steps are

$$R\cdot + O_2 \xrightarrow{k_2} ROO\cdot \tag{10-80}$$

$$ROO\cdot + RH \xrightarrow{k_3} R\cdot + ROOH \tag{10-81}$$

Termination of the reaction occurs by combination of two peroxy radicals; since these are quite a bit more stable than alkyl radicals, their concentration is much greater, and it is therefore more likely that two ROO· radicals will encounter one another than that one ROO· will meet an R· or that two R· radicals will find one another.

$$2ROO\cdot \xrightarrow{k_4} ROOR + O_2 \tag{10-82}$$

The overall rate is equal to the rate of consumption of oxygen, which is used in step 2 but produced in step 4:

$$-\frac{d[O_2]}{dt} = k_2[R\cdot][O_2] - k_4[ROO\cdot]^2 \tag{10-83}$$

Each of the radicals R· and ROO· is a reactive intermediate:

$$\frac{d[R\cdot]}{dt} = k_1[ROOH]^2 - k_2[R\cdot][O_2] + k_3[ROO\cdot][RH] = 0 \tag{10-84}$$

$$\frac{d[ROO\cdot]}{dt} = k_1[ROOH]^2 + k_2[R\cdot][O_2] - k_3[ROO\cdot][RH]$$
$$- 2k_4[ROO\cdot]^2 = 0 \tag{10-85}$$

Adding these two equations yields

$$2k_1[ROOH]^2 - 2k_4[ROO\cdot]^2 = 0 \tag{10-86}$$

which can be simply rearranged to give an expression for the peroxy radical concentration:

$$[ROO\cdot] = \left(\frac{k_1}{k_4}\right)^{1/2}[ROOH] \tag{10-87}$$

Substitution of this expression for [ROO·] in Equation (10-84) leads to

$$k_1[ROOH]^2 - k_2[R\cdot][O_2] + k_3\left(\frac{k_1}{k_4}\right)^{1/2}[ROOH][RH] = 0 \tag{10-88}$$

When this is combined with Equations (10-83) and (10-86), the rate becomes

$$-\frac{d[O_2]}{dt} = k_3\left(\frac{k_1}{k_4}\right)^{1/2}[ROOH][RH] \tag{10-89}$$

This equation agrees with the experimental result that the reaction is first-order in hydrocarbon and also indicates that it is first-order in the initiating material, the peroxide.

Polymerization reactions are in many ways related to the chain reactions we have been describing. They can occur by ionic mechanisms—often catalyzed by acids or by bases—or by radical mechanisms. An example of the latter is the polymerization of styrene initiated by a peroxide or by the radical source, azobisisobutyronitrile:

$$CH_3-\underset{\underset{C\equiv N}{|}}{\overset{\overset{CH_3}{|}}{C}}-N=N-\underset{\underset{C\equiv N}{|}}{\overset{\overset{CH_3}{|}}{C}}-CH_3 \quad \longrightarrow \quad 2(CH_3)_2\dot{C}C\equiv N + N_2$$

The reaction scheme, with R· representing the initiating radical, is typically

$$C_6H_5CH=CH_2 + R\cdot \longrightarrow C_6H_5\dot{C}HCH_2R \tag{10-90}$$

$$C_6H_5\dot{C}HCH_2R + C_6H_5CH=CH_2 \longrightarrow C_6H_5CHCH_2\underset{\underset{C_6H_5}{|}}{\dot{C}}HCH_2R \tag{10-91}$$

The process by which the radical intermediate attacks another monomer molecule continues in successive steps, with the polymer molecule becoming longer and longer, until the radical is lost by a termination process, such as combination with another radical or elimination of a hydrogen atom, which is not capable of continuing the chain, or until the radical is moved to another species by a chain transfer process.

In a chain transfer process, a new radical is created, perhaps by extraction by the polymer radical of a hydrogen atom from another monomer molecule or from a solvent molecule; the growth of the first molecule stops, and that of another molecule begins. A process of this type does not necessarily affect the polymerization rate, because it does not alter the number of radicals, but it does influence the average molecular weight of the polymer. Another possibility of chain transfer is the creation of a new radical somewhere along the backbone of a second polymer molecule by removal of a hydrogen atom. Such an occurrence results in branching of the polymer molecule, for a side unit now grows laterally from the main polymer backbone.

10-6
EFFECT OF TEMPERATURE ON RATE CONSTANT— THE ARRHENIUS EQUATION

The rate of a reaction step that is truly an elementary process increases with increasing temperature. Qualitatively, the explanation of this lies in the existence of an energy barrier which must be crossed by the reacting species. The higher the temperature, the larger the number of molecules possessing sufficient energy of the proper sort to allow the barrier to be crossed.

Empirically it has been found that the dependence of the magnitude of the rate constant for most chemical reactions upon temperature is rather strong. A rule of thumb sometimes stated is that the velocity constant doubles with a 10 K rise in temperature. This is only a very rough approximation, but it gives some idea of the magnitude of the effect that can be expected. In biological work, the ratio of the velocity constants for two temperatures 10 K apart is often referred to as the Q_{10} of a reaction.

For complex reactions, as we have seen, the measured rate constant may well be a composite with factors from several elementary process rate constants appearing in either the numerator or the denominator; accordingly, it is quite possible to find for such reactions apparent rate constants that decrease with increasing temperature. In enzyme-catalyzed reactions, the rate often rises to a maximum with increasing temperature and then decreases; the decrease is attributed to thermal deactivation of the enzyme at higher temperatures.

In his early studies of reaction rates in the nineteenth century, Arrhenius found that the rate constant for the inversion of sucrose could be described by an equation of the form

$$\ln k = -\frac{a}{T} + b \tag{10-92}$$

where a and b are temperature-independent constants. Another version of the same equation is

$$k = Ae^{-a/T} \tag{10-93}$$

Relations of the same form, with empirically determined values of the two constants, A and a, characteristic of the individual reaction, are found to be applicable to many other reactions.

If Equation (10-92) is differentiated with respect to temperature, the result is

$$\frac{d \ln k}{dT} = \frac{a}{T^2} \tag{10-94}$$

This equation can be compared with Equation (4-48) which relates the temperature dependence of an equilibrium constant to the enthalpy change for a reaction. The parallelism of the two equations indicates that the constant a should correspond to an energy change divided by the gas constant R, and Arrhenius therefore suggested that a be replaced by $\Delta E_a/R$, where ΔE_a is the *energy of activation*, the energy required to convert a normal reactant molecule into a molecule ready to undergo the structural changes required to form products.

Arrhenius' explanation for the temperature dependence of the rate constant can also be formulated as follows. There is an equilibrium between normal and activated molecules, the latter denoted by A^*:

$$A \rightleftharpoons A^* \tag{10-95}$$

An equilibrium constant is defined by the equation $K = [A^*]/[A]$, from which the concentration of activated molecules is equal to K multiplied by the concentration of normal molecules, or $K[A]$. The rate of reaction is proportional to the concentration of activated molecules, with a constant of proportionality k' which is independent of temperature:

$$\text{rate} = k'[A^*] = k'K[A] \qquad (10\text{-}96)$$

Since the fraction of molecules in the activated form at any one time is very small, the concentration of normal molecules is the same as the bulk concentration of the reactant.

From Equation (10-96), we can write

$$k = k'K \qquad (10\text{-}97)$$

Differentiation of the logarithmic form of this equation with respect to temperature yields

$$\frac{d \ln k}{dT} = \frac{d \ln k'}{dT} + \frac{d \ln K}{dT} \qquad (10\text{-}98)$$

The first term on the right is zero, since all the temperature dependence of the rate constant is carried by the equilibrium constant K. The usual equation relating the temperature variation of the equilibrium constant to the energy change for the process in question can be substituted for the second term on the right side of the equation:

$$\frac{d \ln k}{dT} = \frac{d \ln K}{dT} = \frac{\Delta E_a}{RT^2} \qquad (10\text{-}99)$$

It is well to point out that, if ΔE_a is understood to be a change in internal energy, this equation should strictly speaking contain an equilibrium constant K_C for molar concentrations. If the rate and equilibrium constants are given in terms of pressures, then it is the enthalpy of activation ΔH_a that should determine the temperature dependence of the rate constant. However, in practice the difference between ΔE_a and ΔH_a is at or below the limits of accuracy of most kinetic experiments, and so the quantity is often simply called "energy of activation" whether it is an enthalpy or an internal energy quantity.

The integrated form of the Arrhenius equation, that is, the form relating values of the rate constants at any two temperatures, T_1 and T_2, is

$$\ln \frac{k_{T_2}}{k_{T_1}} = \frac{\Delta E_a}{R} \frac{T_2 - T_1}{T_2 T_1} \qquad (10\text{-}100)$$

Experimental data are usually best treated by plotting the logarithm of the rate constant against the reciprocal of the absolute temperature. This yields a straight line of which the slope, if logarithms to the base 10 are used, is equal to $-\Delta E_a/2.303R$, from which the value of ΔE_a can easily be calculated.

10-7
TRANSITION-STATE THEORY

A more detailed interpretation of reaction rate constants has been formulated as the *transition-state theory*. An attempt is made to follow the detailed motions of the atoms involved in the reaction from their arrangement in the reactants to their final pattern in the products. Somewhere along the course of this process there is a state of maximum energy for the reacting system, the transition state, which is poised on top of the energy barrier and may go forward to form products or may revert to the original reactants. An important aspect of the transition state is that the bonds to be destroyed in the reactant molecules have been only partially broken but that the new bonds that will appear in the products have already begun to form. The result is that the height of the energy barrier over which the system must pass is not nearly so great as if the bonds in the reactants were required to break completely before the products could begin to form.

If our knowledge of forces holding the atoms in molecules—that is, our knowledge of the nature of valence bonds—were sufficiently detailed, and if extensive computer facilities were available, it would be possible in principle to calculate the energy of any possible configuration of the reacting system from the initial reactants up the energy barrier to the transition state and down the energy slope to the final products, as well as of configurations on either side of the most favorable pathway. From these calculations, both the details of the pathway and the energy of the transition state could be predicted. However, such a calculation from first principles has been found feasible only for very simple reactions. One example, one of the simplest possible reactions, is the displacement reaction in which a hydrogen atom, which we denote as H_A, exchanges places with one of the atoms, H_C, in a hydrogen molecule:

$$H_A\cdot + H_B{-}H_C \longrightarrow H_A{-}H_B + H_C\cdot \qquad (10\text{-}101)$$

It is assumed that this reaction begins with the approach of H_A along the axis of the hydrogen molecule, the justification for this assumption being that an approach along any other direction would lead to a much smaller probability of reaction. It is then possible to represent the configuration of the three atoms in terms of two coordinates, the distance r_{AB} between H_A and H_B and the distance r_{BC} between H_B and H_C:

$$H_A \qquad\qquad H_B{-\!\!-\!\!-}H_C$$
$$\overset{\longleftarrow\; r_{AB}\;\longrightarrow}{\big|}\quad \overset{\leftarrow r_{BC}\rightarrow}{\big|}$$

As the reaction proceeds, r_{AB} decreases and r_{BC} increases:

$$H_A \qquad\qquad\qquad H_B{-}H_C$$
$$H_A \qquad\qquad H_B{-}H_C$$
$$H_A{-\!-\!-\!-\!-}H_B{-\!-\!-\!-\!-}H_C$$
$$H_A{-}H_B \qquad\qquad H_C$$
$$H_A{-}H_B \qquad\qquad H_C$$

Since the atom being removed is identical with the atom entering, the reaction is symmetrical, and the transition state corresponds to the configuration in which $r_{AB} = r_{BC}$.

To represent the calculated variation in the energy of this system with atomic coordinates, it is not sufficient to use a simple two-dimensional graph, since the energy depends upon two variables rather than just one. A fairly good method of presenting the results is by means of a diagram as in Figure 10-4, in which each contour line represents a series of configurations having identical energies. The plot can be visualized as similar to a topographic map in which lines are drawn to represent points of equal elevation, and the *energy surface* consists of two valleys joined by a pass corresponding to the transition state. Points in the upper right region of the diagram correspond to large values of both interatomic distances, in other words, to three separate atoms, and thus the energy in this region is quite high.

Before the reaction, the system of three atoms lies on the floor of one of the valleys. In the course of the reaction, it moves up the valley as the atom approaches the molecule, travels over the pass corresponding to the symmetrical transition state, and then proceeds down the other valley as the displaced atom recedes from the product molecule.

Another way of representing the course of this reaction is to plot the energy of the three-atom system at successive stages in the reaction, taking as the independent variable a measure of the extent of reaction known as the reaction coordinate. The reaction coordinate cannot be unambiguously defined in a quantitative sense, but we can regard it as a measure of the distance traveled by the system from the initial energy level along the dotted line in Figure 10-4 to any specific point. The energy plot is thus a profile, projected into a plane, of the elevation as the system follows this dotted line; an example is shown in Figure 10-5. The activation energy of a reaction is interpreted in this way as the difference in energy between the level at the lowest point in a

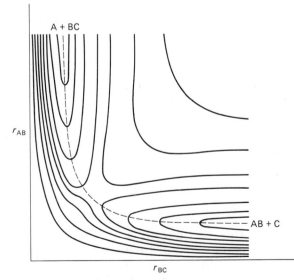

Figure 10-4
Contour diagram of the potential energy surface for a three-atom reaction. Each line passes through points of constant energy. The dashed curve represents the most probable reaction path for a displacement reaction.

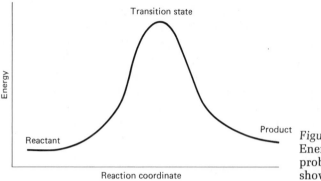

Figure 10-5
Energy profile for the most probable reaction pathway shown in Figure 10-4.

valley and the level at the crest of the pass, where the transition state is located.

QUANTITATIVE FORMATION OF REACTION RATE CONSTANTS

The theory of reaction rates expresses the number of reaction processes per unit time as the product of the "concentration" of the transition state and the probability that this state decomposes in a way so that products are formed. To calculate the concentration of the transition state, one assumes that equilibrium exists between the reactants, which we call A and B, and the transition state, an equilibrium described by the equilibrium constant $K^{\ddagger}$:

$$[TS] = K^{\ddagger}[A][B] \qquad (10\text{-}102)$$

The probability of the transition state forming products is derived by considering that some one mode of vibration of the activated unit forming this state, a mode consisting primarily of the stretching of the bond that is to be broken to form the products, leads to decomposition. This mode is considered distinct from the other vibrational modes of the transition state, and the frequency of decomposition to products is found, on this basis, to be equal to RT/Nh, where R is the gas constant, T is the absolute temperature, N is Avogadro's number, and h is Planck's constant.

The rate of reaction is thus

$$-\frac{d[A]}{dt} = -\frac{d[B]}{dt} = \frac{RT}{Nh}K^{\ddagger}[A][B] \qquad (10\text{-}103)$$

and the rate constant is

$$k = \frac{RT}{Nh}K^{\ddagger} \qquad (10\text{-}104)$$

To be a true equilibrium constant, the quantity $K^{\ddagger}$ should contain the contribution from the vibrational degree of freedom, corresponding to the "falling apart" of the transition complex to form products, which was separated out from it in order to calculate the frequency factor. Despite this theoretical limitation, it appears as a practical matter to be quite satisfactory to treat $K^{\ddagger}$ just like any other equilibrium constant. Another reasonable objection one might advance to this theory of reaction rates is that the transition state never lasts long enough so that equilibrium of the reactants with it can be attained. Despite such objections, the value of the theory is that it works reasonably well and provides a framework within which experimental results, such as the effects of substituents in organic molecules on reaction rates, can be interpreted.

One of the particularly useful ideas that has come from this version of the theory of reaction rates is that of expressing the reaction rate as a function of factors that are thermodynamic, or at least pseudo-thermodynamic, quantities, one factor involving enthalpy effects and the other involving entropy effects. To see how this comes about, let us recall the thermodynamic relationship between the equilibrium constant and the standard-state free energy change:

$$\Delta G° = -RT \ln K \qquad (10\text{-}105)$$

Suppose the equilibrium between the reactants and the transition state is described by an expression like Equation (10-102). For the equilibrium constant in this equation, Equation (10-105) becomes

$$\Delta G°‡ = -RT \ln K‡ \qquad (10\text{-}106)$$

or

$$K‡ = e^{-\Delta G°‡/RT} \qquad (10\text{-}107)$$

Substitution of this in Equation (10-104) leads to an expression for the rate constant:

$$k = \frac{RT}{Nh}e^{-\Delta G°‡/RT} \qquad (10\text{-}108)$$

Next we divide the free energy of activation into the enthalpy and entropy contributions to the activation process:

$$\Delta G°‡ = \Delta H°‡ - T \Delta S°‡ \qquad (10\text{-}109)$$

Combining Equations (10-108) and (10-109) results in the relation

$$k = \frac{RT}{Nh}e^{-\Delta H°‡/RT} e^{\Delta S°‡/R} \qquad (10\text{-}110)$$

The enthalpy of activation is a measure of the energy required to break bonds in the reactant molecules, energy needed to move atoms against the forces holding them in place. Since the enthalpy term carries most of the temperature dependence of k—the effect of T in the exponent far outweighs the effect of the first power of T in the frequency factor—the enthalpy of activation defined by this equation is almost the same as the Arrhenius energy of activation and has approximately the same meaning. The larger its magnitude, the slower the reaction.

The entropy of activation $\Delta S°‡$ measures the gain in disorder accompanying conversion of the reactants to the transition state. For the majority of reactions, this is a negative quantity, meaning that the effect is to reduce the rate. Thus a decrease in entropy is involved

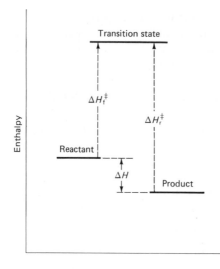

Figure 10-6
Relation between activation enthalpies
for forward and reverse reactions and
the enthalpy change for the overall
reaction.

when two molecules must be brought together into a single unit, and
the more specific the orientation required of the reacting molecules,
the larger the magnitude of the negative entropy of activation.

In a reversible reaction, there is a necessary relation between the
thermodynamic parameters associated with the rate constant for the
forward reaction, those for the rate constant to the reverse reaction,
and the overall changes in thermodynamic properties in going from
reactants to products. For the activation enthalpy, Figure 10-6 illustrates the case in which the overall reaction is exothermic, meaning
that ΔH is negative. It is evident from the diagram that the difference
between the activation enthalpy for the forward process $\Delta H_f^{\ddagger}$ and that
for the reverse process $\Delta H_r^{\ddagger}$ is equal to the enthalpy change for the
overall reaction. Parallel relations apply to the free energy changes
and the entropy changes.

SOME EXAMPLES OF TRANSITION STATES

To give the reader a better insight into the nature of transition states
and to show how the properties of these states can be related to the
activation entropy and activation enthalpy for a reaction, we now discuss a few specific examples. Consider first the reaction leading to the
formation of the intermediate in the acid-catalyzed sucrose hydrolysis
reaction. In this intermediate, a proton is attached to an oxygen atom
of the sucrose molecule. In the transition state for the formation of this
intermediate, the proton is partially transferred from the acid HA to
the oxygen atom of the sucrose. We can picture the proton as approximately halfway between its location when attached to the acid and its
location when attached to the sucrose molecule.

It is tempting to describe a transition state as a "structure," but it is
not a structure in the real sense of the word. It is not static, but its diagram should be imagined as depicting what might be seen if it were
possible to take a very-short-exposure snapshot catching, in the example just described, the proton in the middle of its leap from one bond

to the other. Furthermore, in discussing the course of a reaction, it is important to distinguish between a transition state, which has a lifetime of perhaps 10^{-12} sec, and a reactive intermediate, which may occupy a potential well for perhaps 10^{-6} sec or longer.

In another very common type of reaction, an incoming group may displace a group already attached to a carbon atom. A simple example is the reaction of hydroxide ion with an alkyl halide, such as methyl bromide, displacing the halogen as a halide ion. There is strong evidence, based on the stereochemical configuration of more complex molecules, that the incoming group in a reaction of this sort approaches the carbon atom from the direction opposite the location of the leaving group, just as in the reaction of atomic hydrogen with molecular hydrogen described above. The sequence can be represented as

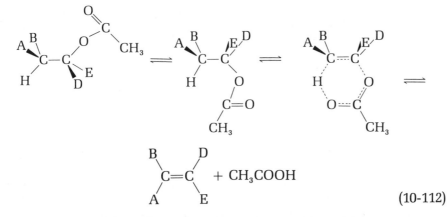

The bonds on the carbon atom that are not directly involved in the reaction undergo an inversion like an umbrella turning inside out. In the transition state, these three bonds are all in a plane perpendicular to the axis formed by the incoming group and the leaving group. This condition represents the point of maximum strain at the top of an energy barrier, from which the system flips quickly to one side or the other.

The steric influence of the transition state on the course of a reaction is illustrated by the pyrolysis reaction of an unsymmetrically substituted acetate. It is proposed that, in order for the reaction to occur, the carbonyl group must come close to the beta hydrogen with which it will combine in the product:

In the diagram of the cyclic transition state, the dotted lines represent bonds that are partially broken or that have been partially formed.

The reaction in which vinyl allyl ether rearranges to form allylacetaldehyde has an entropy of activation of -7.7 cal/deg mol, according to the work of F. W. Schuler and G. W. Murphy, who interpreted this as an indication that the reaction proceeds through a cyclic intermediate in which the freedom of motion of the ends of the molecule is restricted:

$$CH_2{=}CH{-}O{-}CH_2{-}CH{=}CH_2 \rightleftharpoons \begin{matrix} CH_2{\cdots}CH \\ CH_2 \qquad O \\ CH{\cdots}CH_2 \end{matrix} \rightleftharpoons \begin{matrix} & & H \\ & & / \\ CH_2{-}C \\ CH_2 \qquad O \\ CH{=}CH_2 \end{matrix}$$

$$(10\text{-}113)$$

A quite different situation is represented by the thermal denaturation of proteins, which has a positive entropy of activation high enough to offset the effect of a large positive enthalpy of activation, which would otherwise prevent the reaction from occurring under the moderate conditions of temperature at which it takes place. This is interpreted as an indication that bridging bonds between the chains of the protein that stabilize its native structure, chiefly hydrogen bonds, are broken in the formation of the transition state, which is consequently less rigid than the native form of the molecule. Most ring-opening reactions of organic molecules show similar behavior, indicating that the transition state is "looser" than the reactant molecule. Thus, for the unimolecular decomposition of cyclopropane, the activation entropy is about 7 cal/deg mol:

$$\begin{matrix} CH_2 \\ / \quad \backslash \\ CH_2{-}CH_2 \end{matrix} \longrightarrow CH_3{-}CH{=}CH_2 \qquad (10\text{-}114)$$

SALT EFFECTS ON RATES OF IONIC REACTIONS

Reactions occurring between ions in aqueous solution are found to have rate constants that vary considerably with the ionic strength of the medium. This result can be explained by a model in which the transition state has a charge equal to the sum of the charges on the reactant ions. A change in ionic strength affects differently the activity coefficients of ions of different charge, as described in Section 5-2. Therefore, if the transition state is formed by the combination of two ions, a change in ionic strength affects the equilibrium between these reactant ions and the transition state, altering the concentration of the transition state for a given reactant concentration.

At this point, we become involved in a sort of paradox, which puzzled kineticists for some years but which was finally resolved by the transition-state theory. Our intuition leads us to believe that the rate of a reaction—at least a unimolecular reaction—is directly dependent on the concentration of the reacting species. Indeed, this is the assumption made in the transition-state theory, but the *reacting species* must be considered to be the transition state. The reaction rate is proportional to the thermodynamic activity of the initial reactant, then, because the concentration of the transition state depends upon this thermodynamic activity.

That a reaction rate is determined by the thermodynamic activities of the initial reactants rather than by their concentrations is evident if we recall the relation between the equilibrium constant for a reversible reaction and the rate constants for the forward and reverse reactions. For the equilibrium system

$$A + B \underset{k_{-1}}{\overset{k_1}{\rightleftharpoons}} C + D \tag{10-115}$$

the true equilibrium constant is given by an equation involving the activities:

$$K = \frac{a_C a_D}{a_A a_B} = \frac{\gamma_C [C] \gamma_D [D]}{\gamma_A [A] \gamma_B [B]} \tag{10-116}$$

Since the conditions of equilibrium correspond to equal forward and reverse rates, the equilibrium constant must be equal to the ratio of the rate constants k_1/k_{-1} for an ideal solution with activity coefficients equal to unity. This condition permits one to substitute k_1/k_{-1} for K in Equation (10-116):

$$k_1 \gamma_A [A] \gamma_B [B] = k_{-1} \gamma_C [C] \gamma_D [D] \tag{10-117}$$

If this equation is to be satisfied as the ratio of the activity coefficients $\gamma_A \gamma_B / \gamma_C \gamma_D$ is varied by changing the ionic strength, then the ratio k_1/k_{-1} must vary in exactly the opposite way to maintain the value of K constant. If the rate of the forward reaction is represented, in terms of concentrations, by equations such as

$$\text{rate} = k_{obs} [A][B] \tag{10-118}$$

then the observed rate constant includes both the rate constant k_1 and the activity coefficients:

$$k_{obs} = k_1 \gamma_A \gamma_B \tag{10-119}$$

Turning again to the theory of reaction rates, we describe the equilibrium between the reactants and the transition state:

$$K^{\ddagger} = \frac{\gamma_{TS} [TS]}{\gamma_A [A] \gamma_B [B]} \tag{10-120}$$

According to the theory, the rate is proportional to [TS], which is obtained from Equation (10-120), leading to the rate equation

$$\text{rate} = \frac{RT}{Nh} \frac{K^{\ddagger} \gamma_A [A] \gamma_B [B]}{\gamma_{TS}} \tag{10-121}$$

From this, the rate constant is given by

$$k = \frac{RT}{Nh} K^{\ddagger} \frac{\gamma_A \gamma_B}{\gamma_{TS}} \tag{10-122}$$

Inspection of this equation shows that a reduction in γ_A or γ_B compared to γ_{TS} slows the reaction, because of a shift in the concentration equilibrium toward the reactants and away from the transition state. In contrast, a decrease in the activity coefficient of the transition state increases its concentration, accelerating the reaction.

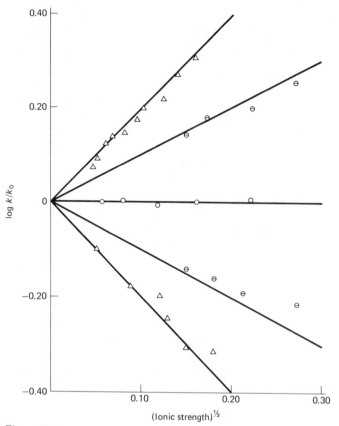

Figure 10-7
Effect of ionic strength of the medium on the rates of reactions involving ions.
The lines are calculated by the Debye–Hückel equation, which should really only
apply to dilute solutions. The reactions represented, in order from top to bottom,
are:

$I^- + S_2O_8^{2-}$
$NO_2NCOOC_2H_5^- + OH^-$
$C_{12}H_{22}O_{11} + H_2O$
$H_2O_2 + H^+ + Br^-$
$Co(NH_3)_5Br^{2+} + OH^-$

Suppose that the rate constant when all the activity coefficients are
unity—the situation approached when the ionic strength approaches
zero in the limit of infinite dilution—is designated k_0. Then any other
rate constant is given by the ratio

$$\frac{k}{k_0} = \frac{\gamma_A\gamma_B}{\gamma_{TS}} \qquad (10\text{-}123)$$

The Debye–Hückel equation provides a means for estimating the
values of the activity coefficients:

$$\log \gamma_i = -AZ_i^2\mu^{1/2} \qquad (10\text{-}124)$$

Using this expression in Equation (10-123) and considering the charge
on the transition state to be the sum of the charges on A and on B, we
obtain

$$\log \frac{k}{k_0} = \log \gamma_A + \log \gamma_B - \log \gamma_{TS}$$
$$= -AZ_A{}^2\mu^{1/2} - AZ_B{}^2\mu^{1/2} + A(Z_A + Z_B)^2\mu^{1/2}$$
$$= 2AZ_A Z_B\mu^{1/2} \qquad\qquad (10\text{-}125)$$

Figure 10-7 shows several sets of experimental results compared to the lines calculated from this equation with the appropriate numerical value of 0.5 for A in aqueous solution at room temperature.

10-8
CATALYSIS

A catalyst is a material that accelerates a reaction without being consumed in the reaction. It participates directly, however, in one or more steps of the reaction. In so doing, it provides an alternate path over which the reaction can proceed with greater velocity than if the catalyst were absent, because this path requires a smaller free energy of activation than the uncatalyzed reaction. Although a catalyst makes it possible for a system to move more rapidly toward a condition of chemical equilibrium, it cannot change the point of the equilibrium. A catalyzed reaction differs in nature from an uncatalyzed reaction only in the special case that one of the participating materials is regenerated during the sequence of steps, so that there is no net change in the amount of this material.

A catalytic reaction proceeds through one or more intermediates in which the other reactants, called substrates, are combined with the catalyst. The intermediates may be well-defined chemical substances stable enough to be isolated or at least to permit their concentration to be readily determined, or they may be very transient, reactive species. If the catalyst is an enzyme, one of the biological catalysts described in more detail in Section 10-9, the substrate is bound at a particular location, called the active site. If the catalyst is a solid, a heterogeneous catalyst, the intermediate is often a surface complex, formed by chemical reaction of the substrate with the atoms in the surface of the solid, as will be described in Chapter 11.

HOMOGENEOUS CATALYSIS
IN THE GAS PHASE

As one example of a gas phase reaction, the oxidation of sulfur dioxide to sulfur trioxide catalyzed by oxides of nitrogen, a process sometimes used on a commercial scale, can be cited. Although there is evidence that nitrosyl sulfates are intermediates in this reaction, the cyclic process can be adequately represented for purposes of illustration by the following scheme:

$$SO_2 + NO_2 \longrightarrow NO + SO_3 \qquad\qquad (10\text{-}126)$$
$$NO + \tfrac{1}{2}O_2 \longrightarrow NO_2 \qquad\qquad (10\text{-}127)$$

The sum of these two steps is the overall reaction

$$SO_2 + \tfrac{1}{2}O_2 \longrightarrow SO_3 \qquad\qquad (10\text{-}128)$$

Another example of a gas-phase catalyzed reaction is that proposed for the depletion of ozone in the atmosphere because of the catalytic effect of hydroxyl radicals and hydrogen atoms:

$$O + OH\cdot \longrightarrow O_2 + H\cdot \tag{10-129}$$

$$H\cdot + O_3 \longrightarrow OH\cdot + O_2 \tag{10-130}$$

The sum of the two steps represents the combination of an ozone molecule with an oxygen atom:

$$O + O_3 \longrightarrow 2O_2 \tag{10-131}$$

ACID–BASE CATALYZED REACTIONS

Many catalyzed reactions of organic molecules in solution are protolytic, or proton-transfer, reactions. The prevalence of these reactions is primarily a consequence of the relative ease of making and breaking bonds involving hydrogen atoms. One acid-catalyzed reaction, the hydrolysis of sucrose to glucose and fructose, has been discussed in Section 10-4 as an example of a complex reaction. The observed rate constant of a reaction such as this is often expressed by an equation:

$$k_{obs} = k_{H_3O^+}[H_3O^+] \tag{10-132}$$

The value of the rate constant k_{obs} is determined for a reaction run starting with a given initial concentration of sucrose and acid. The acid concentration remains constant during the run, while the sucrose disappears according to first-order kinetics. When a series of runs is made in which the initial concentration of a strong catalyzing acid such as hydrochloric acid is varied, the observed rate constant is found to depend upon the concentration of hydrogen ion according to Equation (10-132), and $k_{H_3O^+}$ is termed the catalytic constant for hydrogen ion.

If one wishes to investigate the behavior of an acid- or base-catalyzed reaction at moderate pH values, say in the range of pH 5 to 9, it is not feasible to adjust the hydrogen ion concentration with a strong acid or base, but the pH must be regulated by a buffer mixture. The question then arises as to whether the components of the buffer mixture have an effect on the reaction rate. This can be determined by preparing a series of buffers at the same pH, thus having the ratio of the two components constant, but differing in the total amounts of conjugate acid and base of the buffer pair. Thus a buffer at pH 6.7 can be prepared by mixing together NaH_2PO_4 and Na_2HPO_4 in respective concentrations of 0.15 and 0.05 M, 0.60 and 0.20 M, or 3.00 and 1.00 M. If the rates are about the same in these three buffers, it is concluded that the rate constant depends only on $[H_3O^+]$ or $[OH^-]$ and not on the presence of the other species in the buffers. Such a situation is found for the sucrose hydrolysis reaction, which is therefore referred to as a reaction *specifically catalyzed* by hydrogen ion.

For many reactions studied with this procedure, it is found that the apparent rate constant is *not* independent of the buffer concentration but instead rises as the buffer concentration is increased. These reactions are evidently catalyzed by the acidic or basic component of the buffer, as well as by hydrogen or hydroxide ions. Further study shows that such reactions are catalyzed by any acid or by any base in the broad, Brønsted sense of the term, or perhaps by both these categories, and are thus said to exhibit *general acid* or *general base* catalysis rather than specific hydrogen ion or hydroxide ion catalysis. For phosphate buffers of the kind described above, the observed rate constant of a general-acid catalyzed reaction may require an equation such as

$$k_{obs} = k_{solv} + k_{H_3O^+}[H_3O^+] + k_{H_2PO_4^-}[H_2PO_4^-]$$
$$+ k_{HPO_4^{2-}}[HPO_4^{2-}] \quad (10\text{-}133)$$

Each of the numerical values k_i is termed the catalytic constant for species i. The larger the catalytic constant, the greater the effect of a given concentration of the catalyzing acid on the rate of the reaction.

As a first example of a reaction showing general acid and base catalysis, the mutarotation of glucose is described. The glucose molecule exists in two cyclic forms, designated the α and β forms. These lactone or ring structures differ from one another in stereochemical configuration:

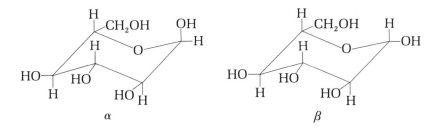

If one starts with an aqueous solution of either of these two forms, there is observed a slow reaction leading to an equilibrium mixture of the two forms, and termed *mutarotation* since the optical rotation approaches a common value from either extreme. For this reaction to occur, the lactone ring must be opened and then reformed. If the pH of the solution is reduced below about 4 by adding an acid such as hydrochloric acid, the rate of the mutarotation reaction is increased, and the increase is found to be proportional to the concentration of hydrogen ions. At pH values above about 6, produced by adding a strong base such as sodium hydroxide, the rate is proportional to the concentration of hydroxide ions. In solutions containing only strong acids and bases, the rate constant is given by the equation

$$k = k_{solv} + k_{H_3O^+}[H_3O^+] + k_{OH^-}[OH^-] \quad (10\text{-}134)$$

The numerical values of the rate constant k_{solv} for the solvent-induced reaction and of the catalytic constants of the hydrogen ion and the hydroxide ion, respectively, all at 25°C, are 0.001, 0.26, and 97×10^2. Within the pH range 4 to 6, and within several units on either side of this range, the catalysis by hydrogen and hydroxide ions is relatively small because neither ion is present in a very large concentration, and catalysis by various other species, such as undissociated acetic acid molecules, acetate ions, ammonium ions, ammonia, sulfate ions, and $Co(NH_3)_5OH^{2+}$ ions, to name only a few from a long list of acids and bases, can be demonstrated. The proof of catalysis by these species consists in carrying out a series of reaction runs at constant pH with varying buffer concentrations as described above. For example, a 1:1 mixture of sodium acetate and acetic acid provides a buffer of pH equal to 4.76. The buffer concentration can be varied from 0.10 to 2.00 M. Since the reaction rate for glucose mutarotation is observed to increase with increasing buffer concentration, the reaction is said to be catalyzed by the acid CH_3COOH or by the base CH_3COO^- or by both, as well as by H_3O^+. As might be expected, other acids show similar effects, to a greater or lesser degree depending upon their strength, as do bases, and thus mutarotation is said to be subject to general acid and general base catalysis. The observed rate constant is now represented by

$$k = k_{solv} + \sum_{acids} k_A[A] + \sum_{bases} k_B[B] \qquad (10\text{-}135)$$

Let us consider a possible mechanism by which an acid catalyst can function in this reaction. The first step may be the transfer of a proton from the catalyst, HA, to the glucose molecule, which we shall here represent as HG:

$$HA + HG \underset{k_{-1}}{\overset{k_1}{\rightleftharpoons}} HGH^+ + A^- \qquad (10\text{-}136)$$

This is assumed to be a reversible process and one in which equilibrium is rapidly established. It is followed by a slower, rate-determining step, in which a proton is removed from the reaction intermediate by a base:

$$HGH^+ + A^- \overset{k_2}{\longrightarrow} GH + HA \qquad (10\text{-}137)$$

This schematic formulation of the reaction does not specifically show the change in structure corresponding to the reaction, but we have written HG and GH to indicate the two configurational isomers of glucose in the lactone form.

To formulate the rate equation, we set the rate equal to the rate of the second step:

$$rate = k_2[HGH^+][A^-] \qquad (10\text{-}138)$$

The concentration of the intermediate is determined by the equilibrium constant

$$K = \frac{k_1}{k_{-1}} = \frac{[HGH^+][A^-]}{[HG][HA]} \qquad (10\text{-}139)$$

Combining these two equations leads to

$$rate = \frac{k_1 k_2}{k_{-1}}[HA][HG] \qquad (10\text{-}140)$$

For the base-catalyzed reaction, the mechanism is probably parallel but somewhat different. The initial step may be the removal of a proton from the glucose molecule to form the ion G^-. This rapid, reversible step would be followed by a rate-determining step involving the transfer of a proton from the conjugate acid to the ion G^- in such a way as to yield the product.

It is to be observed that in each of these reaction schemes for muta-rotation both an acid and a base participate in the overall cycle. The label for the type of catalysis, acid or base, is determined by which species attacks the substrate initially.

Another reaction that has been extensively studied and which shows general acid and general base catalysis is the dehydration of acetaldehyde hydrate. A scheme for the acid-catalyzed reaction is

$$CH_3CH(OH)_2 + HA \underset{k_{-1}}{\overset{k_1}{\rightleftharpoons}} CH_3CH\begin{smallmatrix}OH\\\\OH_2^+\end{smallmatrix} + A^- \tag{10-141}$$

$$CH_3CH\begin{smallmatrix}OH\\\\OH_2^+\end{smallmatrix} + A^- \overset{k_2}{\longrightarrow} CH_3CH\begin{smallmatrix}O^-\\\\OH_2^+\end{smallmatrix} + HA \tag{10-142}$$

$$CH_3CH\begin{smallmatrix}O^-\\\\OH_2^+\end{smallmatrix} \overset{k_3}{\longrightarrow} CH_3C\begin{smallmatrix}O\\\\H\end{smallmatrix} + H_2O \tag{10-143}$$

The last step is rapid, and the second step determines the rate:

$$\text{rate} = k_2[A^-][CH_3CH(OH)OH_2^+] \tag{10-144}$$

If the first step is assumed to be rapid and reversible so that equilibrium is maintained,

$$\text{rate} = k_2[A^-]\frac{k_1}{k_{-1}}[CH_3CH(OH)_2]\frac{[HA]}{[A^-]}$$

$$= \frac{k_1 k_2}{k_{-1}}[CH_3CH(OH)_2][HA] \tag{10-145}$$

Inspection of the scheme shows that the conjugate base A^- has two choices: Removal of a proton from one oxygen leads back to the reactants, and removal from the other leads, via subsequent rapid steps, to water elimination. Increasing the amount of the acid and base in the solution while keeping the pH constant does not shift the equilibrium of Equation (10-141), but it does supply more A^- to participate in the rate-determining second step. The situation is parallel to that for the mutarotation of glucose.

The hydrolysis of esters parallels the hydrolysis of sucrose, since both are specifically catalyzed by hydrogen ion, in the sense that the rate is fixed by the pH in a given buffer system. A mechanism in accord with this finding is

$$CH_3-\overset{O}{\overset{\|}{C}}-O-C_2H_5 + HA \underset{k_{-1}}{\overset{k_1}{\rightleftharpoons}} CH_3-\overset{O}{\overset{\|}{C}}-\underset{H}{\overset{}{O}}-C_2H_5^+ + A^- \tag{10-146}$$

$$CH_3-\overset{\overset{O}{\|}}{\underset{\underset{H}{|}}{C}}-O-C_2H_5^+ + H_2O \overset{k_2}{\rightarrow} CH_3-\overset{\overset{O}{\|}}{C}-O-H + C_2H_5-OH + H^+$$

$$(10\text{-}147)$$

This is followed by a rapid reaction, immaterial to the kinetics of the ester hydrolysis except that the catalyst is regenerated:

$$H^+ + A^- \longrightarrow HA \tag{10-148}$$

The rate of the overall reaction is that of the slow second step:

$$\text{rate} = k_2 \left[CH_3-\overset{\overset{O}{\|}}{\underset{\underset{H}{|}}{C}}-O-C_2H_5^+ \right] [H_2O] \tag{10-149}$$

$$= \frac{k_2 k_1}{k_{-1}} \left[CH_3-\overset{\overset{O}{\|}}{C}-O-C_2H_5 \right] \frac{[HA]}{[A^-]} [H_2O] \tag{10-150}$$

$$= \frac{k_2 k_1}{k_{-1} K_a} \left[CH_3-\overset{\overset{O}{\|}}{C}-O-C_2H_5 \right] [H^+][H_2O] \tag{10-151}$$

Since HA and A^- are members of a conjugate acid–base pair, their concentrations are related to the hydrogen ion concentration of the solution by K_a, the equilibrium constant for the acid dissociation of HA.

Examination of this mechanism to find out why specific hydrogen ion catalysis is indicated shows that, as for sucrose hydrolysis, the hydrogen ion concentration fixes the ratio of acid to conjugate base, determining the concentration of the intermediate. Furthermore, in contrast to the general-acid catalyzed reactions described above, because water reacts with the intermediate in the second step, changing the amount of the conjugate base does not change the rate of this step.

Although· we have presented general and specific catalysis as distinct, there is no sharp dividing line between them. For some reactions, a change in conditions is sufficient to alter the behavior pattern from one to the other.

THE BRØNSTED RELATIONS

A question of some interest is how the strength of an acid or base influences its effectiveness as a general acid or base catalyst. Brønsted proposed equations relating the catalytic constant to the dissociation constant of the acid or base:

$$k_A = g_A K_A{}^\alpha \tag{10-152}$$

$$k_B = g_B K_B{}^\beta \tag{10-153}$$

The quantities K_A and K_B are the conventional acid and base dissociation constants, and the parameters g_A, α, g_B, and β are characteristic of the reaction, solvent, and temperature.

The exponent α or β measures the sensitivity of the reaction to

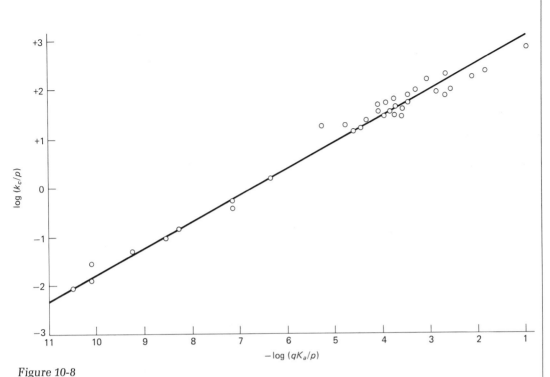

Figure 10-8
Brönsted plot of catalytic constants of acids for the dehydration of acetaldehyde hydrate. Selected data from R. P. Bell and W. C. E. Higginson, *Proc. Roy. Soc.* **A197**, 141 (1949).

change in acid or base strength. To illustrate this, consider the equilibrium

$$\text{HS} + \text{B} \xrightleftharpoons[k_\text{B}]{k_\text{A}} \text{S}^- + \text{HB}^+ \tag{10-154}$$

Imagine the base B to be changed to B′ so that the corresponding equilibrium constant is smaller than that for B by a factor of 10^4. This could correspond to an increase in k_B of 10^2 and a decrease in k_A by a factor of 10^2, in which case the value of β would be 0.5. Alternatively, the same change in K could be brought about by an increase in k_B by a factor of 10^4 without a change in k_A, which would correspond to a β value of unity. The actual situation might range anywhere to the other extreme, at which all of the change is the consequence of reduction in k_A by 10^4, with k_B remaining the same and with a β value of zero.

A Brønsted plot of results for the dehydration of acetaldehyde hydrate, a reaction discussed above, is shown in Figure 10-8. The quantities p and q represent, respectively, the number of equivalent protons in an acid molecule, and the number of equivalent available sites on a base that can accept protons. Thus the statistical correction factor p for oxalic acid is 2. The points plotted are mostly for carboxylic acids and phenols, acids in which the conjugate base differs somewhat from the acid because there is delocalization of the net negative charge in the anion, as in

Other catalysts deviate more widely from the correlation. In nitromethane, the negative charge on the conjugate base is almost entirely on the oxygen, although in the acid the proton is covalently bonded to the carbon atom. Catalysts of this type tend to be much less active, by factors of 30 to 100, than the Brønsted relation predicts. In contrast, molecules such as oximes, where the negative charge stays almost entirely on the atom from which the hydrogen is removed, are more active than expected by a factor of up to 100. Thus it is clear that the Brønsted relation holds only for catalysts of similar structural type.

Furthermore, the Brønsted relations can be expected to hold only over a limited range of acid or base strength. The rate constant for proton transfer has an upper limit, in the vicinity of 10^{10} sec^{-1}, determined by the rate at which the reacting species can diffuse through the liquid medium to meet one another. Rate constants of this magnitude have been measured for certain reactions by techniques discussed in Section 10-11. If the equilibrium constant for an acid–base ionization is very large in either direction, then one rate of transfer or the other is at this limit and is not further increased by structural changes in the proton donor or acceptor. Under these circumstances, all the change in K must be related to the change in one of the rate constants, although in the middle range of equilibrium constants both of the rate constants change when K changes. This is illustrated schematically in Figure 10-9.

The difference between general and specific acid or base catalysis is sometimes attributed to differences in the magnitude of the exponents α and β. Thus, for an acid-catalyzed reaction, a value of α that is large, near unity, reflects great sensitivity of the rate to the strength of the acid catalyst. Since hydrogen ion is the strongest acid that can exist in aqueous solution, a reaction really catalyzed to some extent

Figure 10-9
Effects of approach to the diffusion limit on reaction rate constants.

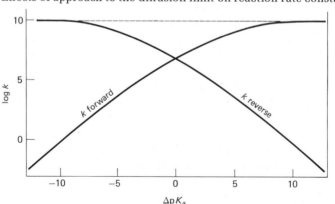

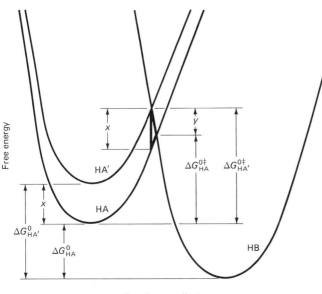

Figure 10-10
Diagram illustrating the linear relationship between change in free energy of
reaction and change in activation free energy for related reactions.

by other acids may appear to have only hydrogen ion catalysis because
the hydrogen ion is so much more effective. However, a reaction characterized by a small value of α is relatively insensitive to the strength
of the acid species and is much more likely to display discernible
general acid catalysis.

It should be pointed out that the Brønsted equations represent only
one example of a group of relationships between equilibrium constants and rate constants referred to as *linear free energy relationships*. To demonstrate the connection between these equations and
free energy changes, there is shown in Figure 10-10 a possible energy
diagram for two acids, HA and HA', either of which can transfer a
proton to base B. The Brønsted equation for each of the acids can be
written in logarithmic form:

$$\ln k_{HA} = \ln g_A + \alpha \ln K_{HA} \tag{10-155}$$

$$\ln k_{HA'} = \ln g_A + \alpha \ln K_{HA'} \tag{10-156}$$

Subtracting these equations yields

$$\ln \frac{k_{HA}}{k_{HA'}} = \alpha \ln \frac{K_{HA}}{K_{HA'}} \tag{10-157}$$

The ratio of the equilibrium constants for the transfers from the two
acids is determined by the free energy difference between the minima
of the curves for HA and for HA', $\Delta G^0_{HA'} - \Delta G^0_{HA}$. The ratio of the rate
constants for proton transfer depends upon the difference between
the crossover points of the curve for HB with the curves for HA and
HA', equal to $\Delta G^{0\ddagger}_{HA'} - \Delta G^{0\ddagger}_{HA}$, for these crossover points give approximations to the free energies for the transition states for the two reac-

tions. If the curves for HA′ and HA have the same shape, that is, if the vertical distance between them represented in the diagram by x is constant, the difference between the crossover points, represented by y, is about the same fraction of the free energy difference between the HA′ and HA curves, regardless of how large this difference is. That this fraction y/x is equal to α can be seen by using the expressions for free energy change in terms of rate and equilibrium constants:

$$\frac{y}{x} = \frac{\Delta G_{HA'}^{0\ddagger} - \Delta G_{HA}^{0\ddagger}}{\Delta G_{HA'}^{0} - \Delta G_{HA}^{0}} = \frac{-RT \ln k_{HA'} + RT \ln k_{HA}}{-RT \ln K_{HA'} + RT \ln K_{HA}}$$

$$= \frac{\ln (k_{HA'}/k_{HA})}{\ln (K_{HA'}/K_{HA})} = \alpha \quad (10\text{-}158)$$

10-9
ENZYME-CATALYZED
REACTIONS

Enzymes are large molecules, mainly protein in nature, which serve as catalysts for processes that occur in living systems. They play a vital part in permitting a wide variety of chemical reactions to take place in a living organism under conditions very mild compared to those a chemist would use in carrying out similar reactions in the laboratory. In many cases, reactions catalyzed by enzymes are much more selective than it has been possible to make their laboratory counterparts.

Enzymes are synthesized by the cells of an organism, but they retain their activity as catalysts in the absence of living material. Although the details of their catalytic function may be quite complex, there is no reason to believe that the explanation of their catalytic mechanism requires any principles beyond the ordinary chemical effects of geometry, structure, and forces of chemical and physical interaction that determine the rates of other reactions.

Some enzymes are simple proteins, but most of them contain in addition to the protein part, called the *apoenzyme,* a second part, termed the *prosthetic group* if it is firmly attached or the *coenzyme* if it is only loosely bound. For certain enzymes, the presence of a particular metal ion is required to maintain activity. Molecular weights of enzymes range from 20,000 to 500,000, and several of them have been isolated as pure crystalline materials.

An enzyme is often very specific in the kind of reaction it catalyzes and in the "substrate" or reactant it accepts. The name of an enzyme is usually an indication of its function: Invertase catalyzes the hydrolysis of sucrose to glucose and fructose, a reaction in which the optical rotation of the sugar is *inverted,* and peroxidases catalyze reactions of hydrogen peroxide. Some enzymes require a particular molecule as substrate; others require one of a small group of molecules with similar functional groups. For example, urease catalyzes the reaction of only one substance, urea. Maltase catalyzes the hydrolysis of α-glucosides but not of β-glucosides. Esterases hydrolyze various esters, but the rate of the reaction depends upon the structure of the ester; this type of reaction selectivity is preferential rather than completely specific.

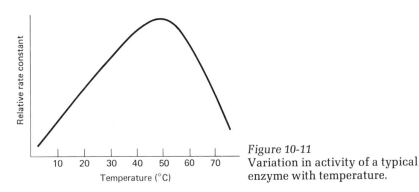

Figure 10-11
Variation in activity of a typical
enzyme with temperature.

The catalytic activity of enzymes is affected by both pH and temperature. There is usually a restricted range of pH in which a particular catalyst is active, and some intermediate pH at which the activity is at a maximum. Like other proteins, enzymes contain functional groups which are weak acids and weak bases, so that both the local charge near the catalytically active site and the overall charge on the molecule are modified by changing the pH. Often a temperature of maximum activity is found for an enzyme, as shown in Figure 10-11. The increase in rate with temperature in the lower temperature range is similar to that expected for any rate process, but an enzyme typically loses its activity by warming to a temperature between 50 and 100°C; it is then said to be denatured. The loss of activity may be reversible or irreversible, depending upon the enzyme, the temperature to which it is taken, and the presence along with the enzyme of other substances which may contribute to its denaturation.

KINETICS OF
ENZYMATIC REACTIONS

In practical investigations of the kinetics of enzyme-catalyzed reactions, it is best to measure the initial velocity of the reaction. That is, a solution of substrate and enzyme in known concentration is prepared, the concentration of the substrate is followed for a short period of time, and the rate of the reaction is extrapolated back to the time of mixing, which is taken to be zero time. Rates are then compared for a series of different initial substrate and enzyme concentrations. This procedure avoids uncertainties caused by such factors as the contribution of a reverse reaction, retardation effects produced by the products of the reaction, or denaturation of the enzyme with lapse of time.

Many enzyme-catalyzed reactions involving a single substrate show a characteristic behavior: The initial velocities in a series of runs with different initial substrate concentrations at a constant enzyme concentration vary linearly with substrate concentration; that is, they are first-order in substrate so long as the substrate concentration is low, but at high substrate concentrations approach asymptotically a limiting value, corresponding to kinetics of zero-order. Illustrated in Figure 10-12, this behavior is in accord with the *Michaelis equation*, in which V_{max} represents the limiting maximum velocity at high substrate con-

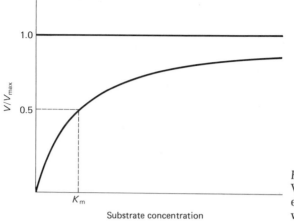

Figure 10-12
Variation in the rate of an
enzyme-catalyzed reaction
with substrate concentration.

centration, [S] is the concentration of substrate, and K_m is termed the Michaelis constant:

$$V = \frac{V_{max}[S]}{K_m + [S]} \tag{10-159}$$

When [S] is very small compared to K_m, it can be neglected in the denominator, and the equation reduces to a form representing the first-order part of the curve:

$$V = \frac{V_{max}[S]}{K_m} \tag{10-160}$$

When [S] is much larger than K_m, the latter can be neglected in the denominator of the Michaelis equation and the rate becomes simply

$$V = V_{max} \tag{10-161}$$

This corresponds to the zero-order limit approached at high substrate concentrations.

The most convenient way of handling enzyme kinetic data is to plot some function of the reaction velocity against a function of the substrate concentration so chosen that a straight line results. A linear plot minimizes the effect of experimental scatter of the data points and facilitates determination of the constants K_m and V_{max} characterizing the reaction. When Equation (10-159) is rearranged by taking the reciprocal of each side, there results

$$\frac{1}{V} = \frac{K_m}{V_{max}[S]} + \frac{1}{V_{max}} \tag{10-162}$$

From this equation, a plot of $1/V$ against $1/[S]$ is predicted to be a straight line with a slope of K_m/V_{max} and with an intercept of $1/V_{max}$ on the $1/[S] = 0$ axis. If the plot is extrapolated to $1/V = 0$, the $1/[S]$ intercept is equal to $-1/K_m$. Any two of the three numerical values (the slope and the two intercepts) read from the graph are sufficient to allow calculation of K_m and V_{max}. This treatment is referred to as the double reciprocal or Lineweaver–Burk method, and the plot appears as shown in Figure 10-13.

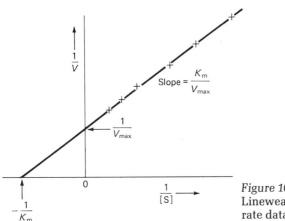

Figure 10-13
Lineweaver–Burk plot of initial
rate data for an enzymatic reaction.

Some workers have preferred alternative ways of plotting enzyme kinetic data. If, for example, both sides of Equation (10-162) are multiplied by [S], the result is

$$\frac{[S]}{V} = \frac{K_m}{V_{max}} + \frac{[S]}{V_{max}} \tag{10-163}$$

Therefore a plot of [S]/V against [S] should be linear with a slope of $1/V_{max}$. Still another alternative is developed by multiplying Equation (10-162) by the product VV_{max}, leading to

$$V_{max} = \frac{K_m V}{[S]} + V \tag{10-164}$$

This can be rewritten

$$V = -\frac{K_m V}{[S]} + V_{max} \tag{10-165}$$

Thus a plot of V against V/[S] should be linear with a slope of $-K_m$. Which of the three methods of plotting is most advantageous depends upon how the experiment has been designed to distribute data points over the range of substrate concentration.

THE MICHAELIS–MENTEN MECHANISM

The treatment associated with names of L. Michaelis and M. L. Menten provides an explanation of the mechanism of an enzyme-catalyzed reaction of a single substrate, which is consistent with the typical kinetic results just described. It is assumed that there is an initial, reversible formation of a complex between enzyme and substrate:

$$E + S \underset{k_{-1}}{\overset{k_1}{\rightleftharpoons}} ES \tag{10-166}$$

The complex may decompose by either of two pathways: It may revert to the enzyme plus substrate as shown in the previous equation, or it may go on to form products:

$$ES \xrightarrow{k_2} E + P \qquad (10\text{-}167)$$

It is assumed that this step is the slow step; the limiting rate observed at a high substrate concentration then corresponds to a situation in which the enzyme is saturated, so that further increase in substrate concentration cannot increase the concentration of complex. The observed limiting rate is therefore equal to k_2 multiplied by the concentration of enzyme added to the solution.

Derivation of the kinetic equation for this mechanism can be based upon the assumption that ES is a *reactive intermediate*. The net rate of change in its concentration is

$$\frac{d[\text{ES}]}{dt} = k_1[\text{E}][\text{S}] - k_{-1}[\text{ES}] - k_2[\text{ES}] \qquad (10\text{-}168)$$

Before this equation can be applied, however, we must distinguish between $[\text{E}]_0$, which is the total concentration of enzyme introduced into the solution initially, and $[\text{E}]$, which is the concentration of *free* enzyme, that part of the total enzyme concentration which is not bound to the substrate nor, in cases to be encountered later, to any other species. The concentration of free enzyme is the quantity needed for use in any equilibrium or rate expression such as Equation (10-168), and here it is equal to the total concentration less the concentration of the enzyme–substrate complex:

$$[\text{E}] = [\text{E}]_0 - [\text{ES}] \qquad (10\text{-}169)$$

On substituting this expression into Equation (10-168) and equating the rate of change of ES concentration to zero, one finds

$$\frac{d[\text{ES}]}{dt} = k_1([\text{E}]_0 - [\text{ES}])[\text{S}] - (k_{-1} + k_2)[\text{ES}] = 0 \qquad (10\text{-}170)$$

From this,

$$[\text{ES}] = \frac{k_1[\text{S}][\text{E}]_0}{k_1[\text{S}] + k_{-1} + k_2}$$

$$= \frac{[\text{S}][\text{E}]_0}{[\text{S}] + \{(k_{-1} + k_2)/k_1\}} = \frac{[\text{S}][\text{E}]_0}{[\text{S}] + K_m} \qquad (10\text{-}171)$$

Since step 2 is the rate-determining step, the rate of the reaction is simply $k_2[\text{ES}]$, or

$$V = -\frac{d[\text{S}]}{dt} = k_2[\text{ES}] = \frac{k_2[\text{S}][\text{E}]_0}{[\text{S}] + K_m} \qquad (10\text{-}172)$$

In these equations, the symbol K_m is employed to show the analogy to Equation (10-159). In the special case in which k_2 is much smaller than k_{-1}, K_m is simply equal to k_{-1}/k_1, which is the *dissociation constant* of the complex ES to form the reactant molecule and free enzyme. Examination of Equation (10-172) also shows that the maximum velocity of the reaction approached at a high substrate concentration is equal to $k_2[\text{E}]_0$, which is in accord with the model of the limiting situation in which the enzyme is working at maximum capability because all of it is in the form of the complex ES.

INHIBITION OF
ENZYMATIC REACTIONS

Certain materials when present in a solution along with an enzyme reduce the activity of the enzyme; these substances are referred to as *inhibitors*. Of course there are circumstances in which a chemical reaction occurs between the added substance and the enzyme that is essentially irreversible, such as the precipitation of enzymes by heavy-metal salts or the combination of iodoacetamide with a sulfhydryl group, but we are concerned here not so much with permanent changes, or poisoning of the enzyme, as with temporary or reversible changes. Indeed, nature can use inhibitors as means for regulating enzyme activity at times when the organism does not require the products of a particular reaction.

The quantitative effects of inhibitors on reaction rates depend upon the concentration of substrate present. In terms of the changes produced in a double-reciprocal plot for a fixed enzyme concentration, as shown in Figure 10-14, inhibition phenomena can be classified into three limiting types, which we describe in turn.

Competitive Inhibition: The intercept of the plot remains constant, but the slope increases, with increasing inhibitor concentration. Thus the degree of inhibition, or the fractional reduction in rate, is greatest when the substrate concentration is least and diminishes as [S] increases, approaching zero at infinite substrate concentration. In terms of the Michaelis–Menten mechanism, this can be explained by binding of the inhibitor at the active site of the enzyme in competition with the substrate. In its ability to be bound, the inhibitor mimics the substrate, but it is not able to undergo the catalyzed reaction. A large concentration of substrate serves to displace the inhibitor from the enzyme.

To develop an equation for competitive inhibition, we add to the Michaelis–Menten scheme an additional equilibrium for the combination of enzyme and inhibitor:

$$EI \rightleftharpoons E + I \tag{10-173}$$

with an equilibrium constant K_I equal to $[E][I]/[EI]$. The total enzyme concentration is now made up of three parts:

$$[E]_0 = [E] + [ES] + [EI] \tag{10-174}$$

In order to determine the concentration of reaction intermediate ES to use in the rate expression, we express [EI] in terms of the equilibrium constant K_I and the free enzyme concentration and then replace the concentration of free enzyme by its equivalent in terms of substrate and complex concentrations, using K_m for the dissociation constant of the complex:

$$\begin{aligned}
[E]_0 &= \frac{K_m[ES]}{[S]} + [ES] + \frac{K_m[ES][I]}{K_I[S]} \\
&= \frac{[ES]}{[S]}\left(K_m + [S] + \frac{K_m[I]}{K_I}\right)
\end{aligned} \tag{10-175}$$

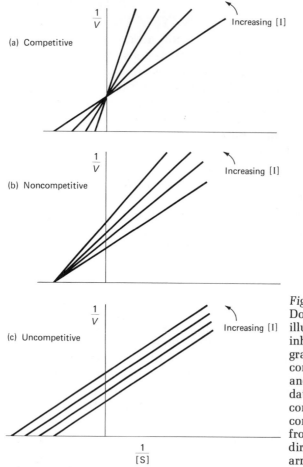

Figure 10-14
Double reciprocal plots
illustrating common types of
inhibition. Each of the three
graphs is for a fixed
concentration of enzyme,
and each line represents
data for a given inhibitor
concentration. The inhibitor
concentration increases
from line to line in the
direction indicated by the
arrow.

This equation is solved for [ES], and the result is used in the rate expression $k_2[ES]$:

$$V = k_2 \frac{[E]_0[S]}{[S] + K_m(1 + [I]/K_I)} \tag{10-176}$$

Examination of this equation shows the V_{max} is equal to $k_2[E]_0$, just as for the uninhibited reaction, shown by Equation (10-172), and is unaffected by the concentration of the inhibitor. The form of the equation useful in interpreting a double reciprocal plot is

$$\frac{1}{V} = \frac{K_m}{V_{max}}\left(1 + \frac{[I]}{K_I}\right)\frac{1}{[S]} + \frac{1}{V_{max}} \tag{10-177}$$

The intercept is $1/V_{max}$, regardless of inhibitor concentration, and the slope is $K_m(1 + [I]/K_I)V_{max}$, increasing with increasing inhibitor concentration.

The study of competitive inhibition in relation to the structure of the inhibitor provides a valuable tool in answering the question as to what detailed structural features in a molecule are important in determining whether it is an acceptable substrate for an enzyme, at least as far as the initial step, the binding event, is concerned.

Noncompetitive Inhibition: The slope and intercept of the double-reciprocal plot both increase by the same factor as the inhibitor concentration is increased. The fractional reduction in velocity produced by a given inhibitor concentration is independent of the substrate concentration. A possible model for this effect is the attachment of the inhibitor to some point on the enzyme other than the active site in such a way as to prevent the enzyme from functioning, but also at a point where the substrate itself cannot be bound and therefore cannot, at high concentrations, displace the inhibitor by an equilibrium effect. This can be represented by

$$ESI \rightleftharpoons ES + I \tag{10-178}$$

Because the substrate and inhibitor bind at different sites, it is reasonable to assume that the presence of the substrate does not influence the binding of the inhibitor. Thus we would expect that the equilibrium of Equation (10-173) is also present, and that the same numerical value of K_I applies to both equilibria:

$$K_I = \frac{[ES][I]}{[ESI]} = \frac{[E][I]}{[EI]} \tag{10-179}$$

For this case, the appropriate form of the reciprocal velocity equation is

$$\frac{1}{V} = \left(1 + \frac{[I]}{K_I}\right)\frac{K_m}{V_{max}[S]} + \frac{1}{V_{max}} \tag{10-180}$$

Uncompetitive or anticompetitive inhibition: The intercept of a double reciprocal plot is increased by increasing the inhibitor concentration, but the slope remains unchanged. In this case, the fractional reduction in velocity increases with increasing substrate concentration. To explain this behavior, it has been proposed that the inhibitor combines with forms of the enzyme that are not themselves active but which are interconvertible with the active form, or that the inhibitor combines only with the enzyme–substrate complex, according to Equation (10-178), but not with free enzyme according to Equation (10-173). The reciprocal rate equation for anticompetitive behavior can be written

$$\frac{1}{V} = \frac{K_m}{V_{max}[S]} + \frac{1}{V_{max}}\left(1 + \frac{[I]}{K_I}\right) \tag{10-181}$$

In many reactions, as studied experimentally, the behavior falls somewhere between the limits described by these three classes of inhibition and is termed mixed. Undoubtedly the reaction pathways are often much more complicated than those that have been described. The models used to explain the types of inhibition apply only to uncomplicated, single-substrate reactions, but the three categories can be used to classify experimental results for more complex, multisubstrate reactions, provided one does not incorrectly infer a pathway from the label that has been applied.

MORE COMPLEX
ENZYMATIC REACTIONS

Suppose that a reaction proceeding according to the Michaelis–Menten mechanism is reversible:

$$E + S \underset{k_{-1}}{\overset{k_1}{\rightleftharpoons}} ES \underset{k_{-2}}{\overset{k_2}{\rightleftharpoons}} E + P \tag{10-182}$$

Using the steady-state assumption for the concentration of ES and assuming that the total substrate concentration is much larger than $[E]_0$, one can show that the rate equation is

$$V = -\frac{d[S]}{dt} = \frac{(V_{max}[S]/K_m) - (V'_{max}[P]/K_p)}{1 + ([S]/K_m) + ([P]/K_p)} \tag{10-183}$$

where K_p is defined as $(k_{-1} + k_2)/k_2$, K_m is $(k_{-1} + k_2)/k_1$, V_{max} is $k_2[E]_0$, the maximum rate of the forward reaction, and V'_{max} is $k_{-1}[E]_0$, the maximum rate of the reverse reaction. Suitable measurements of rates at varying concentrations of enzyme and of substrate permit the four reaction parameters to be evaluated for a given reaction. Of course, if k_2 is very small, then K_p is very large and Equation (10-183) reduces to Equation (10-172). Likewise, if only initial velocity measurements are made so that $[P]$ is always very small, the same simplification applies.

Another sort of complication that may occur is the presence of two successive intermediates in the one-substrate reaction

$$E + S \underset{k_{-1}}{\overset{k_1}{\rightleftharpoons}} ES_1 \underset{k_{-2}}{\overset{k_2}{\rightleftharpoons}} ES_2 \underset{k_{-3}}{\overset{k_3}{\rightleftharpoons}} E + P \tag{10-184}$$

The rate is again given by an expression of the form of Equation (10-183). Indeed, no matter how many sequential intermediates there are in the reaction pathway, the rate equation has the same form. However, the four reaction parameters, K_m, K_p, V_{max}, and V'_{max}, have more complex expressions than in the single-intermediate case, and there are not enough of these measurable parameters to permit individual evaluation of all the rate constants. Of course, the overall equilibrium constant for the reaction, if known, provides one additional relation between the rate constants.

To find the expressions relating the reaction parameters to the rate constants for individual steps, one can solve algebraically a set of simultaneous equations, but E. L. King and C. Altman [*J. Phys. Chem.* **60**, 1375 (1956)] developed, on the basis of the theory of determinants, a short-cut method of obtaining the results. The method proceeds by obtaining a quotient for each enzyme-containing species, which is equal to the fraction of the total enzyme present as that species. The first step is to make a diagram showing all possible pathways by which the various species can be interconverted. For the example in Equation (10-184) with two complexes, this is

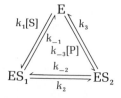

Each arrow corresponding to a reaction step is labeled on the diagram with the quantity that must be multiplied by the concentration of the particular enzyme-containing species to give the rate of that step.

Next one draws diagrams showing the various paths leading to the species for which the fraction is to be evaluated, leaving out one of the possible links in each diagram. If [E] is to be evaluated in the example, the partial diagrams are

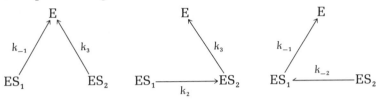

For each of the partial diagrams, the rate constants for the steps involved are multiplied together, there being always in this product one less factor than the number of enzyme-containing species, and all the resulting products are summed to give the numerator of the quotient being sought. For the example, this is

$$k_{-1}k_3 + k_2k_3 + k_{-1}k_{-2}$$

Carrying out a similar process for each of the other enzyme-containing species, one finds

$$k_1k_3[S] + k_1k_{-2}[S] + k_{-2}k_{-3}[P] \qquad \text{(for species ES}_1)$$

and

$$k_1k_2[S] + k_2k_{-3}[P] + k_{-1}k_{-3}[P] \qquad \text{(for species ES}_2)$$

The denominator of the quotient is the sum of the numerators for all the enzyme-containing species present. Thus, for the example,

$$\frac{[E]}{[E]_0} = \frac{k_{-1}k_3 + k_2k_3 + k_{-1}k_{-2}}{\begin{array}{c} k_{-1}k_3 + k_2k_3 + k_{-1}k_{-2} + k_1k_3[S] + k_1k_{-2}[S] \\ + k_{-2}k_{-3}[P] + k_1k_2[S] + k_2k_{-3}[P] + k_{-1}k_{-3}[P] \end{array}} \qquad \text{(10-185)}$$

The concentration of each enzyme-containing intermediate, expressed in terms of $[E]_0$, can then be substituted in the rate equation:

$$V = k_1[E][S] - k_{-1}[ES_1] \qquad \text{(10-186)}$$

If the result is rearranged to correspond to Equation (10-183), the reaction parameters are found to be given by the equations

$$K_m = \frac{k_1k_3 + k_{-1}k_3 + k_{-1}k_{-2}}{k_1(k_{-2} + k_2 + k_3)} \qquad \text{(10-187)}$$

$$K_p = \frac{k_{-1}k_{-2} + k_{-1}k_{-3} + k_2k_3}{k_{-3}(k_{-2} + k_2 + k_{-1})} \qquad \text{(10-188)}$$

$$V_{max} = \frac{k_2k_3[E]_0}{k_3 + k_{-2} + k_2} \qquad \text{(10-189)}$$

$$V'_{max} = \frac{k_{-1}k_{-2}[E]_0}{k_{-1} + k_{-2} + k_2} \qquad \text{(10-190)}$$

If an enzyme-catalyzed reaction involves two substrates rather than only one, there are several alternative sequences by which it may proceed. Suppose the reaction is that of A plus B to give products C and D. The intermediate in the reaction may be a ternary complex EAB which must be formed in a particular order of binding of the two substrates: It may be required that A be bound first to form EA and then B be added, or it may be that B must be combined with the enzyme initially, followed by the addition of A in a second step. An example of such an *ordered* sequence is the combination of succinate and glyoxylate to form isocitrate, catalyzed by the enzyme isocitrate lyase [B. A. McFadden, J. O. Williams, and T. E. Roche, *Biochemistry* **10**, 1384 (1971)], in which the glyoxylate is added first to the enzyme followed by succinate.

Another possibility is that the ternary complex can be formed equally well by either sequence of binding of substrates. An example of this *random* mechanism is the reaction catalyzed by yeast alcohol dehydrogenase involving the oxidation and reduction of NAD on the one hand and the interconversion of an alcohol and an aldehyde on the other. The rate-limiting step is the oxidation–reduction reaction within the complex, and the process can be represented by the scheme

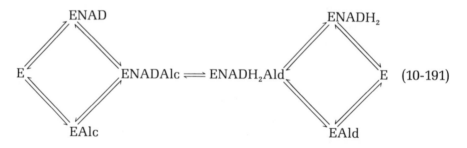

$$\text{(10-191)}$$

Still other two-substrate reactions occur by a pathway that does not involve simultaneous presence of the two reactants on the enzyme; in other words, no ternary complex is ever formed. This sequence is termed a "Ping-Pong" mechanism. It can be represented schematically as

$$A + E \rightleftharpoons AE \rightleftharpoons E' + C \tag{10-192}$$

$$E' + B \rightleftharpoons BE \rightleftharpoons EE + D \tag{10-193}$$

Examples include many group-transfer reactions of which one is the interchange of succinate (succ) and acetylacetonate (acac) between units of coenzyme A (CoA):

$$\text{succ—CoA} + E \rightleftharpoons \text{succ—CoA—E} \rightleftharpoons \text{CoA—E} + \text{succ} \tag{10-194}$$

$$\text{CoA—E} + \text{acac} \rightleftharpoons \text{acac—CoA—E} \rightleftharpoons \text{acac—CoA} + E \tag{10-195}$$

The typical experimental approach for a two-substrate reaction is to hold the concentration of enzyme at some fixed value $[E]_0$ while performing several sets of experiments. In each set, the concentration of one substrate, say B, is held constant at a value substantially larger than $[E]_0$, and the concentration of the other substrate is varied. The results are plotted on a double-reciprocal plot, with each set of experiments giving a curve, which is usually a straight line. For either an ordered or a random mechanism, the various straight lines in the family, when extended, intersect at a point. The Ping-Pong mechanism, however, is unique in that the lines are parallel to one another and do not intersect.

Many additional complexities beyond those already described are encountered in enzyme catalysis. One influence mentioned earlier is the effect of pH, changes in which may modify the ionization of groups in the enzyme and in the complex, and in turn alter the activity, often according to a pattern such as

$$H_2SE^+ \rightleftharpoons HSE \rightleftharpoons SE^- \qquad (10\text{-}196)$$
$$\text{inactive} \qquad \text{active} \quad \text{inactive}$$

General acid and base catalysis, described in Section 10-8, appears to be important in some enzyme mechanisms. This may very well involve the acidic or basic functional groups in the side chains of the amino acid constituents of the enzyme.

Inhibitors can act in various ways in addition to those mentioned. It is possible for the inhibitor to hold on to the substrate rather than to the enzyme. The substrate itself can act as an inhibitor if present in high concentration, often by forming a complex ES_2 which is inactive. In two-substrate reactions, the effects of inhibitors vary with the details of the reaction mechanism and have been used to deduce much of the information available about this mechanism.

10-10
ISOTOPE EFFECTS

It is frequently assumed that the chemical properties of a molecule are substantially unaltered by the substitution of one isotopic species for another in the molecule. This is a reasonable first approximation, but it is not quite true: The physical properties of a material are certainly affected by isotopic substitution, and in many instances both equilibrium and kinetic behavior in chemical processes are modified. Thus the viscosity of D_2O at 25°C is about 23 percent greater than that of H_2O, and acids are weaker in D_2O than in H_2O, with a pK_a difference typically of 0.6 units.

For many chemical reactions, if the mass of an atom near the site of bond breaking or forming is altered, the rate constant is changed measurably. If the substitution involves an atom directly participating in the bond that is broken or formed, the phenomenon is termed the *primary* isotope effect; effects of more remote atoms are termed *secondary*.

The primary isotope effect is fairly well understood, and its rationale can be illustrated for a proton-transfer reaction. In Figure 10-15, we show on the potential energy diagram for the bond to be broken, that is, the bond by which the proton is initially attached to the rest of a molecule, the quantized vibrational levels. The dotted lines apply if the atom is ^1H, and the solid lines apply if the atom is ^2H. Since, at ordinary temperature, most of the molecules in a sample are in the ground vibrational level, the reaction usually starts at the

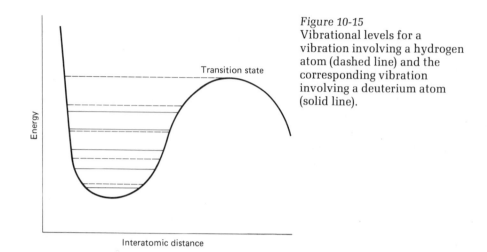

Figure 10-15
Vibrational levels for a
vibration involving a hydrogen
atom (dashed line) and the
corresponding vibration
involving a deuterium atom
(solid line).

lowest vibrational level of the diagram, the vibrational ground state.
This state is above the minimum of the curve by an amount equal to
the zero-point energy, a quantity which is determined by the reduced
mass of the vibrating system, and which is therefore smaller by a factor of $\sqrt{2}$ for deuterium than for hydrogen. If the corresponding hy-
drogen-stretching motion is completely free in the transition state, the
energy of the transition state does not depend upon the isotopic mass.
Consequently, the hydrogen-containing species requires less activation energy than the deuterium-containing species, for it has more
energy to begin with but need only go to the same transition-state energy level, and the hydrogen-containing species reacts 5 to 10 times
more rapidly. If, however, some restriction on the vibration is still
retained in the transition state, then the ratio of the rate constants is
somewhat less, since the energy of the transition state now also varies
with isotopic substitution, although not to so great an extent as does
the ground state energy.

Since substitution of such isotopes as ^{13}C or ^{14}C for ^{12}C, or ^{18}O for
^{16}O, has a much smaller effect on the reduced mass and therefore on
the vibrational zero-point energy, there is observed a change in rate
only of the order of 5 to 20 percent, and then only when the isotope is
directly involved in the bond that undergoes a change in the reaction.

Secondary isotope effects probably also exert their influence by
way of changes in vibrational frequencies, but the situation is much
more complicated than for primary effects. Indeed, both increases and
decreases in rate have been observed upon substitution of heavier
atoms not directly involved in the reacting bond.

Isotope effects can provide information about reaction mechanisms.
For example, the greater strength of acids in water than in D_2O explains the greater rate of hydrolysis of sucrose in D_2O than in H_2O in
the acid-catalyzed reaction described in Section 10-8. The intermediate SD^+ in D_2O is less acidic than SH^+ in H_2O, and accordingly more
SD^+ is formed than SH^+ in the first-step equilibrium. If, however, the
initial step in the reaction, the proton transfer from the catalyst to
sucrose, were rate-determining, the opposite effect would be observed,

for this transfer would be faster for H than for D. The same criterion has been applied to several enzyme-catalyzed reactions to determine whether or not the rate-determining step involves a proton transfer.

10-11
VERY RAPID REACTIONS

One of the most interesting and significant achievements in kinetics in recent years has been the development of methods for measuring the rates of very rapid reactions, extending to those with half-lives as small as 10^{-10} sec. Conventional techniques are limited to reactions with half-lives well over 10 sec.

How are such very rapid rates measured? Several of the methods, the techniques of flash photolysis and pulse radiolysis, will be described in Chapter 14. Nuclear magnetic resonance and electron paramagnetic resonance, suitable for half-lives in the range from seconds to nanoseconds, will be described in Chapter 13. In this section, we devote our attention principally to two groups of methods which have seen extensive application to systems of biological interest, the first group consisting of flow methods and the second of relaxation methods, in which the experimenter follows the return to equilibrium after an abrupt change in temperature or pressure or some other condition affecting the equilibrium.

Historically, flow methods were the first to be extensively developed and applied, their origin dating from the work of F. J. W. Roughton in England in the 1920's. In the *continuous-flow* method, two liquid reactant solutions are forced quickly by pistons into a small mixing chamber, and then the mixed solution flows rapidly down the length of an observation tube. The distance d along the tube from the mixing chamber corresponds to a time of reaction equal to d/u, where u is the flow velocity. Typically, flow velocities can be 10 m/sec, and a distance of 1 cm corresponds to an elapsed time of 1 millisecond. The progress of the reaction can be followed in time by examining the solution by spectroscopic means at various points along the length of the tube, or by varying the velocity of flow. The limiting factor in this, as in any flow method, is the time required for the reactants to be thoroughly mixed, an interval of the order of a millisecond. An advantage of the method is that analytical means requiring a relatively long time to accumulate data can be employed, but a severe disadvantage is the large amounts of reactant solutions required.

To permit study of reactions for which the supply of reactant is limited, the *stopped-flow* method was devised. Here the reactants are again driven quickly into a mixing chamber, but after enough has passed through the chamber to flush it out, flow is suddenly stopped. At this point a recording device such as an oscilloscope, is triggered, and data from a spectrometer or other analytical device is fed into the scope as it sweeps. The trace on the scope can then be photographed. For still speedier processes, the data can be acquired in a storage oscilloscope or a multichannel digital memory and then plotted out in permanent record form later at the convenience of the investigator.

The limitation of mixing time was circumvented and an increase in magnitude of measurable rate constants of about 10^5 was achieved in the 1950's by the use of *relaxation* methods, developed principally by the German scientist, Manfred Eigen, who received the Nobel prize in 1967 for this accomplishment. In a relaxation measurement, a mixture at equilibrium is perturbed by changing one of the conditions that determine the equilibrium constant, and then the changes in concentration are followed as the system "relaxes" or moves toward the equilibrium concentration ratio appropriate to the new conditions. Not only do such methods eliminate the requirement of waiting for mixing to be completed, but they also permit the measurement of rates at or very close to equilibrium.

In the *temperature-jump* method, the source of energy for the increase in temperature is usually the discharge through the solution of current from an electric capacitor which has been charged at high voltage, perhaps 10,000 to 100,000 V. To the solution has been added some inert electrolyte to provide sufficient conductivity. The temperature of an aqueous electrolyte solution can be raised by 5 to 10 K in a microsecond (μsec). Suitable for any polar liquid, and particularly applicable to nonaqueous solvents, is heating by a pulse of microwaves, 1 to 5 μsec in length. The sample is placed in a tuned cavity connected by a waveguide to the microwave source, and the temperature rise attainable is about 1 K. A third possibility is irradiation of the solution with a pulse from a laser, a device which will be described in Chapter 14. The radiation may be of a wavelength at which the solvent absorbs, or a dye may be added to absorb the energy, or one of the reactants may have an absorption band that can be irradiated.

A *pressure jump* can be applied to a reaction mixture by suddenly rupturing a diaphragm through which a pressure of up to 100 atm has previously been applied to the solution. Electric field jumps have also been used for reactions involving ions.

The concentration change after a perturbation can be followed by an optical method, such as absorption spectroscopy, in the same way as described for very rapid reactions measured by the flow technique, using a triggered oscilloscope or a multichannel memory.

To learn how the rate constants are related to the observations made in a relaxation experiment, one can consider a reversible process, monomolecular in both directions, such as an isomerization:

$$A \underset{k_{-1}}{\overset{k_1}{\rightleftarrows}} C \qquad (10\text{-}197)$$

Before the impulse is applied, the reaction is at equilibrium and the concentrations are $[A]_0$ and $[C]_0$. Following the perturbation, a new numerical value of the equilibrium constant K applies; let us suppose that the new value of K is larger than the previous value. The final equilibrium concentrations toward which the system moves are $[A]_e$, which is smaller than $[A]_0$, and $[C]_e$, which is larger than $[C]_0$.

The net rate at which the system changes is

$$-\frac{d[A]}{dt} = +\frac{d[C]}{dt} = k_1[A] - k_{-1}[C] \qquad (10\text{-}198)$$

Letting x be a variable equal to the deviation of the concentration from the final equilibrium values, so that $x = [A] - [A]_e = [C]_e - [C]$, one can write the rate equation

$$-\frac{dx}{dt} = k_1(x + [A]_e) - k_{-1}([C]_e - x) \tag{10-199}$$

However, the equilibrium concentrations are related by

$$k_1[A]_e = k_{-1}[C]_e \tag{10-200}$$

Therefore

$$-\frac{dx}{dt} = (k_1 + k_{-1})x = kx \tag{10-201}$$

Thus the return to equilibrium is first-order, with a rate constant k equal to the sum of the two opposing rate constants. If the value of the equilibrium constant, $K = k_1/k_{-1}$, can be established by independent measurements, the numerical values of the two individual rate constants can be obtained.

The reciprocal of the rate constant k is often called the *relaxation time* and symbolized by τ. It is equal to the time in which the deviation from equilibrium is reduced to $1/e$ of its original value and is simply $1/0.693$ or 1.44 times the half-life for the return to equilibrium.

Another type of reaction that can be treated formally in a similar way is that in which a bimolecular process is opposed by a unimolecular process:

$$A + B \underset{k_{-1}}{\overset{k_1}{\rightleftharpoons}} C \tag{10-202}$$

If the system is not too far from equilibrium, the equation for the return to equilibrium is

$$-\frac{dx}{dt} = (k_1[A]_e + k_1[B]_e + k_{-1})x \tag{10-203}$$

The quantity in parentheses is a constant for a given temperature and can be regarded as the apparent first-order rate constant for the return to equilibrium. The relaxation time τ is, as in the simpler case, the reciprocal of the rate constant. For this reaction type, however, the return to equilibrium is first-order only if the displacement from equilibrium x is small.

For more complex reactions, the results can be analyzed to obtain a series, or spectrum, of relaxation times. However, as for the case just illustrated, each relaxation time is a function of all the rate constants and equilibrium concentrations. To find values for individual rate constants, it is necessary to measure relaxation times as a function of concentration.

By relaxation measurements, it has been possible to show that protolytic reactions often occur with rate constants so large that the step determining the rate must be a diffusion process, rather than a process in which a chemical bond is broken. Thus, in aqueous solutions, a proton is transferred from H_3O^+ to a base such as CH_3COO^-, NH_3, imidazole, or $CuOH^+$ with a rate constant of about 10^{10} liters/mol sec. The model proposed for these reactions is the formation of a hydrogen bond between the acid and the base, followed by the jump of the proton along the axis of the bond. The rate is controlled by the steps in which the hydronium ion diffuses to the base and the products diffuse apart from one another. Some of the consequences of diffusion-limited rates were indicated in Figure 10-9.

Rates of proton transfer less than the diffusion limit have been found in certain circumstances. In the ionization of carbon acids and the recombination of their conjugate base with hydrogen ion, it appears that rearrangement of the structure of the proton donor may be required and that this is the slow step. Such is certainly the case with nitromethane, for which the rate constant for reaction with hydroxyl ion is only about 10^2 liters/mol sec. For a series of organic carboxylic acids with a variety of structures, the carboxylate–hydronium ion recombination rate constants are near the diffusion limit, with values of 1×10^{10} liters/mol sec or more, and all the differences in acid strength, corresponding to a range of K_a values of almost 10^3, are reflected in the variation in dissociation rate constant from about 10^8 sec^{-1} for the strongest acid to 3×10^5 sec^{-1} for the weakest.

Relaxation methods have been applied to follow the changes in conformation of polynucleotides and polypeptides in solution in an effort to understand corresponding changes in nucleic acids and proteins, the nature of which will be discussed further in Chapter 12. For example, in the formation of a double-stranded helix from two short polynucleotide chains, polyadenine and polyuracil, it has been found that there is only a single relaxation time. The rate constant is of the order of 10^6 liters/mol sec, and the activation enthalpy value of -20 kcal/mol suggests that three hydrogen bonds must be formed in a preliminary equilibrium if the approach of the two chains is to result in the formation of a double helix.

The kinetics of enzyme-catalyzed reactions have been extensively studied by both rapid-mixing and relaxation techniques. Relaxation studies are particularly suited to measurement of the rates of binding and removal of substrate and product molecules. In addition, they have contributed to a growing body of results which indicate that conformational changes in an enzyme, possibly induced by substrate molecules, are of frequent occurrence along reaction pathways, and that most reaction pathways have several intermediates.

As a first example of the results for enzymatic reactions obtained from relaxation experiments, we describe the work on bovine pancreatic ribonuclease (RNase) carried out by G. G. Hammes and coworkers [Acc. Chem. Res. 1, 321 (1968)]. This enzyme has a molecular weight of only 13,683, and its structure has been well worked out by various methods, including x-ray diffraction. RNase catalyzes the breakdown of ribonucleic acid (RNA) in two successive reactions. In the first step, an oxygen–phosphorus bond in a nucleotide containing cytosine or uracil is cleaved, forming a 2',3'-cyclic phosphate derivative. In the second step, this derivative is hydrolyzed to yield a terminal pyrimidine 3'-monophosphate:

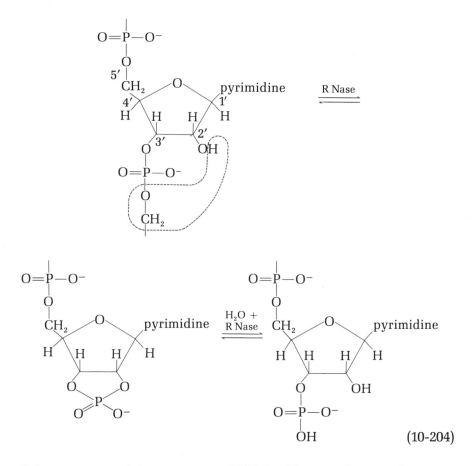

(10-204)

It is not very satisfactory to use RNA itself as a substrate, because, as the enzyme breaks the polymer into sections of various kinds, each section becomes a reactant for subsequent stages of breakdown, and the kinetic analysis is very difficult. Rather, model compounds have been used, such as cytidylyl-3′,5′-cytidine, to serve as a reactant for the first step, yielding cytidine 2′,3′-cyclic phosphate which in turn is a substrate for the hydrolysis step, being transformed to cytidine 3′-phosphate.

In the presence of a dye which serves as a pH indicator, but in the complete absence of substrate, temperature-jump studies show that RNase undergoes a change with a relaxation time τ_1 of 10^{-3} to 10^{-4} sec. From the pH dependence of the relaxation, it is concluded that the conformational change, which this evidently must be, involves an ionizable group with a pK_a of about 6. Other evidence has suggested that this group is in the histidine residue designated 48 in the peptide sequence, a unit located so that it may be part of a hinge region involved in the opening and closing of the molecule about the active site. The rate constants, k_1 and k_{-1}, for the conformational change are 780 and 2470 sec^{-1}.

When either a dinucleotide, a pyrimidine 2′,3′-cyclic phosphate, or a pyrimidine 3′-phosphate is allowed to interact with the enzyme, two relaxation processes are observed, which can be explained by a mechanism in which a bimolecular combination is followed by an isomerization:

$$E + S \underset{k_{-2}}{\overset{k_2}{\rightleftharpoons}} ES_1 \underset{k_{-3}}{\overset{k_3}{\rightleftharpoons}} ES_2 \qquad (10\text{-}205)$$

Neglecting the isomerization of the free enzyme and assuming that step 2 is rapid compared to step 3, Hammes writes equations for the relaxation times:

$$\frac{1}{\tau_2} = k_2([E]_0 + [S]) + k_{-2} \qquad (10\text{-}206)$$

$$\frac{1}{\tau_3} = \frac{k_3}{1 + k_{-1}/([E]_0 + [S])} + k_{-3} \qquad (10\text{-}207)$$

Equation (10-206) is found for all the systems studied to be well obeyed by τ_2, but τ_3 is sometimes difficult to measure as a function of concentration, although it appears to follow Equation (10-207). The magnitude of k_2 is 10^7 liters/mol sec, of k_{-2} is 10^4 sec^{-1}, and of k_3 and k_{-3} is 10^3 sec^{-1}. Thus it appears that any of the substrates binds extremely rapidly, and the binding is always followed by a slower isomerization or conformational change. The pH dependence of k_2 indicates involvement of ionizable groups with a pK_a of 5.4 and 6.4, probably the imidazole portions of histidine residues 12 and 119 in the peptide sequence, and the pH dependence of τ_3 is similar to that of τ_1, indicating that the conformational change in the presence of substrate has some similarity to that in the absence of substrate. It is seen, then, that the behavior of the reacting system is similar qualitatively, no matter which of the three species is added to the enzyme, differing only quantitatively in the magnitude of the rate constants. Thus both the first and second steps of the reaction as represented in Equation (10-204) operate in much the same way.

What remains to be observed in this reaction are the actual proton transfers between the enzyme in its active conformation and the bound substrate. It is presumed that the reason these steps have not been found in relaxation investigations is that the reaction intermediates are present in too low a concentration.

One interesting point of technique in the study of the RNase reaction is the combination of stopped-flow and temperature-jump methods. The temperature-jump, or any other relaxation method, is inapplicable if a reaction goes very far toward completion; Hammes overcame this difficulty by applying the temperature impulse immediately after stopping the flow in a rapid-mixing apparatus, so that the overall reaction did not have the opportunity to go to equilibrium, which might be nearly to completion.

Another of the enzymes about which relaxation measurements have given significant information is glyceraldehyde-3-phosphate dehydrogenase, a catalyst for one of the steps in the glycolysis sequence, which was described in Section 4-10:

$$\text{glyceraldehyde-3-phosphate}^{2-} + NAD + H_2PO_4^- \rightleftharpoons$$
$$\text{1,3-diphosphoglycerate}^{4-} + NADH_2 + H^+ \qquad (10\text{-}208)$$

The enzyme is different from RNase, but characteristic of a large group of enzymes in that it consists of four similar units. Each of the units has a molecular weight of about 35,000, and each has a catalytically active site. However, although the units are stable individually, they have no catalytic activity unless they are assembled into the tetramer. The individual units can have either of two different structures, the enzyme from yeast consisting of four units of one type and that from rabbit muscle of four units of the other type. Several tetramers consisting of combinations of the two structures have also been found to be active.

In studying this enzyme, as well as other dehydrogenases for which NAD–NADH$_2$ is a cofactor, the fluorescence of NADH$_2$ has been used as an analytical method in the temperature-jump experiment. The enzyme-bound NADH$_2$ gives a fluorescence spectrum with a maximum at 435 nm, compared to 462 nm for the free form, and with double the intensity of the spectrum of the free form.

Temperature-jump experiments on the yeast enzyme show five different relaxation times, of which three can be readily interpreted. One of these, having a magnitude in the range of seconds, has been identified with an interconversion between two forms of the enzyme, probably differing in conformation, an active form designated R, and an inactive form designated T, which, despite its inactivity, binds both substrate and NAD. The rate of binding of NAD to the R form of the enzyme is characterized by a relaxation time of the order of 100 μsec and to the T form by a time of milliseconds. In the absence of NAD, most of the enzyme is in the inactive state, but the binding of NAD induces the transformation of the T state to the active R state. Indeed, when the enzyme is dissociated into individual units, NAD accelerates recombination to form the tetramer.

EXERCISES

1. The initial concentration of reactant for a second-order reaction involving only one reactant is 0.378 M. In 17 min, the concentration is reduced to 0.189 M. Calculate k and determine the concentration of the reactant after 34 min.

2. The gaseous reaction

$$2N_2O_5 \longrightarrow 2N_2O_4 + O_2$$

is first-order with a rate constant at 378 K of 0.122 sec^{-1}. Assuming the reaction is carried out at constant volume, make a plot showing how the partial pressures of N$_2$O$_5$ and O$_2$ vary with time, starting with an N$_2$O$_5$ pressure of 10 torr.

3. A solution of sucrose is mixed with hydrochloric acid and placed in a polarimeter tube which is then maintained at 30°C. The optical rotation is read at various times. Using an appropriate plot, determine the rate constant from the following results.

Minutes	Degrees
0	+13.07
10	+10.33
20	+8.10
30	+6.12
40	+4.50
50	+3.16
60	+2.01
70	+1.04
80	+0.23
90	−0.45
100	−1.02
∞	−4.08

4. Suppose that the reaction between nitric oxide and chlorine proceeds by the following pathway, with $k_2 \ll k_1 \ll k_{-1}$.

$$NO + Cl_2 \underset{k_{-1}}{\overset{k_1}{\rightleftharpoons}} NOCl_2$$

$$NOCl_2 + NO \overset{k_2}{\longrightarrow} 2NOCl$$

Determine whether the resulting rate equation would be consistent with the experimental result that the reaction is third-order.

5. Consider the acid catalysis of the sucrose hydrolysis reaction in Equations (10-33) to (10-35). What are the species HA and A^- when hydrochloric acid is the catalyst? When acetic acid is the catalyst?

6. Derive the relations between the half-life of a first- or second-order reaction and the rate constant for each type of reaction.

7. J. W. Moore and R. C. Anderson [*J. Am. Chem. Soc.* **66**, 1476 (1944)] studied the reaction

$$2Ce^{4+} + As^{3+} \longrightarrow 2Ce^{3+} + As^{5+}$$

In a run at 30°C with an initial concentration of Ce^{4+} equal to 0.0234 mol/liter and of As^{3+} equal to 0.0117 mol/liter, the following results were obtained.

Time (min)	Concentration of Ce^{4+} (mol/liter)
0	0.0234
70	0.0193
130	0.0171
272	0.0139
335	0.0130
399	0.0122

By plotting suitable functions of concentration against time, test the data to determine the order of the reaction and evaluate the rate constant. Note that the initial concentrations represent equivalent amounts of the reactants, so that this can be treated as a reaction between like molecules.

8. A chemical reaction is studied at temperatures of 27° and 37°C. The forward rate constants for the two temperatures are 3.4 min^{-1} and 8.5 min^{-1}, respectively. Calculate the Arrhenius energy of activation for the reaction. Calculate the enthalpy of reaction from the temperature dependence of the equilibrium constant, which is found to be 2.3×10^{-6} at 27°C and 5.1×10^{-6} at 37°C. Assuming the Arrhenius activation energy is equal to the enthalpy of activation, draw an enthalpy contour for the reaction.

9. Derive an equation for the rate of thermal decomposition of acetaldehyde to give methane and carbon monoxide according to the following scheme, with the methyl radical as a reactive intermediate.

$$CH_3CHO \overset{k_1}{\longrightarrow} CH_3 \cdot + CHO \cdot$$

$$CH_3CHO + CH_3 \cdot \overset{k_2}{\longrightarrow} CH_4 + CO + CH_3 \cdot$$

$$2CH_3 \cdot \overset{k_3}{\longrightarrow} C_2H_6$$

10. A reaction is found to have an Arrhenius activation energy of 25.8 kcal and a rate constant of 15 min^{-1} at 30°C. Predict the rate constant at 10°C.

11. Show how Equation (10-59) can be obtained from the assumption that the total number of radicals is constant and that the only radical source and sink are steps 1 and 5 of the hydrogen–bromine chain reaction.

12. From the definition of equilibrium as a condition in which forward and reverse reaction rates are equal, derive the relation between K and the rate constants given in Equation (10-36).

13. The following half-lives for the first-order hydrolysis of pantothenate in acid solution were found by D. V. Frost and F. C. McIntire [*J. Am. Chem. Soc.* **66**, 425 (1944)]. From a suitable plot, determine the apparent activation energy:

Temperature (°C)	$t_{1/2}$ (days)
100	0.304
75	1.7
55	7.8
39	50.6
23	224
10	1250

14. For the reaction of H_2 and Br_2, the rate during the initial period is given by

$$k_{app} = 2k_2(k_1/k_5)^{1/2}$$

Show how ΔE_{app}, the overall activation energy as measured, is related to the activation energies of the individual steps 1, 2, and 5. (Hint: Substitute the Arrhenius expression for each of the rate constants, take logarithms, and differentiate with respect to temperature.)

15. The following data are obtained for a simple enzyme-catalyzed reaction of substrate S, using an enzyme concentration of 2×10^{-5} mol/liter.

Concentration of S (mol/liter)	Rate (mol/liter sec)
1×10^{-3}	0.063×10^{-8}
2×10^{-3}	0.118×10^{-8}
4×10^{-3}	0.211×10^{-8}
6×10^{-3}	0.286×10^{-8}
8×10^{-3}	0.348×10^{-8}
1×10^{-2}	0.400×10^{-8}

Determine the constants of the Michaelis–Menten equation.

16. Show that an activation energy of 12,000 cal/mol corresponds near room temperature to a Q_{10} of approximately 2.

17. Data have been obtained for the destruction of thiamine by an enzyme [R. R. Sealock and R. L. Goodland, *J. Am. Chem. Soc.* **66**, 507 (1944)], both with and without added o-aminobenzyl-4-methylthiazolium chloride as an inhibitor. By means of a suitable plot of the following initial rates, determine the type of inhibition.

Substrate concentration (mol/liter) $\times 10^4$	Initial velocity (mol/liter min) $\times 10^6$
Runs with no inhibitor:	
1.0	0.746
2.5	0.992
5.0	1.12
10.0	1.26
20.0	1.36
Runs with a 0.2×10^{-4} M inhibitor concentration:	
1.0	0.136
2.5	0.285
5.0	0.618
10.0	0.758
20.0	1.29

18. For the reversible reaction of substance A to form substance B, the enthalpy of activation is $+30$ kcal/mol and the entropy of activation at 300 K is -20 cal/mol deg. The enthalpy change for the formation of B from A is -13 kcal/mol. The entropy change for the formation of A from B is $+30$ cal/mol deg. Make diagrams showing the relative enthalpies, entropies, and free energies of A, the transition state, and B, and predict whether the rate constant for the forward or the reverse reaction is the larger.

19. The reaction rate for the iodination of tyrosine is given by

$$-\frac{d[\text{tyrosine}]}{dt} = k_2[\text{tyrosine}][\text{iodine}]$$

Measured values of k_2 for this catalyzed reaction in phosphate buffers at 25°C are:

pH	$[\text{HPO}_4{}^{2-}] \times 10^2$	k_2
6.60	4.02	10.7
	6.03	14.0
	8.06	17.4
6.40	1.50	5.1
	3.00	7.8
	5.40	11.8

Assuming that $k_2 = k_{\text{OH}}[\text{OH}^-] + k_{\text{HA}}[\text{HPO}_4{}^{2-}] + k_{\text{H}_2\text{A}}[\text{H}_2\text{PO}_4{}^-]$, estimate k_{OH}, k_{HA}, and $k_{\text{H}_2\text{A}}$ from the data.

20. If the termination step in the thermal decomposition of hydrocarbons were the combination of two hydrogen atoms to form H_2, rather than that given in Equation (10-71), what would be the predicted kinetic equation for a long-chain reaction?

21. Derive Equation (10-50) for the reaction of NO with Cl_2. Describe an experimental procedure to test whether equilibrium is maintained up to the rate-determining step in this reaction.

22. Alkyl bromides can be added to olefins in reactions catalyzed by benzoyl peroxide, which thermally decomposes, forming benzoyl radicals as a principal product. For the following scheme, assuming a long-chain reaction, deduce the kinetic equation.

$$(\text{C}_6\text{H}_5\text{CO})_2\text{O}_2 \xrightarrow{k_1} 2\text{C}_6\text{H}_5\text{COO}\cdot$$

$$\text{C}_6\text{H}_5\text{COO}\cdot + \text{CCl}_3\text{Br} \xrightarrow{k_2} \text{CCl}_3\cdot + \text{C}_6\text{H}_5\text{COOBr}$$

$$\text{CCl}_3\cdot + \text{C}_6\text{H}_5\text{CH}{=}\text{CH}_2 \xrightarrow{k_3} \text{C}_6\text{H}_5\overset{\cdot}{\text{C}}\text{HCH}_2\text{CCl}_3$$

$$\text{C}_6\text{H}_5\overset{\cdot}{\text{C}}\text{HCH}_2\text{CCl}_3 + \text{CCl}_3\text{Br} \xrightarrow{k_4} \text{C}_6\text{H}_5\text{CHBrCH}_2\text{CCl}_3 + \text{CCl}_3\cdot$$

$$2\text{CCl}_3\cdot \xrightarrow{k_5} \text{C}_2\text{Cl}_6$$

23. For the reaction

$$\text{Br}_2 + \text{CH}_3\text{COCH}_3 \xrightarrow{\text{catalyst}} \text{CH}_2\text{BrCOCH}_3 + \text{HBr}$$

general acid catalysis is found, while the rate is independent of the concentration of bromine and is first-order in ketone concentration. Show that these results can be explained by the following mechanism, together with certain assumptions about the relative magnitudes of the rate constants. HA stands for any acid, while A^- is the conjugate base.

$$HA + CH_3-\overset{\displaystyle O}{\underset{\displaystyle CH_3}{C}} \underset{k_{-1}}{\overset{k_1}{\rightleftharpoons}} \left[CH_3-\overset{\displaystyle OH}{\underset{\displaystyle CH_3}{C}} \right]^+ + A^-$$

$$\left[CH_3-\overset{\displaystyle OH}{\underset{\displaystyle CH_3}{C}} \right]^+ + A^- \underset{k_{-2}}{\overset{k_2}{\rightleftharpoons}} CH_3-\overset{\displaystyle OH}{\underset{\displaystyle CH_2}{C}} + HA$$

$$CH_3-\overset{\displaystyle OH}{\underset{\displaystyle CH_2}{C}} + Br_2 \overset{k_3}{\rightarrow} CH_3-\overset{\displaystyle O}{\underset{\displaystyle CH_2Br}{C}} + HBr$$

24. According to the Michaelis equation, what concentration of substrate would yield a rate equal to one-half the maximum rate V_{max}?

25. For p-aminosalicylic acid, the rate of protonation is much less than the diffusion limit. Suggest an explanation.

26. The following scheme applies to the oxidation of glucose to yield a gluconolactone and hydrogen peroxide, catalyzed by glucose oxidase.

$$E + glucose \longrightarrow E'\text{-lactone}$$
$$E'\text{-lactone} \longrightarrow E' + lactone$$
$$E' + O_2 \longrightarrow E-H_2O_2$$
$$E-H_2O_2 \longrightarrow E + H_2O_2$$

Classify this as an ordered, random, or Ping-Pong mechanism.

REFERENCES

Books

I. Amdur and Gordon G. Hammes, *Chemical Kinetics: Principles and Selected Topics*, McGraw-Hill, New York, 1966. Intermediate level; especially good for relaxation methods and enzymatic reactions.

R. P. Bell, *The Proton in Chemistry*, 2nd ed., Cornell University Press, Ithaca, New York, 1973. Very readable; includes kinetics of proton-transfer and proton-catalyzed reactions.

Myron L. Bender, *Mechanisms of Homogeneous Catalysis from Protons to Proteins*, Wiley-Interscience, New York, 1971. Thorough treatment, at a moderately advanced level, of all varieties of catalyzed reactions in solution.

Myron L. Bender and L. J. Brubacher, *Catalysis and Enzyme Action*, McGraw-Hill, New York, 1973. Introductory level.

Paul D. Boyer, Ed., *The Enzymes*, Vols. 2 and 3, Academic Press, New York, 1971. Detailed accounts of mechanisms of various enzymatic reactions are included.

E. F. Caldin, *Fast Reactions in Solution*, Wiley, New York, 1964. A clear, readable account of measurement methods for rapid reactions. Intermediate.

E. F. Caldin and V. Gold, *Proton Transfer Reactions*, Halsted Press, New York, 1975. A thorough, up-to-date treatment.

Joseph B. Dence, Harry B. Gray, and George S. Hammond, *Chemical Dynamics*, W. A. Benjamin, Menlo Park, Calif., 1968. Elementary introduction.

M. Dixon and E. C. Webb, *Enzymes*, Academic Press, New York, 1964. A comprehensive survey.

E. M. Eyring and B. C. Bennion, "Fast Reactions in Solution," in *Annual Review of Physical Chemistry*, Vol. 19, Annual Reviews, Palo Alto, Calif., 1968.

Arthur A. Frost and Ralph G. Pearson, *Kinetics and Mechanism*, 2nd ed., Wiley, New York, 1961. A general, intermediate-level textbook in reaction kinetics.

H. Gutfreund, *Enzymes: Physical Principles*, Wiley, New York, 1972. A very good general account, including kinetics.

Gordon G. Hammes, "Relaxation Spectrometry of Biological Systems," in *Advances in Protein Chemistry*, Vol. 23, Academic Press, New York, 1968.

Gordon G. Hammes, Ed., *Investigation of Rates and Mechanisms of Reactions*, Wiley-Interscience, New York, 1973. Up-to-date descriptions of experimental methods and equipment.

W. P. Jencks, *Catalysis in Chemistry and Enzymology*, McGraw-Hill, New York, 1969. An advanced, comprehensive volume.

Keith J. Laidler and Peter S. Bunting, *The Chemical Kinetics of Enzyme Action,* 2nd ed., Oxford University Press, New York, 1973. Probably the best single volume on the subject. Fairly advanced, but quite readable.

H. R. Mahler and E. H. Cordes, *Biological Chemistry*, 2nd ed., Harper and Row, New York, 1971. Chapter 6 has a very detailed account of enzyme kinetics.

Dennis Piszkiewicz, *Kinetics of Chemical and Enzyme-Catalyzed Reactions*, Oxford University Press, New York, 1977. Quite good, intermediate-level treatment.

Kent Plowman, *Enzyme Kinetics*, McGraw-Hill, New York, 1972. Intermediate level.

I. H. Segel, *Enzyme Kinetics*, Wiley, New York, 1975. Comprehensive, including expressions for a variety of reaction schemes.

G. B. Skinner, *Introduction to Chemical Kinetics*, Academic Press, New York, 1974. Intermediate level.

A. G. Sykes, *Kinetics of Inorganic Reactions*, Pergamon Press, Elmsford, N.Y., 1966. Particularly useful for reactions involving ions, including complexes, in solution.

P. R. Wells, *Linear Free Energy Relationships*, Academic Press, New York, 1968.

Finn Wold, *Macromolecules: Structure and Function*, Prentice-Hall, Englewood Cliffs, N.J., 1968. Includes section on enzyme kinetics.

Journal Articles

Addison Ault, "An Introduction to Enzyme Kinetics," *J. Chem. Educ.* **51**, 381 (1974).

R. R. Baker and D. A. Yorke, "Theories on the Slow-Gas-Phase Oxidation of Hydrocarbons," *J. Chem. Educ.* **49**, 351 (1972).

Myron L. Bender, Ferenc J. Kezdy, and Fred C. Wedler, "α-Chymotrypsin: Enzyme Concentration and Kinetics," *J. Chem. Educ.* **44**, 84 (1967).

Jacob Bigeleisen, "Chemistry in a Jiffy," *Chem. Eng. News,* p. 26 (April 25, 1977).

J. P. Birk, "Mechanistic Implications and Ambiguities of Rate Laws," *J. Chem. Educ.* **47**, 805 (1970).

J. D. Bradley and G. C. Gerrans, "Frontier Molecular Orbitals—A Link between Kinetics and Bonding Theory," *J. Chem. Educ.* **50**, 463 (1973).

Edward Caldin, "Temperature-Jump Techniques," *Chem. Brit.* **11**, 4 (1975).

G. V. Calder, "The Time Evolution of Drugs in the Body," *J. Chem. Educ.* **51**, 19 (1974).

W. W. Cleland, "What Limits the Rate of an Enzyme-Catalyzed Reaction?" *Acc. Chem. Res.* **8**, 145 (1975).

Michael R. J. Dack, "The Influence of Solvent on Chemical Reactivity," *J. Chem. Educ.* **51**, 231 (1974).

J. O. Edwards, E. F. Greene, and J. Ross, "From Stoichiometry and Rate Law to Mechanism," *J. Chem. Educ.* **45**, 381 (1968).

Larry Faller, "Relaxation Methods in Chemistry," *Sci. Am.* **220**, 30 (May 1969).

J. E. Finholt, "The Temperature-Jump Method for the Study of Fast Reactions," *J. Chem. Educ.* **45**, 394 (1968).

Joseph S. Fruton, "The Active Site of Pepsin," *Acc. Chem. Res.* **7**, 241 (1974).

V. Gold, Application of Isotope Effects," *Chem. Brit.* **6**, 292 (1970).

W. Keith Hall, "Catalytic Function of Hydrogen Bound to the Surfaces of Oxides," *Acc. Chem. Res.* **8**, 257 (1975).

Gordon G. Hammes, "Relaxation Spectrometry of Enzymatic Reactions," *Acc. Chem. Res.* **1**, 321 (1968).

George Hammons, F. H. Westheimer, K. Nakaoka, and R. Kluger, "Proton-Exchange Reactions of Acetone and Butanone. Resolution of Steps in Catalysis by Acetoacetate Decarboxylase," *J. Am. Chem. Soc.* **97**, 1568 (1975).

Kenneth R. Hanson and Irwin A. Rose, "Interpretations of Enzyme Reaction Stereospecificity," *Acc. Chem. Res.* **8**, 1 (1975).

Emil Thomas Kaiser and Bonnie Lu Kaiser, "Carboxypeptidase A: a Mechanistic Analysis," *Acc. Chem. Res.* **5**, 219 (1972).

Keith J. Laidler, "Unconventional Applications of the Arrhenius Law," *J. Chem. Educ.* **49**, 343 (1972).

I. R. Lehman, "DNA Ligase: Structure, Mechanism, and Function," *Science* **186**, 790 (1974).

Gustav E. Lienhard, "Enzymatic Catalysis and Transition-State Theory," *Science* **180**, 149 (1973).

M. Morton, "Polymerization as a Model Chain Reaction," *J. Chem. Educ.* **50**, 740 (1973).

Ramesh C. Patel, Gordon Atkinson, and R. J. Boe, "Fast Reactions: Rapid Mixing and Concentration Jump Experiments," *J. Chem. Educ.* **47**, 800 (1970).

Ralph M. Pollack and Thomas C. Dumsha, "Imidazole-Catalyzed Hydrolysis of Anilides, Nucleophilic Catalysis or Proton-Transfer Catalysis?" *J. Am. Chem. Soc.* **97**, 377 (1975).

M. D. Porter and G. B. Skinner, "The Steady-State Approximation in Free-Radical Calculations," *J. Chem. Educ.* **53**, 366 (1976).

Chong Wha Pyun, "Steady-State and Equilibrium Approximations in Chemical Kinetics," *J. Chem. Educ.* **48**, 194 (1971).

R. R. Rando, "Mechanisms of Action of Naturally Occurring Irreversible Enzyme Inhibitors," *Acc. Chem. Res.* **8**, 281 (1975).

William F. Sheehan, "Along the Reaction Coordinate," *J. Chem. Educ.* **47**, 254 (1970).

L. E. H. Smith, L. H. Mohr, and M. A. Raftery, "Mechanism for Lysozyme-Catalyzed Hydrolysis," *J. Am. Chem. Soc.* **95**, 7497 (1973).

Robert M. Stroud, "A Family of Protein-Cutting Proteases," *Sci. Am.* **231**, 74 (July 1974).

Ronald P. Taylor, S. Berga, V. Chau, and C. Bryner, "Bovine Serum Albumin as a Catalyst. III. Conformational Studies," *J. Am. Chem. Soc.* **97**, 1943 (1975).

Richard Wolfenden, "Analog Approaches to the Structure of the Transition State in Enzyme Reactions," *Acc. Chem. Res.* **5**, 10 (1972).

William T. Yap, Barbara F. Howell, and Robert Schaffer, "Determination of the Kinetic Constants in a Two-Substrate Enzymatic Reaction," *J. Chem. Educ.* **54**, 254 (1977).

Eleven
Adsorption
and Surface
Effects

The properties of many systems, including a large number that are of biological importance, are materially affected by the presence of a surface or interface between phases. The greater the extent of the surface in relation to the amount of material, the larger the contribution the nature and structure of the surface region make in determining the properties of the system.

One of the factors leading to the existence of a relatively large amount of surface is the presence, in the system, of units of very small dimensions. Thus fine particles of solid or fine droplets of liquid have, by their very geometry, a relatively large surface/volume ratio. Further, needlelike geometry, with one large and two small dimensions, or a disclike shape, with one small and two large dimensions, is each associated with appreciable surface area per unit mass, although the interfacial areas are not as large as when all three dimensions are small. Another kind of structure in which a relatively large area can be developed is that of a porous solid which contains a spongelike arrangement of many channels having sufficiently small diameter so that a large total surface of the solid is exposed.

Interfaces are often characterized by a particular molecular ordering or by a composition different from that of the bulk phase, effects that are termed adsorption. In this chapter, we first present a general principle governing the extent of adsorption at any type of interface, the Gibbs equation. We then consider the surfaces of pure and mixed liquids and show how molecular behavior at these surfaces is parallel to the formation of insoluble monomolecular films on liquid substrates. Next we describe some of the aspects of adsorption on solids, both from the gas phase and from the liquid phase, and show how adsorption plays a part in surface-catalyzed reactions.

Not only are adsorption and surface effects of practical importance in themselves but, in addition, the forces and influences involved are intimately related to those encountered in more complex systems, such as colloidal dispersions and biological membranes, to be discussed in Chapter 12.

11-1
ENERGY RELATIONS
AND ADSORPTION

The free energy of a surface is equal to the work required to produce that surface. As seen in Chapter 1, the amount of this work is equal to the product of the surface tension and the surface area of the system. Thus

$$G_{\text{surface}} = \gamma A \qquad (11\text{-}1)$$

The general tendency to minimize the free energy is manifested for surfaces as a reduction in the area or a decrease in surface tension or both. In a liquid, the freedom of the surface to change in shape provides an opportunity for the area to be reduced, as illustrated by the formation of spherical droplets and by the rise of liquid in a wetted capillary tube. The other factor in the product, the surface tension, can be decreased if the molecules of a liquid are unsymmetrical and can orient themselves in a particular way at the surface, or if the liquid is composed of several components so that preferential migration of one of these to the surface can make the concentration of the surface different from the concentration in the bulk of the liquid.

Any difference existing between the concentration at the surface of a phase and the concentration in the bulk of the phase constitutes a form of *adsorption*. If a solute in a liquid reduces the surface tension of the solvent by its presence, the solute tends to migrate to the surface and its equilibrium concentration in the surface will be higher than the bulk concentration. If the solute increases the surface tension, it is found that the liquid surface is depleted in solute, compared to the body of the liquid. When a liquid solution is in contact with a solid, adsorption at the liquid–solid interface is modified considerably by the interaction of the solute with the solid, and material from the solution may tend to concentrate in relatively large amounts at the solid surface, as do many organic molecules on the surface of activated carbon. When a solid and a gas phase are in contact with one another, molecules of the gas may adhere to the solid surface. An example is the binding of molecules of water by silica gel. In either of these two instances, the process of adsorption lowers the surface free energy of the system below that of the pure solid.

The amount of adsorption at a surface depends upon the amount by which the surface tension changes with increased concentration of the adsorbed material. A generally applicable quantitative expression is the Gibbs equation,

$$\Gamma \frac{\text{mol}}{\text{cm}^2} = -\frac{C}{RT} \frac{d\gamma}{dC} = -\frac{1}{RT} \frac{d\gamma}{d \ln C} \qquad (11\text{-}2)$$

The amount adsorbed at concentration C is represented by the symbol Γ, Greek capital gamma, and is defined as the *excess* of solute per unit area of surface over that which would be present if the concentration near the surface were the same as in the bulk of the phase. The rate of variation in surface tension with concentration is given by the derivative $d\gamma/dC$.

Example: The surface tension of solutions of butyric acid at 15°C is 49.0 dyn/cm at 0.25 M concentration and 45.5 dyn/cm at 0.35 M concentration. Calculate the excess surface concentration of butyric acid in a 0.30 M aqueous solution according to the Gibbs equation.

Solution: Take the average rate of change in surface tension with concentration between 0.25 M and 0.35 M as that at 0.30 M. From Equation (11-2),

$$\Gamma = -\frac{(0.30\ M)(49.0 - 45.5)\ \text{dyn/cm}}{8.314 \times 10^7\ \text{ergs/mol deg}\ (288\ \text{K})(0.35\ M - 0.25\ M)}$$
$$= 4.4 \times 10^{-10}\ \text{mol/cm}^2$$

It is possible to understand fairly simply the qualitative basis for the Gibbs equation. Consider the case of a liquid-phase solution. The driving force for adsorption is the lowering of the surface tension by the solute. Suppose that Δn moles of solute transferred to a region of A square centimeters of surface causes a reduction in surface tension $d\gamma$. The change in free energy is then

$$dG = A\ d\gamma \tag{11-3}$$

However, accumulation of solute at the surface is contrary to the tendency for uniform concentration to be maintained throughout the liquid, the osmotic tendency, and transfer of solute to the surface against a concentration difference dC corresponds to an increase in free energy of magnitude

$$dG = (\Delta n)RT\ d\ln C \tag{11-4}$$

At equilibrium, the sum of these two changes in free energy must be zero:

$$A\ d\gamma + (\Delta n)RT\ d\ln C = 0 \tag{11-5}$$

Rearranging,

$$\frac{\Delta n}{A} = \Gamma = -\frac{d\gamma}{RT\ d\ln C} \tag{11-6}$$

We emphasize that this is not a rigorous derivation of the Gibbs equation, but it is presented to emphasize the two physical effects that play a part in determining how much adsorption occurs at equilibrium.

The Gibbs equation has a thermodynamic basis, and there is much circumstantial evidence in support of its soundness, so that it is certainly valid, with the one qualification that activities should be employed in place of concentrations for the best results. However, it has been very difficult to devise a measurement technique suitable for testing the equation experimentally. Among the methods that have been tried are slicing off a thin layer of the surface with a device like a microtome and analyzing this layer, or labeling the surface-active material with radiotracers emitting radiation of very short path length and evaluating the concentration of molecules near the surface by counting the radioactivity. Because of various complications in the experiments and because of the very small magnitude of the total amount of material adsorbed, none of these experiments has been thoroughly conclusive. Nevertheless, the Gibbs equation is used with confidence to calculate the amount of material adsorbed in a given system from the measured dependence of surface tension on concentration.

It has been shown that the surface tension is equal to the surface free energy per unit area. What about the enthalpy of a surface? The answer to this can be found by the general method for obtaining values of an enthalpy change, which is by studying the effect of temperature on a phenomenon. Using the thermodynamic relation, Equation (4-43), for the entropy in terms of the derivative of free energy with respect to temperature at constant pressure, $S = -(dG/dT)_P$, we can write the general equation

$$H = G + TS = G - T\left(\frac{dG}{dT}\right)_P \qquad (11\text{-}7)$$

For the surface, this becomes the special equation

$$H_{\text{surface}} = \gamma - T\left(\frac{d\gamma}{dT}\right)_P \qquad (11\text{-}8)$$

Is the difference between the free energy of the surface and the enthalpy of the surface significant? This can be answered using the example of water at 25°C. The surface tension is 72.0 ergs/cm² as compared to the surface enthalpy which is 118.7 ergs/cm², indicating a substantial contribution from the entropy term.

When aqueous solutions of various solutes are examined, there are found to be two general classes of behavior. Most inorganic solutes produce a slight increase in surface tension with increasing concentration, and therefore *negative* adsorption is predicted by the Gibbs equation, but the magnitude is too small to be significant. Many soluble organic molecules lower the surface tension of water rather drastically, large *positive* adsorption occurs, and they are said to be *surface-active*. Most of these substances are composed of molecules containing a polar, or even ionic, functional group. Examples of neutral groups are the hydroxyl, amino, carboxyl, amide, ester, and nitro groups, and typical ionized groups are the carboxylate, sulfonate, and ammonium groups. The polar part of the molecule confers water solubility, whereas the hydrocarbon part has little affinity for water and thus provides the driving force for adsorption at the surface of the aqueous phase. Of course, if the hydrocarbon portion of the molecule is sufficiently large, the substance is essentially insoluble and can only form a film on the surface of the water, as discussed in Section 11-3.

11-2
LIQUID SURFACES

ORIENTATION IN
PURE LIQUIDS

In a pure liquid sample of a compound consisting of molecules with a polar head and a hydrocarbon tail, there is the possibility of orientation of the molecules at the liquid surface. Irving Langmuir, one of the pioneers in the study of liquid and solid surfaces, compared the properties of octane and octyl alcohol. Substitution of a hydroxyl group for a hydrogen in the hydrocarbon increases the enthalpy of evaporation

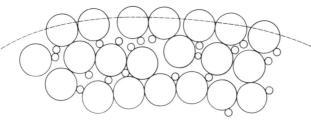

Figure 11-1
Preferential orientation in a liquid surface. The large spheres represent the hydrocarbon tails, and the small spheres are the polar functional groups, or heads. The dashed line represents the effective surface of the liquid.

by about 45 percent but leaves the surface enthalpy substantially unchanged. This can be explained by the tendency of the alcohol molecules to turn their polar groups inward. The liquid then exposes a surface essentially hydrocarbon in nature with the appropriate surface energy for a hydrocarbon. When evaporation occurs, the hydroxyl group of a molecule must be torn away from the hydroxyl groups of other molecules against strong attractive forces, and thus the increased enthalpy of evaporation is reasonable. The cause for orientation of the molecules in the surface can be described in either of two ways: (1) the polar groups have a greater mutual attraction than do the hydrocarbon parts of the molecules and so are pulled together as closely as possible; (2) the surface containing hydroxyl groups has a higher surface tension than the surface containing hydrocarbon chains, and therefore the latter is preferred. Figure 11-1 shows schematically the arrangement of molecules in a liquid surface of this sort.

SOLUBLE FILMS

As mentioned above, surface-active derivatives of hydrocarbons concentrate near the surface of an aqueous solution and may lower the surface tension of the solution much below that of water. The decrease in surface tension is easily and directly measured, and the existence of adsorption is deduced from the values of surface tension lowering. Just as for orientation in the surface of a pure liquid, it can be considered that the driving force for adsorption at the surface is lowering of the surface tension; or, with equal justification, it can be considered that the surface tension is lowered as a consequence of replacement of the aqueous surface by a hydrocarbon surface resulting from the lack of affinity of the hydrocarbon parts of the molecules for the solvent. It is immaterial which of the two effects is viewed as the cause and which as the result, for they are inextricably linked.

In Figure 11-2 are shown the measured values of the surface tension for aqueous solutions of the lower aliphatic acids. For the lower aliphatic acids, as well as for several other homologous series, I. Traube found that the surface activity increases regularly with chain length and that the concentration of carboxylic acid required to reach a given surface tension is reduced by a factor of three for each added methylene group. *Traube's rule* has been explained by picturing the molecules as curled up in a spherelike shape, as depicted in Figure 11-1, in order to present a minimum of hydrocarbon surface to the water. A molecule floating in this configuration in the surface produces an effect roughly proportional to its own surface area, and the observed molecular volumes of the hydrocarbons indicate that the surface area of the spherical molecule varies linearly with the number of carbon atoms in the chain.

It is possible to show that, for low concentrations of surface-active materials, the adsorbed layer behaves in such a way that it can be characterized as a two-dimensional gas. Imagine a dilute solution placed next to a portion of pure liquid solvent, with a barrier impenetrable to solute molecules separating the two surfaces. The force per unit length exerted by the surfaces on the barrier, and perpendicular to it, is equal to the difference in the surface tensions of the two liquids:

$$\frac{F}{l} = \gamma_1 - \gamma_2 = \Delta\gamma \tag{11-9}$$

The barrier is pulled more strongly by the pure solvent, for this is the liquid with the greater inclination to shrink its surface. To compute the surface tension of the solution, we apply the Gibbs equation, Equation (11-2). In dilute solutions, γ is nearly linear from zero concentration up to the concentration of the solution in question, so that $-d\gamma/dC$ can be evaluated as $(\gamma - \gamma_0)/(C - 0)$ or $\Delta\gamma/C$. Further, the adsorption Γ is equal to n moles of solute per A square centimeters of surface. Then

$$\frac{n}{A} = -\frac{C}{RT}\frac{d\gamma}{dC} \simeq \frac{C}{RT}\frac{\Delta\gamma}{C} = \frac{\Delta\gamma}{RT} \tag{11-10}$$

Finally, the surface tension difference is considered to be a kind of two-dimensional surface pressure, and is represented by ϕ. Equation

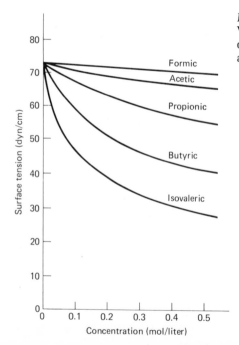

Figure 11-2
Variation in surface tension with concentration for aqueous solutions of aliphatic carboxylic acids.

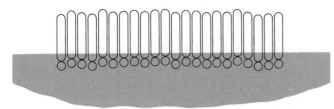

Figure 11-3
Probable arrangement of molecules in a saturated surface film.

(11-10) can then be rearranged to a form resembling the ideal gas equation:

$$\phi A = nRT \qquad (11\text{-}11)$$

As solutions of surface-active materials are made more concentrated, the surface tension curves flatten out and a point is reached beyond which a further increase in concentration in the bulk phase has no further effect. This limit apparently corresponds to a condition of surface saturation, in which practically the entire surface is covered with a layer of solute molecules. The three-carbon to six-carbon carboxylic acids show about the same molar surface concentration at saturation, with an area of about 27 Å^2 occupied by each molecule, an area which is apparently related to the cross-sectional area of the hydrocarbon chain but is independent of the chain length. In these concentrated films, the molecules are evidently forced to uncoil and become oriented parallel to one another with the carboxyl heads in the aqueous phase and the hydrocarbon tails standing upward, as shown in Figure 11-3. The consistency in the surface concentration at saturation calculated by the Gibbs equation is one of the points of confirmation of the validity of this equation.

Aliphatic alcohols behave much like the carboxylic acids. However, for short-chain alcohols, the minimum area per molecule approached at high concentrations is appreciably greater than the cross section of the hydrocarbon chain, indicating the possibility that the alcoholic hydroxyl groups bind rather firmly water molecules which consequently occupy a portion of the surface. Many detergent and soaplike materials exhibit similar surface tension-lowering effects on water when present in sufficiently dilute solution. Examples are sodium oleate, sodium dodecyl sulfate, and cetylpyridinium bromide. These substances, however, form aggregates or micelles above a critical concentration, behavior that will be discussed more fully in Chapter 12.

INTERFACIAL TENSION

If two immiscible liquids are in contact with one another, the surface between then has associated with it an interfacial tension corresponding to the energy required to produce unit area of that surface. Interfacial tensions can be measured by suitable modifications of the capillary rise or du Nouy tensiometer methods.

The more alike the two liquids touching one another, the lower the interfacial tension. The value of an interfacial tension is almost always less than the liquid–gas surface tension of either one of the two liquids involved. The interfacial tension between water and n-octyl alcohol at 20°C is 8.5 dyn/cm compared to 27.5 dyn/cm for the surface tension of n-octyl alcohol. This is an indication of special structure in the water–alcohol surface. Undoubtedly many alcohol molecules are oriented so that their polar groups are surrounded by water molecules; possibly these project into the water phase and so serve to bridge the gap between the two unlike liquids.

11-3
INSOLUBLE FILMS
ON LIQUIDS

There are several different types of molecules that are insoluble in water but which can be induced to spread over a water surface to form layers or films only one molecule in thickness. These molecules are again characterized by a dual nature, containing a hydrocarbon portion which reduces water solubility and a polar portion which confers affinity with the aqueous phase. Many of them are of the same general structure as the surface-active molecules described earlier, but simply have longer hydrocarbon chains. Thus an alcohol with a chain length greater than 14 carbon atoms is substantially insoluble in water but forms a film on the surface of water. Molecules of somewhat shorter chain length can be spread on aqueous salt solutions, in which the solubility of the surface-active material is less than in pure water. Other film-forming molecules are of much greater molecular weight and include synthetic polymers as well as proteins and other naturally occurring polymers.

It is interesting that the first recorded observation of an insoluble surface film was made by Benjamin Franklin in 1774. He reported an experiment in which he spread less than a teaspoonful of olive oil on a pond on Clapham Common in England and observed it to spread to an area of "perhaps half an acre," making the surface of the water "as smooth as a looking glass." However, no one realized for over a hundred years that the thickness of Franklin's spread film, roughly 25 Å, was related to the dimensions of individual molecules.

EXPERIMENTAL METHODS
Some films can be formed by simply pouring the material on water, but usually it is desirable to use a spreading solvent, a material that evaporates rapidly after the solution is poured on the aqueous phase, leaving the solute spread over the surface. Suitable liquids include benzene, ethyl ether, chloroform, and hexane. In some systems the nature of the solvent seems to affect the properties of the resulting film, and care must then be taken in the choice of solvent and attention paid to the question of reproducibility of results from one solvent to another.

Surface films can most conveniently be studied by using a shallow, rectangular trough filled to the brim with water. The edges are waxed or made of Teflon to retain the water in the trough. Many film formers are very sensitive to the presence of metal ions, and thus an uncoated

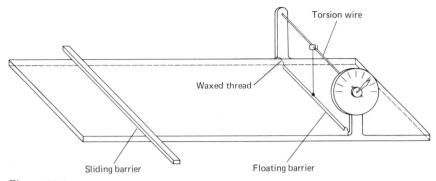

Figure 11-4
Diagram of a Langmuir surface pressure balance.

metal tray cannot safely be used. Quantitative measurements require great care as to cleanliness and removal of impurities. Contaminants on the surface of the substrate liquid are customarily removed by using bars to sweep repeatedly across the surface toward the edge, carrying off both dust and soluble surface-active materials.

The surface pressure balance developed by Langmuir for determination of the force–area relationships for a film consists of a shallow trough with a floating barrier near one end. The barrier is pivoted on a torsion wire by means of which the force exerted against it can be measured. The ends of the barrier are linked to the sides of the trough by waxed threads to keep molecules of the film former from escaping past the barrier. A diagram of the apparatus is shown in Figure 11-4.

By bringing a movable bar successively closer to the floating strip, the force exerted on the float can be determined as a function of the area available for the film. The "pressure," or force per unit length ϕ on the float is equal to the difference in surface tension between the film on one side and the pure solvent on the other side. The situation is quite like that described above for a soluble surface-active material with a barrier dividing its film from a surface of pure solvent, except that in the present case the insoluble molecules really cannot escape under the float. In fact, if curves are plotted of the pressure–area product against pressure, as in Figure 11-5, the results from both methods for a homologous series show a continuous gradation; those for the C_{12} carboxylic acid obtained from surface tension lowering and the Gibbs equation are in good agreement with those from the surface pressure balance.

Spreading pressures of stearic acid films have been extensively studied, beginning with the work of Langmuir. This acid, with the formula $C_{17}H_{35}COOH$, forms a coherent film. Even when there is a large area available per molecule, the molecules cluster together to form islands, and the spreading pressure, or escaping tendency of individual molecules, is relatively small. As the area per molecule is reduced to the vicinity of 21 Å^2, there is a sudden increase in pressure, as shown in Figure 11-6, corresponding to nearly complete coverage of the surface by the film. Further compression of the film results in a pileup of the molecules to give a multilayer structure with erratic values of the spreading pressure.

In the compact, monomolecular film, the molecules can be visualized as arranged vertically, with the hydrocarbon chains parallel to one another and directed upward, and the polar heads in the water surface. The area occupied per molecule seems to be the cross-sectional area of the hydrocarbon chain, since about the same value is found for other aliphatic straight-chain acids, amines, amides, and methyl ketones. As the length of the carbon chain is increased by the addition of a methylene group, the film thickness increases by about 1.4 Å.

The presence of a single methyl branch in isostearic acid, $(CH_3)_2CH(CH_2)_{14}COOH$, is sufficient to raise the area requirement from that of stearic acid to 31.6 Å^2, presumably by interfering with the compact arrangement of the parallel chains. Oleic acid, $CH_3(CH_2)_7CH{=}CH(CH_2)_7COOH$, with the same chain length as stearic acid, but with a kink at the double bond, requires 46 Å^2 per molecule. Other molecules with single, saturated hydrocarbon chains, such as esters, polyphenols, and α-bromo acids, apparently have their area requirements determined by the size of the head group, which varies from 25 to 35 Å^2. Glyceryl tristearate, with three hydrocarbon chains, has a molecular area of 66 Å^2, corresponding approximately to three times the area of a single chain.

Other physical properties of monolayers can be investigated. The viscosity can be qualitatively observed by dusting a small amount of talc onto the film and blowing on it. If the film is rigid, the powder particles resist attempts to move them. Viscosity can be estimated in a more quantitative way by allowing the film to flow through a channel in the confining barrier on the surface pressure trough or by utilizing

Figure 11-5
Pressure–area versus pressure curves for carboxylic acids of various chain lengths. [After N. K. Adam, *The Physics and Chemistry of Surfaces*, Oxford University Press.] For C_4 to C_{12}, values were calculated from the surface tension by the Gibbs equation; for C_{12} and C_{15}, spreading pressures were directly measured.

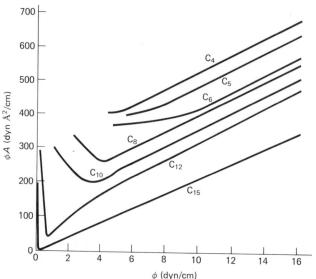

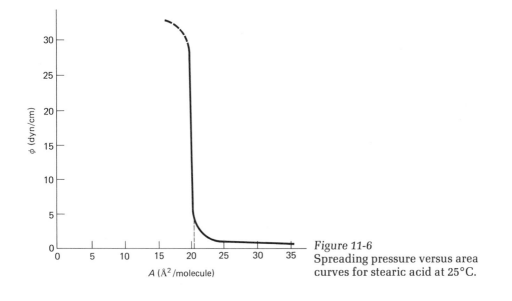

Figure 11-6
Spreading pressure versus area curves for stearic acid at 25°C.

a circular disk mounted horizontally and fastened to the bottom of a vertical shaft so that the disk just touches the film-covered surface of the liquid. The resistance of the film to deformation is measured either by determining the torque required to rotate the disk at some constant velocity, or by measuring the damping effect upon the disk when it is set into torsional oscillation.

The vertical electric potential difference across the film gives additional clues to the molecular arrangement. In the measurement of this potential, a reference electrode, such as a calomel electrode, is located in the liquid phase. In one method, a small metal probe carrying a bit of radioactive material which ionizes the air so that it becomes conducting is placed above the liquid, and the potential difference between it and the probe is measured. The film potential depends primarily upon the concentration and orientation of the electric dipoles in the molecules forming the film. The small probe of the radioactive method can be used to explore the surface and look for discontinuities in the film, showing clearly for some substances such as myristic acid an arrangement of patches of condensed film with intervening regions of a dilute "gaseous" layer. A change in the film potential contribution per molecule with changing concentration can be taken to indicate a change in the tilt of the molecules with respect to the surface or a rearrangement of the distribution of counterions in the solution.

FILMS OF MOLECULES OF SMALL AND MODERATE SIZE

We turn now to the description of some of the properties of specific types of monomolecular films. Depending on the nature of the material, the amount of film former applied per unit area of surface, and the temperature, there may be observed any of a variety of phase behaviors, from very dilute gaseous films with independently moving molecules at one extreme, to the closely packed, rigid films at the other

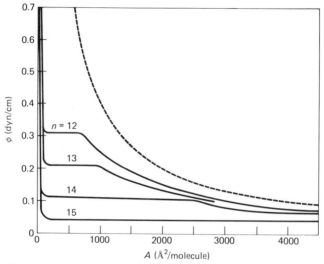

Figure 11-7
Spreading pressure versus area curves for carboxylic acids, $C_nH_{2n+1}COOH$, of varying chain length: $n = 12$, tridecylic acid; $n = 13$, myristic acid; $n = 14$, pentadecylic acid; $n = 15$, palmitic acid.

extreme. The properties of the film of a specific substance depend upon a delicate balance between the forces of van der Waals attraction between the hydrocarbon chains of adjoining molecules, on the one hand, and repulsive or steric effects which prevent these chains from approaching one another closely, on the other hand. The latter effects include the repulsions of charged head groups, as well as the presence of multiple polar groups which influence a molecule to lie flat on the surface. For example the bromide salt of trimethylammonium ion with a twenty-carbon chain attached forms a gaseous film despite the length of the chain, presumably because of the effects of electrostatic repulsion. The dibasic ester $C_2H_5OOC(CH_2)_{11}COOC_2H_5$ forms good gaseous films because both ends of the molecule are held to the surface.

Evidence for the effect of van der Waals forces is found in the phase behavior of a homologous series of compounds such as carboxylic acids. Those of lower molecular weight form gaseous films, those of higher molecular weight form condensed films, and those of intermediate molecular weight form either type as well as a series of intermediate phases, depending on conditions.

Figure 11-7 shows the pressure–area curves of several carboxylic acids of varying chain length. Following each curve from low pressure to high pressure, that is, large area to small area, one sees first the behavior of a very compressible gas, nearly ideal, followed by a horizontal line corresponding to condensation to a liquidlike phase. Along this line of constant force there are two phases present, just as when a three-dimensional gas condenses to liquid at a given temperature and pressure.

Pentadecylic acid, in the vicinity of room temperature, forms a variety of condensed phases. At the highest pressure is the solid phase, which is rigid and has very low compressibility: The chains and heads are both packed together as tightly as possible. In a *liquid-condensed* phase, the film is more compressible than the solid, but the viscosity is still high. Possibly the heads are close-packed, but the chains are not quite completely touching; further compression then either forces the polar groups to move upward or downward to give a staggered arrangement, or causes the polar groups to be stripped of some solvent molecules. For a *liquid-expanded* phase, the viscosity is somewhat lower, but the compressibility is too low for there to be much space between the molecules, and no patches of condensed phase as distinct from a gaseous phase are detectable by surface potential measurements. One model advanced for this phase is a distribution of molecular configurations differing in the fraction of the hydrophobic chain that is lifted above the surface.

Films consisting of several kinds of molecules may give information about the nature of the interactions between the different molecules. One of the components may initially be in solution beneath a film of the other material. The solute may penetrate the film, as sodium cetyl sulfate penetrates a cetyl alcohol monolayer, increasing the spreading pressure but having little effect upon the surface potential. Sometimes, however, the two substances form a complex film, equivalent to two superimposed monolayers. Thus hexadecylamine in solution attaches itself to the bottom of a cholesterol monolayer. Formation of this type of film does not materially alter the pressure, but it is indicated by a change in the surface potential.

FILMS OF PHOSPHOLIPIDS

It is possible to extract from the membranes of cells several types of molecules, termed phospholipids, containing two long hydrocarbon chains attached to glycerin by two ester linkages, with the third hydroxyl of the glycerin linked through phosphate to a polar group:

$X = -CH_2CH_2NH_3^+$, phosphatidylethanolamine

$X = -CH_2CH_2\overset{+}{N}(CH_3)_3$, lecithin, or phosphatidylcholine

$$\begin{array}{l} O \\ \parallel \\ R_1C-O-CH_2 \\ O \\ \parallel \\ R_2C-O-CH \\ \qquad\qquad\quad O^- \\ \qquad\quad CH_2-O-\overset{\displaystyle O^-}{\underset{\displaystyle O}{P}}-O-X \end{array}$$

$X = -CH_2-CH-COO^-$
$\qquad\qquad\quad | $
$\qquad\qquad\ NH_3^+ \qquad$, phosphatidylserine

$X = -C_6H_6(OH)_5$, phosphatidylinositol

The alkyl groups, R_1 and R_2, have chain lengths varying from 12 to 24 carbon atoms and contain from zero to six double bonds per chain. The double bonds are usually unconjugated and in the cis configuration.

The structures of membranes containing phospholipids will be discussed in Chapter 12. However, because membranes can be expected to contain these molecules in some sort of regular pattern or ordered arrangement such as they might have in a surface film, phospholipids provide particularly interesting subjects for monolayer studies.

As a point of comparison, the isotherms of three glycerin derivatives are shown in Figure 11-8. In 1,2-dipalmitin, the first and second hydroxyl groups of glycerin are esterified with palmitic acid, the sixteen-carbon straight-chain acid. The film properties resemble closely those of 1,3-dipalmitin in which the two terminal hydroxyls are esterified, except that the area per molecule occupied by the latter is a bit larger. However, at least at the temperature shown here, both these films are considerably more compact than the film of 1,2-dimyristin, in which two adjacent hydroxyls are esterified with the fourteen-carbon acid, demonstrating again the effect of the greater dispersion forces of attraction between longer hydrocarbon chains.

The general behavior of phospholipid films is much like that of carboxylic acid films: With sufficiently long straight chains the monolayer is condensed, whereas shortening of the chains or introduction of points of unsaturation or branching results in a gaseous monolayer. Somewhere in the middle of a homologous series, there are molecules that can form a variety of film types, depending upon surface concentration and temperature, and it is the manner in which these monolayers, usually of the liquid-expanded or liquid-condensed type, respond to changing conditions that supplies the greatest amount of the information being sought—that about the details of molecular interactions.

A step closer to the situation in real membranes is a film in which several different components are present together. It is found that two lecithins that individually form films of the same type mix in a nearly ideal fashion and that the properties of the mixed films are simply additive. However, if the conditions of the measurement are such that the two components would individually exist in different monolayer phases, the mixture is nonideal or, in the extreme case, the components are immiscible.

Figure 11-8
Spreading pressure versus area curves for 1,2-dipalmitin (A), 1,3-dipalmitin (B), and 1,2-dimyristin (C) at 23°C on solutions of pH = 5.5. D. A. Cadenhead, *J. Chem. Educ.* **49**, 152 (1972).

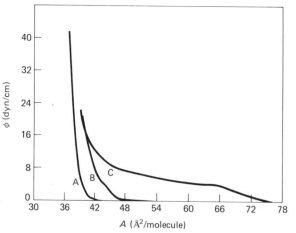

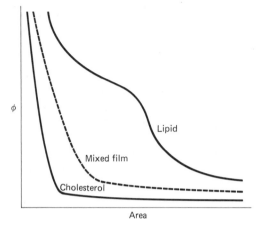

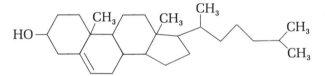

Figure 11-9
Typical surface pressure isotherms of a lipid, of cholesterol, and of a mixed film containing both lipid and cholesterol.

Along with phospholipids, another abundant component of cell membranes is cholesterol, a steroid:

When cholesterol is mixed with a lipid and a monolayer is formed, the resulting film has rather interesting properties. If the lipid by itself forms an expanded film, the cholesterol reduces greatly the area per molecule and thus serves to bind together the molecules of the other component. Typical surface pressure curves are shown in Figure 11-9. If the mixed film is measured, instead, under conditions in which the lipid would when alone form a condensed film, there is little effect on the surface pressure. The rather striking effect, however, is that both categories of mixed films containing cholesterol have an unexpectedly low viscosity, and it has been suggested that the importance of cholesterol in living membranes may be the result of its ability to hold together other molecules without forming a rigid structure.

FILMS OF PROTEINS

Many proteins spread at the water–air interface to form monomolecular films in which the complex native structure, including regions in which the peptide backbone is in a helical conformation, is lost and the molecules are extensively or completely unfolded. As a consequence of this, the protein loses its biological activity, often irreversibly. To illustrate the drastic change in geometry a spread protein may undergo, globular proteins which are normally approximately spherical in shape with diameters of perhaps 45 Å form layers on the surface only 8 to 10 Å thick, a thickness about that required to accommodate polypeptide chains with their axes parallel to the surface. The films behave as if many of the polar groups in the side chains are directed into the aqueous phase and the hydrocarbon side chains are exposed to the air.

It is also observed that proteins that are soluble in water initially tend gradually to diffuse to the surface and become denatured. The protein molecule in solution is folded up into essentially a spherical form, with the hydrocarbon portions principally tucked inside and the outside composed largely of polar groups which can thus be in contact with the solvent. In a sense, the protein loses its solubility when it forms a surface film, since the pull of the polar groups downward into the water prevents the nonpolar portion from curling back on itself. Diffusion to the surface with accompanying denaturation is aided by agitation of a protein solution and, in handling such solutions, particular care must be taken to avoid surface denaturation. Many proteins are also strongly adsorbed on glass surfaces, from which they can be removed only with difficulty, so that this is another potential source of loss of material in handling solutions.

Protein films under most conditions have rather low spreading pressures. They do not undergo very distinct phase changes but are transformed gradually as the pressure is changed. On compression, the films have been described as slowly becoming rigid or gelling. Spreading on a subphase of dilute acid or concentrated ammonium sulfate makes the films gaseous at higher concentrations than when they are spread on water.

It has been found that the nature of protein films can be modified by the conditions of spreading. Two standard procedures for obtaining isotherms have been developed:

(1)   If a fixed amount of protein is placed on a large area of surface and measurements taken as the area is gradually reduced, the resulting isotherm is termed a ϕ-A (or Π-A) isotherm. It is under these conditions that the protein is most likely to be fully spread.

(2)   If increments of protein are added to a fixed area of liquid surface, as by adding successive aliquots of a solution, a ϕ-C (or Π-C) isotherm is obtained. In this experiment, much of the protein is forced to spread against the pressure of the molecules already present on the surface and the protein is less likely to be unfolded.

When bovine serum albumin and β-casein were spread on a phosphate buffer of pH 7 at 25°C by the first method, both gave films that were essentially completely unfolded, judging by the dimensions (M. T. A. Evans, J. Mitchell, P. R. Mussellwhite, and L. Irons, in *Surface Chemistry of Biological Systems,* Martin Blank, Ed., Plenum Press, New York, 1970). When bovine serum albumin was spread by method 2, the spreading pressure observed on the ϕ-C isotherm was smaller than that of the ϕ-A isotherm obtained by method 1, indicating incomplete unfolding.

The behavior of gaseous films of a protein has been used as a means of determining the molecular weight of the protein. The films are found to obey, when they are dilute enough to be gaseous, the equation

$$\phi(A - nA_2) = nRT \tag{11-12}$$

The total area of the film is represented by the quantity A, and of this the area A_2 is that occupied by the molecules themselves, as distin-

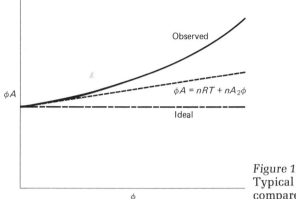

Figure 11-10
Typical behavior of a gaseous monolayer compared with two equations of state.

guished from the empty space between them. The equation can be rewritten as

$$\phi A = nRT + nA_2\phi \tag{11-13}$$

It is apparent that a plot of ϕA against ϕ should be linear with an intercept at zero pressure of nRT. A typical example of such a plot is shown in Figure 11-10. At 300 K, the numerical value of ϕA per mole at zero pressure is 411 if ϕ is expressed in dynes per centimeter and A is expressed as square angstroms per molecule. Another set of units is used in the following illustration.

Example: A solution containing 1 mg of protein is spread on a 5 percent solution of ammonium sulfate at 25°C. A plot of ϕA against ϕ is made, with the area expressed in square centimeters and the "pressure" in dynes per centimeter. The intercept at $\phi = 0$ is found to be 6.0×10^2. Calculate the molecular weight of the protein.

Solution: The number of moles of protein is 0.001 g divided by M, the molecular weight. Then

$$6.0 \times 10^2 = \frac{0.001}{M}RT$$

or

$$M = \frac{(0.001)(8.31 \times 10^7)(298)}{6.0 \times 10^2} = 41,300$$

SOME PRACTICAL APPLICATIONS OF MONOLAYERS

In this section, two applications of monolayer films are mentioned. The first is the buildup of multilayer films on solids by the transfer of monolayers from a liquid surface. Barium stearate, for instance, is formed by spreading stearic acid on dilute barium hydroxide or chloride. The film is then kept under pressure by a piston oil, such as castor oil, which spreads at a very nearly constant pressure; the piston oil is separated from the stearate film by a waxed silk thread. When a metal or glass slide is dipped through the film and withdrawn, the slide carries with it one layer of barium stearate molecules. Additional layers can be deposited on top of the first layer by repeatedly inserting and withdrawing the slide.

The deposition and orientation of the layers depends upon the film material and the slide material. If the slide is clean and the film is something like barium stearate or calcium palmitate, no layer is deposited on the first excursion down; a layer with hydrocarbon tails extended outward and polar heads attached to the slide is then deposited on the excursion up, and a second layer, with hydrocarbon tails toward the hydrocarbon tails of the first layer, is deposited on the next trip down. Each successive raising or lowering of the slide adds another layer, and layers alternate in direction of the molecule. As a result, a slide with an odd number of layers has a hydrophobic surface, and one with an even number of layers has a hydrophilic surface.

Multilayers of any desired thickness can be built up by the method described. Coating a glass surface to a thickness of one-fourth the wavelength of light greatly reduces the reflection of light at the surface and this method has been used in optical instruments such as microscopes.

Another, quite different, way in which monolayer films have been used is to retard evaporation from large bodies of water, such as ponds or reservoirs, to conserve water supplies. The fact that a monolayer is so very thin permits a small amount of material to cover a relatively large surface area.

11-4
SOLID ADSORBENTS

The surface of a solid has associated with it a surface energy. Since the surface of the solid is fixed in shape, it is not possible to measure the surface energy by the method suitable for a liquid, that is, measurement of surface tension by deformation of the surface. However, there is no doubt that a solid has a surface tension, and for most solids the value of the surface tension, and of the surface energy, is much greater than that for liquids. Since the solid surface is fixed in shape and concentration, the free surface energy cannot be reduced by changing the shape of the surface, or by molecular orientation, or by a change in concentration. However, the surface can be covered with a layer of *adsorbate*, a layer of gas or liquid molecules held to or adsorbed on the surface. The effect of adsorption is to replace the surface of the solid with two other interfaces and, since the surface energy of the solid is fairly large, this reduces the total surface energy. At the same time, since most solid surfaces are rough and the best adsorbents are those that are especially porous, a solid exposed to a gas phase may have its area reduced upon adsorption as the adsorbate fills in cracks, pores, and other irregularities.

Solids may adsorb materials from either the liquid phase or the gas phase; the effective and practical adsorbents for the two cases are similar or even identical materials, and possess certain characteristics in common. Since adsorption occurs at the surface, the greater the area of surface, the greater is the capacity of the adsorbent. Adsorbents of large specific surface area are either finely divided or very porous. Particles of small diameter possess relatively much more external surface, since the volume is proportional to the cube of the radius while the surface area is proportional to the square of the radius. Thus, for a given mass of material, the area is inversely proportional to the radius of the individual particles. Porous materials contain much internal surface, as much as 500 m²/g; for some charcoals the surface is so large that almost every atom is exposed to a pore.

Porous solids are not extremely stable in the form in which they function as adsorbents; if circumstances permit—for example, if the temperature is raised—the solid tends to aggregate into larger units, or "sinter," and the surface tends to become smooth and inactive. Adsorbents are extremely sensitive to the method of preparation and past history and, indeed, the history of the sample and method of activation may be fully as important as the chemical composition in determining the adsorption capacity.

Some representative adsorbents are activated carbon, silica gel, activated aluminum oxide, platinum black, finely divided nickel, and magnesium oxide. Each must be specially prepared in active form. Adsorbent charcoal can be derived from organic matter by a suitable process of activation. Volatile materials are driven from the original substance, such as wood or bone, by careful heating in the presence of a suitable atmosphere, which may be air, steam, or carbon dioxide. Before heating, the source material may be impregnated with an activating agent, such as a zinc salt. As the volatile materials leave, they produce a very fine-pored, open structure. Along with many other uses, the adsorption of gases in gas masks and the decolorization of sugar solutions in the refining process are significant applications of active charcoal as an adsorbent.

Active carbon can also be produced by incomplete combustion of fuel gas. "Channel black" is this sort of carbon, made by allowing a low-oxygen flame to impinge on metal channels on which a low-density, high-area, sootlike material collects.

Silica gel is prepared by precipitation of silicic acid by the addition of hydrochloric acid to a solution containing sodium silicate, followed by washing of the gelatinous precipitate to remove electrolytes. The precipitate is highly hydrated, and careful removal of most of the water leaves an open structure. The activity of the resulting solid depends upon the temperature and solution concentrations during the precipitation process, as well as upon subsequent treatment of the precipitate.

Aluminum hydroxide is precipitated from a solution of an aluminum salt by addition of a base. The precipitate is highly hydrated and gelatinous; careful drying and heating at 350 to 500°C converts the hydroxide to the active oxide. Again, as for silica, conditions of precipitation are important in determining the physical state and adsorptive properties of the final product. Alumina is commercially available with a variety of pore sizes and with a variety of surface treatments, including impregnation with acid or base, which serve to modify its properties for the adsorption of organic molecules containing various functional groups.

Molecular sieves, or synthetic zeolites, are solids of a rather unusual type. These materials have an arrangement of pores built into the crystal structure, rather than simply having openings and channels which were produced incidentally in the history of a particular sample and thus would tend to disappear on heating, as is true for most other high-area solids. There are some naturally occurring zeolitic minerals, but the advent of their synthetic counterparts has made possible the design of pore structures for special purposes.

Zeolites consist of an aluminum silicate framework covalently bonded together, but with net negative charge, along with some cations, which may be hydrogen, ammonium, or metal ions, to give electrical neutrality. Indeed, they are ion-exchange materials, for one type of cation can readily be replaced by another. The synthetic materials are made by crystallization from an alkaline aqueous medium which is in contact with an aluminosilicate gel. As the zeolite crystals precipitate from the solution, the gel dissolves, maintaining the concentrations in the solution. The precipitate is then filtered off, dried, and activated by heating to from 400 to 800°C. By varying the conditions of precipitation and the cations present, it is possible to control the nature of the crystal structure and the size of the pores. Thus material can be produced with pores so small that no molecule larger than water can enter and be adsorbed, and this product serves as a selective drying agent for liquids. Under other conditions, the pore size obtained can be adjusted to admit water and methanol without allowing ethanol or larger molecules to enter.

Zeolites have been found to be quite active as catalysts for heterogeneous reactions such as are described in Section 11-7, and in many cases to promote selectively reactions that are desirable in the refining of petroleum. The bulk of their use is, however, as adsorptive and drying agents, particularly in the purification of organic compounds.

11-5
ADSORPTION OF GASES ON SOLIDS

Adsorption of a gas on a solid surface can occur in either one of two ways. In one form of adsorption, the gas molecules condense on the solid in much the same manner as they would condense to form a liquid in the absence of the solid; this is called *physical adsorption* or van der Waals adsorption, for the forces leading to it are similar to the van der Waals forces by which gas molecules attract one another. In contrast to this is *chemical adsorption*—chemisorption, for short—in which valence bonds in the molecule can be broken and new bonds to the solid formed. Chemisorbed molecules are held very strongly, and the enthalpy evolved in the adsorption process is from 25 to 75 kcal/mol of adsorbate taken up, as compared with an enthalpy loss of perhaps 3 to 10 kcal/mol of adsorbed gas for physical adsorption, an amount quite like the heat evolved when the gas condenses to the liquid phase.

Physical adsorption is a generally occurring phenomenon: If the temperature is low enough, any gas will be adsorbed on any solid surface. Relative amounts of different gases physically adsorbed under the same conditions on the same solid are roughly parallel to the ease of condensation of the gases, as judged by the normal boiling point or the critical temperature. Chemical adsorption, in contrast, is quite specific to the particular combination of gas and solid, just as the possibility of a chemical reaction is specific to the nature of the reactants.

The rate at which physical adsorption occurs is usually quite rapid. Furthermore, it is usually readily reversible, and equilibrium can be attained easily from either the adsorption or desorption direction. Chemical adsorption is rapid in a few instances, but it is much more commonly a slow process, requiring an activation energy similar to that for an ordinary chemical reaction. Indeed, it is sometimes termed *activated* adsorption. At low temperatures, it may occur too slowly ever to become measurable.

The behavior of oxygen on charcoal illustrates the differences between these two types of adsorption. If the gas is admitted to the adsorbent at 0°C and the two phases are then allowed to come to equilibrium, much of the oxygen can later be removed from the charcoal by evacuation at the same temperature with a vacuum pump. However, a portion of the oxygen cannot be recovered at this temperature, no matter how long the sample is pumped or how low the gas pressure is reduced. If the temperature is increased, gas is evolved by the solid, but it is now a mixture containing some carbon monoxide and possibly some carbon dioxide in addition to part of the original oxygen. It can therefore be concluded that most of the oxygen is physically adsorbed and that this portion is easily removed by simply reducing the pressure above the solid, but that some of the

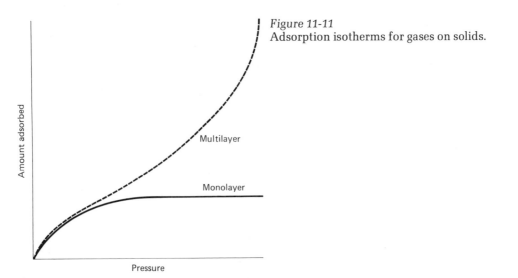

Figure 11-11
Adsorption isotherms for gases on solids.

oxygen undergoes a chemical reaction with the solid, becoming at-tached by some type of valence bond to carbon atoms in the solid.

The amount of gas adsorbed in a given gas-adsorbent system at a given temperature depends upon the pressure of the gas and always increases with increasing pressure. A relation between the amount adsorbed and the pressure of adsorbate in the gas phase, as shown in Figure 11-11, is known as an adsorption isotherm.

Because chemical adsorption requires the formation of valence bonds with atoms in the surface of the solid, it is limited to the forma-tion of a single layer of molecules completely covering the available surface. Complete coverage corresponds in some instances to the situation in which all available physical space is occupied, and there is no room for additional molecules in the first layer of adsorbate; and in other instances it corresponds to the condition in which each available position on the surface that can form the required bond is occupied, although there is still some vacant physical space between the adsorbate molecules, a condition that is only possible when the adsorbate molecules are rather small.

An isotherm equation for monolayer adsorption was derived by Langmuir. He proposed that adsorption equilibrium be described as the situation when the rate at which molecules are desorbed from the surface is equal to the rate at which molecules striking the surface become bound to it, and he assumed that molecules striking any part of the surface already occupied by adsorbate simply bounce off. At equilibrium a certain fraction θ of the total number of sites is occupied by adsorbed molecules, and a fraction $(1 - \theta)$ of the sites is vacant. The rate of adsorption is proportional to the *rate* at which the gas strikes the surface—which is in turn proportional to the gas pressure —to the fraction of the surface sites that are *vacant*, and to the chance that a molecule striking a vacant site will *stick* to that site.

Combining all the constants of proportionality into one constant, we can write

$$\text{rate of adsorption} = k_a(1 - \theta)P \tag{11-14}$$

The rate of evaporation from the surface, or the rate of desorption, is proportional to the fractional number of sites occupied:

$$\text{rate of desorption} = k_d\theta \tag{11-15}$$

When the equation obtained by equating these two rates is solved for θ, there is obtained one form of the Langmuir isotherm:

$$\theta = \frac{k_a P}{k_d + k_a P} \tag{11-16}$$

The ratio of the constants k_a/k_d can be thought of as the equilibrium constant K for distribution of the adsorbate between the gas space and the surface of the solid.

If the numerator and the denominator of the Langmuir equation are each divided by k_d, there is obtained

$$\theta = \frac{k_a P/k_d}{1 + k_a P/k_d} = \frac{KP}{1 + KP} \tag{11-17}$$

Since the value of θ cannot be directly measured, the amount adsorbed is expressed as the fraction x/x_m, where x is the amount adsorbed at any given pressure and x_m is the amount in the same units required to saturate completely the sample of adsorbent. Equation (11-17) then becomes

$$x = \frac{x_m KP}{1 + KP} \qquad (11\text{-}18)$$

To obtain a function linear in P, so that data treatment can be simplified, we divide both sides by P and then invert both sides:

$$\frac{P}{x} = \frac{1 + KP}{x_m K} = \frac{1}{x_m K} + \frac{P}{x_m} \qquad (11\text{-}19)$$

According to this equation, a plot of P/x against P should yield a straight line with a slope of $1/x_m$ and an intercept of $1/x_m K$. From these two numerical values, the constants x_m and K can be calculated.

It is interesting and significant to observe that the form of this equation, as well as the behavior of the amount adsorbed at the extremes of pressure, parallels that of the Michaelis equation for the kinetics of an enzyme-catalyzed reaction discussed in Chapter 10. In that case, first-order kinetics at low concentrations turned into zero-order kinetics as the enzyme became saturated.

Since the Langmuir isotherm is limited to monomolecular adsorption, it is especially suited to describe chemisorption. However, its derivation included an implicit assumption that the equilibrium constant for adsorption at a particular location on the surface is the same as that for any other site, and thus is independent of the amount of surface already covered. This assumption may fail for two reasons: The surface is likely to be nonuniform, having different adsorption enthalpies at different places, and, second, there are undoubtedly forces of attraction or repulsion between neighboring molecules on the surface, which cause the equilibrium constant to depend upon the number of sites adjacent to the one in question that are already occupied by adsorbate molecules. Despite these limitations, the equation is found to be experimentally applicable to many systems over considerable ranges of pressure.

When adsorption occurs by a physical process rather than as chemisorption, it is no longer limited to a single layer, but multilayer adsorption can occur as well. S. Brunauer, P. H. Emmett, and E. Teller extended the Langmuir model to describe a series of equilibria for a series of layers of molecules on a surface, so that the rate of desorption from any one of the layers is equated to the rate of condensation upon the layer beneath. They assumed that K for adsorption has one value for the first layer, and that it has a different but uniform value for all succeeding layers. Multilayer adsorption is not characterized by saturation as an upper limit, but rather by a very rapid rise in amount adsorbed as the condensation pressure of the adsorbate gas is approached. This pressure, often denoted by P_0, is the pressure at which the gas would condense to a liquid in the absence of any adsorbent. A typical multilayer isotherm was shown in Figure 11-11 along with an isotherm of the Langmuir type.

The equation for a multilayer isotherm resulting from the Brunauer–Emmett–Teller (BET) approach is

$$\frac{P}{x(P_0 - P)} = \frac{1}{x_m c} + \frac{(c-1)P}{x_m c P_0} \tag{11-20}$$

The constant c depends on the difference between the equilibrium constant for adsorption in the first layer and that for higher layers, and the quantity x_m is the amount of adsorbate corresponding to a single layer covering the entire surface.

The data to which the BET isotherm is most commonly applied is that for the physical, nonspecific adsorption of nitrogen on a solid near the boiling point of liquid nitrogen. Further, nitrogen is small enough in molecular dimensions to penetrate into fine pores. The value of x_m, then, corresponds to the amount of nitrogen just sufficient to cover the entire surface of the solid, including all the walls of pores and cracks, with a monolayer. Of course, the point of using this elaborate isotherm equation is that physically there is never obtained a situation in which this layer exists alone; rather some parts of the solid are covered by two or three layers before the first layer is completed. However, if the isotherm equation is valid, its application permits the calculation of x_m. Experience has indicated that each nitrogen molecule occupies about 16.2 Å^2 on the surface, so that knowledge of the amount required to form a monolayer permits the true surface area of the solid to be calculated. This technique has had particular application in evaluation of the surface areas of solid adsorbents for commercial use, of solid catalysts, of pigments, and of other finely divided solids.

The BET isotherm has also been applied to interpretation of the adsorption of vapors on solid proteins. Figure 11-12 shows an isotherm of water vapor on bovine serum albumin. It is interesting that

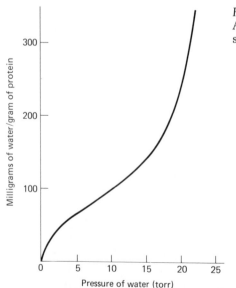

Figure 11-12
Adsorption isotherm of water vapor on serum albumin.

calculation of the apparent area from a water vapor isotherm gives a value for the "area" of the protein several times larger than that obtained by the use of nitrogen in the usual way. The explanation for this is that the water molecules bind to the protein in such a way that they can penetrate *within* the solid protein molecule as a result of their ability to form hydrogen bonds. Since the nitrogen molecules cannot do this, they give a better measure of the "surface area" of the protein, to the extent that this term has a valid meaning, whereas the x_m value from the water isotherm corresponds to the condition in which the polar groups in the protein that are able to form hydrogen bonds are saturated.

**11-6
ADSORPTION FROM
SOLUTION**

Adsorption on a solid surface from a solution is a fairly specific process, depending in rate and amount upon the nature of the solute and the adsorbent. It is also a selective or competitive process, in which a given solute is adsorbed in competition with other solutes and with the solvent.

More complex molecules tend to be adsorbed in preference to less complex molecules. This is the basis for the use of charcoal in removing colored material in the purification of organic compounds, since usually the colored impurities are of higher molecular weight. Examples of the use of adsorption for purification are the concentration of alkaloids by adsorption on aluminum silicate followed by the washing away of impurities and regeneration with dilute alkali, and the use of charcoal as an adsorbent to concentrate penicillin from the dilute solutions in which it is made.

Since the solvent can be adsorbed as well as the solutes, it is possible for "negative adsorption" to occur, with the solvent present at the solid–solution interface in a concentration greater than in the bulk of the solution so that the solution is more concentrated in solutes than if adsorption had not taken place.

The amount of material adsorbed is decreased somewhat by increasing the temperature, but the effects of temperature and pressure are usually small. The nature of the solid adsorbent is important. A general classification can be made of adsorbents into polar and nonpolar, although this classification is only approximate and relative. More highly polar adsorbents include quartz, magnesium oxide, titanium dioxide, silica gel, and aluminum oxide; these generally bind polar molecules or ions more tightly than do nonpolar adsorbents such as sucrose, most charcoals, and graphite. However, properties of adsorbents can be considerably modified by the method of activation, or treatment before use.

When the order of adsorption from different solvents is compared, it is found that, the greater the solubility of a material in the solvent, the smaller the extent of its adsorption. Thus, for organic substances adsorbed on charcoal, the order of decreasing adsorbability is from water, ethyl alcohol, pyridine, benzene, acetone, ether, carbon disulfide, and petroleum ether. The situation, in fact, is much like that for the distribution of a solute between two immiscible liquids; here we can consider the solute as distributed between the solvent and the adsorbed layer according to its relative affinity for the solvent and for the adsorbent.

If one places a solid adsorbent material in a vertical column constricted at the bottom and then passes down through the column a solution containing a mixture of materials adsorbed to different extents by the solid, the components of the mixture are found to move down the column at a rate that varies inversely with the strength of adsorption: The material held most weakly can be found in a band farthest down the column. This separation is the result of a series of successive transfers from the liquid phase to the adsorbed layer and back again. If more of the same solvent or some other solvent of increased polarity is then poured down the column, the bands of adsorbate continue to move downward and can be completely washed through the column or *eluted*, leaving in succession from the bottom of the column. Since this method was first employed by the Russian botanist M. Tswett to separate the colored components of leaves, it was given the name *chromatography*. Other separation procedures in which one phase moves relative to another and variations of the distribution coefficients of components between phases are utilized to separate the constituents of a mixture, have come to be included in the term chromatographic separations. Some of these procedures involving selective distribution between immiscible liquids or between a liquid and a gas were mentioned in Chapter 2.

One version of adsorption chromatography that has been extensively used for analysis or separation is *thin layer chromatography*. A slurry of an adsorbent material such as silica gel is spread in a uniform layer from 0.1 to 2.0 mm thick on a glass plate or a polyester film and then dried. The solution to be analyzed is applied to the surface near one end; this is followed by a "developing" or eluting solvent which flows through the adsorbent, carrying the several solutes various distances from the starting point. The solutes, if colorless, are located by treating them with a color-developing reagent or possibly by examining the adsorbent for fluorescence or phosphorescence under ultraviolet light. Sections of the adsorbent can be removed separately, and the solutes eluted.

Compared to gas–liquid partition chromatography and thin-layer chromatography, column adsorption chromatography is time-consuming and yields relatively broad bands of the materials to be separated. Consequently, it was not used extensively during the 1960's. However, the recent development of *high-pressure liquid chromatography* has overcome these disadvantages. By use of a mechanical pump, a solution is forced at high pressure through a metal column packed with adsorbent. The resulting high flow rate substantially decreases the time required, and also improves the sharpness of the separation, for there is less time for diffusion to broaden the advancing front of a material.

In addition to its significance as a separation method, adsorption is also important because of its role in many chemical and biochemical phenomena, such as precipitation of ionic materials, the process of dying fabrics, and the stabilization of colloidal suspensions and emulsions, to be discussed in Chapter 12.

11-7
HETEROGENEOUS
REACTIONS AT
SOLID SURFACES

Many chemical reactions of considerable importance involving gas or liquid phase reactants occur at the surface of solids that act as catalysts to promote the reactions. Thus the hydrogenation of unsaturated hydrocarbons or fats can be carried out, either in the laboratory or on a commercial scale, with the aid of a platinum or nickel catalyst. Petroleum fractions are converted into economically more desirable compounds by cracking, reforming, and isomerization reactions carried out over solid catalysts, which may be aluminosilicates, zeolites, or platinum metals. Synthetic ammonia is produced from nitrogen and hydrogen over solid catalysts containing iron and aluminum oxides, together with small amounts of other materials, called promoters, which increase the catalytic activity of the principal components. In recent years, there has been intensive work on the development of catalytic converters to reduce the amount of unburned material discharged into the atmosphere by internal combustion engines in automobiles.

KINETICS OF CATALYZED REACTIONS

The course of a reaction occurring at a solid surface involves a series of events, beginning with the diffusion of molecules in the gas or liquid reactant toward the surface, followed by chemisorption on the surface. The reaction of the chemisorbed molecule may then occur as a unimolecular process while the molecule is held on the surface, or it may be a bimolecular process involving either another molecule on the surface or a molecule from the gas phase which happens to strike the adsorbed molecule. Following the reaction, the products must then desorb and diffuse away from the surface in order to allow the reaction to proceed. A typical energy profile is shown in Figure 11-13.

Under most circumstances, the diffusion processes are rapid enough so that they are not the rate-limiting steps. The slow steps may be the adsorption process, the surface reaction, or the desorption process, but most commonly the reaction on the surface is the slow step. It is then possible to assume equilibrium in the adsorption process, and the rate of the reaction is set equal to a rate constant, involving the usual exponential activation energy factor, multiplied by the concentration of the reactant on the surface.

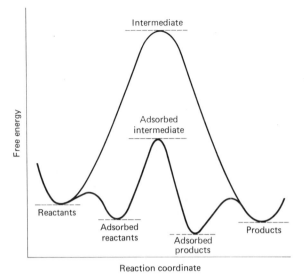

Figure 11-13
Free energy profile for a typical gaseous reaction catalyzed by a solid (dark line)
and the same reaction when uncatalyzed (light line). The apparent overall free
energy of activation for the catalyzed reaction includes positive and negative
contributions from various adsorption and desorption steps, as well as from the
reaction itself.

To express the concentration of the reactant on the surface, the
Langmuir isotherm is often employed. If the reactant is fairly weakly
adsorbed or if its concentration in the gas phase is low, then the concentration on the surface is directly proportional to the concentration
in the gas phase, and the reaction is first-order in the partial pressure of
the reactant in the gas. If the reactant is strongly adsorbed, then the
reaction may appear to be zero-order in the reactant pressure, since
the concentration on the surface is independent of this pressure.

Sometimes a molecule is chemisorbed with dissociation into atoms.
Suppose, for instance, that hydrogen is adsorbed according to the
equation

$$H_2(gas) \rightleftharpoons 2H \cdot (surface) \tag{11-21}$$

The equilibrium constant for the adsorption–desorption equilibrium
is

$$K = \frac{[H \cdot]^2}{[H_2]} \tag{11-22}$$

From this equation, it is found that the concentration of hydrogen
atoms on the surface is proportional to the *square root* of the concentration of molecular hydrogen in the gas phase. Provided adsorption
is reasonably weak, so that the concentrations correspond to the
low-pressure portion of the isotherm, the kinetics of a reaction in
which hydrogen participates may well be half-order in the pressure
of hydrogen.

Another situation frequently encountered is that in which two
kinds of reactant molecules are both adsorbed on the same sites on

the surface. If the adsorbate molecules are labeled A and B, the Langmuir isotherms for this situation are

$$\theta_A = \frac{K_A P_A}{1 + K_A P_A + K_B P_B} \qquad (11\text{-}23)$$

$$\theta_B = \frac{K_B P_B}{1 + K_A P_A + K_B P_B} \qquad (11\text{-}24)$$

If the rate of reaction is proportional to the surface concentration, the rate equation is then

$$\text{rate} = \frac{k K_A K_B P_A P_B}{(1 + K_A P_A + K_B P_B)^2} \qquad (11\text{-}25)$$

The mathematical form of this equation is such that the rate is a maximum for some intermediate value of the composition ratio P_A/P_B and falls off when P_A/P_B is either very large or very small.

An example will illustrate the application of these relations. The hydrogenation of ethylene over a copper catalyst is found experimentally to obey the equation

$$\text{rate} = \frac{k P_{H_2} P_{C_2H_4}}{(1 + K P_{C_2H_4})^2} \qquad (11\text{-}26)$$

From the fact that the denominator is squared, it is deduced that reaction occurs on the surface between adsorbed ethylene and adsorbed hydrogen. However, the absence of a term in the denominator for the concentration of hydrogen indicates that ethylene is much more strongly adsorbed than hydrogen.

SOME EXAMPLES OF CHEMISORPTION

In recent years, the application of several instrumental techniques, particularly infrared spectroscopy, has made it possible to say with a bit more assurance than was possible in the past what kinds of bonds hold chemisorbed molecules to a solid. It is possible that these adsorbed species are intermediates in the path of catalyzed reactions.

In Figure 11-14 are shown some of the ways in which acetylene has been found to be adsorbed. When the adsorbent is made by depositing a thin layer of metallic nickel on a silica support, there is observed an infrared absorption spectrum characteristic of the C—H stretching vibrations of the ethyl group, indicating that some of the acetylene has been converted to $CH_3CH_2\cdot$ units. Where has the acetylene obtained the extra hydrogen atoms required in this reaction? Apparently from other acetylene molecules which have been dehydrogenated, for, when the system is treated with hydrogen gas, the spectrum of the ethyl groups grows in intensity, an effect that can only be explained as the result of the reaction of hydrogen with the skeletons of dehydrogenated acetylene molecules left on the surface. This surface "carbide" has not been directly detected, for no spectral absorptions have been found that can be specifically assigned to it.

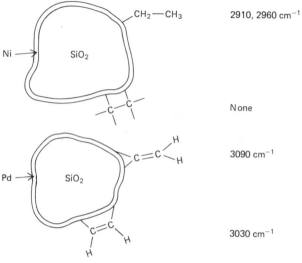

Figure 11-14
Several ways in which acetylene is chemisorbed on metals supported by silica, and the corresponding infrared absorption frequencies for the C—H stretching vibration.

If acetylene is admitted to the presence of an adsorbent made by depositing metallic palladium on silica, no spectrum of saturated hydrocarbon appears. Instead the vibrational absorptions observed correspond to those assigned to olefinic hydrogens. Two different frequencies are found, and the two structures shown in the figure have been proposed to account for them.

EXERCISES

1. A solution containing 10.0 mg of a protein is spread on ammonium sulfate solution at 25°C. The ϕA intercept at zero pressure is 1120 dyn cm². Calculate the molecular weight of the protein.

2. An adsorbent consists of spherical particles of 1 μm radius. If the density of the solid is 2.4 g/cm³, what is the surface area of the material in square meters per gram?

3. If the particles in Exercise 2 are stacked together in straight lines in all three dimensions—in a cubic array—what is the apparent or bulk density of the solid? If you can devise a more compact mode of packing the spheres, describe it.

4. A surface film containing 0.058 mg of n-hexadecylamine spreads under a pressure of 20 dyn/cm to an area of 600 cm². Calculate the apparent cross section of the amine molecules in the film.

5. The surface tension of a solution containing 5 percent by weight of acetone in water at 25°C is 55.5 dyn/cm. The surface tension of pure water at this temperature is 72.0 dyn/cm. Estimate the amount of acetone adsorbed per square centimeter of the surface of the solution.

6. A 0.025 M aqueous solution of a surface-active material is found to have a surface excess of solute of 1.85×10^{-9} mol/cm² at 25°C. Estimate the surface tension of the solution.

7. Calculate the length of a molecule of dodecyl alcohol from the fact that an amount of 6.22×10^{-5} g spread on water in a close-packed monomolecular layer occupies an area of 500 cm². The density

of the alcohol is 0.831 g/cm³. Describe how you would conveniently introduce a known quantity of this magnitude of the alcohol onto the surface of a Langmuir surface-pressure apparatus.

8. When a protein molecule from the interior of a solution moves to the surface and is denatured, would each of the quantities ΔH and ΔS be expected to be positive or negative? Why?

9. Estimate the temperature coefficient of the surface tension of water at 25°C from the data given in this chapter.

10. The surface tension of solutions of butyric acid at 15°C is 49.0 dyn/cm at 0.25 M and 45.5 dyn/cm at 0.35 M. Calculate the excess surface concentration of butyric acid in a 0.30 M aqueous solution according to the Gibbs equation.

11. The following data are obtained for the adsorption of oxalic acid on a sample of charcoal. Plot the results according to the Langmuir equation, using concentration instead of pressure, and determine the constants a and K.

Concentration of acid in solution at equilibrium (M)	Moles acid adsorbed per gram of charcoal
0.434	0.153
0.208	0.126
0.0638	0.087
0.0209	0.069
0.00736	0.053

12. From Equation (11-20) devise a suitable plot on which data obeying the BET relation should be linear, and plot the following results for the adsorption of nitrogen on 0.548 g of aluminum oxide at 77 K. Using the value of 16.2 Å² for the area occupied by one molecule of nitrogen, calculate the surface area per gram of the solid.

Pressure (atm)	N₂ adsorbed (mmol)
0.1063	0.964
0.1382	1.038
0.1699	1.109
0.2022	1.175
0.2238	1.222

13. Draw an analogy between the kinetic behavior of a reaction occurring at the surface of a solid heterogeneous catalyst and one catalyzed by an enzyme.

14. In choosing a material to form a monolayer film on a lake or reservoir to retard surface evaporation, certain practical considerations enter. List some of the properties required of the film former to make such an arrangement feasible.

15. Show how the relation of amount adsorbed to pressure (or concentration) behaves in the limits of high pressure and of low pressure when the Langmuir equation applies.

16. The exchange of deuterium with ammonia over iron is found to follow the equation

$$\text{rate} = \frac{k[D_2]^{1/2}[NH_3]}{(1 + K[NH_3])^2}$$

What deductions about the course of the reaction can be drawn from this equation? Why should this reaction be of interest to chemists engaged in the practical application of catalysts?

17. Show that the adsorption of a gas A in the presence of gas B which also can be adsorbed on the same sites on the surface is given by the following equation, if each gas follows the Langmuir model of adsorption:

$$\theta_A = \frac{K_A p_A}{1 + K_A p_A + K_B p_B}$$

18. Describe how one could determine the heat of hydration of a dry protein by water vapor from adsorption measurements without the use of a calorimeter.

REFERENCES

Books

Arthur W. Adamson, *Physical Chemistry of Surfaces,* 3rd ed., Wiley, New York, 1976. A comprehensive treatment at the intermediate to advanced level.

Robert E. Baier, Ed., *Applied Chemistry at Protein Interfaces,* American Chemical Society, Washington, D.C., 1975. Surfaces of proteins and the behavior of proteins at surfaces.

G. C. Bond, *Catalysis by Metals,* Academic Press, New York, 1962. An excellent account of heterogeneous catalysis and chemisorption by metals.

Henry B. Bull, *An Introduction to Physical Biochemistry,* Davis, Philadelphia, 1964. Chapter 9 emphasizes liquid surfaces.

Farrington Daniels and Robert A. Alberty, *Physical Chemistry,* 4th ed., Wiley, New York, 1975. Chapter 8 describes the thermodynamics of surfaces and adsorption.

J. T. Davies and Sir Eric Rideal, *Interfacial Phenomena,* 2nd ed. Academic Press, New York, 1963. An advanced treatment of surfaces and interfaces in liquid systems.

G. L. Gaines, *Insoluble Monolayers at Liquid-Gas Interfaces,* Wiley, New York, 1966. Very readable, comprehensive account.

Norman L. Gershfeld, "Physical Chemistry of Lipid Films at Fluid Interfaces," in *Annual Review of Physical Chemistry,* Vol. 27, Annual Reviews, Palo Alto, Calif., 1976.

E. D. Goddard, Ed., *Monolayers,* American Chemical Society, Washington, D.C., 1975. Collection of reports on various monolayer systems and methods of studying them.

S. J. Gregg and K. S. W. Sing, *Adsorption, Surface Area and Porosity,* Academic Press, New York, 1967. Properties of solid adsorbents.

Barry L. Karger, Lloyd R. Snyder, and Csaba Horvath, *An Introduction to Separation Science,* Wiley, New York, 1973. An excellent intermediate-level treatment, including chromatography.

Hans Kuhn, Dietmar Möbius, and Hermann Bücher, "Spectroscopy of Monolayer Assemblies," in *Techniques of Chemistry,* A. Weissberger and B. W. Rossiter, Eds., Vol. 1, Part IIIB, Chapter VII, Wiley-Interscience, New York, 1972.

L. H. Little, *Infrared Spectra of Adsorbed Species,* Academic Press, New York, 1966.

Walter J. Moore, *Physical Chemistry,* 4th ed., Prentice-Hall, Englewood Cliffs, N.J., 1972. Chapter 11 is an advanced treatment of interfaces and adsorption.

S. Ross and J. P. Oliver, *On Physical Adsorption,* Wiley, New York, 1964. The phenomena and theories of physical adsorption.

G. A. Somorjai, *Principles of Surface Chemistry,* Prentice-Hall, Englewood Cliffs, N.J., 1972.

W. J. Thomas and J. M. Thomas, *Introduction to the Principles of Heterogeneous Catalysis,* Academic Press, New York, 1967. An excellent, intermediate-level summary.

Journal Articles

Frederick J. Almgren, Jr., and Jean E. Taylor, "The Geometry of Soap Films and Soap Bubbles," *Sci. Am.* **235**, 82 (July 1976).

Jeffrey C. Buchholz and Gabor A. Somorjai, "The Structure of Adsorbed Gas Monolayers," *Acc. Chem. Res.* **9**, 333 (1976).

D. A. Cadenhead, "Film Balance Studies of Membrane Lipids and Related Molecules," *J. Chem. Educ.* **49**, 152 (1972).

J. G. Dash, "Two-Dimensional Matter," *Sci. Am.* **228**, 30 (May 1973).

K. H. Drexhage, "Monomolecular Layers and Light," *Sci. Am.* **222**, 108 (March 1970).

J. C. Giddings, "The Conceptual Basis of Field-Flow Fractionation," *J. Chem. Educ.* **50**, 667 (1973).

Vladimir Haensel and Robert L. Burwell, "Catalysis," *Sci. Am.* **225**, 46 (December 1971).

Paul C. Hiemenz, "The Role of van der Waals Forces in Surfaces and Colloid Chemistry," *J. Chem. Educ.* **49**, 164 (1972).

Barry L. Karger, "Resolution in Linear Elution Chromatography," *J. Chem. Educ.* **43**, 47 (1966).

L. S. Lobo and C. A. Bernardo, "Adsorption Isotherms and Surface Reaction Kinetics," *J. Chem. Educ.* **51**, 723 (1974).

Peter F. Lott and Robert J. Hurtubise, "Instrumentation for Thin-Layer Chromatography," *J. Chem. Educ.* **48**, A437 (1971).

W. J. Popiel, "Adsorption by Solids from Binary Solutions," *J. Chem. Educ.* **43**, 415 (1966).

Harold H. Strain and Joseph Sherma, "M. Tswett: Adsorption Analysis and Chromatographic Methods—Application to the Chemistry of the Chlorophylls," *J. Chem. Educ.* **44**, 238 (1967).

Harold H. Strain and Joseph Sherma, "Modifications of Solution Chromatography Illustrated with Chloroplast Pigments," *J. Chem. Educ.* **46,** 476 (1969).

Kenzi Tamaru, "New Catalysts for Old Reactions," *Am. Sci.* **60,** 474 (1972).

H. Veening, "Recent Developments in Instrumentation for Liquid Chromatography," *J. Chem. Educ.* **50,** A481 (1973).

J. A. Wood, "Colloidal Surfactants," *J. Chem. Educ.* **49,** 161 (1972).

Twelve
Macromolecules and Molecular Aggregates

Of great interest and importance, to both the chemist and biochemist, are the properties of molecules in the range of size corresponding roughly to molecular weights between several thousand and several million. These include not only synthetic polymers, such as polystyrene, polytetrafluoroethylene, elastomers, and many others, but also a great variety of high-molecular-weight molecules, such as proteins and nucleic acids, synthesized within living organisms. These molecules differ from smaller ones in many properties, for example, their ability to scatter light, merely by virtue of their size. In addition, investigation of their behavior shows that it turns upon such factors as conformation and shape, extent of solvation, and amount and distribution of electric charges in ways more complicated than does the behavior of "ordinary" molecules.

Before the existence of the covalently bonded macromolecules mentioned above was recognized, chemists developed techniques for studying other particles in the same size range, often called the colloidal range. The colloidal particles recognized at an early stage were mostly those of inorganic substances including such materials as sulfur, phosphorus, metals, metal oxides, and even some salts, substances that are otherwise insoluble. It is possible to produce dispersions, called *sols,* of such materials in an aqueous medium, but these dispersions are metastable at best; the particles have an innate tendency to grow spontaneously and fuse together with other particles until a precipitate is formed. Because the disperse or solute phase has no great affinity for the medium or "solvent," it is termed *lyophobic,* or "solvent-hating." Most inorganic colloids also differ from true macromolecular materials in that the size of the dispersed particles depends to a great extent on the circumstances of the formation of the dispersion since these colloids do not involve a particular pattern of covalent bonds; indeed, most of them are formed by ionic rather than directed covalent bonds.

In contrast to lyophobic materials, many macromolecules disperse spontaneously in the solvent to form sols and are termed *lyophilic,* or "solvent-loving." Most proteins go readily into water, at least if the pH is suitably adjusted, and synthetic organic polymers are dispersed by solvents such as benzene and acetone.

In addition to large organic molecules in which the components are more or less irreversibly bound together, we shall have occasion to consider systems in which smaller organic molecules aggregate reversibly. Some of these components are much like the surface-active molecules we dealt with in Chapter 11, but here we consider especially those that, although they exist as individual molecules in dilute solution, come together into large units at higher concentrations. We also devote particular attention to molecules that, because of their anisotropic geometry or special arrangement of functional groups, array themselves into regular patterns in solution or in the membranes constituting parts of living cells.

12-1
SYNTHETIC AND
NATURAL POLYMERS

Synthetic polymers are formed by chemical reactions in which successive building blocks, or monomer units, are added to a backbone or main chain of atoms held together by covalent bonds. For many monomers, branches in the main chain of atoms may also be produced as well as, in some cases, cross links between chains. The kinetics of one type of reaction by which polymers are formed were described in Chapter 10.

Some polymerization reactions involve *addition* across a double bond in the monomer. This is the type of reaction by which poly(methyl methacrylate) and Teflon are formed:

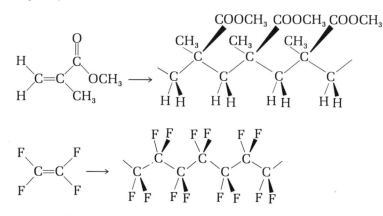

In condensation polymerization, a small molecule, usually water, is eliminated between successive units. Examples are polyamides, such as nylon, formed from $H_2N(CH_2)_p NH_2$ and $HOOC(CH_2)_q COOH$:

$$\left(HN-(CH_2)_p-NH-\underset{O}{\overset{}{C}}-(CH_2)_q-\underset{O}{\overset{}{C}}\right)_n$$

and polypeptides, such as poly(L-alanine):

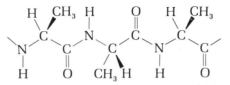

Poly(methyl methacrylate) and polyalanine also provide illustrations of another aspect of importance to the properties of polymers—the presence of asymmetric or chiral centers. For the polypeptide, the asymmetric carbon atom has its configuration determined by the configuration of the amino acid from which the polypeptide was made. In most naturally occurring proteins and peptides, the amino acids have L configuration. For the synthetic polymer, the situation is different in that the bonds to the asymmetric carbon are formed in the course of the polymerization reaction, and thus the configuration of the center depends upon the nature of this reaction. The formula drawn above for the methacrylate polymer places all the ester groups in front of the plane of the paper and all the methyl groups behind the plane. This implies that the configurations of the asymmetric carbons are all the same. Such a molecule is termed *isotactic* and is an example of a stereoregular polymer. It is found that isotactic polymers are much more readily crystallized than random polymers, because the regularity of the chains allows them to pack more closely together in the solid state. Under other reaction conditions, polymers such as methacrylate are obtained with configurations of successive centers alternating regularly from one to the other; these are termed *syndiotactic*.

Synthetic polymer samples usually contain molecules having a variety of numbers of monomer units and therefore a distribution of molecular weights. When various experimental methods, some of which we describe in succeeding sections of this chapter, are applied in an effort to determine the molecular weight of the material, only an *average* value is obtained, and the exact significance of this average differs from method to method. For instance, measurement of a colligative property of a solution of the polymer yields a *number average,* defined by the equation

$$M_n = \frac{\sum_i N_i M_i}{\sum_i N_i} \tag{12-1}$$

The symbol $\sum_i$ represents a summation over species of various sizes each designated by a particular value of the index i. N_i is the number of i molecules and M_i is the molecular weight of the ith molecule. In this sort of average, each particle contributes, regardless of its mass, an equal share to the overall average.

Other measurements, such as those involving the average kinetic energy of a molecule or its ability to scatter light, yield a *weight-average* molecular weight:

$$M_w = \frac{\sum_i W_i M_i}{\sum_i W_i} = \frac{\sum_i N_i M_i^2}{\sum_i N_i M_i} \qquad (12\text{-}2)$$

The weight average is based upon the weight fraction analysis of the sample, rather than on the number fraction which determines M_n.

If it happens that all the molecules in a macromolecular sample have the same molecular weight, the two averages are then equal. The greater the range of molecular weights in a sample, the more the values of the averages deviate from one another; this deviation can be used as a measure of the width of the mass distribution. M_w is always equal to or *larger than* M_n and is of the order of $2M_n$ for typical synthetic polymers.

Naturally occurring proteins are much like polypeptides in structure but, rather than having a single type of amino acid, each consists of definite sequences of a variety of amino acid residues. The peptide units form a chain, terminated at one end by an amino group and at the other end by a carboxyl group, and from this chain there extend out side chains, some entirely hydrocarbon in nature, others including polar functional groups, as determined by the particular amino acids forming the protein. In any given protein, such as RNase, or lactoglobulin, or insulin, there is a fixed sequence of amino acid constituents, and a definite molecular weight, evidenced by the equality of M_n and M_w and by the ability to form crystalline solids. From the study of the structure of these crystals, it has been possible to learn much about the chemical sequence as well as about the geometric arrangement of amino acids in proteins.

Nucleic acids consist of nucleotides joined together in polymeric chains. A nucleotide is a molecule in which a purine or pyrimidine base has attached to it a pentose sugar ring, in which there is in turn substituted a phosphate group; an example is adenosine 5′-monophosphate, in which the base is adenine, as represented in the formula of ATP on page 142. The nucleic acid structure can be regarded as a backbone of alternating sugar groups and phosphate linkages; each sugar ring has as its side chain a purine or pyrimidine ring. In ribonucleic acid (RNA) the sugar is ribose, and in deoxyribonucleic acid (DNA) it is deoxyribose. The other common purine molecule, along with adenine, is guanine:

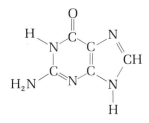

The common pyrimidine bases are thymine and 5-methylcytosine, which are mostly found in DNA, uracil, mostly occurring in RNA, and cytosine, found in both DNA and RNA:

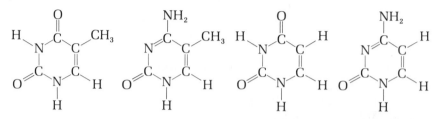

There are several less common components which appear in low concentrations, mostly in what is called *messenger* RNA. Nucleic acids are found in the chromosomes of the cell, and the genetic information borne by the chromosomes is coded in sequences of the nucleotides.

Other natural polymers include rubber, which is polyisoprene, and polysaccharides, such as starch. For the polysaccharide cellulose, the structure is

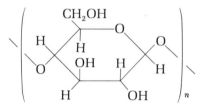

12-2
X-RAY DIFFRACTION
AND POLYMERS

A beam of x rays is diffracted by a regular array of atoms in the same way that a light beam is diffracted by a ruled grating, as described in Chapter 8. Pulling a polymeric fiber tends to align the long axes of the component macromolecules in the direction in which the fiber is stretched, and an x-ray diffraction experiment performed on such a fiber containing oriented molecules may give results that are relatively easy to interpret in terms of repeat distances in the structure. Figure 12-1 shows an arrangement in which an x-ray beam is incident at an angle of 90° on a vertically oriented fiber and produces a pattern of horizontal layer lines on a photographic plate. Figure 12-2 is a cross section of the same arrangement, showing that the angle of maximum intensity, the diffraction angle θ, is related to the repeat distance d along the length of the fiber and to the wavelength λ of the x rays by the equation

$$n\lambda = d \sin \theta \qquad (12\text{-}3)$$

From this equation, in which n is an integer, it is seen that large displacements of layer lines, that is, large values of $\sin \theta$, are associated with small values of the spacing d.

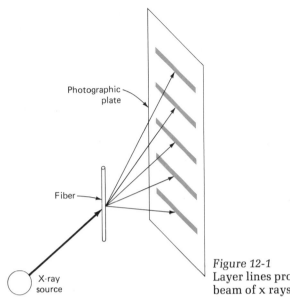

Figure 12-1
Layer lines produced by diffraction of a
beam of x rays by a stretched fiber.

Structure perpendicular to the fiber axis is indicated by division of
the layer lines into spots, the lateral positions of which give information about repeated spacings in a second dimension.

One of the outstanding triumphs of the x-ray diffraction of polymers
was elucidation of the structure of the α helix found in polypeptides
and in some proteins. A characteristic repeat distance d along the fiber
axis of 5.1 Å is observed for this conformation. Linus Pauling showed
that a particular helical structure is consistent with this observation
and with the geometric requirement that all the atoms in the group

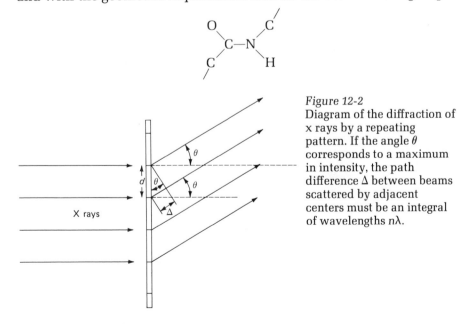

Figure 12-2
Diagram of the diffraction of
x rays by a repeating
pattern. If the angle θ
corresponds to a maximum
in intensity, the path
difference Δ between beams
scattered by adjacent
centers must be an integral
of wavelengths $n\lambda$.

be in a plane, as a consequence of electron delocalization over the N—C—O atomic sequence and sp^2 hybridization of the nitrogen and carbon atoms in this sequence. Furthermore, the α helix is stabilized by hydrogen bonds between the NH in one residue and the carbonyl group three residues farther along the chain. These hydrogen bonds have a nearly linear arrangement for the atoms —N—H---O—C— and have a direction approximately parallel to the axis of the helix. The helix has 3.6 amino acid residues per turn. Based on the usual bond distances and angles, this would require 5.4 Å per turn. The fact that the observed repeat distance is slightly smaller is explained by a gentle spiraling of the entire helix. Figure 12-3 shows one of the two possible helix arrangements, right-handed or left-handed; in most proteins the helix is present with the right-handed sense.

In another conformation, found in some proteins and termed the

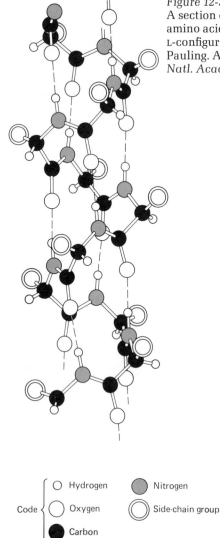

Figure 12-3
A section of right-handed α helix, with amino acid residues in the
L-configuration, as worked out by Pauling. After Pauling and Corey [*Proc. Natl. Acad. Sci.* **37**, 205 (1951)].

Code
- ○ Hydrogen
- ◯ Oxygen
- ● Carbon
- ◉ Nitrogen
- ◎ Side-chain group

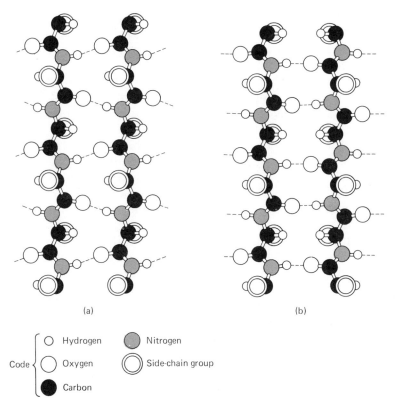

(a)   (b)

Code

○ Hydrogen   ◉ Nitrogen

○ Oxygen   ◎ Side-chain group

● Carbon

Figure 12-4
Diagrams of the parallel (a) and antiparallel (b) pleated sheet arrangement of
polypeptides.

β-chain structure, the protein molecules are more extended than in the
α helix. A series of β chains can lie side by side, with hydrogen bonds
between the NH of one chain and a carbonyl oxygen of the neighboring chain. The resulting arrangement contains a series of alternating
ridges and valleys because of the zig-zag arrangement of the chains and
is described as a *pleated* sheet. The two possible versions of the β
pleated-sheet structure, one with the chains parallel and the other with
alternate chains antiparallel, are shown in Figure 12-4. Side-chain
groups project from the ridges, alternately on one side of the sheet and
the other and, when the sheets are stacked, fit into the valleys of the
adjacent sheets.

In discussing the conformation of a protein backbone, either that in
the solid or that in solution, which we consider later, it is convenient to
describe the chain geometry in terms of standard angles. As indicated
in Figure 12-5, the angle ϕ describes rotation about the N—C$^\alpha$ bond

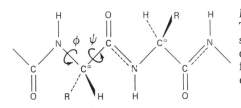

Figure 12-5
The fully extended form of a short
section of a polypeptide or protein
chain, showing the conventions defined
for describing the conformation in terms
of rotational angles.

and is zero when the C^{α}—C bond is trans to the NH and 180° when the two bonds are cis, which is the situation represented in Figure 12-5. The alpha carbon in the amino acid residue is represented by C^{α} and the carbonyl carbon by an unlabeled C. For L acids the value of ϕ is taken to be $+120°$ when C^{α}—R is trans to NH and $-120°$ when C^{α}—H is trans to NH. The angle ψ describes rotation about the C^{α}—C bond axis and is zero when C^{α}—N is trans to the carbonyl group and 180° in the fully extended conformation in Figure 12-5. It is $+120°$ when C^{α}—H is trans to the carbonyl and $-120°$ when the group trans to the carbonyl is C^{α}—R. In the right-handed α helix of L acids, typical values of ϕ and ψ are $-57°$ and $-47°$, respectively, and for the antiparallel pleated sheet, $-139°$ and $+135°$, respectively.

Many protein molecules, including cytochrome, myoglobin, lysozyme, and a variety of others, exist in a relatively compact, globular conformation, which for a given protein is quite specific and reproducible. By applying the methods of x-ray diffraction as they are extended to three dimensions, along with complex mathematical calculations and the skill and experience of crystallographers, it has been possible to locate the positions of individual atoms in the crystals of several of these proteins, as had been done much earlier for simple inorganic compounds, and thus to establish their complete molecular structure. It turns out that the peptide chain in globular proteins typically has α-helical segments alternating with less regular sections, as well as β-chain segments. The relative amounts of the different conformations vary among proteins, with hemoglobin and myoglobin having 70 to 75 percent α helix, insulin and serum albumin about 40 to 50 percent, and RNase only about 15 percent. For some enzyme–substrate systems, x-ray diffraction has been successful in showing in detail how

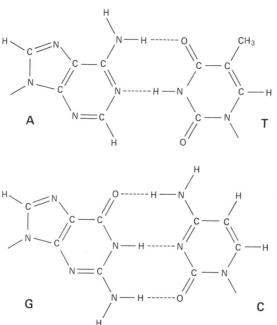

Figure 12-6
Hydrogen-bonding schemes for adenine–thymine and guanosine–cytidine pairs in the helix of nucleic acids. The bonds left free are attached to ribose or deoxyribose units in the dual chains.

substrates, as well as inhibitors, are bound to the active site, and thus in contributing information about the mechanism of the reaction.

Another achievement of x-ray crystallographers was the unraveling of the structure of DNA, a material in which two strands are intertwined in a *double helix,* a pattern proposed by J. D. Watson and F. H. C. Crick. The arrangement is such that the planes of the purine and pyrimidine bases are parallel to one another and perpendicular to the axis of the helix. A base from one chain is linked by two or three hydrogen bonds to a base from the other chain, as shown in Figure 12-6. The geometric requirements of the hydrogen bonds limit the possible base pairs to adenine with thymine, or guanine with cytosine. The repeat distance along the spiral is 34 Å, and the distance between successive base pairs is 3.4 Å, so that there are 10 units in each chain per turn of the spiral.

12-3
CONFORMATION OF MACROMOLECULES IN SOLUTION

Linear polymers in solution can exist in any of a great variety of conformations ranging from, at one extreme, forms in which the backbone is more or less completely surrounded by solvent to, at the other extreme, structures of a compact globular shape in which the molecule is folded back upon itself in such a way as to exclude most of the solvent. Polymer conformation is affected by the nature and magnitude of the forces of attraction or repulsion between various parts of the macromolecule as compared to the interaction of the macromolecule with the solvent. If the solvent–solute attractive forces are large—the solvent is then termed a *good* solvent—or if there are large repulsions between successive segments of the polymer, the solute molecule tends to be in the extended shape; but if the segments or side groups of the polymer attract one another more strongly than they do the solvent—the solvent is *poor*—the solute molecule tends to fold up into the compact conformation.

A polymer is said to exist in a *random coil* when it is in an extended form in which the orientation of successive segments of the chain is random with respect to the orientation of previous segments. In other words, the angle of rotation about each bond that is free to turn is independent of the conformation about other bonds up and down the chain. Since one segment of the chain cannot occupy a region of space already occupied by another segment, there is necessarily always a certain volume *excluded* from the space available to a given segment. For a molecule to be in a random coil, the excluded volume must be counterbalanced by at least a small attractive force between neighboring segments in excess of the solvent–solute attraction.

A convenient way of characterizing the conformation of a linear polymer is in terms of the rms end-to-end distance $\sqrt{\overline{h^2}}$. From statistical considerations, this distance for a random coil is equal to $(ZB^2)^{1/2}$, where Z is the number of rigid links in the chain and B is the length of each link. Chain elongation can also be expressed in terms of a somewhat more basic quantity, the radius of gyration R_G, which can be used for branched as well as straight-chain molecules. The radius of gyration is defined by the relation $MR_G{}^2 = I$, where I is the moment of inertia of a body and is equal to $(\overline{h^2}/6)^{1/2}$ for a random coil; the bar denotes the average value of h^2. For a sphere of radius r, which might be an approximation to a globular protein molecule, R_G is equal to $(3r^2/5)^{1/2}$ and, for a rod of length L, it is equal to $(L^2/12)^{1/2}$.

Polypeptides exist in solution under some conditions in the form of an α helix, but they can undergo what is termed a helix–coil transition, produced by either a change in solvent or an increase in temperature; in this process the helix becomes unraveled, and the molecule assumes a form somewhere near that of a random coil, with no regular conformational pattern from one amino acid residue to the next.

Proteins undergo changes somewhat analogous to the helix–coil transition. Since the environment of proteins in a living system resembles to some extent an aqueous solution, it is instructive to consider the conformation of proteins in an aqueous medium. Many proteins, including enzymes, have a particular three-dimensional structure in which they are able to perform their biological functions. In this arrangement, which is usually a compact one with only a small amount of water in the interior as well as with the hydrophobic side chains turned toward the interior, they are said to be in the native state. The native or compact conformation in solution probably resembles rather closely the arrangement of the same molecule in the crystalline state.

If, by a change in solvent composition, addition of a reagent such as urea or guanidine, or an increase in temperature, the native structure of a protein is destroyed, as evidenced by a change in properties and particularly by loss of biological activity, the protein is said to be denatured. Some processes of denaturation are irreversible, and the activity of the protein is completely lost; others are less drastic and can be reversed by restoring the original conditions. Denaturation undoubtedly resembles the helix–coil transition of polypeptides, because the sections of the denatured molecule are indeed found to have much greater freedom of motion than they do in the native protein. The much greater complexity of the protein structure causes the denaturation process to be more complicated and less readily definable than the helix–coil transition.

In an effort to estimate the helical content of proteins for which the structure has not been established by x-ray analysis, as well as to establish the conformation of protein molecules in solution and to follow the changes occurring during denaturation, several spectroscopic methods sensitive to conformation have been employed. One such method is the study of absorption in the ultraviolet region, where there are observed bands in the range of 250 to 300 nm associated with the π–π^* transitions of the aromatic groups in tryptophan, tyrosine, and phenylalanine, and in the range 190 to 200 nm, and more weakly at 225 nm, resulting from amide group transitions.

The intensity of the amide absorption depends upon conformation and is only about one-half as great for the α helix as for the random coil or β chain. The helix–coil transition of a polypeptide or the denaturation of a protein is thus usually accompanied by *hyperchromism*— a term for increase in absorption intensity—for this band.

The absorption maximum for the aromatic groups shifts slightly to shorter wavelength—only about a nanometer—when a protein is denatured. This effect is attributed to a change from a hydrophobic environment to a hydrophilic environment as the group moves from a position within the molecule that is protected from the solvent to a condition of exposure to solvent, and is parallel to changes in spectra of model compounds produced by a change in solvent. The effects are so small that they are normally determined from *difference spectra* in which two samples of the protein, one native and one to be compared, are placed simultaneously in the two beams of a spectrometer and differences in their absorption are measured directly.

The ability to rotate the plane of polarization of light is another property of macromolecules that is sensitive to conformation. Native proteins have values of specific rotation $[\alpha]$, a quantity defined on page 256, of between -30 and $-80°$. The observed rotation is the sum of that from the chiral centers of the residues, which contribute in the negative sense, and the overall asymmetry of the helix, which contributes in the positive sense. On denaturation, the values increase in magnitude to the range between -80 and $-120°$, comparable to the average value of about $-100°$ for the constituent amino acids.

Optical rotation of proteins is often expressed as the reduced mean residue rotation $[m']$ according to the equation

$$[m'] = \frac{3}{n^2 + 2} \frac{M_0[\alpha]}{100} \tag{12-4}$$

The factor $3/(n^2 + 2)$ corrects the value to that which would be measured in a medium of unit refractive index n, and M_0 is the average molecular weight of an amino acid residue. For some samples, the wavelength dependence of $[m']$ is represented adequately by the simple Drude equation [like Equation (9-27)]

$$[m'] = \frac{a_0\lambda_0^2}{\lambda^2 - \lambda_0^2} \tag{12-5}$$

where a_0 and λ_0 are adjusted to fit the data.

A better representation of experimental data for proteins, especially when the helical content is large, is given by an equation derived by W. Moffitt, which takes into account coupling between the oscillating electric dipoles of the functional groups situated close together along the helix:

$$[m'] = \frac{a_0\lambda_0^2}{\lambda^2 - \lambda_0^2} + \frac{b_0\lambda_0^4}{(\lambda^2 - \lambda_0^2)^2} \tag{12-6}$$

In using the Moffitt equation for polypeptides or proteins, it is usually assumed that a λ_0 value of 212 nm applies. For an α-helical polypeptide, right-handed, and with no optically active R groups, it is found that b_0 has a value of $-630°$, so that an observed value of $-315°$, for example, corresponds to 50 percent helix and 50 percent random coil. Of course, if a left-handed helix or a β-chain structure, each of which gives positive b_0 values, is also present, the interpretation of numerical results becomes less clear-cut.

If we follow the wavelength dependence of optical rotation to wavelengths short enough to reach the vicinity of an absorption band, it is possible to observe the Cotton effect, as described on page 330. An α helix shows a characteristic trough, or negative extremum, at 233 nm, corresponding to the amide $n-\pi^*$ absorption band at 225 nm. At 233 nm, $[m']$ for the α helix is $-12,700°$ compared to 1800° for a protein in the absence of the helix. At still shorter wavelengths, a positive Cotton effect is observed for the $\pi-\pi^*$ transition in the α helix, with a peak at about 198 nm, compared to a weak trough at 204 nm and a peak at 190 nm for the corresponding random coil.

The circular dichroism of polypeptides is also indicative of conformation, and CD measurements have been used to estimate amounts of various conformations in proteins. Thus the characteristic α-helix $\pi-\pi^*$ transition near 200 nm leads, as predicted by Moffitt for coupled oscillators, to two different CD bands, a negative one near 207 nm and a positive one at 190 nm. There is also a quite strong negative band at 222 nm, probably for the $n-\pi^*$ transition. This transition has a high rotational strength, although the absorption band to which it corresponds is moderately weak. In contrast, the random coil has a single strong negative peak near 200 nm for the $\pi-\pi^*$ transition and a weak positive band near 220 nm, whereas the antiparallel β sheet has a strong positive band near 195 nm and a moderately strong negative band near 217 nm.

12-4
FILTRATION
AND PARTICLE SIZE

Filter media and membranes can be prepared in a form in which macromolecules and colloidal particles above a certain size limit are retained while neutral molecules or ions in the ordinary range of molecular weight, including the solvent molecules, are permitted to pass. Such media can be used to characterize roughly the size of a macromolecule by determining the pore size required for it to pass; they can also be employed to carry out separations of molecules of different sizes.

A large range of pore sizes is available. Filter paper has pores of diameter 1 to 5 μm and readily passes most macromolecules. Porous earthenware filters that retain particles of the size of bacteria have openings down to about 100 nm. Viruses were first characterized as "filterable viruses" because it was found that the capacity to produce disease could be passed through filters that retained bacteria. Very fine pores, with diameters of less than 1 nm, can be obtained with films made of regenerated cellulose or hardened gelatin. To provide mechanical support against pressure, films of this kind can be supported against a fine wire screen or deposited in the pores of filter paper or earthenware. In an *ultrafilter,* pressure is applied to force solvent and small solute molecules through such a supported barrier, while leaving behind a more concentrated solution of macromolecules.

Methods used in the preparation of membranes from collodion illustrate the way in which pore size can be controlled. These membranes are formed by spreading a thin layer of a solution of cellulose nitrate in an alcohol–ether solvent upon a support, such as the inside of a test tube, and allowing the solvent to evaporate. The pore size depends upon the alcohol/ether ratio in the solvent, the presence of additives such as glycerol in the solution, and the treatment received by the membrane after it has been cast. Soaking it in an alcohol–water mixture of high alcohol content tends to increase the pore size. Filters with various pore sizes made of cellulose derivatives or of polycarbonate polymers are commercially available under several trade names, such as Millipore and Uni-Pore. Alternatively, natural membranes, such as those prepared from animal bladders, can be used, but their pore size is less readily controllable.

In the process of *dialysis,* a concentration gradient is the driving force for the transfer of smaller ions or molecules through a membrane that retains colloidal particles. As an example, a freshly prepared aqueous colloidal suspension can be placed in a collodion sack and a stream of running water allowed to flow over the outside of the sack until the concentration of the electrolyte within the colloid has been reduced to the desired level. For lyophobic colloids, such treatment is often necessary to keep them from coagulating within a short time because of the relatively high concentration of ions left from the process of preparation. Dialysis is employed in the study of protein solutions to adjust the concentration of electrolytes before electrophoresis experiments, as will be described later.

If electrolytes are to be removed by dialysis, the process is accelerated by applying an electric potential across the membrane. This is conveniently accomplished in a three-chamber apparatus in which the solution is contained in the middle compartment between two membranes. One electrode is placed in each of the end compartments, and anions are drawn out toward the anode and cations move in the opposite direction toward the cathode.

12-5
OSMOTIC PRESSURE

In Section 2-9, osmotic pressure was discussed as one of the colligative properties of solutions. For a solution that behaves ideally, the colligative properties depend merely upon the number of independent particles present in a unit volume of solution, and thus the measurement of osmotic pressure for a solution containing a known weight of solute can be utilized to evaluate the molecular weight of the solute, using the ideal equation

$$\Pi = \frac{RTc}{M} \qquad (12\text{-}7)$$

where c is the number of grams of solute per unit volume of solution. In a nonideal solution system, extrapolation to infinite dilution allows the molecular weight to be estimated, and at the same time deviations from ideality provide additional information about solvent–solute interactions and the structure of the solution.

MOLECULAR-WEIGHT DETERMINATION

For molecules in the molecular-weight range of 10,000 to 500,000, the osmotic pressure was for many years the most satisfactory of the colligative properties to measure because of experimental considerations. In this size range, the molar concentrations of macromolecules easily attainable are inconveniently small for conventional methods of measurement of freezing point depression or boiling point elevation— aside from the hazard of changes in the nature of the solute under these conditions—but correspond to osmotic pressures of centimeters of mercury or decimeters of water or of an organic liquid, small enough for columns of liquid to be handled in the laboratory, yet large enough to measure with some precision. At the same time, it is much easier for the experimenter to obtain membranes that pass the solvent but restrain the macromolecular solute than to find membranes impermeable to small molecules. However, recently developed instrumentation such as the vapor pressure osmometer described in Chapter 2 has made it possible to determine very accurately small changes in vapor pressure, freezing point, or boiling point. One of the difficulties in colligative property measurements of the molecular weight of polymers is that the result is a number average, and a small total weight of impurities of low molecular weight may have a large effect on the result.

OSMOTIC PRESSURE OF NONIDEAL SOLUTIONS

According to Equation (12-7), the ratio Π/c should have a constant value, equal to RT/M, independent of the concentration of the solution. In many real solutions of macromolecules, this is only approximately true, and the experimental results are better characterized by a power series in the concentration, such as

$$\Pi = RT\left(\frac{c}{M} + Bc^2 + Cc^3 + \cdots\right) \qquad (12\text{-}8)$$

This is termed a *virial* expansion, since it is a power series. Usually data are adequately represented by the first two terms, and higher terms need not be included. The quantity B, termed the *second virial coefficient* or the *interaction constant,* can then be evaluated by fitting the results to the equivalent equation:

$$\frac{\Pi}{c} = RT\left(\frac{1}{M} + Bc\right) \qquad (12\text{-}9)$$

It is instructive to compare the virial equation to the van der Waals equation after multiplying out the latter to give a polynomial in n/V, dropping the cubic term, and substituting c/M for n/V. One finds that

B is then given by $(b - a/RT)/M^2$. It will be remembered that, for the gas phase, a is a quantity measuring the forces of interaction between molecules and b is an "excluded-volume" term, representing the volume occupied by the molecules themselves. With slight modifications, we can interpret these quantities similarly for the solution case.

In solution, the intermolecular force term is a measure of the excess of attractive forces between molecules of the solute over the forces binding solute to solvent. Thus if a solvent is *good,* which means that it has a large affinity for the solute, usually because it is chemically or physically similar, the numerical value of a is small. At the opposite extreme of behavior, when a is large, the solute molecules can be drawn so tightly together that phase separation occurs at certain temperatures. The excluded-volume parameter b is related to the shape of the macromolecule; for molecules of fairly large size, an elongated rod keeps other molecules from occupying a larger region of space than is excluded by a compact sphere of the same volume.

For a good solvent, the value of the interaction constant B is usually positive at all temperatures, because the small value of a makes the influence of temperature small. For a poor solvent, B is negative at low temperatures, goes through zero at some temperature, referred to as the *theta point,* and is positive at higher temperatures. The significance of the theta point is that it is a condition in which the polymer is in a truly random conformation. The term *theta solvent* is applied to the composition of a mixed solvent corresponding to a value of B of zero at a given temperature.

THE DONNAN
MEMBRANE
EQUILIBRIUM

One of the limitations of osmotic pressure measurements applies to those macromolecules, such as proteins, that ionize as electrolytes. Known as the Donnan membrane effect, the phenomenon comes into play whenever charged colloidal particles are retained by a barrier that allows the counterions to pass.

Let us consider first the general equilibrium condition for the ions of any electrolyte that can pass freely from one solution to another, taking as an example sodium chloride. Suppose the ions of sodium chloride are distributed between solutions 1 and 2 which are separated by a membrane permeable to both sodium ions and chloride ions. When a mole of sodium ions is transferred from solution 1 to solution 2, the change in free energy is

$$\Delta G_{Na^+} = (G^0 + 2.303RT \log [Na^+]_2) - (G^0 + 2.303RT \log [Na^+]_1)$$

$$= 2.303RT \log \frac{[Na^+]_2}{[Na^+]_1} \tag{12-10}$$

In order to preserve electroneutrality, the transfer of a mole of sodium ions is accompanied by the transfer of the same number of chloride ions, which is associated with a free energy change:

$$\Delta G_{Cl^-} = 2.303RT \log \frac{[Cl^-]_2}{[Cl^-]_1} \tag{12-11}$$

If the two solutions are at equilibrium, the total free energy change, the sum of ΔG_{Na^+} and ΔG_{Cl^-}, must be zero. Combining this condition with Equations (12-10) and (12-11) leads to the result

$$\frac{[Na^+]_2}{[Na^+]_1} = \frac{[Cl^-]_1}{[Cl^-]_2} \tag{12-12}$$

To illustrate the application of this equation to equilibria involving ions that cannot pass through a membrane, we treat the case in which a salt containing a nondiffusible anion, represented by NaR, is placed on one side of a membrane, and a diffusible salt, say NaCl, is placed on the other side of the membrane. For simplicity, the volumes of solution on the two sides of the membrane are considered equal and transfer of solvent through the membrane is assumed to be prevented by application of suitable pressure. Since the membrane is permeable to NaCl, Equation (12-12) describes the equilibrium finally reached. Let the initial concentration of NaR be C_1 and that of NaCl be C_2, and let the amount of NaCl transferred through the membrane at equilibrium be x moles per liter. The equilibrium concentrations in the NaCl solution are $C_2 - x$ for each of the ions. In the colloid, the concentration of sodium ion is $C_1 + x$ and of chloride ion is x.

The equilibrium condition then is

$$\frac{C_2 - x}{C_1 + x} = \frac{x}{C_2 - x} \tag{12-13}$$

Solving this equation for x yields

$$x = \frac{C_2{}^2}{C_1 + 2C_2} \tag{12-14}$$

The ratio x/C_2 is the fraction of the sodium chloride that diffuses through the membrane. If C_1 and C_2 are equal, x/C_2 is 1/3. If C_2 is very much larger than C_1, the latter can be neglected in the denominator in comparison with C_2, and x/C_2 is equal to 1/2; this corresponds to a uniform distribution of sodium chloride through the two solutions.

Similar equilibrium equations can be derived for the case in which a colloid containing a nondiffusible cation is placed on one side of the membrane and a salt with a common anion is placed on the other side, or the case in which the electrolyte has neither of its ions in common with the colloid. The general equation which applies to any case can be written

$$\frac{[cat^+]_2}{[cat^+]_1} = \frac{[an^-]_1}{[an^-]_2} = \left(\frac{[cat^{2+}]_2}{[cat^{2+}]_1}\right)^{1/2} = \left(\frac{[an^{3-}]_1}{[an^{3-}]_2}\right)^{1/3} = \cdots = \lambda \tag{12-15}$$

where λ is the Donnan distribution coefficient.

The Donnan equilibrium affects osmotic pressure results for solutions of macromolecules because of the contribution from the unequal concentrations of counterions on the two sides of the membrane. The Donnan effect is small at the isoelectric point of a protein or in the presence of a large salt concentration, so that osmotic measurements should be made under one of these conditions.

The Donnan equilibrium has been applied to the distribution of ions between the red blood cells and the plasma. The equilibrium ratios of chloride and bicarbonate ion concentrations, determined by the presence of cations that cannot penetrate the cell walls, should be equal:

$$\frac{[HCO_3{}^-]_{plasma}}{[HCO_3{}^-]_{cell}} = \frac{[Cl^-]_{plasma}}{[Cl^-]_{cell}} \tag{12-16}$$

Thus transfer of bicarbonate ion between the cell and the plasma should be accompanied by the transfer of an equivalent amount of chloride ion. When carbon dioxide is absorbed by the blood from the tissues, it is not immediately converted to bicarbonate ion; this conversion requires the presence of an enzyme and occurs much more rapidly in the cells than in the plasma, since the enzyme is present in greater concentration in the cell interior. Now if carbon dioxide is absorbed by the blood, the bicarbonate ion concentration inside the cells builds up immediately, with two initial consequences: Chloride ion diffuses in from the plasma to satisfy the Donnan equilibrium, and bicarbonate ion diffuses out to reach equilibrium with the plasma. Diffusion of the bicarbonate outward leads to accompanying outward transfer of chloride ion after the initial inward surge. The reverse of this cycle occurs when carbon dioxide is lost from the blood in the lungs. The *chloride shift* between plasma and cells is therefore a continuously occurring process.

12-6
DYNAMIC PROPERTIES

It is possible to obtain significant information about the shape and size of macromolecules by studying the dynamics of their behavior in a liquid medium under various influences. The basis of several dynamic effects is the principle of equipartition of energy, which, as we saw in Chapter 3, states that the total translational energy of molecules is $\frac{3}{2}RT$ per mole of material. This is applicable to large as well as to small molecules, and to particles in colloidal dispersions as well as to gaseous molecules. However, because of the much greater mass of a colloidal particle, the same amount of kinetic energy represents a much smaller velocity of motion than it does for a typical gaseous molecule.

BROWNIAN MOTION

In 1827, Robert Brown, a Scotch botanist, noticed that pollen grains in suspensions are in rapid and random motion. In many colloidal dispersions, similar motion can be observed by use of an ultramicroscope, as described in Section 12-9. Lyophobic particles can be more readily observed in the ultramicroscope than lyophilic particles, but Brownian motion takes place in any macromolecular solution.

Brownian motion is the counterpart of the movement postulated by the kinetic molecular theory for gaseous molecules. Colloidal particles are sufficiently small that the molecules of the medium carry

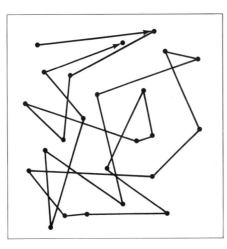

Figure 12-7
Path of a particle undergoing Brownian
motion. Dots show locations of the
particle at successive times.

enough energy to move them upon collision and that the effect of
collisions upon them is not fully averaged to zero, but are sufficiently
large so that motional velocity is much less than that of solvent molecules. The result of the sharing of kinetic energy between the solvent
and the solute is the movement of colloidal particles that can be observed as they are buffeted about, first in one direction, and then in
another. An example is shown in Figure 12-7.

In deriving an equation describing Brownian motion, consideration
is limited for simplicity to the component of motion in one direction,
say the x direction. The motion may be assumed to be a *random-walk*
process, one that takes place by successive steps of uniform distance,
with equal probabilities that any step will be in one direction or in the
other. For a group of particles initially all at one location, a random
walk leads to a Gaussian distribution of concentration at some later
time; as time elapses the region over which the distribution extends
becomes larger and larger. For a given time interval, the average displacement from the starting point reached at the end of the interval is
proportional to the number of steps taken during the interval. If now
the principle of equipartition of energy is assumed valid for the kinetic
energy of the particles, this model leads to an equation, similar to one
derived by Einstein, for the average of the square of the displacement
$\overline{\Delta x^2}$ during time interval τ:

$$\overline{\Delta x^2} = \frac{2kT}{\phi}\tau \tag{12-17}$$

The quantity ϕ is the *frictional coefficient,* defined as the negative
ratio between the frictional force on a moving particle and the velocity
u with which it moves:

$$\text{force} = -\phi u \tag{12-18}$$

An equation derived by G. G. Stokes gives the frictional coefficient
of a rigid spherical particle of radius r which is relatively large compared to the size of the particles of the medium but small compared

to the distance from other like particles and to the distance from the wall of the vessel:

$$\phi = 6\pi\eta r \qquad (12\text{-}19)$$

The viscosity coefficient η is that of the medium. On substitution of this expression, the Brownian motion equation becomes

$$\overline{\Delta x^2} = \frac{kT\tau}{3\pi\eta r} = \frac{RT\tau}{3\pi\eta rN} \qquad (12\text{-}20)$$

For spherical particles, the results of J. Perrin, a French scientist working in the first decade of this century, and of subsequent investigators, are quite consistent with this equation. If a molecule is solvated, the radius r must include the strongly held solvent which travels with the particle, as well as the "dry" size. For nonspherical molecules, ϕ must be multiplied by a shape factor, which is always less than unity since ϕ is larger for any other shape of the same volume than for the Stokes' law sphere. In other words, the frictional ratio $\phi_{meas}/\phi_{sphere}$ is always equal to or greater than unity.

DIFFUSION

Closely related to Brownian motion is the phenomenon of diffusion, the transfer of a substance from a region in which its concentration is larger to a region in which its concentration is smaller. Indeed, it was the study of diffusion rates that led Thomas Graham in 1861 to coin the term *colloid,* meaning *gluelike,* to apply to the materials he observed to diffuse relatively slowly as compared to the rapidly diffusing *crystalloids.*

To avoid the complexities of a full three-dimensional treatment, we again consider only the case in which diffusion occurs in one dimension, to which we assign the x coordinate. The basic rule governing the rate of material transfer by diffusion is known as *Fick's first law:*

$$\frac{dn}{dt} = -DA\frac{dc}{dx} \qquad (12\text{-}21)$$

The quantity of material dn is transferred in the time interval dt. The driving force for transport is the concentration gradient dc/dx and A is the cross-sectional area of the path or channel available for movement. The *diffusion coefficient D* depends upon the size and shape of the diffusing particles, the solvent medium, and the temperature. It is this quantity in which we are primarily interested, because of the information it gives about the solute molecule. Dimensions of the diffusion coefficient are length^2/time. If the amounts of material in dn and dc are expressed in the same units and the lengths in D, dc, and dx are in identical units, only relative concentrations need be measured.

One method for measuring the diffusion coefficient involves the use of a porous barrier, usually a porous disk of alundum or sintered glass, mounted in a glass tube, initially with pure solvent on one side of the barrier and a solution on the other. As solute diffuses through the disk, the concentration gradient across the disk diminishes and diffusion slows correspondingly. Both liquids are vigorously stirred so that

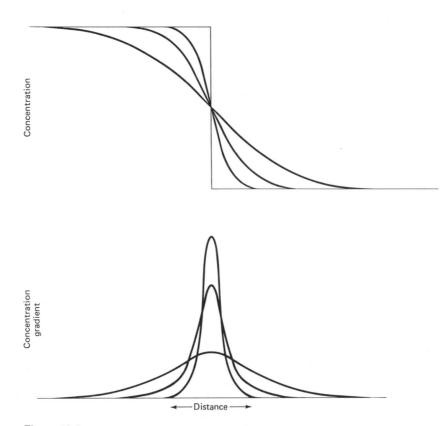

Figure 12-8
The concentration and concentration gradient variation with time as a result of diffusion across an initially sharp boundary.

there is no concentration difference within either liquid phase. The rate of transfer is then evaluated as a function of concentration difference between the two solutions, and the gradient *within the disk* is assumed to be linear, so that Equation (12-21) can be written

$$\frac{dn}{dt} = -DA\frac{\Delta c}{l} \tag{12-22}$$

where l is the thickness of the porous disk and Δc is the difference in concentration between the two liquids, one on either side of the disk. The ratio A/l can be evaluated for the particular measuring system by calibration with a material of known D.

A second experimental method follows the spreading with time of a boundary formed initially by contact between a solution and the pure solvent. This can be done conveniently with equipment used for sedimentation or electrophoresis measurements, both to be described later. In contrast to the first method, care is taken that the liquid on either side of the boundary is not disturbed by mechanical agitation or convection. Figure 12-8 shows the concentration and concentration gradient functions at successive times for an ideal system. Mathematical analysis of the situation shows that the concentration gradient can be expressed at any time t by the Gaussian function

$$\frac{dc}{dx} = -\frac{c_0}{(4\pi DT)^{1/2}} e^{-x^2/4Dt} \tag{12-23}$$

From this equation and the measured results, D can be evaluated.

How are we to relate the value of D to the size of the molecule? The approach is quite like that for Brownian motion, in that the diffusion coefficient is expressed as a function of the viscous resistance to motion of the particle passing through the medium, and one can indeed set up a model in which diffusional transfer along a tube occurs by Brownian motion, from which it is possible to show that

$$D = \frac{\overline{\Delta x^2}}{2\tau} \tag{12-24}$$

Combining this equation with Equation (12-17) leads to the relation

$$D = \frac{kT}{\phi} \tag{12-25}$$

which is Einstein's law of diffusion.

For a spherical particle, Stokes' law permits calculation of the particle radius from a knowledge of the value of the diffusion coefficient. For particles of other shape, values of D can be combined with sedimentation velocity results, described below, to obtain particle size information independently of the shape of the particle.

RATE OF SEDIMENTATION

Application of a downward mechanical force to solute molecules causes them to settle to the bottom of their container. Particles with diameters of the order of 0.1 μm or larger settle under the influence of gravity when a suspension is allowed to stand for several hours or days, and differential settling rates are used to separate particles of various sizes in samples of soil, sand, or minerals.

The apparatus used to study the sedimentation of macromolecules quantitatively is termed an *ultracentrifuge*. Not only is an ultracentrifuge designed to apply a large centrifugal force by spinning at high speed a rotor containing the solution, but it also has additional special features. Disturbances of the sedimentation process, either by mechanical vibration or by convection currents, are minimized. To eliminate vibrational effects, the rotor can be suspended on a vertical wire, or even held in a magnetic field, free of direct mechanical drive. To reduce convection currents, the interior of the chamber housing the rotor is partially evacuated to avoid frictional heating, and it is refrigerated at a temperature near the temperature of maximum density of water at 4°C. The liquid is protected from evaporation by using a sealed cell, and the cell is sector-shaped, with a larger cross section at the bottom than at the top, so that particles can move radially without striking the sides of the cell. And perhaps the most important feature is that there is provision for continuous analysis of the cell contents at various levels without interrupting the centrifugation process.

The Spinco-Beckman Model E is a widely used ultracentrifuge capable of operating at speeds up to about 70,000 revolutions per minute (rpm). Figure 12-9 shows the whole system and the rotor, about 190 cm in diameter, which when spinning is enclosed in an armor-plated chamber. The cell is placed in a hole in the rotor, along with a counterbalance or solvent-filled cell on the opposite side, and the cell windows are transparent, to permit passage of the light used for analysis of the contents.

The optical analytical method is commonly any of three types: absorption of visible or ultraviolet light, measurement of refractive index by the interference fringe pattern of an interferometer, or determination of refractive index gradient by a schlieren system. The interferometric method is quite precise, but the schlieren method is adequate to follow the movement of the boundary at the top of a sedimenting layer of solute.

The schlieren system depends upon the principle that parallel rays of light passing through a region of refractive index gradient are bent in the direction of this gradient. Everyone has observed this phenomenon in the shimmering distortions produced by warm air currents rising from a chimney or from the surface of the earth on a hot day. In a macromolecular solution, the refractive index gradient is proportional to the concentration gradient, and the angular deviation of a beam of light thus is proportional to the *concentration gradient*. Now if one imagines oneself as an observer looking toward the cell with a light source on the opposite side, one can visualize seeing a lesser intensity directly transmitted by the region with the refractive index gradient, but a correspondingly greater intensity at some other place in the image where the ray of light that has been bent overlaps light passed by another level in the cell without deviation. It is difficult to measure the deviated beam originating at one level of the cell in the presence of the undeviated beam originating at another level. In one of the commonly used optical systems, a diagonal slit is introduced into the optical path in such a way as to convert the light ray deviation

Figure 12-9(a)
A Spinco-Beckman ultracentrifuge, Model E, equipped with a photoelectric scanner which provides chart recordings while a run is in progress.

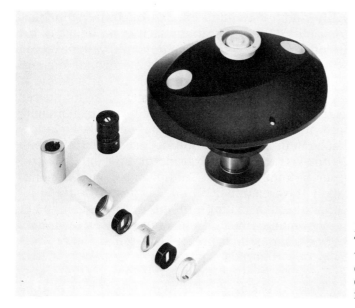

Figure 12-9(b)
The centrifuge rotor with an assembled cell on the left, and cell components in front.

parallel to the concentration gradient into a deviation perpendicular to this gradient.

To understand how this is done, we can follow the rays of light through the schlieren optical system in Figure 12-10. Light from a source is rendered parallel by a collimating lens and then passes through the cell. That section of the light beam which passes through a region where there is a refractive index gradient is bent in the direction of larger refractive index, which is downward if the cell is in a vertical position. The schlieren lens brings the parallel rays from the cell together in the plane of the diagonal slit, and the undeviated rays are simply narrowed horizontally to a pencil of light, which later

Figure 12-10
Schematic diagram of a schlieren optical system. The paths of three rays are shown. One of these passes through the region of changing refractive index at the boundary and is bent downward; because of the inclination of the diaphragm, only that part of the ray toward the front can pass, and this is deviated backward by the cylindrical lens.

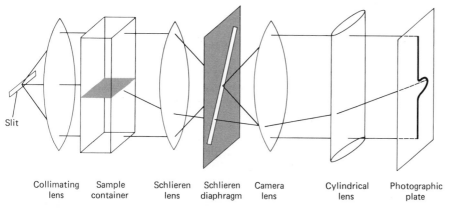

Slit							
Collimating lens	Sample container	Schlieren lens	Schlieren diaphragm	Camera lens		Cylindrical lens	Photographic plate

spreads apart vertically, and passes unaffected through the cylindrical lens, having been focused by the camera lens so that an image of the cell is formed on the photographic plate. The deviated ray, however, strikes a lower portion of the plate containing the diagonal slit, and thus only a portion of it which is a certain distance to one side of the center is transmitted. Passing through the cylindrical lens off-center, it is bent by this lens laterally and reaches the photographic plate with a horizontal deviation proportional to the original vertical deviation.

Figure 12-11 is a schematic representation of a sedimentation pattern. The ordinate is the refractive index gradient, and the peak corresponds to the upper boundary of the solute. For a well-behaved macromolecular system, the area under the peak is proportional to the concentration of the solute. An example of how differences in the sedimentation velocity of macromolecules in a mixture can be used to show that more than one component is present is given in Figure 12-12. The boundaries of various materials separate as the materials sediment with differing velocities. As illustrated in the figure, blood serum from humans with certain diseases, such as nephritis, shows characteristic changes in the sedimentation pattern from that of a normal individual. It is also interesting to realize that it was the work of The Svedburg, who developed the ultracentrifuge in the 1920's and 1930's, that first clearly demonstrated that each protein is a homogeneous substance with a definite molecular weight and therefore does truly consist of large molecules.

One notes also in Figure 12-12 the effect of diffusion on the nature of the solute boundary: As the boundary moves downward with the passage of time, it becomes broader, somewhat as the boundary spreads in the free-diffusion measurement of diffusion coefficient.

To treat quantitatively observations of sedimentation velocity, we

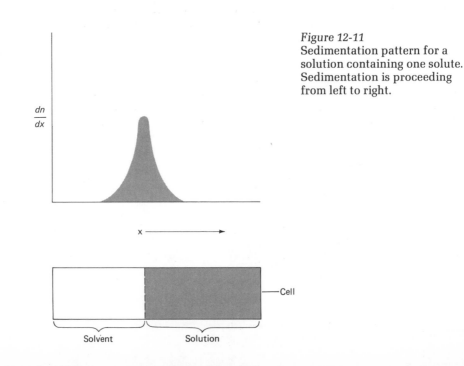

Figure 12-11
Sedimentation pattern for a solution containing one solute. Sedimentation is proceeding from left to right.

$\dfrac{dn}{dx}$

x ⟶

Cell

Solvent   Solution

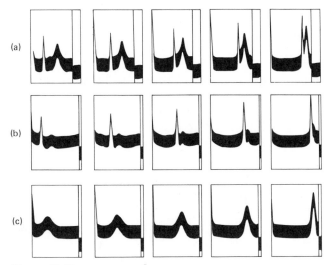

Figure 12-12
Sedimentation diagrams of abnormal human serum from a patient with multiple myeloma (a), of the electrophoretically separated globulins from the same serum (b), and of the ultracentrifugally homogeneous urinary protein from the same patient (c). Sedimentation proceeds from right to left in the diagrams. The photographs were taken at intervals of 16 min for the serum proteins and 32 min for the urinary protein, at a rotor speed of 59,780 rpm, corresponding to a force about 250,000 times that of gravity. The serum and the urinary proteins were the same specimens for which electrophoretic diagrams are shown in Figure 12-15. (Courtesy of Dr. F. W. Putnam, Distinguished Professor of Molecular Biology and Zoology, Indiana University, Bloomington.)

apply the fact that molecules in a cell in a spinning untracentrifuge rotor are quickly accelerated to a terminal velocity at which the frictional retarding force is equal to the centrifugal force. As we saw in Equation (12-18), the frictional force is equal to the frictional factor times the velocity of the particle:

$$F = -\phi \frac{dx_r}{dt} \qquad (12\text{-}26)$$

The velocity is expressed in terms of x_r, the radial distance from the axis of rotation of the centrifuge to the molecule in question. The centrifugal force of motion in a circular path is mv^2/x_r, where m is the mass of the moving object and v is the linear velocity. Since the linear velocity is the product of the radius and the angular velocity ω, the centrifugal force is $m\omega^2 x_r$.

Because the molecules are suspended in a liquid medium, their effective mass is less than their total mass. The buoyancy correction factor multiplying the total mass can be written as $(1 - \rho/\rho_p)$, where ρ_p is the density of the particle and ρ is the density of the medium. A more exact version is $(1 - \overline{v}\rho)$, in which $\overline{v}$ is the partial specific volume of the solute, the increase in volume of the solution when 1 g of solute is added. The density of the solvent is often used in this expression as an approximation to the density of the solution ρ.

Equating the magnitudes of the centrifugal and viscous forces and introducing the buoyancy factor, we find

$$\frac{dx_r}{dt} = \frac{m(1 - \bar{v}\rho)\omega^2 x_r}{\phi} \tag{12-27}$$

Integration of this equation from level x_{r_1} at time t_1 to level x_{r_2} at time t_2 leads to

$$\ln\frac{x_{r_2}}{x_{r_1}} = \frac{m(1 - \bar{v}\rho)\omega^2}{\phi}(t_2 - t_1) \tag{12-28}$$

The quantity $m(1 - \bar{v}\rho)/\phi$ is referred to as the sedimentation coefficient s and is the velocity of fall under unit force. It can be calculated from experimental data by the equation

$$s = \frac{\ln(x_{r_2}/x_{r_1})}{\omega^2(t_2 - t_1)} \tag{12-29}$$

Values of s are often given in svedburg units, S, equal to 10^{-13} sec. Typical values range from 1.2 S for insulin to 185 S for tobacco mosaic virus.

Provided the particles are spherical, Stokes' law can be used to express ϕ. With the buoyancy factor in the form $(1 - \rho/\rho_p)$, the sedimentation coefficient is equal to

$$s = \frac{4}{3}\pi r^3 \frac{\rho_p(1 - \rho/\rho_p)}{6\pi\eta r} = \frac{\frac{2}{9}r^2(\rho_p - \rho)}{\eta} \tag{12-30}$$

Thus it is possible to evaluate r from the sedimentation coefficient. If the particle is not spherical, then sedimentation rate measurements alone do not yield a definite value for the particle size. However, sedimentation velocity plus diffusion coefficient values, taken together, provide this information, since the frictional factor ϕ, which depends upon shape and solvation, cancels out when the results are combined. For a solute with a continuous distribution of molecular weights, such as a typical synthetic polymer, sedimentation rates give an average molecular weight that is a complicated function of the weight distribution rather than any one of the simple averages.

There are several experimental complications in sedimentation rate measurements. The presence along with macromolecules of counterions—small ions of opposite charge—affects the results, because the counterions sediment at rates much lower than those of the macromolecules. Addition of excess inert electrolyte at 0.1 to 1.0 M concentration, as well as maintenance of pH near the isoelectric point of the macromolecule, tends to minimize this effect.

Another anomaly results from the dependence of the sedimentation coefficient on total concentration. Ideally, there should be no concentration effect, but in real solutions there is a variation, and the best values are accordingly obtained by extrapolation to infinite dilution. Indeed, for a one-component system, the effect of concentration may be beneficial, because the greater sedimentation rate at lower concentration speeds up laggard molecules behind the boundary and reduces diffusional broadening of the boundary.

When a collection of particles such as the molecules of gas in the atmosphere is subjected to a force field, a gradient of concentration is set up in the direction of the force. A description of the concentration gradient can be obtained by applying the thermodynamic concept of free energy. The molar free energy difference between two elevations in the atmosphere because of the force field of gravity is equal to the work required to bring about the reversible transfer of a mole of gas from the lower altitude to the higher altitude. This work is equal to the force on the molecules, or Mg, where g is the gravitational constant multiplied by the distance, say dh for an incremental transfer. At the same time, the free energy varies with molar concentration according to the relation $d_G = RT\, d \ln c$. At equilibrium, the excess free energy because of a greater h must be exactly compensated for by a smaller free energy because of concentration:

$$RT\, d \ln c + Mg\, dh = 0 \qquad (12\text{-}31)$$

Applied to two points in the atmosphere, with concentrations c_2 at h_2 and c_1 at h_1, this equation becomes

$$\ln \frac{c_2}{c_1} = -\frac{(h_2 - h_1)Mg}{RT} \qquad (12\text{-}32)$$

An equivalent form is

$$c_2 = c_1 e^{-(h_2 - h_1)Mg/RT} \qquad (12\text{-}33)$$

One sees that this is a form of the Boltzmann distribution equation presented as Equation (1-19).

The same sort of distribution is found in a suspension of colloidal particles, and this provides another analogy between the behavior of these particles and that of gaseous molecules.

To attain measurable concentration gradients of smaller colloidal particles, it is necessary, as for sedimentation rate measurements, to use an ultracentrifuge. To describe the situation after a macromolecular solute in a centrifuge has come to equilibrium, we replace the gravitational factor g in Equation (12-31) by $\omega^2 x_r$, introduce the buoyancy correction, and change the sign because force and distance are now measured in the same sense:

$$RT\, d \ln c = (1 - \bar{v}\rho)M\omega^2 x_r\, dx_r \qquad (12\text{-}34)$$

Integrating between two levels in the cell, one obtains

$$\ln \frac{c_2}{c_1} = \frac{(1 - \bar{v}\rho)M\omega^2}{RT} \frac{x_{r_2}^2 - x_{r_1}^2}{2} \qquad (12\text{-}35)$$

From a plot of $\ln c$ against x_r^2, the molecular weight of a macromolecule can be calculated. For a sample with a continuous distribution of molecular sizes, the result is a weight average. In calculations based upon sedimentation equilibrium experiments, there is no uncertainty about the frictional factor, such as there is in interpretation of rate data. One can regard the equilibrium condition as that in which the rate of sedimentation downward because of centrifugal force is exactly equal to the rate of diffusion upward because of the concentration gradient, with the frictional factor entering in the same way in both dynamic processes.

Example: A sample with a density of 2.831 is suspended in water at 5°C. The ratio of concentrations is found to be 1.72 at 10.10 cm from the axis of the centrifuge to 1.00 at 10.00 cm when the rotor is spun at 4000 rpm. Estimate the molecular weight.

Solution: To calculate the buoyancy factor, we assume the solution is ideal and use the density of the solvent as 1.000 at 5°C. A speed of 4000 rpm corresponds to $4000 \times 2\pi/60$ or 419 radians/sec. Substituting in Equation (12-35),

$$2.3 \log \frac{1.72}{1.00} = \frac{[1 - (1.000/2.831)]M(419)^2(10.10^2 - 10.00^2)}{2(8.314 \times 10^7)(278)}$$

From this, $M = 1.10 \times 10^5$.

Sedimentation equilibrium runs have the disadvantage, compared to sedimentation velocity measurements, that they take much longer to carry out because the centrifuge must be operated at a considerably lower speed to avoid sending all the solute to the bottom of the cell. Recently methods have been developed to minimize this difficulty: The rotor speed is programmed, starting with a high velocity and decreasing as equilibrium is approached, and column heights of a millimeter or less are used instead of lengths of a centimeter or more.

In a nonideal solution, the equilibrium distribution depends upon the concentration. To obtain valid results for molecular weight, one can extrapolate to zero concentration, or apply a correction factor:

$$RT \, d \ln c \left(1 + \frac{\partial \ln \gamma}{\partial \ln c} \right) = (1 - \bar{v}\rho)M\omega^2 x_r \, dx_r \qquad (12\text{-}36)$$

Sedimentation equilibrium can be utilized as a separation process by carrying it out in a medium that has a density gradient. A preformed gradient is sometimes prepared by layering sucrose solutions of decreasing concentration on top of one another with the macromolecular solution at the top. This is more easily done in a centrifuge with a swinging-bucket rotor, where the sample tubes are initially vertical and swing out as the rotor is accelerated, so that the force is always directed toward the bottom of the sample container. Alternatively, the centrifugation process itself can be used to generate a density gradient in a high concentration of salt placed in the medium, such as 7.7 M cesium chloride for nucleic acids or potassium bromide for proteins. In a density gradient, the macromolecules form a band in the cell at the level where the effective density of the macromolecules and associated species equals the density of the medium.

Density gradient ultracentrifugation is more effective with nucleic acids than with proteins, since the former have greater density differences. A high content of guanine–cytosine base pairs leads to higher density than a high content of adenine–thymine pairs. It has also been possible to label DNA with ^{15}N and to follow the path of the label through successive generations of cells by separating the ^{14}N and ^{15}N fractions of the DNA, as well as mixed fractions, by sedimentation in a density gradient.

12-8
ELECTROPHORESIS

Particles of colloidal dimensions dispersed in an aqueous medium almost always carry electric charges on their surface. These charges may arise from the way in which the disperse phase is formed, as illustrated by positively charged silver iodide microcrystallites produced when excess silver salt is mixed with a soluble iodide, or by the presence of ionized impurities, as when ferric oxide is formed by the hydrolysis of ferric chloride but the reaction stops short of completion, leaving some ferric ions unreacted within the colloidal particles. And for proteins, it is the ionization of weak acid and base functional groups, along with a small amount of specific ion binding, that imparts a surface charge. For both hydrophobic and hydrophilic colloids, the surface charge is a very important factor in stabilizing the dispersion, both because it increases binding of the solvent and because like-charged particles repel one another. Of course, the charges on the colloidal particles are always balanced by charges of opposite sign on the counterions.

An electric potential gradient applied to a dispersion of charged particles in a liquid medium causes the particles to migrate toward the electrode with charge opposite that of the net charge on the particle. This type of transport is termed *electrophoresis*. If a semipermeable membrane is interposed so that the flow of solute particles is prevented, application of an electric field results in *electroosmosis,* or flow of the liquid medium, including counterions as well as solvent, in the direction opposite that in which the macromolecular particles would move if unimpeded by the barrier.

If a mechanical force is applied so as to cause a shearing motion of two phases on opposite sides of an interface, the consequent separation of charge leads to electric potentials; the converse of electrophoresis occurs in the sedimentation of macromolecules described above, and the resulting potential is often called a *sedimentation potential.* The converse of electroosmosis results when a liquid is constrained to move relative to a stationary solid phase: A *streaming potential* is set up when liquid is forced through small-diameter capillaries of glass or through a porous diaphragm which behaves as a bundle of capillaries. The glass walls of the capillary have been negatively charged by adsorption of hydroxyl ions.

TECHNIQUES OF
ELECTROPHORESIS

Studies of electrophoresis may be directed toward the goal of understanding the pattern of charges present at the interface, or toward the separation and analysis of components of mixtures that differ in

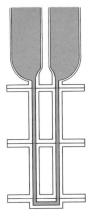

Figure 12-13
Tiselius electrophoresis cell. Sharp boundaries are formed by
sliding the sections horizontally with respect to one another.

charge because of slight differences in molecular structure. The technique has been particularly valuable in the characterization and purification of biochemical materials such as proteins, polysaccharides, and nucleic acids.

The apparatus and techniques for the moving-boundary method of electrophoresis were developed by Arne Tiselius in the 1930's and improved by L. G. Longsworth. The Tiselius apparatus consists of a U-shaped pathway made of optically plane glass plates arranged so that the cross section of the solution is rectangular, about 3×25 mm. The U-tube is divided into several sections with glass plates forming the two ends of each section. The sections can be moved laterally with respect to one another by sliding them with end plates in contact and, by bringing the channels into alignment, sharp initial boundaries can be formed between the liquid in one section containing the macromolecules and the liquid containing only solvent medium in another section. The arrangement is illustrated in Figure 12-13.

As we shall see below, factors such as pH and ionic strength modify the charges on macromolecules, particularly on proteins. To avoid a sudden change as the protein moves from its initial environment into a new phase, the liquid placed above the solution in the electrophoresis cell should be as nearly like the macromolecular solution as possible. It is desirable to have pH, ionic strength, and electric conductivity each be the same on both sides of the boundary, but not all three conditions can be met simultaneously. A compromise is made by adding a portion of buffer to the protein solution and then dialyzing this mixture against another portion of the buffer in order to equalize the activities of the counterions. The same two solutions are then used to form the boundary.

To minimize external disturbances of the liquid in the electrophoresis cell, electrodes are chosen that do not form gas bubbles, usually silver–silver chloride electrodes, and the experiment is conducted in a refrigerating thermostat near 4°C. Typically, several hundred volts potential difference is applied across the electrodes, and the solution is noticeably warmed by the current that flows. The movement of the boundaries in the electrophoresis apparatus is followed by the schlieren method described above for the ultracentrifuge. A pattern is obtained in which a peak corresponds to a maximum in the refractive index gradient and thus to the boundary of a particular solute. The boundaries in the two limbs of the U-tube move in opposite directions: In one side, at the *descending boundary,* the colloidal particles move out of the buffered medium, while at the *ascending boundary* in the other side, the particles move into the buffer solution. There is always found in the schlieren pattern an "anomalous" peak corresponding to the original position of each boundary before migration started.

For quantitative measurements, the voltage drop across the solution is calculated from a measurement of the current flowing and a separate measurement of the specific conductance of a sample of the macromolecular solution. The mobility of the solute molecules, which is their velocity under unit potential gradient, is calculated by dividing the observed velocity of the boundary, in centimeters per second, by the potential drop, in volts per centimeter.

THE ELECTRICAL DOUBLE LAYER

It would be very useful if electrokinetic mobilities could be interpreted in a way so as to give information about the concentration and arrangement of charges at the surface of a colloidal particle, as well as about the spatial distribution of counterions in the nearby solution. However, the situation is complex, and most of the deductions that can be made are qualitative at best.

A result of the mutual interactions of the charges is the presence of an electrical double layer at the solid–liquid interface. Not only is the structure of the double layer a matter of some uncertainty, but the ionic arrangement in it is transient. L. F. Helmholtz assumed that the double layer is simply a layer of charges of one sign on the surface of the particle, with an equivalent layer of ions of opposite sign at a uniform distance away in the solution. This picture has been modified and elaborated by G. Gouy and O. Stern, so that the pattern of counterions on the solution side is considered to be diffuse rather than compact, extending some distance out into the solution, with the concentration of ions decreasing with increasing distance from the surface. The Helmholtz double layer and the diffuse double layer are represented in Figure 12-14.

The situation of the counterions in the diffuse double layer is somewhat like that postulated by Debye and Hückel for the ionic atmosphere around a single ion. A principal distinction is that the surface of the colloidal particle is practically planar on the size scale of ions, in contrast to the central point charge to which the Debye–Hückel equations apply. An important consequence of the diffuse nature of the double layer is that some counterions near the surface of the particle are present in the solvent which moves *with* the particle, and these ions are forced to accompany the particle when it moves in a field. Thus the effective or net charge on the particle is its own charge

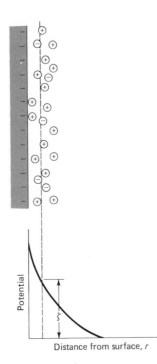

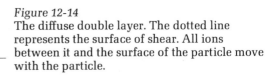

Figure 12-14
The diffuse double layer. The dotted line represents the surface of shear. All ions between it and the surface of the particle move with the particle.

less the charge carried by counterions located inside the boundary between bound solvent and free solvent, the boundary called the *surface of shear*. Moreover, an increase in ionic strength of the medium has the same effect as it does in the Debye–Hückel model—the counterions move closer to the surface. But in this case, those counterions that happen to move from outside the surface of shear to within the surface of shear become effectively bound and decrease the net charge on the electrokinetic moving unit. Of course, if the particle is a protein or other weak electrolyte, a change in ionic strength changes the activity coefficients of the ions and consequently the dissociation of functional groups, and thus, *in addition to* the effect on the dimensions of the ionic atmosphere, has an additional influence on the charge of the particle.

When account is taken of the presence of the "atmosphere" of counterions, other features complicating the interpretation of electrokinetic mobilities appear. As the macromolecule moves through the solution, the atmosphere must dissipate in the region vacated by the particle and be built anew at the location to which the particle is moving; in other words, the atmosphere must *relax*. Furthermore, the counterions tend to flow in a direction opposite the direction of motion of the macromolecules, carrying some solvent with them. Both factors reduce the observed particle mobility below that which would be calculated for a simple charged particle moving in a neutral fluid medium.

Some workers express the properties of the double layer in a particular colloidal system in terms of the electrokinetic or *zeta* potential ζ, which is visualized as the electric potential difference between the bulk of the medium and the surface of shear of the particle, in contrast to the electrochemical potential, which is the total potential measured from the bulk of the solution all the way to the particle itself. The zeta potential is considered to be the potential which, if it existed across the two plates of an electric capacitor equivalent to the two sides of the Helmholtz double layer, would yield electrokinetic effects similar to those actually measured. A simple derivation, using the laws of electrostatics and the principles of flow in a fluid medium of viscosity η, leads to the equation

$$\zeta = \frac{4\pi\eta u}{D} \tag{12-37}$$

where D is the dielectric constant of the medium and u is the electrokinetic mobility. (It is to be noted that mobilities as measured and reported are usually in units of square centimeters per second per practical volt, with 300 practical volts corresponding to 1 electrostatic volt). In the light of all the complications of interpretation mentioned, the zeta potential is best regarded as simply another scale on which electrokinetic mobility can be expressed.

RESULTS WITH PROTEIN SOLUTIONS

In addition to the magnitude of the mobility under given conditions, the pH value at which the macromolecules have zero mobility is also characteristic of a particular protein. On the acid side of this isoelectric point the protein migrates toward the cathode, whereas on the basic side it migrates toward the anode. Isoelectric points, as well as mobilities, vary slightly with the ionic strength of the medium and with the specific composition of the buffer employed.

Mixtures of proteins with different mobilities can be analyzed by electrophoresis. Each constituent of a mixture produces its own boundary, and the relative areas under the schlieren pattern are indicative of relative concentrations. The existence of several globulins in the blood was established by electrophoresis, and some typical patterns for plasma proteins are shown in Figure 12-15. It is seen that the boundaries in the ascending and descending patterns are not quite equivalent because of slight differences in the ionic environment. For a protein of questionable homogeneity, the formation of only one boundary on electrophoresis leads to a strong presumption that the material consists of only one component.

Distinctions made by electrophoresis between various forms of hemoglobin serve as a classic example of the power of the method. Hemoglobin is a dimer consisting of two symmetrical protein components. Each half-molecule contains an α and a β chain of peptides. Individuals with sickle-cell anemia, a disease in which red blood cells change from a spherical to a sickle shape when the blood oxygen level is reduced, because the hemoglobin becomes less soluble, have a form of hemoglobin, designated Hb S, with a mobility greater than that of normal hemoglobin, Hb A, and with an isoelectric point about 0.2 units higher. This difference has been traced to the replacement of a particular glutamic acid residue, which has a negative charge, in each half of the dimer by a valine residue in hemoglobin Hb S, adding two additional positive charges to the whole molecule. The abnormality is

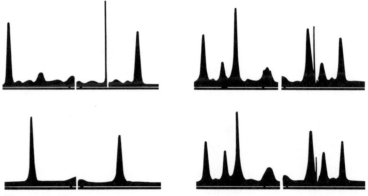

Figure 12-15
Tiselius electrophoretic diagrams. The left half of each of the four diagrams is
the pattern of the ascending limb of the solution in the U-tube, that to the right,
of the descending limb. The photographs were taken after migration at 1°C for
200 min in a Veronal buffer of pH 8.6 with a ionic strength of 0.1, using a current
of 16 mA. The gap in the white line superimposed on each pattern indicates the
origin. The diagram at the upper left is that of normal human serum; the
components in order of decreasing mobility are albumin, α_1-globulin, α_2-globulin,
β-globulin, and γ-globulin. There are also "anomalous" peaks at the origin which
do not correspond to a solute. The diagram at the upper right is that of abnormal
serum from a patient with multiple myeloma, that at the lower left is of the
Bence–Jones protein found in the urine of the same patient, and that at the lower
right is of a mixture of this protein and the serum of the patient. The last pattern
indicates clearly that the Bence–Jones protein is homogeneous with one of the
constituents of the abnormal serum. (Courtesy of Dr. F. W. Putnam.)

located in the sixth amino acid position from the N terminus of the β
chain. About 50 other genetically controlled variants of hemoglobin
have been found, such as Hb C which has lysine, bearing a positive
charge, at the same sixth position.

ELECTROPHORETIC SEPARATIONS

Separations can also be carried out by electrophoresis, allowing the
faster-moving component to migrate ahead and then physically removing this portion of the solution. If the pH can be adjusted to a value
between the isoelectric points of two components in a mixture, the
resulting migration in opposite directions facilitates separation.

When it is not necessary to measure mobilities quantitatively, and
when separation is the primary aim, it is simpler and more satisfactory
to conduct an electrophoresis experiment with a rigid supporting
medium, such as paper or a gel. In paper electrophoresis, a horizontal
strip of paper is moistened with buffer solution, and the paper is laid
horizontally on top of a glass plate support with the two ends of the
paper dipping in vessels each containing buffer solution and an electrode. The sample is placed as a spot or stripe in the middle of the
paper, and the solutes separate into distinct zones as they migrate
toward the electrodes in the flow of current along the paper. The strip
can be treated with reagents to locate the protein molecules or can be

cut apart after the experiment to achieve their separation. In gel electrophoresis, the procedure is very similar, with the solution placed in a groove made across the surface of a horizontal strip of gel and perpendicular to the direction of current passage. Starch, agar, or polyacrylamide gels can be used.

There is no difficulty with convection in this procedure, but there is always the possibility that the support will influence the mobility. Thus proteins tend to be adsorbed, and therefore retarded, by paper. Electroosmosis, the bulk flow of solvent, occurs when migration of the macromolecules is hindered, and this is observed with a paper or agar medium. In some circumstances, however, it is possible to utilize differential retardation of mobility according to the size of the macromolecules, along with charge differences, to achieve better separations than unimpeded electrophoresis can provide.

12-9
OPTICAL PROPERTIES
AND LIGHT SCATTERING

One of the characteristics that distinguishes colloidal dispersions from true solutions is the scattering of light by the colloidal particles. A system of either type appears clear if the viewer looks through it toward a source of light, but a colloidal dispersion appears cloudy if viewed in a direction transverse to a light beam passing through it. Scattering of light, termed the *Tyndall effect,* can be observed visually from a beam passing through a lyophobic colloidal dispersion in a beaker or through a cloud of smoke in a darkened room, as in the case of a projection beam in a movie theatre.

In an ultramicroscope, scattering of light is utilized for the observation of individual colloidal particles. This instrument is merely an ordinary microscope provided with a dark field: The illuminating system is arranged so that the rays of light come into the sample from the side rather than being transmitted directly through the sample to the eye of the observer. Individual particles scatter photons into the optical system of the microscope and thus, like stars in the sky, can be seen despite the fact that they subtend an angle in the field of view so small that their dimensions cannot be determined. Thus the Tyndall effect permits the particles in a dispersion to be counted, although their size and shape cannot be determined directly. Unfortunately, many lyophilic macromolecules in solution do not differ sufficiently from the solvent in index of refraction, a property significant in determining the extent of light scattering, to permit them to be observed as discrete particles in the ultramicroscope.

The quantitative theory of light scattering by small isolated particles was worked out by Lord Rayleigh whose interest in the subject was aroused by the problem of the color of the sky. For particles with dimensions less than one-tenth the wavelength of the light λ, the ratio of the intensity scattered at an angle θ from the direction of the transmitted beam to the intensity of incident, unpolarized, monochromatic radiation is given by

$$\frac{I_\theta}{I_0} = \frac{8\pi^4 a \alpha^2}{\lambda^4 r^2}(1 + \cos^2 \theta) \qquad (12\text{-}38)$$

In this equation, a is the number of scattering particles per unit volume, r is the distance of the observer from the sample, and α is the electrical polarizability of the particle. Since the particle scatters light in three dimensions, and the total energy must be constant, the intensity varies inversely with distance in the same way that the total surface of a sphere varies directly with the radius: according to the second power. The polarizability enters because the incident light induces vibrating dipoles in the particle which then act as sources and radiate light in all directions; the more polarizable are the electric charges in the particle, the greater is the magnitude of these oscillating dipoles. An outstanding feature of the Rayleigh equation is the very strong dependence of the scattered intensity upon the wavelength, which appears to the inverse fourth power. The fact that blue light is scattered more than red light was used by Rayleigh to explain the blue color of the sky when viewed in any direction but toward the sun.

The dependence upon the angle θ arises because an observer at any one location looking toward the scattering molecules can see only a component of the oscillations of the dipoles. Figure 12-16 shows how the dipoles appear to an observer: Oscillations perpendicular to the line of sight are seen in full magnitude and give rise to the term unity in the last factor in Equation (12-38). Those in any other direction are seen with reduced amplitude corresponding to their projection on the line of sight. As the observation point is moved around in a circle in any plane containing the incident light beam, the projection factor varies as $\cos \theta$; since the energy of a wave depends on the square of the amplitude, the angular factor in the intensity equation is $\cos^2 \theta$, varying from unity at $\theta = 0°$ to zero at $\theta = 90°$ and back to unity at $\theta = 180°$.

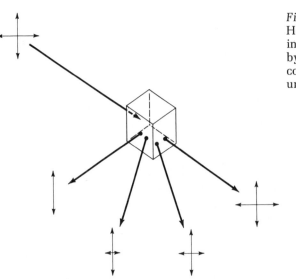

Figure 12-16
How an observer views the induced dipoles produced by the two perpendicular components of incident unpolarized radiation.

To apply Equation (12-38) in practice, it is necessary to use the refractive index n of the scattering molecules as a measure of their polarizability. Optical theory leads to an equation for the refraction per unit volume, the specific refraction:

$$\frac{n^2 - 1}{n^2 + 2} = \frac{4}{3}\pi a \alpha \tag{12-39}$$

If the refractive index is approximately unity, this simplifies to

$$\alpha = \frac{n^2 - 1}{4\pi a} \tag{12-40}$$

However, for solutions, it is the difference between the polarizability of the solution and that of the pure solvent that is significant; using zero to denote the solvent,

$$\alpha - \alpha_0 = \frac{n^2 - n_0^2}{4\pi a} \tag{12-41}$$

The refractive index for a solution of concentration c grams per liter can be written approximately as

$$n = n_0 + c\frac{dn}{dc} \tag{12-42}$$

Squaring both sides of this equation leads to

$$n^2 = n_0^2 + 2n_0 c\frac{dn}{dc} + c^2\left(\frac{dn}{dc}\right)^2 \tag{12-43}$$

The last term is small and can be neglected, so that

$$n^2 - n_0^2 = 2n_0 c\frac{dn}{dc} \tag{12-44}$$

We now express the concentration a as cN/M, where N is Avogadro's number, and combine Equations (12-41) and (12-44), yielding

$$\alpha - \alpha_0 = \frac{n_0(dn/dc)M}{2\pi N} \tag{12-45}$$

Substitution of the square of this quantity for α^2 in Equation (12-38) leads to

$$\frac{I_\theta}{I_0} = \frac{2\pi^2 n_0^2(dn/dc)^2}{N\lambda^4}\frac{1 + \cos^2\theta}{r^2}Mc \tag{12-46}$$

The first factor on the right, containing the wavelength of the light and information about the refractive index, is termed the optical constant and is represented by the symbol K. The symbol R_θ represents the intensity ratio corrected for the geometry of the system by multiplying by $r^2/(1 + \cos^2\theta)$; R_θ is independent of the value of θ. Thus the equivalent of Equation (12-46) is

$$R_\theta = KMc \tag{12-47}$$

From Equation (12-47), the molecular weight of a solute can be calculated. If the solute consists of molecules such as synthetic polymers or a mixture of proteins, all of essentially the same chemical nature so that dn/dc is constant, the value of M obtained is a weight-average molecular weight.

If a solution in which the solute is scattering light is not ideal—in other words, if the solute particles are not independent of one another—an approach developed by Einstein and by Debye can be used. The basic idea is that no light will be scattered by the solution as a whole if the solute is uniformly distributed on a microscopic scale throughout the solution, because light from one particle will interfere destructively with light from a neighbor, just as no light is scattered by a transparent crystal so long as there are no crystal imperfections. Fluctuations in concentration result from random motion such as Brownian motion, and their extent varies inversely with the tendency to restore uniform concentration when a deviation from uniformity occurs—in other words, inversely with the osmotic pressure developed by a given concentration difference. The equation derived from this theory is

$$\frac{Kc}{R_\theta} = \frac{1}{RT}\frac{d\pi}{dc} \tag{12-48}$$

In the ideal case, $\pi = cRT/M$, and this equation reduces to the equivalent of Equation (12-47). For a nonideal solution, we can use Equation (12-9) for osmotic pressure in terms of the interaction constant B, and Equation (12-48) then becomes

$$\frac{Kc}{R_\theta} = \frac{1}{M} + 2Bc \tag{12-49}$$

From the slope and intercept of a plot of Kc/R_θ against concentration, the values of B and M can both be evaluated.

The treatment up to this point has been valid for particles smaller than the wavelength of light. If the dimensions of the particles are

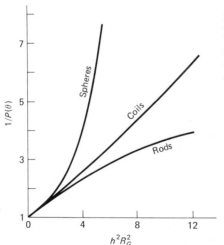

Figure 12-17
Effect of the shape of a large particle on the variation in the reciprocal particle scattering factor for light of wavelength of the same order as the particle dimension.

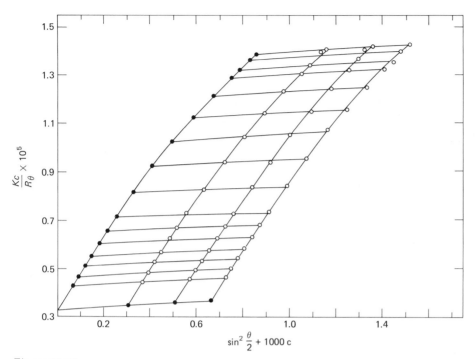

Figure 12-18
A Zimm plot, as described in the text, based on the data of Doty for collagen in a citrate buffer at pH 3.7 and 15°C. The molecular weight is 310,000 and the length of the rod-shaped molecules is 300 nm. Reproduced with permission from H. Boedtker and P. Doty, *J. Am. Chem. Soc.* **78**, 4267 (1956). Copyright by the American Chemical Society.

sufficiently large to be appreciable in comparison to the wavelength, then interference occurs between light scattered from different parts of the same particle. The ratio of the scattered intensity at any angle θ to that which would be observed if the particle had the same molecular weight but dimensions small compared to the wavelength of the light is termed the *particle scattering factor* $P(\theta)$. This factor depends upon the shape of the particle and can in principle be evaluated by integrating over all pairs of scattering elements in a particle of any assumed shape. The results for spheres, rods, and coils are shown in Figure 12-17 as a function of h, defined as $(4\pi/\lambda) \sin (\theta/2)$, and the radius of gyration R_G, which was defined in Section 12-3. Often measurements are made at θ values of 45 and 135°, and the ratio of these two intensities is plotted. At low scattering angles or longer wavelengths, the function becomes independent of particle shape, and the radius of gyration can be evaluated, without knowledge of the shape:

$$\lim \frac{1}{P(\theta)} = 1 + \frac{h^2}{3} R_G^2 \qquad (12\text{-}50)$$

Of course, conversion of the R_G value to a particle weight requires information about the shape.

A method of plotting experimental data that takes into account both solution nonideality and the effects of particle size and shape according to the approximate equation

$$\frac{Kc}{R_\theta} = \frac{1}{M}\left(1 + 2Bc + \frac{h^2}{3}R_G{}^2\right) \qquad (12\text{-}51)$$

has been developed by B. H. Zimm. The experimental values of Kc/R_θ are plotted against $\sin^2(\theta/2) + kc$. The first term is proportional to h^2, and the factor k in the second term is an arbitrary constant chosen to make the data points spread conveniently over the graph. On the plot, lines are drawn through points at constant angle and extrapolated to zero concentration, and lines are drawn through points at constant concentration and extrapolated to zero angle. The two sets of limiting values are each then extrapolated to a common intercept at zero value of the abscissa, yielding $1/M$. An example of a Zimm plot is shown in Figure 12-18.

12-10
GELS AND
MOLECULAR
EXCLUSION
CHROMATOGRAPHY

Gels are colloidal dispersions that possess physical properties much like those of a solid, including substantial resistance to flow, although they contain a rather large amount of one component which is normally a liquid at the same temperature. The setting of a concentrated gelatin sol upon cooling to a semisolid mass is an example of gel formation. Clotting of the blood is gelation of the plasma under the influence of fibrin. Formation of a gel from a colloidal sol differs from coagulation or precipitation of the dispersion in that the gel contains all the liquid component present in the solution.

STRUCTURE
AND PROPERTIES OF GELS

The structure of gels obtained by solidification of a sol is still imperfectly understood. It is evident that the molecules constituting the disperse phase in the corresponding sol have somehow been linked together to give the whole a rigidity that cannot be contributed by the solvent. In some systems the fibers of the disperse phase may be extensively branched and may become entangled with one another, perhaps in a way analogous to that in which branches in a brush pile become interlaced to form a skeleton structure of some rigidity without occupying a very large fraction of the total volume and without being firmly bonded together as a single unit. The mutual attraction of solvent and solute molecules must somehow play a role in stabilizing the gel, but experimental evidence indicates that the rate of diffusion of small solute molecules and the capacity of ionic solutes to conduct electric current are reduced only very slightly, if at all, by gelation. Thus the transformation of a sol to a gel does not involve loss of continuity of the liquid phase nor much loss of mobility, but leads to a structure with two continuous intermeshing phases.

Most inorganic gels are inelastic: When the gel is dehydrated, as by gentle warming, the mass shrinks only to a limited extent, after which the liquid phase is replaced by air. If a partially dried gel is exposed to water, some water is reabsorbed but not as much as has been removed, and if too much water is removed, the gel loses the capacity to absorb water. Silica gel, which can be prepared by adding acid to a sodium silicate solution of suitable concentration, is typical of this sort of material.

Many organic gels can be formed by absorption of solvent by the dry material. A dry protein such as gelatin or agar undergoes *imbibition* or *swelling* in liquid water or even in the presence of water vapor. Gelatin swells to a limit, whereas egg albumin swells indefinitely and eventually is dispersed to a solution. Elastic gels can be pictured as composed of molecules rather loosely held together with the contacts able to rearrange as the liquid content or the temperature is varied. Organic polymers swell in organic solvents, as does rubber in benzene.

Some gels can be reversibly converted to sols by the application of a shearing force. If an electrolyte that would cause a dilute sol to coagulate is added to a concentrated sol of a material such as ferric oxide or aluminum oxide, the sol sets to a *thixotropic* gel, one that can be liquefied temporarily by shaking the container. Thixotropy is possessed by aqueous suspensions of some clays such as bentonite; these are used in muds for petroleum drilling because the mud sets when the drill stops but liquefies easily when the drill is started. A thixotropic substance is an example of a non-Newtonian material, for its flow rate is not proportional to the applied force, and thus there is no definite value for the coefficient of viscosity. It seems probable that thixotropy is characteristic of the protoplasm in living cells.

GEL CHROMATOGRAPHY

We have mentioned several methods for the analysis and separation of mixtures of macromolecules. Another method, recently developed and widely used, is termed *gel filtration* or *gel permeation* or *molecular exclusion* chromatography. Like other forms of chromatographic separation and analysis, this technique utilizes a series of many successive contacts between two environments to accentuate the effect of a preferential distribution between them.

In this version of chromatography, small beads or particles of organic polymer are swollen with solvent to form a porous gel which is packed into a vertical tube or column. A small volume of solution containing the solute to be separated is placed above the gel, and then solvent is added continuously to produce flow through the column. If the size of the gel pores has been suitably chosen so that they are accessible to smaller molecules in the solute mixture, these smaller molecules wander into the pores and are restricted in their progress down the column compared to the larger molecules which cannot enter the pores and which pass rapidly down the column through the void space, exiting from the bottom ahead of the smaller particles. A differential refractometer or an ultraviolet transmission device can be used to indicate the appearance of solute molecules in the effluent from the column.

If the gel contains a more-or-less continuous distribution of pores of differing size, then the volume accessible to molecules of various sizes also varies continuously, and molecules will come out of the column sequentially according to their size. To cover a wider range of sizes than can be accommodated with a single gel, several columns with gels of different porosity can be used in tandem.

Gel permeation chromatography was first developed for biochemical separations of proteins, using the polysaccharide dextran (Sephadex) and polyacrylamide (BioGel). Agarose gels have been employed for the separation of bacteria, cell nuclei, ribosomes, and other larger particles. For synthetic polymer molecular-weight determinations or separations, polystyrene with a varying extent of cross-linking to control pore size has been found effective.

12-11
MICELLES AND EMULSIONS

In this and the following sections we describe several systems in which molecules arrange themselves into particular structures based on amphiphilic (or amphipathic) properties—the same duality in nature associated with the presence of polar groups and a nonpolar hydrocarbon chain which leads to surface activity—often accompanied by asymmetry in shape.

MICELLES AND ASSOCIATION COLLOIDS

Certain amphiphiles exist as single molecules or ions in true solution at low concentrations in water but associate at higher concentrations into aggregates termed *micelles*. An example is sodium dodecyl sulfate (also called sodium lauryl sulfate) for which extrapolation to infinite dilution of measurements of colligative properties at low concentrations indicates an average particle weight of 144, a value half the formula weight and one consistent with ionization into sodium ion and dodecyl sulfate ion. At concentrations above about 25 millimolar, the ratio of osmotic pressure to weight of solute falls substantially over a narrow range of concentration, indicating a growth in particle size. The aggregates formed at higher concentrations are called micelles, and the solution composition at which aggregation sets in is termed the *critical micelle concentration* or *cmc*. Of course, above the cmc there are also monomers present in equilibrium with the micelles. Other examples of micelle formers are sodium stearate, $C_{17}H_{35}COO^-Na^+$, and cetyltrimethylammonium bromide, $C_{16}H_{33}(CH_3)_3N^+Br^-$. Detergents and water-soluble dyes are association colloids of practical importance.

Many micelle formers are electrolytes, and measurements of electric conductivity are an aid in characterizing the extent of their ionization. The equivalent conductance drops as the concentration increases through the cmc, but not as much as do the colligative properties. A typical micelle of sodium dodecyl sulfate contains about 60 lauryl

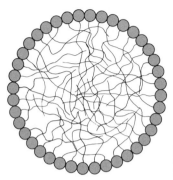

Figure 12-19
Possible arrangement of molecules in a spherical micelle.

sulfate ions and about 40 sodium ions, leaving about 20 sodium ions free as counterions, with a net negative charge of about 20 units on the micelle.

The sizes of micelles have been measured by light-scattering techniques, and further information about their shape and size has been obtained by hydrodynamic methods, such as sedimentation experiments. Apparently those formed by many systems in the concentration region not too far above the cmc are very nearly spherical in shape and have a diameter of the order of twice the length of an individual molecule. It seems clear that the arrangement of molecules must be one in which the polar groups are in a spherical array facing the solvent on the outside, with the hydrocarbon tails somehow interlaced in the interior of the micelle. Why do these structures appear rather abruptly at a critical concentration? Probably the critical concentration is the lowest one at which molecular geometry permits a large enough loss in the solvent–hydrocarbon interface when the hydrocarbon chains are brought into a compact unit isolated from the solvent to overcome the repulsive forces between neighboring polar groups and the unfavorable entropy of association.

Spectroscopic results indicate that the interiors of spherical micelles are almost liquid in nature, with considerable freedom of motion of the hydrocarbon chains still preserved. It has also been found that some 10 or more water molecules are incorporated into a micelle for each amphiphilic molecule present. Figure 12-19 indicates the possible structure of a spherical micelle.

A most significant feature of micellar behavior is the phenomenon of solubilization, by which organic molecules such as benzene or a water-insoluble dye are brought into solution as a result of the presence of micelles. Examination by x-ray diffraction shows that the characteristic spacings in micelles are increased by the presence of solubilized material. Solubilization is one of the principal mechanisms by which detergents and soaps are effective in removing organic soils. In some instances, the solubilized species helps to stabilize the micelle, for the cmc is lowered in its presence. The interior of the micelle, hydrocarbon in nature, provides a favorable environment for the solubilized hydrocarbon species, and perhaps a similar effect is involved in the solubilization of fats in the digestive system.

At higher concentrations of amphiphile, the structures of association colloids change: Spheres or nearly spherical ellipsoids give way to cylindrical, rod-shaped units, which can grow in length indefinitely without changing the surroundings of the individual amphiphilic molecules. At still higher concentrations, the amphiphilic species may be found packed in bilayers in a leaflet arrangement. More is said about such larger micelles in Section 12-12.

In general it is found that the preferred micelle geometry, both the shape and the size, depends upon the length of the hydrocarbon chain, the size and solvation of the polar groups, the temperature, and the ionic strength of the medium, which affects electrostatic interactions between the polar groups and ions, as well as upon the concentration.

EMULSIONS

An emulsion is a dispersion of droplets of one liquid in another liquid with which it is immiscible. The commonly encountered examples consist of water in an organic liquid—an "oil"—or an organic liquid in water. Pharmaceuticals are often prepared in the form of emulsions, and both plant latexes and commercial latex paints are emulsions.

The droplets in an emulsion may range in size from colloidal dimensions up to those visible in the microscope or to the naked eye. Such dispersions are stable only in the presence of a third material, an emulsifying agent, situated at the interface between the phases. On standing, even if an emulsifying agent is present, they may in time *cream,* with droplets concentrating in an upper layer if the disperse phase is less dense, or even separate into two bulk phases. In the absence of an emulsifying agent, the interfacial tension between the two liquids is sufficiently great so that separation into two layers occurs in order to minimize the area of contact. The stabilizing species is typically an amphiphile, such as a metal carboxylate soap, an alkylsulfonic acid, or an alkyl sulfate. Obviously the polar group is on the aqueous side of the interface, and the hydrocarbon chain is on the oil side. Hydrophilic colloids such as gelatin and certain gums, as well as solids such as silica and carbon black, also serve as emulsifying agents.

The nature of the emulsifying agent is important in determining which phase becomes the disperse or discontinuous phase in an emulsion and which phase becomes the continuous phase. Alkali-metal soaps favor oil-in-water emulsions in contrast to the corresponding salts of polyvalent cations such as zinc, aluminum, and the alkaline earths which favor water-in-oil emulsions. The larger ratio of hydrocarbon cross section to polar group cross section in the latter apparently influences the hydrocarbon to be on the outside of the spherical shell.

To see the relation in structure between emulsions and micelles, one may imagine a process of solubilization in which a spherical micelle absorbs organic molecules, becoming larger and larger until eventually the interior forms a droplet of organic liquid. In practice, emulsions are not often formed this way, but they are made by mechanical agitation ranging in severity from vigorous shaking to forcing a mixture through a jet under pressure or spraying the liquid against a rotor turning at a high speed, methods used in homogenizing milk by dispersing the globules of butterfat into smaller droplets in the aqueous medium.

**12-12
LIQUID CRYSTALS AND
LARGE MICELLES**

The physical properties of ordinary molecular liquids, as well as those of colloidal dispersions, solutions of macromolecules, and spherical micelles, are isotropic. This means that the properties are the same in every direction, in distinction to the anisotropy exhibited by many crystalline solids for which the properties vary with the direction in which they are measured. The refractive index and the extinction coefficient for absorption of light are readily measured properties which may be anisotropic; less readily measurable characteristics such as compressibility and thermal conductivity also vary with direction in anisotropic solids.

Liquid crystals represent a somewhat unusual type of matter in which the substance flows like a liquid but has anisotropy in its optical properties much like that observed in solids. Certain substances, most of which have fairly rigid, rod-shaped molecules containing suitable polar functional groups, as well as polarizable units such as aromatic rings joined by an unsaturated linkage, melt from the solid to form an anisotropic liquid crystalline phase, from which they undergo another transition to an isotropic liquid at higher temperatures. Liquid crystals formed in this way, by pure materials undergoing a change in temperature, are termed *thermotropic*. Because the liquid crystal phase is intermediate in properties and lies between solid and liquid on the phase diagram, it is often referred to as *mesomorphic*. A mesomorphic phase retains something of the order present in the solid phase of the same material.

Liquid crystals can be classified according to the nature of the order in their molecular arrangements, as illustrated in Figure 12-20. The *smectic*, or "soaplike," type contains an array of elongated molecules lying side by side like logs in a woodpile and stratified into layers or sheets able to move past one another. The direction of a sheet is perpendicular to the long axis of the molecule. Within a sheet the molecules may be randomly arranged or may be located in a regular pattern.

Smectic

Nematic

Figure 12-20
Arrangement of axes of
rod-shaped molecules in small
regions of smectic and
nematic liquid crystals.

A molecule that forms a smectic phase is ethyl-p-azoxy cinnamate:

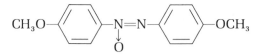

$$CH_3CH_2OOC-CH=CH-\langle\ \rangle-N=N-\langle\ \rangle-CH=CH-COOCH_2CH_3$$

Some substances form several different smectic phases which vary in appearance and optical properties; possible structural differences among them include tilting of the molecules away from exact perpendicularity to the layer, as well as differing arrangements with respect to neighbors within the layer. The smectic phase has a relatively high degree of order, and the overall order is not very sensitive to environmental conditions, such as temperature or magnetic fields.

In a *nematic* liquid crystal, the molecules are also parallel to one another, but there are no layers: There is no periodicity in the direction parallel to the long axis of the molecule. The term "nematic" comes from the threadlike appearance of structures observed in this kind of material in the microscope. An example is p-azoxyanisole, which is nematic in the range 84 to 150°C:

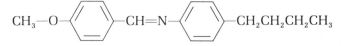

$$CH_3O-\langle\ \rangle-N=N-\langle\ \rangle-OCH_3$$

Some substances form both smectic and nematic phases, with a transition temperature between the stability ranges of the two mesomorphic forms. Since the smectic phases correspond to a higher degree of order, they are generally stable at lower temperatures than the corresponding nematic phases. Recently efforts have been made to design compounds for which the liquid crystal phase is stable down to ambient temperatures, taking into account the delicate balance between lateral intermolecular forces of attraction on the one hand and steric repulsions or failure of the molecules to achieve a good mutual fit on the other hand. An example of such a compound is the Schiff base derivative, 4-methoxybenzylidene-4'-n-butylaniline (MBBA):

$$CH_3-O-\langle\ \rangle-CH=N-\langle\ \rangle-CH_2CH_2CH_2CH_3$$

This substance forms a nematic phase stable from 21 to 47°C, and the corresponding ethyoxy compound, EBBA, forms a nematic phase stable from 36 to 80°C.

In a nematic phase, the ordering seems usually to extend over only a limited range of distance and is sensitive to external forces; a rather uniform order can be established by mechanical treatment, such as rubbing the material over a surface or rotating the container, or by the application of an external electric or magnetic field. Of course, thermal motion produced by increasing the temperature works to disrupt the order.

Small molecules can be dissolved in many thermotropic liquid crystals, and when these solute molecules are geometrically anisotropic, they assume some of the order of the medium. The resulting ordered solutions make it possible to study the polarization properties of spectroscopic transitions of the solute molecules.

A third type of liquid crystal order is termed *cholesteric*, because the molecules exhibiting it are chiefly fatty acid esters of cholesterol, although cholesterol itself does not form a liquid crystalline phase. Some workers regard this type as a special case of the nematic phase, and the molecules are arranged in a somewhat similar fashion. The difference is that, as a cholesteric phase is traversed in one of the directions perpendicular to the long axis of the molecules, each successive layer of molecules is twisted in a regular way compared to the previous layer. The pitch of the spiral—the distance for a 360° turn in direction—is between 0.2 and 20 μm. Optically active molecules are necessary to form a cholesteric phase, and the resulting phase is optically active. From the circular dichroism associated with optical activity (see Section 9-10), bright iridescent colors result. Since the color the material assumes depends upon the pitch of the helix, and the latter varies with temperature, cholesteric liquid crystals have been used as sensitive temperature indicators. One application is in the measurement of skin temperature in order to detect malignancies by virtue of the higher temperature associated with their higher metabolic rate.

As distinguished from thermotropic phases, *lyotropic* liquid crystals are those formed in certain concentration ranges, usually of solutions of an amphiphile in water. Indeed the most common sort is closely related to the association colloids, or micelle formers, described in Section 12-11. If one increases the concentration of a micelle former, such as sodium dodecyl sulfate, well above the cmc, there appears a second fairly abrupt change, corresponding to an increase by a factor of about 10 in the equivalent molecular weight of the micelles. The best evidence seems to indicate that the micellar structure has changed from spherical or ellipsoidal to cylindrical or rod-shaped; the rodlike micelle can grow to any length without changing the environment of the individual molecules. Cylindrical micelles can be oriented by mechanical motion of the suspension or, if they are present in sufficiently high concentration, they can stack together in a close-packed hexagonal array.

At still higher concentrations of amphiphiles, there is a further transition to leaflet or lamellar structures, in which the molecules are arranged as stacked bilayers, hydrocarbon tail to hydrocarbon tail and polar head to polar head, as shown in Figure 12-21. The heads can still be surrounded by water molecules or separated from the heads of adjacent layers by a region of water and counterions. The lamellar micelles may shrink or expand, to a limited extent, by the transfer of material between this aqueous region and the bulk of the solution as influenced by changes in ionic concentrations in the solution. A lamellar micelle corresponds to the smectic structure described for thermotropic liquid crystals.

Representative substances that form lyotropic liquid crystals are alkyl trimethylammonium ions, hexadecylpyridinium chloride, and n-octylhexaoxyethylene glycol mono ether [$CH_3(CH_2)_7O(C_2H_4O)_6H$].

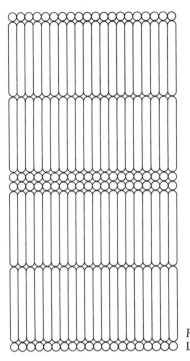

Figure 12-21
Lamellar structure of a micelle.

The doubly refracting, or birefringent, parts of striated muscle fiber, the axons of nerve cells, and the cilia of lower organisms can all be classified as liquid crystals because of their optical properties. Finally, a particularly interesting class of liquid crystal formers is the group of phospholipids discussed in Section 12-13.

12-13
BIOLOGICAL MEMBRANES
AND PHOSPHOLIPID
BILAYERS

Biological membranes are thin layers composed mostly of proteins and of less polar or fat-soluble *lipids*. In addition to enclosing the entire cell contents, membranes in cells of higher organisms provide internal structure for the organelles within the cell. Thus there is a membrane surrounding the nucleus, membranes forming the mito-chondria, and membranes constituting the endoplasmic reticulum, a network of channels throughout the cell. The myelin sheath sur-rounding the axon of a nerve cell and the discs of rod cells in the retina of the eye are membranes specialized for particular functions.

The lipid fraction of membranes varies from about one-fourth to three-fourths of the total material, while the remainder is protein. There is also some carbohydrate, but it is usually in the form of a sugar attached to a protein or lipid molecule, a glycoprotein or a glycolipid. The principal lipid material consists of molecules with two long hydro-carbon chains such as the phosphatidyl compounds described on page

415. Sphingolipids represent another class of lipids of which sphingomyelin is an example:

$$CH_3(CH_2)_{12}CH=CH-\underset{\underset{\displaystyle OH}{|}}{CH}-\underset{\underset{\displaystyle NH}{|}}{CH}-CH_2-O-PO_2-O-CH_2-CH_2-N(CH_3)_3$$

$$\underset{\underset{\displaystyle R_1}{|}}{C}=O$$

$\ominus \qquad \oplus$

Cholesterol is also found in the membranes of higher animals, comprising, for example, 25 percent by weight of the membranes of human erythrocytes, but is not found in bacterial membranes.

There is considerable evidence that the lipid components of a membrane are arrayed in a bilayer structure, with the polar ends of the molecules forming the inner and outer surfaces of the membrane, much like one layer of a smectic lyotropic liquid crystal. Various ways have been developed to assemble lipid bilayer structures outside the cell in order to study their properties to provide a model for cell membranes. One of these procedures begins with a solution of a lipid in a hydrocarbon solvent between octane and dodecane in molecular size. A drop of this solution is painted with a fine brush across a small hole drilled in a piece of Teflon. It is best to have the Teflon support immersed in an aqueous solution. The resulting film first shows colored interference patterns, but in a short time, if the experiment is successful, excess material drains out of the film and its two surface layers of lipid molecules come together, tail to tail, forming a single bilayer often called a *black lipid* membrane because of its unusual appearance.

This membrane can be used to demonstrate the effect of the cyclic dodecapeptide valinomycin in facilitating the transfer of potassium ion. Initially, there is a large electric resistance across the membrane, but this is reduced by a factor of about 10^5 when a trace of valinomycin is added to the solution if potassium ion is also present. However, there is no effect if only sodium ion is present. D. W. Urry has proposed that valinomycin can bind a potassium ion in a kind of pore in its center, and that the complex may then undergo a conformational change which moves the ion through the pore to the other side of the peptide. Hydrated sodium ion is evidently too large for this process to occur. An interesting point is that valinomycin can apparently insert itself into the membrane so that the transfer can take place through the barrier.

When lipid molecules, except for cholesterol, are introduced into aqueous solutions, they behave very much like other amphiphiles. Those with two hydrocarbon chains tend not to form spherical or cylindrical micelles, however, because of geometric considerations —the polar groups are too small in proportion to the volume of the hydrocarbon chains—but they do readily form stacked bilayer sheets which are smectic liquid crystals. Lipids are not easily dispersed by simply dropping them into water, but liquid crystalline dispersions are more readily prepared by carefully moistening the anhydrous material. X-ray diffraction studies of stacked bilayers of dipalmitoyl phosphatidylcholine show a repeat distance of 43 Å, corresponding to the distance between head groups on opposite sides of a single bilayer. A second repeat distance, 16 Å in the dry state and increasing to 25 Å when water is added, corresponds to the distance between head groups in adjacent bilayers. In the direction parallel to the bilayer surface, a reflection corresponding to 4.2 Å is observed; this is characteristic of the separation between extended hydrocarbon chains which are closely packed.

Ultrasonic irradiation of a smectic phase lipid solution causes the planar layers to fold into spherical vesicles which entrap in their interior a portion of the aqueous medium. Each vesicle consists of one or more concentric bilayers in which the hydrocarbon tails of the two halves of the bilayer adjoin each other in the interior of the bilayer. A cross section through a vesicle with two bilayers is shown in Figure 12-22. By suitable separation procedures, such as gel permeation chromatography, a fraction of small vesicles consisting of only one bilayer shell plus the interior liquid can be obtained. From egg yolk phosphatidylcholine, the vesicles obtained are fairly uniform in size, with a radius of about 250 Å containing about 2700 molecules. Materials trapped within the vesicle can remain there for relatively long periods of time, and rates of permeation through the vesicle can be measured. The process of transfer of a lipid molecule forming part of one half of the bilayer to become part of the other side of the bilayer is very slow, with a half-time of the order of hours or longer; but, at least above a transition temperature which will be described below, molecules within one layer can move around quite rapidly in that layer.

X-ray diffraction and electron microscope studies indicate that cell membranes are quite similar in structure to spherical vesicles. Although there has been some dispute about the interpretation of the electron micrographs, which must be obtained by staining a section of cell with a material like OsO_4 or UO_2^+ which is opaque to electrons, these pictures invariably show for a membrane a sandwichlike arrangement of two dark surfaces with a light layer between. The

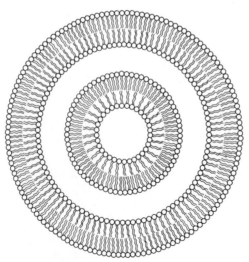

Figure 12-22
Section through a spherical phospholipid vesicle containing two concentric bilayers.

outer surfaces are evidently the polar groups, and the light region is the dual layer of hydrocarbon chains. X-ray diffraction shows a spacing parallel to the bilayer plane of between 4 and 5 Å, similar to that in the vesicles. It has also been possible to interpret intensities of scattered x rays as indicating two layers of high electron density on the membrane about 40 Å apart, corresponding to the polar head groups, with a trough in electron density midway between, corresponding to the methyl group locations. The protein components of a membrane constitute a great variety of molecular types and, further, they are not arranged in any regular structural pattern, so that x rays give no information about their location in the membrane. Despite the fact that proteins may form the major portion of the membrane material, it seems clear that the general arrangement is governed by the properties of the lipids and not by the proteins.

A process of fusion of two membranes occurs when cells are infected with lipid-enveloped viruses, and corresponding changes have been found in model systems. H. M. McConnell and co-workers demonstrated that vesicles of phosphatidylcholine of 250 Å diameter spontaneously fuse together in pairs to form larger vesicles, as indicated by a mixing of their contents and by a decrease in the motional constraints upon the inward-facing molecules. McConnell was also able to incorporate labeled lipids, as described in Chapter 13, into membranes reconstituted from the components of the sarcoplasmic reticulum, which controls the flow of calcium ions to the muscles, while these membranes retained their functional integrity, as well as into the membranes of intact living cells of the organism *Mycoplasma laidlawii*.

Thermal transitions—types of phase changes—are observed both in cell membranes and in model phospholipid systems. In anhydrous phospholipids, a transition is found much below the normal capillary melting point and is evident from differential thermal analysis (see Section 3-6), as well as from an increase in interchain spacing from 4.2 to 4.7 Å as measured by x-ray diffraction. In the hydrated liquid crystalline state, similar transitions are observed, but at yet lower temperatures—for the C_{18} molecule, 58°C compared to 115°C for the same substance when anhydrous. Measurements using both nuclear magnetic resonance and electron paramagnetic resonance techniques (see Chapter 13) show a change in mobility of various parts of the molecule associated with the transition; they indicate that the hydrocarbon chains become fluid, consistent with the x-ray findings, but that the internal motion is more restricted for parts of the hydrocarbon chain closer to the head group. The head groups, while somewhat less rigid, are still constrained to a regular array even above this transition temperature. For model systems composed of several phospholipids mixed together, the thermal transition occurs over a range of temperature, so that the solid phase and the liquid crystal phase can coexist throughout a range of temperature rather than merely at a single temperature. The phase behavior of a mixed lipid system is often typical of that of a mixture of substances that are miscible or partially miscible in the liquid phase but insoluble in the solid phase. It has been proposed that the degree of unsaturation in membrane lipids varies because of the need for each organism or organ to have the transition temperature of its cell membranes at an optimum value for physiological function at the usual temperature of its environment.

The model of membrane structure accepted for some time was that proposed by H. Davson and J. F. Danielli; in this model the proteins are pictured as forming a coating on both sides of the lipid bilayer, thus being associated with the polar groups. Recent evidence indicates that this picture is incorrect, or at least incomplete, and that probably most of the proteins are included in one half of the bilayer or pass completely through the bilayer. Considerable information about protein location, as well as about the effects of lipid phase equilibria on the membrane, has been obtained by the freeze-fracture method of electron micrography.

In the freeze-fracture technique, a system containing membranes is frozen very rapidly by immersion in liquid nitrogen or chilled Freon and then split with a very sharp knife. The fracture often occurs along the plane between the hydrocarbon portions of a bilayer. The surface is sometimes then etched by raising the temperature to $-100°C$ in a very high vacuum and subliming off some of the ice that may have been formed. A carbon–platinum deposit, which is opaque to electrons, is then laid down. The beam of atoms is allowed to impinge on the surface at a shallow angle, giving a shadow pattern which serves to make visible the surface contours in the subsequent electron micrograph. What appears when a membrane system is treated in this way is a smooth surface in which are embedded a number of knobs or protruding structures, perhaps 50 to 80 Å in diameter. Each protrusion is associated with a membrane molecule—either it is the molecule itself or it represents some disturbance in the structure resulting from the presence of a protein at that point. If the temperature of the membrane before freezing is above the upper transition temperature, the protrusions are scattered fairly uniformly about the surface. In contrast, when the initial temperature is below the upper limit of the chain melting transition for the particular film, the protrusions are found to be segregated into localized patches with large areas of vacant lipid surface between. Evidently the bare regions constitute that part of

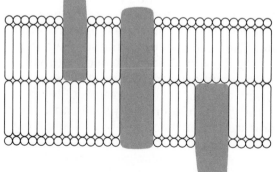

Figure 12-23
Possible structure of a typical cell membrane. The larger units are proteins which may be on one or the other side of the bilayer or may pass entirely through.

lipid which is in the solid phase, and the proteins are localized in still-liquid sectors.

From these freeze-fracture electron micrographs, we conclude that the protein pattern is quite mobile. The presence of a protein molecule probably requires modification of the conformation of the surrounding lipid molecules, and this may be why proteins are found only in the "liquid" phase. Figure 12-23 represents a scheme that may correspond to the structure, at least at one moment, of a typical cell membrane. In addition to the determination of membrane geometry, however, understanding of membrane function requires the unraveling of complex effects of selective transport, that which is passive as well as the active transport which takes place against a concentration gradient.

EXERCISES

1. A particle for which the frictional coefficient is 65 dyn cm/sec is pulled through a solution by a force of 30.0 dyn. At what velocity does it travel?

2. Calculate the rms displacement over a period of 30 sec of colloidal particles of radius 0.05 μm undergoing Brownian motion in an aqueous solution at 30°C.

3. In an electrophoretic experiment, a particle moves at a rate of 4.5×10^{-2} cm/sec under the influence of an electric field gradient of 80 V/cm. What is its electrokinetic mobility?

4. Estimate the frictional coefficient of a sphere of radius 10 μm in water at 25°C. Calculate the rms displacement of the particle as a result of Brownian motion for a time interval of 2.0 sec.

5. An x-ray beam of wavelength 1.539 Å is incident at an angle of 90° on a stretched fiber. The most intense diffracted beam is deviated from the original beam direction by an angle of 21°. What is the repeat distance in the fiber?

6. A sample of polymer has five molecules of molecular mass 10,000, 12,000, 25,000, 30,000, and 100,000, respectively. Calculate the number-average and the weight-average molecular mass.

7. The electrokinetic mobility of a globulin at pH 8.6 is 7.5×10^{-5} cm²/(V sec) at 30°C in an aqueous medium. The dielectric constant of water at this temperature is 76.8, and the mobility is given in practical volts. Estimate the zeta potential in practical volts.

8. Compute the terminal velocity of fall in an aqueous suspension of a particle of density 6.5 and radius 500 nm under the influence of gravity.

9. What is the sedimentation velocity of particles of radius 0.100 μm and density 1.65 in water at 5°C in a centrifuge of 5.00-cm path radius rotating at 100 revolutions per second?

10. What angular velocity is required to give a concentration ratio of 2:1 for a 1.00-mm difference in position in a centrifuge in which the sample has a density of 5.0 and is suspended in water, the radius of the path is 10.0 cm, the molecular weight is 350,000, and the temperature is 25°C?

11. From a plot of the following data for bovine serum albumin, calculate the molecular weight and virial coefficient or interaction constant B.

$(KC/R_\theta) \times 10^5$	Concentration, g/cm³
5.20	0.00240
3.95	0.00160
2.82	0.00095
1.90	0.00034

12. The current flowing through an electrophoresis tube of cross section 0.75 cm² containing an aqueous solution of specific conductance 0.225 ohm⁻¹ cm⁻¹ is 500 mA. What is the potential gradient in the tube?

13. Derive the equation for the equilibrium distribution of molecules in the atmosphere under the influence of the force field of gravity, Equation (12-32), by treating a thin, horizontal layer of gas in a vertical column of air and considering that the force at the bottom of the layer is larger than that at the top by an amount corresponding to the force of gravity on the material within the layer.

14. A linear polymer consists of 250 units of the monomer, each of which is 45 Å in length. Calculate the rms end-to-end distance and the radius of gyration of the random coil form of the polymer.

15. The zeta potential of a protein in a suspension at 25°C is 0.150 ordinary volts. Calculate the corresponding electrokinetic mobility of the protein particles.

16. A solute diffuses through a long cylindrical tube of 2.00-mm diameter at 27°C at a rate such that 0.025 mol crosses during each second a plane at a location where the concentration gradient is 0.050 mol/(liter cm). What is the diffusion coefficient? What is the frictional coefficient of the particle?

17. In a suspension of particles of density 4.00 in water at room temperature, the concentration ratio at equilibrium under the force of gravity is found to be 1.45:1.00 at a difference in level of 10.0 cm. Calculate the particle weight.

18. The following values are found for the osmotic pressure of cellulose nitrate in solution in acetone at 27°C. Estimate the molecular weight of the cellulose nitrate by a suitable extrapolation to zero concentration, and estimate the interaction constant for the polymer in solution.

Concentration, g/liter	Osmotic pressure $\times 10^4$, atm
3.00	4.70
6.00	13.35
9.00	25.86
12.0	42.4

19. A protein molecule has an isoelectric pH of 7.8. (a) What effect does a change in solution pH from 6.5 to 7.0 at constant ionic strength have on the electrokinetic mobility of the protein? (b) Explain how an increase in ionic strength at a constant pH of 6.5 affects the mobility.

20. The sedimentation coefficient for bovine serum albumin is 5.0×10^{-13} sec at 25°C. The partial molar volume of this protein is 0.734 cm³/g and the density of water at this temperature is 1.0024 g/cm³. Assuming the molecules are spherical, estimate the molecular weight of the protein.

21. A 0.010 M solution of a colloidal electrolyte NaX is placed on one side of a membrane which is permeable to Na^+ but not to X^-. Calculate the equilibrium distributions of ions if on the other side of the membrane there is placed (a) a 0.050 M NaCl solution (b) a 0.025 M Na_2SO_4 solution. Assume the total volumes of solutions on the two sides of the membrane are equal.

22. A solution of 1.07 g of protein in 150 cm³ of water has an osmotic pressure of 328 mm of water at 25°C. Estimate the molecular weight of the protein.

23. A portion of 100 cm³ of a 0.0125 M solution of a colloid RCl containing ionizable chloride is placed on one side of a semipermeable membrane with 100 cm³ of water on the other side. Estimate the equilibrium pH reached by the water in light of the requirement that the product $[H_3O^+][Cl^-]$ be the same on both sides of the membrane.

24. Calculate the sedimentation velocity in a centrifuge of radius 7 cm, rotating at 2500 rpm, for particles of radius 3.0 μm and density 1.220 in an aqueous medium at 7°C.

25. The following data were obtained in a light-scattering experiment on a cellulose nitrate fraction in solution in acetone at 25°C [H. Benoit, A. M. Holtzer, and P. Doty, *J. Phys. Chem.* **58**, 635 (1954)]. Prepare a Zimm plot and evaluate the molecular weight and interaction constant B. The values tabulated are $(Kc/R_\theta) \times 10^7$.

	Concentration $\times 10^3$			
$\sin^2(\theta/2)$	0.45	0.56	0.75	0.89
0.07	17.0	18.5	21.0	22.2
0.14	23.0	24.4	26.3	27.5
0.25	29.8	31.5	33.3	34.9
0.41	40.6	42.4	44.8	45.7
0.59	53.0	55.0	55.6	56.6
0.75	63.4	65.0	65.6	67.1

REFERENCES

Books

K. H. Altgelt, *Gel Permeation Chromatography,* Dekker, New York, 1971. A very good summary of the principles and applications of this method.

Robert L. Baldwin, "Intermediates in Protein Folding Reactions and the Mechanism of Protein Folding," in *Annual Review of Biochemistry,* Vol. 44, Annual Reviews, Palo Alto, Calif., 1975.

T. L. Blundell and L. N. Johnson, *Protein Crystallography,* Academic Press, New York, 1976. Contains extensive diagrams; begins at the introductory level but develops the subject to an advanced level of practical application.

Frank A. Bovey, *Polymer Conformation and Configuration,* Academic Press, New York, 1969. Describes various physical methods of characterizing polymer structure; very readable.

Benjamin Carroll, Ed., *Physical Methods in Macromolecular Chemistry,* Dekker, New York, 1972. Volume 2 has particularly good chapters on gel permeation chromatography and thermal methods.

Dennis Chapman, Ed., *Biological Membranes, Physical Fact and Function,* Academic Press, New York, 1968. Describes methods of study and models of membranes.

R. E. Dickerson and I. Geis, *The Structure and Action of Proteins,* Harper and Row, New York, 1969. Excellent account with outstanding structural diagrams.

Carl Djerassi, *Optical Rotatory Dispersion,* McGraw-Hill, New York, 1960. Includes descriptions of a variety of pioneering applications.

M. Edidin, "Rotational and Translational Diffusion in Membranes," in *Annual Review of Biophysics and Bioengineering,* Vol. 3, Annual Reviews, Palo Alto, Calif., 1974.

D. Freifelder, *Physical Biochemistry,* Freeman, San Francisco, 1976. An up-to-date textbook covering a number of the areas described in this chapter.

H. Gasparoux and J. Prost, "Liquid Crystals," in *Annual Review of Physical Chemistry,* Vol. 27, Annual Reviews, Palo Alto, Calif., 1976.

P. G. de Gennes, *The Physics of Liquid Crystals,* Oxford University Press, New York, 1974. An excellent general account.

Carlos Gitler, "Plasticity of Biological Membranes," in *Annual Review of Biophysics and Bioengineering,* Vol. 1, Annual Reviews, Palo Alto, Calif., 1972.

G. W. Gray and P. A. Winsor, *Liquid Crystals and Plastic Crystals,* Vols. 1 and 2, Wiley, New York, 1974. A fairly extensive and detailed account.

Roger Harrison and George G. Lunt, *Biological Membranes: Their Structure and Function,* Wiley, New York, 1975. A very good review of the subject.

Rudy H. Haschemeyer and Audrey E. V. Haschemeyer, *Proteins. A Guide to Study by Physical and Chemical Methods,* Wiley, New York, 1973. An intermediate-level description of conformation and of various techniques of studying physical properties of macromolecules.

Edward D. Korn, Ed., *Methods in Membrane Biology,* Vol. 1, Plenum Press, New York, 1974. Chapter 1 on membrane models, Chapter 2 on lipid monolayers, and Chapter 3 on circular dichroism of membranes are particularly recommended.

Sydney J. Leach, Ed., *Physical Principles and Techniques of Protein Chemistry,* Academic Press, New York, 1969. Has excellent, intermediate-level chapters on x-ray methods, electrophoresis, and ultraviolet absorption. Part B, published in 1970, has relevant chapters on ultracentrifugal analysis, light scattering, and infrared methods.

Leo Mandelkern, *An Introduction to Macromolecules,* Springer-Verlag, New York, 1972. Introductory level; treats both synthetic and natural polymers.

A. D. McLachlan, "Protein Structure and Function," in *Annual Review of Physical Chemistry,* Vol. 23, Annual Reviews, Palo Alto, Calif., 1972.

D. L. Melchior and J. M. Steim, "Thermotropic Transitions in Biomembranes," in *Annual Review of Biophysics and Bioengineering,* Vol. 5, Annual Reviews, Palo Alto, Calif., 1976.

Harold J. Morowitz, *Entropy for Biologists,* Academic Press, New York, 1970. Includes a more advanced treatment of diffusion and Brownian motion.

Roger S. Porter and Julian F. Johnson, Eds., *Ordered Fluids and Liquid Crystals,* American Chemical Society, Washington, D.C., 1967. Accounts of investigations of specific systems.

Howard K. Schachman, *Ultracentrifugation in Biochemistry,* Academic Press, New York, 1959. Authoritative treatment by a leader in development of the field.

Charles Tanford, "Protein Denaturation," in *Advances in Protein Chemistry,* Vol. 24, Academic Press, New York, 1970.

Charles Tanford, *The Hydrophobic Effect: Formation of Micelles and Biological Membranes,* Wiley, New York, 1973. A detailed exposition of the principles on which membrane structure and function may be based, including the structure of water, solubility, and thermodynamic principles.

S. N. Timasheff and G. D. Fasman, Eds., *Biological Macromolecules,* Vol. 2, *Structure and Stability of Biological Macromolecules,* Dekker, New York, 1968.

Kensal Edward Van Holde, *Physical Biochemistry,* Prentice-Hall, Englewood Cliffs, N.J., 1971. Introductory level; sections on diffusion, sedimentation, electrophoresis, and optical properties of macromolecules are specially recommended.

A. J. Verkleij and P. H. J. Th. Ververgaert, "The Architecture of Biological and Artificial Membranes as Visualized by Freeze Etching," in *Annual Review of Physical Chemistry,* Vol. 26, Annual Reviews, Palo Alto, Calif., 1975.

Alan G. Walton and John Blackwell, *Biopolymers,* Academic Press, New York, 1973. Structural information plus extensive accounts of the physical methods by which it is obtained.

James D. Watson, *Molecular Biology of the Gene,* 3rd ed., W. A. Benjamin, Menlo Park, Calif., 1976. An interesting book by one of the leaders in nucleic acid research.

Finn Wold, *Macromolecules: Structure and Function,* Prentice-Hall, Englewood Cliffs, N.J., 1971. A very good introductory book.

Journal Articles

Scot D. Abbott, "Size Exclusion Chromatography in the Characterization of Polymers," *Am. Lab.,* 41 (August 1977).

C. B. Anfinsen, "Principles That Govern the Folding of Protein Chains," *Science* **181,** 223 (1973).

R. Bittman and L. Blau, "Kinetics of Solute Permeability in Phospholipid Vesicles," *J. Chem. Educ.* **53,** 259 (1976).

Donald D. Bly, "Gel Permeation Chromatography," *Science* **168,** 527 (1970).

Allen E. Blaurock and Glen I. King, "Asymmetric Structure of the Purple Membrane," *Science* **196,** 1101 (1977).

Sir Lawrence Bragg, "X-Ray Crystallography," *Sci. Am.* **219,** 58 (July 1968).

Glenn H. Brown, "Liquid Crystals and Their Roles in Inanimate and Animate Systems," *Am. Sci.* **60,** 64 (1972).

David E. Burge, "Molecular Weight Determination by Osmometry," *Am. Lab.,* 41 (June 1977).

Roderick A. Capaldi, "A Dynamic Model of Cell Membranes," *Sci. Am.* **230,** 26 (March 1974).

Raymond Chang and Lawrence J. Kaplan, "The Donnan Equilibrium and Osmotic Pressure," *J. Chem. Educ.* **54,** 218 (1977).

Carolyn Cohen, "The Protein Switch of Muscle Contraction," *Sci. Am.* **233,** 36 (November 1975).

Anthony R. Cooper and David P. Matzinger, "Aqueous Gel Permeation Chromatography," *Am. Lab.* 13 (January 1977).

John T. Edward, "Molecular Volumes and the Stokes-Einstein Equation," *J. Chem. Educ.* **47,** 261 (1970).

Paul J. Flory, "Spatial Configuration of Macromolecular Chains," *Science* **188,** 1268 (1975).

C. Fred Fox, "The Structure of Cell Membranes," *Sci. Am.* **226,** 31 (February 1972).

R. D. B. Fraser, "Keratins," *Sci Am.* **221,** 86 (August 1969).

Gopinath Kartha, "Picture of Proteins by X-Ray Diffraction," *Acc. Chem. Res.* **1,** 374 (1968).

Sidney A. Katz, Charles Parfitt, and Robert Purdy, "Equilibrium Dialysis," *J. Chem. Educ.* **47**, 721 (1970).

Milton Kerker, "Brownian Movement and Molecular Reality Prior to 1900," *J. Chem. Educ.* **51**, 764 (1974).

William J. Lennarz, "Studies on the Biosynthesis and Functions of Lipids in Bacterial Membranes," *Acc. Chem. Res.* **5**, 361 (1972).

C. D. Linden and C. F. Fox, "Membrane Physical State and Function," *Acc. Chem. Res.* **8**, 321 (1975).

William N. Lipscomb, "Structure and Mechanism in the Enzymatic Activity of Carboxypeptidase A and Relations to Chemical Sequence," *Acc. Chem. Res.* **3**, 81 (1970).

J. A. McCammon and J. M. Deutch, "'Semiempirical' Models for Biomembrane Phase Transitions and Phase Separations," *J. Am. Chem. Soc.* **97**, 6675 (1975).

A. L. McClellan, "The Significance of Hydrogen Bonds in Biological Structures," *J. Chem. Educ.* **44**, 547 (1967).

E. T. McGuinness, "Estimation of Protein Size, Weight, and Asymmetry by Gel Chromatography," *J. Chem. Educ.* **50**, 826 (1973).

Robert Olby, "The Macromolecular Concept and the Origins of Molecular Biology," *J. Chem. Educ.* **47**, 168 (1970).

Edward P. Otocka, "Modern Gel Permeation Chromatography," *Acc. Chem. Res.* **6**, 348 (1973).

Bhinyo Panijpan, "The Buoyant Density of DNA and the G + C Content," *J. Chem. Educ.* **54**, 172 (1977).

David C. Phillips, "The Three-Dimensional Structure of an Enzyme Molecule," *Sci. Am.* **215**, 78 (November 1966).

M. C. Porter and A. S. Michaels, "Membrane Ultrafiltration," *Chemtech,* 56 (January 1971).

C. C. Price, "Some Stereochemical Principles from Polymers," *J. Chem. Educ.* **50**, 744 (1973).

A. Rudin, "Molecular Weight Distributions of Polymers," *J. Chem. Educ.* **46**, 595 (1969).

Henry M. Sobell, "How Actinomycin Binds to DNA," *Sci. Am.* **231**, 82 (August 1974).

Deborah H. Spector and David Baltimore, "The Molecular Biology of Poliovirus," *Sci. Am.* **232**, 25 (May 1975).

Michael E. Starzak, "Ion Fluxes through Membranes," *J. Chem. Educ.* **54**, 200 (1977).

Roger Steinert and Bruce Hudson, "The Helix-Coil Transition of DNA," *J. Chem. Educ.* **50**, 129 (1973).

Serge N. Timasheff, "Protein-Solvent Interactions and Protein Conformation," *Acc. Chem. Res.* **3**, 62 (1970).

D. R. Uhlmann and A. G. Kolbeck, "The Microstructure of Polymeric Materials," *Sci. Am.* **233**, 96 (December 1975).

G. R. Van Hecke, "Thermotropic Liquid Crystals," *J. Chem. Educ.* **53**, 161 (1976).

Peter H. von Hippel and Thomas Schleich, "Ion Effects on the Solution Structure of Biological Macromolecules," *Acc. Chem. Res.* **2**, 257 (1969).

Thirteen
Magnetic Resonance Spectroscopy

The methods of magnetic resonance, developed for the most part very recently, have supplied extensive information about molecular structure and have provided chemists and biochemists valuable means for identifying molecules as well as for studying molecular structure and dynamics. These techniques are based upon the fact that electrons, protons, and neutrons all behave as if each were a small magnet, producing a magnetic field in its surroundings and therefore interacting with other magnetic fields. We have seen in Chapter 8 some effects of electronic magnetic moments—attributed to electron *spin*—upon the spectra of atoms. Wherever an unpaired electron occurs, the effect of its spin becomes evident; furthermore, many nuclides also have magnetic moments which are the resultants of contributions from their constituent particles. Unpaired electrons can, in addition, generate magnetic fields through their orbital motion. In this chapter, we examine briefly the bulk magnetic properties of matter and then describe how the resonant absorption and emission of energy by nuclei and by electrons in a magnetic field has been developed into a powerful structural tool.

13-1 MAGNETIC SUSCEPTIBILITY

If a species such as O_2 or NO or $CH_3\cdot$ having unpaired electrons is placed in an external magnetic field, the moments arising from electrons in the molecules or radicals tend to be aligned parallel to the field direction. Viewed from the outside, the sample of matter is magnetized in a way so that its magnetic field adds to that externally applied. Such a material is termed *paramagnetic*. The molar susceptibility of a material χ_M is defined as the ratio of the amount of magnetization produced in a mole of material by an applied field to the strength of that field. The paramagnetic contribution to the magnetic susceptibility of the material is positive in sign, since the induced field is in the same direction as the applied field, and it decreases with increasing temperature because thermal motion disorders the molecules; if μ_M is the magnetic moment of one particle, the molar susceptibility is given by $N\mu_M{}^2/3kT$. A paramagnetic sample tends to move *into* a magnetic field, and its susceptibility can be determined by measuring the force pulling it into the field.

Next we ask how the magnitude of a permanent dipole moment μ_M is related to the number of unpaired electrons in a paramagnetic species. Under some circumstances, the moment is given by the expression

$$\mu_M = g\sqrt{J(J+1)}\mu_B \tag{13-1}$$

where J is the total electronic quantum number described in Section 8-9, g is a complicated function of the quantum numbers L, S, and J, and μ_B is a unit of magnetic moment, equal to $eh/4\pi mc$, called the Bohr magneton. However, in a condensed phase, μ_M is often closely represented by

$$\mu_M = 2.00\sqrt{S(S+1)}\mu_B \tag{13-2}$$

an expression obtained from Equation (13-1) by setting $J = S$. The g value is very close to the value of 2, more precisely 2.0023, which is found for a free electron—one alone in space and not in an atomic orbital. What has become of the contribution to the magnetic moment from orbital motion? The situation is as if the random buffeting of the radical by other molecules causes the orientation of the orbit to change so rapidly that it has essentially a random orientation. The orbital contribution to the moment is then said to be quenched, and Equation (13-2) is termed the spin-only approximation.

In transition metal complexes, the number of unpaired electrons can be evaluated from the magnetic susceptibility by use of Equation (13-2). For the enzyme catalase, which contains a heme–iron complex, susceptibility measurements show a moment of 5.6 μ_B for the enzyme, 5.5 μ_B for the enzyme–azide complex, and 2.3 μ_B for the enzyme–cyanide complex. Thus the first two contain high-spin iron and the last contains low-spin iron, each in the ferric oxidation state.

A further effect of an applied magnetic field occurs for all matter, whether or not there are permanent magnetic moments present in the molecules. Moments can be *induced* by the field, which causes small changes in the orbital motion of the electrons. The induced moments are produced in a direction governed by *Lenz's law*: A magnetic field applied to a conductor results in a current so directed that it produces a magnetic field partially offsetting the applied field. In a molecule, the current can be visualized as a speeding up or slowing down of the electron circulation; since there is no resistance to electron flow in an atom or molecule, the effect persists so long as the applied field is present. Since the induced moments reduce the field inside the sample below that outside, fewer lines of magnetic force pass through the sample per unit area than through empty space in its vicinity, and the sample tends to move out of the applied field. A material having only this type of magnetic polarization has a negative susceptibility and is said to be *diamagnetic;* when a permanent electronic moment is present in a molecule or radical, its diamagnetism is pretty well obscured by the far larger paramagnetism.

The total magnetic susceptibility per mole is given by the equation

$$\chi_M = N\left(\alpha_M + \frac{\mu_M{}^2}{3kT}\right) \tag{13-3}$$

where α_M is a magnetic polarizability factor, indicating how easily dipoles can be induced. Measured susceptibilities of paramagnetic systems can be readily corrected for the diamagnetic contribution by utilizing the additivity of the diamagnetic term in characteristic contributions for various atoms and bonds in a molecule.

13-2
PRINCIPLES
OF NUCLEAR MAGNETIC
RESONANCE
SPECTROSCOPY

Just as electronic magnetic moments are ascribed to "spin," so can be those of protons and neutrons. Either of these two particles has a spin quantum number I of $\frac{1}{2}$, and a corresponding spin angular momentum of $\sqrt{(\frac{1}{2})(\frac{3}{2})}\,(h/2\pi)$. Nuclei heavier than hydrogen, which is merely a proton, can be regarded as composites of protons and neutrons, with quantum numbers obtained by combining the quantum numbers of the constituent particles in a way similar to that in which atomic quantum numbers are obtained from the quantum numbers of individual electrons. Some empirical rules have been developed to correlate the quantum numbers of nuclear ground states, the only states involved here because excited states lie at very high energies: (1) nuclei with even mass numbers and even atomic numbers (for example, ^{12}C, ^{16}O, and ^{32}S) have zero spin and zero magnetic moment; (2) nuclei with odd mass numbers have half-integral spin; (3) nuclei with even mass numbers but odd atomic numbers have integral spin. Common nuclei with

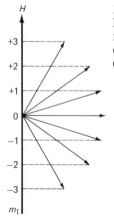

Figure 13-1
Possible orientations of the magnetic moment-angular momentum vector for spin quantum number $I = 3$ in an external magnetic field H. The projections on the magnetic field direction are labeled by the quantum numbers m_I.

spins of $\frac{1}{2}$ are ^{13}C, ^{19}F, and ^{31}P. Examples with spins of 1 are ^2H and ^{14}N, and with spins of $\frac{3}{2}$ are ^{11}B, ^{23}Na, ^{35}Cl, and ^{63}Cu.

Following the conventions of Section 8-9, the angular momentum of a particle can be represented by a vector parallel to the axis of spin, with a length proportional to the magnitude of the angular momentum, $\sqrt{I(I + 1)}(h/2\pi)$. In an external magnetic field, the particle is limited by quantum rules to any one of a certain series of orientations in which the component of the angular momentum vector along the field direction, designated $m_I(h/2\pi)$, has one of the values in the series: $+I(h/2\pi)$, $(+I - 1)(h/2\pi)$, . . . , $(-I + 1)(h/2\pi)$, $(-I)(h/2\pi)$. An example for the case $I = 3$ is shown in Figure 13-1. Since the magnetic dipole moment always lies along the spin axis, its spatial orientation is tied to the orientation of the angular momentum vector.

Frequently the magnetic properties of a nucleus are expressed in terms of the *magnetogyric ratio* γ, the ratio of the maximum component μ of the magnetic moment to the maximum component of the angular momentum along the field:

$$\gamma = \frac{\mu}{Ih/2\pi} \qquad (13\text{-}4)$$

For an electron as well as a neutron, the angular momentum vector is opposite in direction to the magnetic moment of the particle, as expected for a magnetic field generated by circulation of *negative* charge, and γ is said to be negative; for the proton and for many nuclei, the magnetic moment and the angular momentum vector are directed in the same sense and γ is positive.

Nuclei with spin quantum numbers of 1 or more have, in addition to a magnetic dipole, an electric quadrupole, corresponding to a deviation of the charge cloud from spherical symmetry. The charge may be concentrated along the spin axis, forming a prolate ellipsoid, or it may bulge out transverse to the axis, forming an oblate ellipsoid. An electric quadrupole interacts with the gradient of a spatially varying electric field and, since electric fields are largely determined by the electron distribution in a molecule, quadrupole interactions give information about electronic bonding, such as an indication of the type of hybridization used by an atom in forming a covalent bond.

THE NUCLEAR MAGNETIC RESONANCE EXPERIMENT

The nuclear moments in a sample placed in a magnetic field come to equilibrium according to the Boltzmann distribution, with a slight excess of population in the lower-energy states. However, because the nuclear moments are a thousand or more times smaller than electronic moments, the net magnetization is too small to be measured. Nevertheless, a great amount of information can be obtained by measuring the relative ability of radiation of various frequencies to bring about transitions from one quantized spin level to another.

Figure 13-2 shows the NMR spectrum for the hydrogen atoms in the compound benzyl acetate, $CH_3COOCH_2C_6H_5$; there is also a peak in the spectrum for the reference material tetramethylsilane, from which the positions of the other peaks are measured. The benzyl acetate

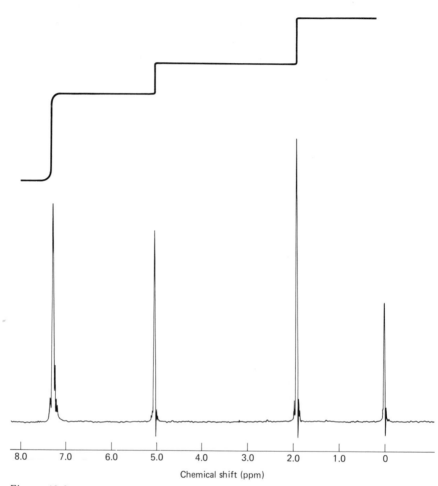

Figure 13-2
Hydrogen NMR spectrum of benzyl acetate, $CH_3COOCH_2C_6H_5$, with the reference tetramethylsilane (TMS) appearing at high field—at the right-hand end of the spectrum. The curve above the spectrum is the integral presentation, with the height of the step at each peak proportional to the area of the peak.

spectrum contains three peaks, for which the relative areas from left to right, deduced from the integrals plotted above the absorption peaks, are 5:2:3. Since areas in a given nuclear magnetic resonance (NMR) spectrum are proportional to the numbers of nuclei responsible for the peaks, it is possible to assign the three absorptions to the phenyl, methylene, and methyl groups, respectively. We now describe how a spectrum of this type is obtained and then see why the hydrogens in different groups produce distinctive absorption peaks.

In an NMR spectrometer, the sample is placed in a tube of ordinary Pyrex glass supported in a magnetic field of strength H_0 between the poles of a magnet. For a spin-$\frac{1}{2}$ nucleus, such as a hydrogen atom, there are two possible states in the magnetic field: one of lower energy with m_I of $+\frac{1}{2}$ and the angular momentum vector more nearly parallel to the field, and the other of higher energy with m_I of $-\frac{1}{2}$ and with the vector antiparallel to the field. A nucleus in either of these states is not able to align its moment exactly along the field direction, and con-

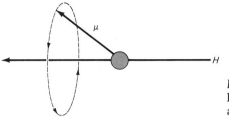

Figure 13-3
Precession of a magnetic moment about
an applied magnetic field.

sequently it is subjected to a torque from the field. The direction of the torque is perpendicular both to the magnetic moment and to the field, causing the axis of the moment to precess at a frequency of $\gamma H_0/2\pi$ about the field as shown in Figure 13-3, in the same manner that the axis of a spinning top tilted away from the vertical precesses about the vertical direction of the gravitational field.

The energy of interaction of the magnetic field H_0 with the magnetic moment is equal to $-H_0\mu_z$, where μ_z is the component of the moment in the direction of the field. Since μ_z equals $\mu(m_I/I)$, the difference in energy between the parallel and antiparallel states is $2H_0\mu$ or $2H_0\gamma Ih/2\pi = H_0\gamma h/2\pi$, using Equation (13-4) with $I = \frac{1}{2}$. This difference depends, obviously, on the magnitude of the magnetic field, and the dependence is represented by the solid lines in Figure 13-4. The frequency of radiation required to induce transitions is equal to the energy difference divided by h:

$$\nu = \frac{\Delta E}{h} = \frac{\gamma H_0}{2\pi} \tag{13-5}$$

For nuclei with spins greater than $\frac{1}{2}$, spectroscopically allowed transitions correspond to changes in m_I of ± 1, that is, changes between neighboring states. Thus the frequency for transitions of these nuclei is also given by Equation (13-5). Furthermore, the frequency of radiation required to induce an NMR transition of a nucleus turns out to be

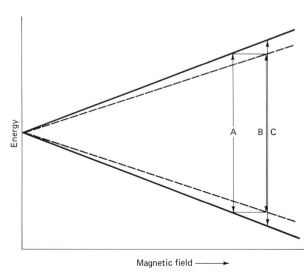

Figure 13-4
Energy difference of the two states of a spin-$\frac{1}{2}$ magnetic moment as a function of the external magnetic field. The dashed line represents the situation when the nuclear moment is shielded by the electron cloud surrounding it. Transitions A and B represent the spin flips of the unshielded and shielded nuclei at the same frequency and different fields; transitions B and C are those of the shielded and unshielded nuclei at the same field and different frequencies.

equal to the frequency with which the axis of spin of that nucleus precesses around the direction of the magnetic field.

The radiation required to excite NMR transitions in the magnetic field of a typical spectrometer lies in the short-wave portion of the spectrum and is generated by a radio transmitter, from which it reaches the sample by way of a coil of wire mounted in the probe holding the sample tube. This coil, serving as a transmitting antenna, is usually arranged as a split solenoid, part in front and part in back of the sample tube, with its axis horizontal. The receiver coil, surrounding the sample and serving as an antenna to pick up the signal, is arranged with its axis vertical so that there is minimum leakage to it from the transmitter coil in the absence of a signal from the sample. Some spectrometers use the same coil both as transmitter and receiver coil. Spin flips of nuclei caused by the radio signal correspond to motion of magnets within the closed loop of the receiver coil and thus induce in that coil a voltage that is amplified, detected, and sent to an oscilloscope or recorder. Figure 13-5 shows the experimental arrangement.

Many NMR spectra require resolution of 1 part in 10^8 or better to show chemically significant features; consequently, the magnetic field must be uniform in space over the volume of the sample within the receiver coil, as well as constant in time, both to this degree of accuracy. Spatial uniformity is made possible by sets of shim coils on the faces of the magnet poles, carrying currents which can be adjusted to eliminate effects of magnet inhomogeneity. In addition, homogeneity is improved by spinning the sample about the vertical axis with an air turbine. To minimize variations in the magnetic field with time, the spectrometer is provided with circuits which regulate the current through the magnet, and a final "lock" circuit regulates the field very precisely by requiring it to track a resonance signal of a substance placed either within the sample tube—an *internal* lock—or in a second container close to the sample—an *external* lock.

Typical operating fields for commercial NMR spectrometers vary from 14 to 84 kilogauss (kG). The corresponding frequencies for hydro-

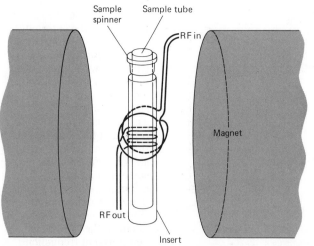

Figure 13-5
Schematic diagram
of the sample region
of an NMR
spectrometer.

Sample spinner   Sample tube

RF in

Magnet

RF out

Insert

gen nuclei are 60 and 360 megahertz (MHz), respectively, and for carbon nuclei, with nuclear moments about one-fourth those of hydrogen, 15 and 90 MHz. A common field for research instruments is 23 kG, which corresponds to a frequency of 100 MHz for hydrogen, 40 MHz for ^{31}P, and 25 MHz for ^{13}C, with other frequencies in proportion. For fields higher than 23 kG, adequate field homogeneity cannot be obtained from iron-core magnets, and there are employed superconducting solenoids, made of alloys such as niobium–tin which have essentially zero resistance when kept at the boiling point of liquid helium. Since the sensitivity of the instrument increases about as the three-halves power of the frequency employed for a given nucleus, it is usually advantageous to operate at as high a field as possible.

Nuclei of a given species which are in different environments absorb energy at different points in the spectral range, because it is the net magnetic field reaching a nucleus that determines its transition energy, and the external field is modified slightly by the effects of the electrons in the molecule, as well as possibly by the spin states of other nuclei, effects that provide for the chemist extensive information about molecular structure. A spectrum can be obtained experimentally by scanning either the magnetic field or the frequency of the radio transmitter. The energies of nuclei in two environments were shown in Figure 13-4, the dotted lines corresponding to the nucleus *shielded* to the greater extent from the external magnetic field. If the shielding or screening parameter is represented by σ, the effective magnetic field reaching the nucleus is $H_0(1 - \sigma)$. A larger shielding parameter results in a smaller effective field which, for a fixed operating frequency, requires H_0 to be larger in order to reach the resonance condition. An experiment at constant frequency corresponds to constant vertical distance on the diagram. For a given H_0, a larger shielding parameter causes the energy difference and therefore the resonance frequency to be smaller, corresponding to a variation along a vertical line in the diagram.

In the conventional presentation of an NMR spectrum, magnetic field increases to the right and frequency increases to the left. Thus the resonance of a more shielded nucleus lies to the right of that of a less shielded one. It is generally immaterial to the appearance of a spectrum whether the field or the frequency was swept in the actual experiment.

CHEMICAL SHIFTS

We turn now to the question of how shielding by electrons occurs, so that various nuclei resonate at different positions in a spectrum, or have different *chemical shifts*. The answer is that the applied field affects the electron cloud in a molecule, and the resulting change in electron circulation produces magnetic fields which aid or oppose the applied field. The induced fields are proportional in magnitude to the applied field, and therefore the chemical shift is conveniently expressed as a fraction of the applied field, usually in parts per million (ppm). Becauses changes in frequency and changes in field are proportional to one another, each being a relatively small fraction of the absolute magnitude, it is equally satisfactory to calculate chemical shift differences as the ratio, in parts per million, of the frequency

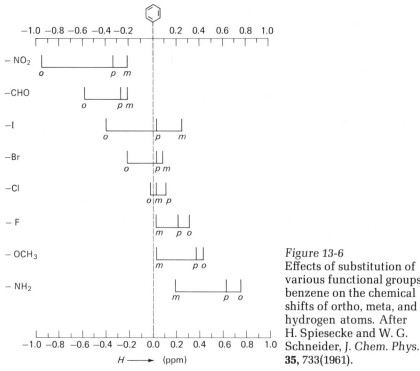

Figure 13-6
Effects of substitution of various functional groups in benzene on the chemical shifts of ortho, meta, and para hydrogen atoms. After H. Spiesecke and W. G. Schneider, *J. Chem. Phys.* **35**, 733(1961).

difference to the operating frequency. Thus in Figure 13-2, the peak for the phenyl hydrogens appears at 725 Hz from the peak for the chemical shift reference tetramethylsilane (TMS), and, since the spectrometer frequency was 100 MHz, the chemical shift is 7.25 ppm.

One part of the mechanism by which electrons shield nuclei corresponds to the diamagnetic effect described in Section 13-1: The external field induces local currents which produce magnetic fields offsetting the applied field. The more electrons there are surrounding a nucleus, the more it is shielded, other things being equal. An example of the effect of electron density upon chemical shift is observed in the resonances of hydrogen atoms in aromatic molecules as influenced by substituents. Such substituent effects in monosubstituted benzenes are shown diagramatically in Figure 13-6. The large downfield shift observed when a nitro group is present, for example, is the consequence of electron withdrawal from the ring by the group through both inductive and resonance effects.

There is also in some systems a *paramagnetic* effect, leading to unshielding of nuclei. This arises when the applied field causes a mixing into the ground state of excited electronic states having energies not too far above the ground state. This mixing occurs in such a way that the induced magnetic field adds to the applied field. Carbonyl compounds and complexes of transition metals are examples of systems with low-lying excited electronic states for which a paramagnetic shift can be observed with a magnitude related to the energies of these states.

Some molecules, of which benzene is an example, have an aniso-tropic magnetic susceptibility. This means that the electron circulation induced by an external magnetic field is greater when the field is in one direction—for aromatic molecules, in the direction perpendicular to the plane of the ring—than it is in other directions. The reason for this property of aromatic molecules is that electrons in π orbitals can circulate much more easily parallel to the plane of the ring than in any other direction. From the *ring current* induced by the applied field, there results an induced magnetic field which has a maximum in a direction perpendicular to the aromatic ring. This field shields nuclei lying above or below the ring, but unshields nuclei that lie outside the ring near its plane, including the hydrogen atoms of the aromatic molecule itself. Porphyrin rings constitute extended systems of delocalized π electrons, and hydrogen atoms lying near the center of a porphyrin ring are found to resonate at unusually high magnetic fields.

In acetylene, electron circulation is induced most readily around the direction of the C–C bond, leading to a shielding of the acetylenic hydrogen so that it resonates at a field several parts per million higher than olefinic or aromatic hydrogens. The π electrons in olefinic and carbonyl double bonds contribute antisotropic magnetic fields, as do even the σ electrons in C—H bonds; however, for these bonds the geometric characteristics are not so well defined and the magnitudes of

Figure 13-7
Typical chemical shift ranges of hydrogen atoms in various chemical structures.

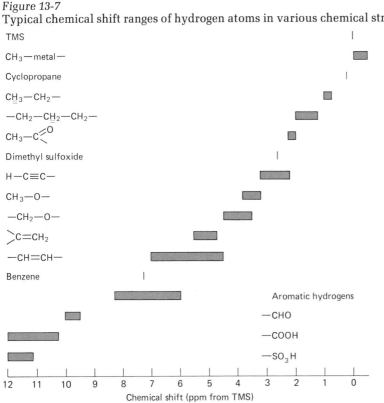

the effects are smaller. In cyclohexane derivatives, a hydrogen in the axial position is shielded by about 0.2 to 0.4 ppm more than if it is brought into the equatorial position by having the ring undergo a chair-to-chair conformational interconversion. This has been attributed to anisotropy of the C—C bonds.

Figure 13-7 is a diagram of typical chemical shift regions for hydrogen atoms. The presence of an electronegative atom, such as oxygen in an ether or alcohol, or nitrogen in an amine, produces a downfield shift, whereas an electropositive metal causes a shift to high field. Cyclopropyl hydrogens occur at higher fields than corresponding hydrogens in straight-chain compounds, possibly because of a ring current in the three-membered ring.

SPIN–SPIN SPLITTING

It is frequently found that the NMR resonance for nuclei of a certain chemical shift does not appear as a single line but is instead a multiplet. For example, the fluorine and hydrogen spectra of 5-fluorouracil, a substance used to treat certain types of cancer, as shown in Figure 13-8, contain doublets for the fluorine in position 5 and the hydrogen in position 6. These nuclei are *spin–spin coupled* to one another, and

Figure 13-8
NMR spectra of 5-fluorouracil. The proton spectrum shows, from left to right, the hydrogens on nitrogen, the position-6 hydrogen, the residual hydrogens in the deuterated dimethyl sulfoxide solvent, and TMS. The inset in the center is an expansion of the position-6 hydrogen resonance, and the inset at the upper left is the ^{19}F spectrum of the molecule; the two insets are on the same horizontal scale.

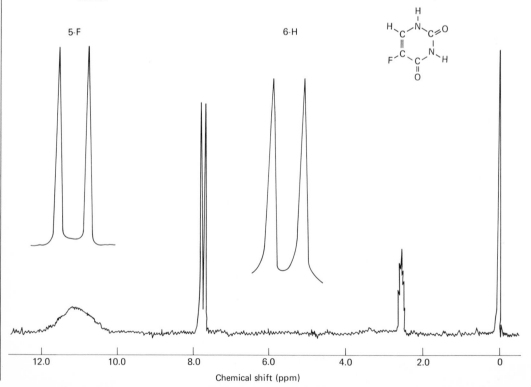

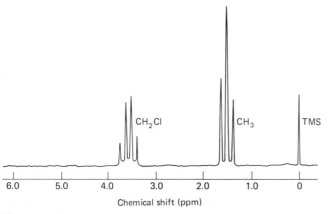

Figure 13-9
Hydrogen NMR spectrum of ethyl chloride at 60 MHz. Because the chemical shift difference between the methyl and methylene hydrogens is only about 15 times as large as the coupling constant between the two kinds of hydrogen, the multiplets are not exactly symmetrical, each having the intensity of its components increasing slightly in the direction of the other.

the magnitude of the coupling constant is 4.4 Hz, the separation between the two peaks in each of the two doublets.

Spin–spin coupling arises when the energy levels of one nucleus depend upon the spin orientation of another nucleus. In 5-fluorouracil, the fluorine nucleus, which like hydrogen has a spin quantum number of $\frac{1}{2}$, can have either of two states in a magnetic field. The circulation of the electrons near the fluorine is slightly different for the two orientations of the nucleus, and this magnetic polarization is transmitted through the electron cloud of the molecule until it arrives at the hydrogen nucleus. Thus the magnetic field at the hydrogen nucleus is slightly different, dependent on the fluorine nuclear spin state. One line of the hydrogen doublet then corresponds to those molecules in which the orientations of the hydrogen and fluorine are opposite, and the other line corresponds to those molecules in which the orientations are the same. Similarly, the fluorine transition energy depends upon the direction of the hydrogen spin.

Because the spin–spin coupling information is transmitted from one nucleus to another through electrons, the magnitude of the interaction generally falls off with increasing number of bonds between the nuclei involved. However, a double bond usually transmits information more readily than does a single bond. Coupling can sometimes also be transmitted through lone-pair electrons between nuclei sufficiently close together to be in physical contact although not directly bonded. Spin–spin coupling transmitted through electrons is called *indirect* coupling; furthermore, it is described as *isotropic,* which means that it does not change as a molecule tumbles in the liquid phase.

One of the features of spin–spin splitting is that the magnitude of J is independent of the operating frequency, in contrast to a frequency separation produced by a chemical shift difference, which is proportional to the operating frequency.

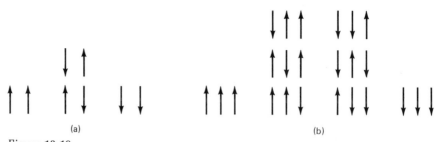

Figure 13-10
The several ways of orienting nuclei of spin 1/2 in groups of (a) two or (b) three.

As a more complicated example of indirect spin–spin coupling, consider the spectrum of ethyl chloride in Figure 13-9. There is a quartet for the methylene group at 3.57 ppm from TMS and a triplet for the methyl group at 1.48 ppm. The three components of the triplet correspond to the four ways in which two spins of the two methylene hydrogens can be arranged, yielding three different values of the total spin as shown in Figure 13-10a; the two possible combinations with zero total spin are reflected in the double intensity of the middle peak of the triplet. The four components of the quartet result from the eight ways three hydrogen spins in the methyl group can be combined, yielding four values of the total spin, as in Figure 13-10b. The general rule, applicable if the chemical shift difference between the coupled multiplets is very much greater than the coupling constant, is that n nuclei of spin $\frac{1}{2}$ split another coupled resonance into $n + 1$ components with spacings equal to the coupling constant and with relative intensities proportional to the coefficients of the binomial expansion. In ethyl chloride, the methyl–methylene chemical shift difference is not quite large enough for this rule to apply strictly, and the intensities of the components of each multiplet are peaked slightly toward the other multiplet.

An important point is that the spectral selection rules forbid the appearance in the spectrum of any transition that would indicate the magnitude of the coupling constant between equivalent nuclei. Thus, although it is certain that the two nuclei in the methylene group of ethyl chloride are spin–spin coupled to one another, there is no way in which this can be established by examination of the spectrum of the molecule unless the equivalence of the two nuclei is removed by substitution of deuterium for one of them. This substitution does not appreciably change the electronic structure of the molecule, and the value of the measured H—D coupling constant can be converted to the corresponding H—H value by multiplying by γ_H/γ_D, which is about 6.5. Other illustrations of this point are the single-line spectra of the hydrogens in molecules such as t-butyl chloride, tetramethylsilane, or benzene.

Table 13-1 includes some typical values of coupling constants. The magnitudes of coupling constants are intimately related to the geometric relationship of the bonds over which the coupling is transmitted. For example, the H–H coupling constant in the unit H—C—C—H,

Table 13-1
Representative magnitudes of nuclear spin-spin coupling constants, in hertz

One-bond		
H—H		276
H—F		521
H—C		120–260
C—F		150–300
C—C		35–175

Two-bond

		0–3
(H₂C=)	in cyclopropanes	2–5
C=C—H	in olefins	1–12
C—C—H	in aromatics	1–4
(F₂C<)		10–80
(FHC<)		40–80

Three-bond

(H—C–C—H)	freely rotating	0–12
(H\C=C/H)		5–14
(H\C=C/H cis-trans)		11–19
H(ax)—H(ax)	in cyclohexane	13
H(eq)—H(eq)	in cyclohexane	4
(H—C=C—H)	in benzenes	6–10
H—C—C—F		7–13
F—C—C—F		0–20
(F—C=C—F)	in substituted benzenes	18–20
(H—C=C—C)	in benzenes	6–10

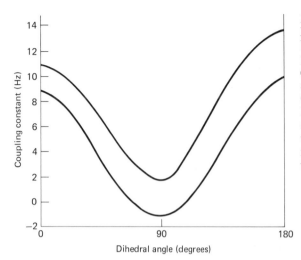

Figure 13-11
Karplus-type relations between the H—C—C—H proton–proton coupling constant and the dihedral angle between the two C—H bonds. Most values lie between the two curves, and, for a given molecular type, the variation can be represented by a curve of the shape of those shown.

termed the vicinal coupling constant and illustrated in the case of ethyl chloride in Figure 13-9, depends upon the dihedral angle between the two C—H bonds as shown in Figure 13-11. The dihedral angle is the angle between the projections of the C—H bonds as they are viewed along the C—C bond axis. The exact magnitude of the vicinal coupling depends upon the nature of the other substituents on the two carbons, particularly their electronegativity, which seems to shift the curve in Figure 13-11 up or down without altering its shape. A similar relationship has been applied to the vicinal coupling constant in the unit H—C—N—H in peptides in solution, for which the angle ϕ described in Chapter 12 can be estimated from the J value. Of course, two different values of ϕ may correspond to a given value of J, but one angle can often be eliminated as unreasonable on the basis of other evidence. Magnitudes of vicinal coupling constants have been used to work out the exact conformation of several furanose and pyranose rings in sugar derivatives.

Coupling to nuclei other than hydrogen can often be observed. A nucleus such as fluorine or phosphorus, with a spin of $\frac{1}{2}$, leads to a doubling of the coupled proton peak. If the coupling of a hydrogen is to deuterium or to nitrogen-14, the hydrogen resonance appears as a triplet, with all three components of equal intensity, because the nucleus of spin 1 has an equal probability of being in any one of its three possible spin states. Similarly, coupling to boron causes a proton pattern to be a quartet of lines of equal intensity.

Carbon-13 has a spin of $\frac{1}{2}$, and the 1 percent of hydrogen atoms attached to the ^{13}C nuclei present in natural abundance in any organic sample give rise to doublets approximately centered on the "main" peak, which represents protons attached to ^{12}C. The separation of the two ^{13}C satellites is equal to the one-bond carbon–hydrogen coupling constant. Because the coupling constant is transmitted to a nucleus to an extent determined by the magnitude of the electronic wave function right at the nucleus, the value of the C—H coupling can be related to the hybridization of the carbon atom, for it is only s orbitals that have nonzero values at the nucleus. The results are values of roughly 120 Hz for sp^3 hybridized carbons, 180 Hz for sp^2 carbons, and 240 Hz for sp carbons, so that the J value increases with the fractional s character of the bonding orbitals.

COMPLEX SPECTRA

Many spectra are more complicated than the simple *first-order* patterns we have seen to this point. One contributing factor is the presence of peaks for which the ratio of the chemical shift difference from neighboring peaks to the coupling constant is very small. Selection rules valid under the condition of a large ratio of chemical shift difference to J no longer apply, and therefore the number of transitions observed is greater, each transition may involve several nuclei, and the intensities deviate from simple integral ratios.

To describe various spin systems, an alphabetical code is employed. Letters near one another in the alphabet are assigned to nuclei with near-equal chemical shifts. Thus an A_2BXY_2 system has two A nuclei, one B nucleus with a shift close to that of the A nuclei, and one X and two Y nuclei with chemical shifts far removed from A and B but close

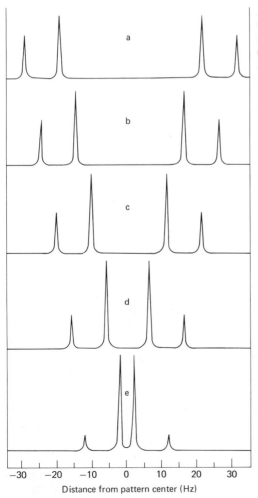

Figure 13-12
Calculated AB spectra for $J = 10$ Hz and the ratio of coupling constant to chemical shift of (a) 0.20, (b) 0.25, (c) 0.33, (d) 0.50, and (e) 1.00. After Bovey, *Nuclear Magnetic Resonance Spectroscopy.*

−30   −20   −10   0   10   20   30
Distance from pattern center (Hz)

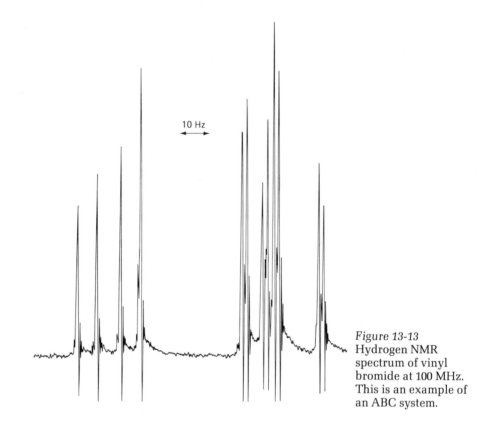

10 Hz

Figure 13-13
Hydrogen NMR
spectrum of vinyl
bromide at 100 MHz.
This is an example of
an ABC system.

to one another. The simplest system, two coupled nuclei with different shifts, as illustrated above for the hydrogen and fluorine nuclei in fluorouracil, is described as an AX system. If the shifts are close together, the two-spin system is termed an AB system, and typical spectra for this case are shown in Figure 13-12, with the two inner peaks having higher intensities than the two outer peaks. The spacing between either pair of peaks gives the coupling constant directly, as in the AX system, but the chemical shift difference can only be established by a calculation involving all the peak positions.

Figure 13-13 shows the ABC pattern of vinyl bromide. The peaks associated with the hydrogen geminal to the bromine are downfield, and the peaks of the two slightly different hydrogens in the $=CH_2$ group are upfield. Examination of the pattern shows three spacings each repeated four times, but these spacings are not exactly equal to the coupling constants; and extensive mathematical analysis is required to obtain the coupling constants and chemical shifts from the spectrum. This pattern is first-order in terms of the number of peaks observed, but it again illustrates the manner in which the intensities of the coupled multiplets peak toward one another. Interpreting this spectrum as nearly first-order, we describe the CH resonance downfield as doubled by a large trans coupling, approximated by the distances from the first to the third and from the second to the fourth peaks, and doubled again by a cis coupling of smaller magnitude given roughly by the distances from the first to second and third to fourth peaks.

Another circumstance leading to spectral complexity is the presence of two or more nuclei which have the same chemical shift—and are indistinguishable chemically—but which differ in their coupling to one or more other nuclei. An example is benzene containing two different para substituents, as represented in Figure 13-14a by p-iodoanisole. The two hydrogens ortho to the iodine atom are chemically equivalent and, by analogy with other iodobenzenes, should appear downfield; they are designated A nuclei. The two hydrogens ortho to the methoxy group are likewise chemically equivalent, appearing at a higher field, and are labeled B nuclei. The spectrum might then be expected to be simply a four-peak AB pattern, doubled in intensity because there are two sets of AB hydrogens, with splitting characteristic of ortho coupling between the A and B hydrogens, and at first glance it appears to be this. Closer inspection, however, shows additional peaks, for neither the two A nuclei or the two B nuclei are *magnetically* equivalent: The coupling constant of a given B hydrogen to one A hydrogen is that for para coupling and to the other A hydrogen is that for ortho coupling, the two values being quite different, and the same is true for coupling of a given A hydrogen to the two B hydrogens. The spin system in this molecule is designated AA′BB′ to denote the magnetic inequivalence.

The spectrum of o-iodoanisole in Figure 13-14b illustrates by comparison with that of the para compound how NMR spectra can be used to determine the relative positions of substituents in an aromatic ring. This pattern can be interpreted on a nearly first-order basis, remembering that the ortho H–H coupling is the largest. The four hydrogens are all nonequivalent. The high-field multiplet consists of a doublet superimposed on the two lower components of a triplet and has an area approximately twice as great as that of each of the other two multiplets. Thus there is a large doublet for each of the two hydrogens having only one other hydrogen adjacent, and a widely spaced triplet

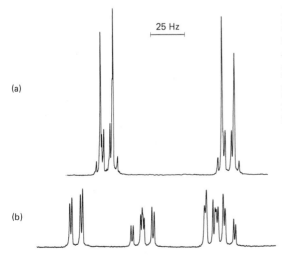

(a)

(b)

25 Hz

Figure 13-14
Ring hydrogen regions of NMR spectra of (a) p-iodoanisole, an AA′BB′ system, and (b) o-iodoanisole, an ABCD system. The resonances of the methoxy hydrogens appear at higher field, far to the right on the scale of these spectra.

for each of the two hydrogens having two neighbors. The low-field doublet is for the hydrogen ortho to the iodine and the high-field doublet is that for the hydrogen ortho to the methoxy—and meta to the iodine. The high-field triplet must be for the hydrogen meta to the iodine and para to the methoxy group, still in accord with the electron-releasing effects of this group, and the triplet in the middle is the hydrogen meta to the methoxy and para to the iodine.

In Figure 13-15 is shown the ^{19}F spectrum of the molecule CHFCl—CHFCl. Since ^{19}F has a spin of $\frac{1}{2}$, as does ^1H, it contributes to a spectrum just like a hydrogen atom of vastly different chemical shift. The two hydrogens in the molecule are magnetically inequivalent, for their coupling constants to a given fluorine nucleus are different. Likewise the two fluorine atoms are inequivalent, having differing coupling to a given hydrogen atom. The system is described as an AA′XX′ system. A further complication in this molecule is the presence of two chiral centers, so that one spectrum is observed for the dl and ld forms and a second spectrum for the dd and ll forms. Because of the larger chemical shift difference in ^{19}F spectra, the patterns for the two forms are resolved, whereas in the hydrogen spectrum, the two patterns, each of which is identical with the ^{19}F pattern, overlap. It is important to realize that, for a molecule with only a single chiral center, the spectra of the two enantiomers are identical.

The spectrum of crotonaldehyde, CH_3—CH=CH—CHO, is shown in Figure 13-16. The peak for CHO is at low field and is doubled by coupling to the nearest olefinic hydrogen. The methyl resonance occurs at high field and is split into four components of approximately equal intensity by the coupling of different magnitudes with each of the two olefinic hydrogens. The resonance pattern of the olefinic hydrogens is an AB pattern—two pairs of lines with the inner member of each more intense than the outer one. The separation in each portion of the pattern is equal to the coupling constant between A and B, and the value indicates that this is a trans olefin. The upfield olefinic hydrogen resonance is further doubled by coupling with the aldehyde hydrogen, and each part of the doublet is split into a quartet by coupling with the methyl hydrogens. The low-field olefinic hydrogen is

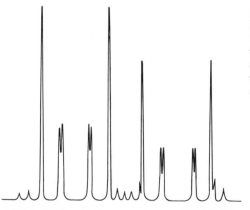

Figure 13-15
Fluorine-19 NMR spectrum of CHFCl-CHFCl. The ten lines to the left represent the resonances of one optical form, the ten lines to the right represent the resonance of the other form.

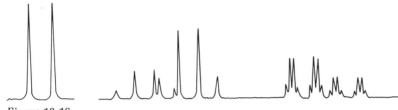

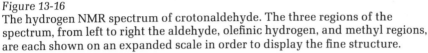

Figure 13-16
The hydrogen NMR spectrum of crotonaldehyde. The three regions of the
spectrum, from left to right the aldehyde, olefinic hydrogen, and methyl regions,
are each shown on an expanded scale in order to display the fine structure.

apparently that closest to the methyl group and is coupled to this group
by a larger amount, which causes its resonance to be split into a quartet much larger than that for the hydrogen giving the upfield resonance.

13-3
DYNAMIC EFFECTS
IN NMR

In addition to chemical shift and spin–spin coupling, spectral parameters that aid directly in the assignment of molecular structure, NMR
spectroscopy also provides a substantial amount of information about
molecular dynamics and rapid chemical reactions through investigation of the shape of absorption lines and of the quantities called nuclear relaxation times. In addition, a knowledge of relaxation behavior
of nuclei in various environments aids in obtaining spectra properly
and in their correct interpretation.

LINE SHAPES

Suppose a nucleus can exist in either of two environments and can
move, either by a physical process or by a chemical reaction, from one
to the other. Will the NMR spectrum show two distinct peaks or only
a single peak? The rule-of-thumb answer is that two peaks will appear
if the nuclei have a lifetime in one environment that is at least as great
as a reciprocal of the difference between resonance frequencies in the
two environments. As an example, suppose that there are two resonance peaks which at low temperature, where the transfer between
locations is slow, are observed at a separation of 50 Hz. As the interchange is accelerated by raising the temperature, they remain distinct
so long as the rate of transfer between environments does not exceed
50 times per second, which is equivalent to the statement that the lifetime in one environment must be of the order of $\frac{1}{50}$ sec or longer. At
rates much in excess of this limit, only a single line is seen, at a position
that is a concentration-weighted average of the chemical shifts of the
two environments. As the rate of the process increases from the lower
extreme at which two narrow lines are seen, the first effect observed is
a broadening of each line; then the region between the lines fills in, and
finally the merged peak narrows and becomes tall. From the line shape
at any stage in the regions of successive broadening, merging, and narrowing, it is possible to calculate the rate of the process.

Figure 13-17 shows how the line shape for the resonance of the two methyl groups in dimethylnitrosamine, a carcinogenic material, changes as the temperature increases. At low temperatures, the two methyl groups are nonequivalent because the partial double-bond character of the N—N bond keeps one of the groups cis to the oxygen and the other trans. This barrier is only about 23 kcal/mol, small enough so that an increase in temperature to about 200°C speeds up rotation about the N—N bond to the point where the two methyl groups give only a single, although still somewhat broadened, peak.

A representative chemical exchange process evident in an NMR spectrum is the transfer of a hydroxyl hydrogen in aqueous ethanol solution between ethanol molecules or between an ethanol and a water molecule; concurrently, exchange occurs from water molecule to water molecule. Under conditions where the exchange of hydroxyl hydrogens in the ethanol molecules is slow, the hydroxyl resonance appears as a triplet because of coupling to the methylene hydrogens, and the methylene resonance is roughly a quartet doubled. The addition of a trace of acid or an increase in water concentration accelerates the exchange, so that the hydroxyl peak becomes a singlet, the methylene resonance collapses to the more familiar quartet, and spin–spin coupling between the two is no longer evident. At still higher exchange rates, the resonances of the hydroxyl hydrogen in the water and the hydroxyl hydrogens of the alcohol, which are distinct at lower rates, merge into a single peak, the position of which is the concentration-weighted average of the two components in the solution, taking into account the effects of hydrogen bonding described in Section 13-6.

When the motions of molecules are greatly restricted, as in solids or very viscous liquids, or adsorbates on surfaces, NMR bands are typically very broad. This is a result of the magnetic fields produced by nuclei adjacent to the one under observation. The effect occurring is not the indirect spin–spin interaction previously described, which is transmitted by the electrons, but a direct interaction through space between one magnetic dipole and another. The interaction energy of two magnetic dipoles as seen in the NMR experiment is given by the equation

$$E = \mu_1 \mu_2 \frac{3 \cos^2 \theta - 1}{r^3} \tag{13-6}$$

where θ is the angle between the magnetic field and the vector joining the nuclei and r is the distance between the nuclei, as shown in Figure 13-18. The effect of rapid motion, as in a typical organic solvent or aqueous solution of ordinary viscosity, is to eliminate the effect of this interaction, for if θ is averaged over all orientations of the two nuclei with every orientation equally probable, the value of $\cos^2 \theta$ is $\frac{1}{3}$, yielding a value of zero for E. Thus in the usual high-resolution spectrum

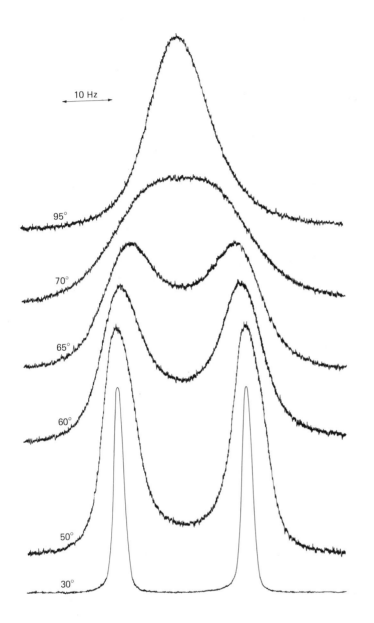

10 Hz

95°

70°

65°

60°

50°

30°

Figure 13-17
Hydrogen NMR spectrum of N,N-dimethylnitrosamine in the presence of
CF$_3$COOH, as a function of temperature. The two methyl groups give distinct
peaks at room temperature as a result of restricted rotation about the N—N bond
in (CH$_3$)$_2$NNO; increase in temperature increases the rate at which they
exchange environments by increasing the rate of rotation about the bond.
Presence of the strong acid substantially reduces the temperature of merging of
the two peaks below that in the absence of acid. The total peak areas in each
of the spectra should be the same; however, the vertical scale has been varied
to fit the patterns into the figure.

of a liquid, one sees no effect of the direct dipolar interaction. This interaction is of course anisotropic, implying that its magnitude depends upon the orientation of the molecule in the magnetic field, and if the spectrum of a single crystal, in which molecules have a particular orientation, is examined, it is found to vary with orientation of the crystal in the spectrometer.

If molecular motions occur at a suitable rate, they tend to average out the direct dipolar interaction to a greater or lesser degree and thus produce narrowing of the very broad line characteristic of the solid in which there is no motion. For example, the proton resonance of solid ice is of the order of 10^5 Hz in width at typical spectrometer frequencies. Here the broadening mechanism is primarily an intramolecular one, resulting from the field of one proton at the location of the other proton in the same molecule. Water molecules bound to dry, solid proteins or adsorbed at the surface of adsorbents such as silica gel or alumina show line widths of the order of 10^2 to 10^4 Hz; with increasing temperature or increasing coverage of the surface, the molecules become more liquidlike and the line narrows. From studies of the temperature dependence of these line widths, it is possible to estimate the activation energy for reorientation of the water molecules. The width of a line is conventionally measured at points corresponding to one-half the maximum intensity, as shown in Figure 13-19; this value can be taken as inversely proportional to the rate of the motion producing narrowing.

Studies have also been made of the "wide-line" resonance absorption of solid polymeric materials. In poly(methyl methacrylate), for example, it is possible to observe the onset of methyl rotation at very low temperatures; and then at higher temperatures, there are transitions corresponding to the beginning of rotation of polymer segments about the axis of the main chain; all this occurs far below the bulk melting point of the polymer.

Similar effects of line-width changes may be observed in the spectra of macromolecules in solution as the temperature is changed. The transition from the helical arrangement of a protein, in which the units of the macromolecule are firmly held in place so that the only motion is the slow rotation of the entire molecule, to a random coil, in which there is much freedom of motion of various segments of the peptide backbone as well as of the side chains, is evident in a marked sharpen-

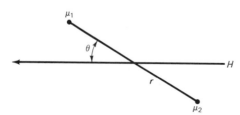

Figure 13-18
Diagram of the direct dipole interaction between two magnetic nuclei.

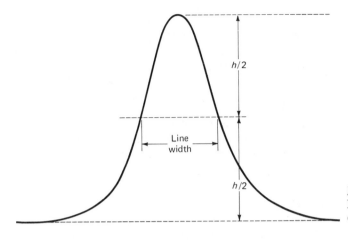

Figure 13-19
Definition of the width of an absorption band.

Figure 13-20
Portion of the hydrogen NMR spectrum of lysozyme in D_2O showing the aromatic resonances (a) at 61°C; (b) at 75°C, which is above the thermal transition temperature; (c) at 61°C in the presence of the denaturant dimethylnitrosamine.

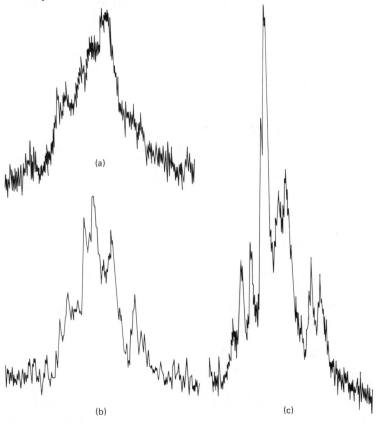

ing of the spectrum, as illustrated in Figure 13-20 for the protein lysozyme. Similarly, studies of the proton spectra of molecules in lipid vesicles have been made, in which the mobilities of the hydrocarbon chains were estimated from line widths.

RELAXATION

The observed NMR spectral lines result from absorption of energy during spin transitions induced by the radiofrequency signal from the spectrometer transmitter. Magnetic fields fluctuating at the resonance frequency can also arise at the nucleus from motion or change in quantization of other magnetic nuclei or of electrons in the vicinity of the nucleus, and these fields are also able to produce transitions, which can be seen in the spectrum as influences on the line intensities or shapes. Such transitions involve the transfer of energy between the nucleus and its immediate environment and are termed spin–lattice relaxation processes. The *spin–lattice relaxation time* T_1 is the reciprocal of the rate constant for the first-order kinetic process in which these transitions occur.

Spontaneous transitions within the sample are indeed necessary if the NMR signal is to be observed, for the applied radiofrequency field induces transitions in both directions and thus tends to equalize the populations of the spin states. Since the spectrometer is sensitive only to the *difference* between upward and downward transitions, rates that depend upon the populations of the initial states, equalization of these populations *saturates* the signal, causing it to disappear.

Local motions in the sample are effective in the spin–lattice relaxation of a nucleus to the extent that they produce fluctuating fields at that nucleus which have a component at the proper frequency to induce transitions at the operating magnetic field. Thus motions that are very much faster, such as rotations in a nonviscous liquid with an average frequency of perhaps 10^{12} per second, are not very effective in producing T_1 relaxation. Likewise, very slow motions in a solid, with time constants of milliseconds or longer, are also ineffective. However, motional or diffusion processes within colloidal systems frequently have a large component of frequency of the order of 10^8 per second and are thus effective in relaxation.

One can express motion in terms of a *correlation time,* which can be regarded as an average time over which the system retains its geometric arrangement. For a rotation, the correlation time is defined as the average time required to turn through a radian; for diffusion, it is approximately the average time required to migrate a distance equal to the molecular diameter. Relaxation becomes more effective and T_1 becomes shorter as the correlation time approaches the reciprocal of the spectrometer frequency from either direction, a relation represented in Figure 13-21.

Other important mechanisms for relaxation include the effects of fluctuating electric fields on nuclei that have electric quadrupole moments. These moments, which are present in nuclei with spins of 1 or more, provide a handle by which a changing electric field can exert a torque tending to produce a spin transition. Since there are electrons moving all about a nucleus in a molecule, quadrupole relaxation often

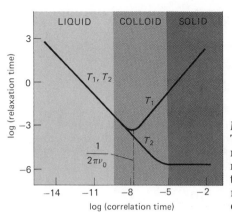

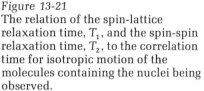

Figure 13-21
The relation of the spin-lattice relaxation time, T_1, and the spin-spin relaxation time, T_2, to the correlation time for isotropic motion of the molecules containing the nuclei being observed.

occurs quite rapidly, and the result may be difficulty in observing the resonance of a nucleus such as ^{14}N, because the spin state changes too rapidly. A second consequence is that rapid relaxation washes out spin–spin coupling of a nucleus of this type to hydrogen, so that, for example, we never observe spin–spin multiplets that might be produced in a hydrogen spectrum by chlorine nuclei, which have a spin of $\frac{3}{2}$.

Another mechanism by which spin–lattice relaxation occurs is that in which spin–spin coupling to another nucleus or to an electron spin varies with time. Paramagnetic species containing unpaired electrons produce nuclear relaxation by a combination of direct dipole coupling and time-variable spin–spin coupling. It is often impossible to obtain an NMR spectrum of the nuclei in an organic free radical because the odd electron makes the nuclear relaxation times so short that the lines are very, very broad.

Spin–lattice, or T_1, relaxation can be viewed on a macroscopic basis as associated with the way in which an NMR spectrum can be obtained. Suppose that one starts with an ordinary sample of matter and then places it in the fixed large field of the spectrometer magnet, with the field in the z direction. The relaxation processes, occurring under the influence of the external fixed field, lead to a state in which an excess of nuclear spins is in the lower-energy state, for hydrogen, parallel to the magnetic field. This means that the sample is magnetized in the z direction, parallel to the field. The time constant for development of this magnetization is T_1 and, if the magnetic field is suddenly removed, the time constant for decay of the magnetization is T_1.

Although the sample is magnetized in the z direction, there is still no signal in the spectrometer receiver, which is sensitive to the motion of moments in the xy plane, transverse to the fixed field direction. Applying a radio signal to the sample causes the nuclear spins, which have been precessing around the external field direction with random orientation, to bunch up together, as shown in the second part of Figure 13-22, so that the sample has a macroscopic fluctuating magnetization transverse to the field. It is the transverse magnetization that the receiver senses to produce a spectrum.

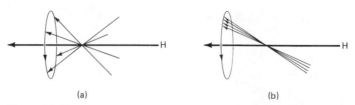

(a)   (b)

Figure 13-22
(a) Nuclear magnetic moments precessing about an external fixed magnetic field. Only the excess moments parallel to the field over those antiparallel are shown. (b) Nuclear moments precessing in phase about the field, bunched together as a result of the effect of the application of an rf field at the resonant frequency, and following the oscillations of that field.

There are relaxation processes that cause the precessing nuclei to get out of phase with one another; these are called *spin–spin relaxation* processes and are characterized by a time constant T_2. Thus a fixed magnetic field applied to some of the nuclei can cause them to precess faster than the others and thus get out of step and no longer contribute to the transverse magnetization, although a constant field cannot contribute to T_1 relaxation at all. In a solid, where there is little motion, T_2 is very short, whereas T_1 is quite long, in contrast to the situation in a liquid where T_1 is equal to T_2; these relations can be seen in Figure 13-21. The lineshape function for what is called a Lorentzian line, which is typically observed for liquid phase spectra, is

$$g(\nu) = \frac{KT_2 H_0}{1 + T_2{}^2 4\pi^2(\nu - \nu_0)^2} \tag{13-7}$$

provided there is no saturation. The maximum value of this function is obtained when $\nu = \nu_0$, that is, right on resonance. To reduce the amplitude to one-half of the maximum, the denominator must be equal to 2, or $T_2{}^2 4\pi^2(\nu - \nu_0)^2$ be equal to 1, from which we deduce that the line width at half-maximum intensity is inversely proportional to the value of T_2. These relations are sometimes expressed in terms of ω, the angular frequency in radians/sec, which is equal to $2\pi\nu$.

13-4
SPECTRA OF
OTHER NUCLEI

NMR spectra of any of a variety of nuclei that have magnetic moments can in principle be obtained. The spectra of different nuclei cannot be confused, because the chemical shift range of any one nucleus is very small compared to the difference between the resonance frequencies of two nuclides. Spectra are most easily obtained for the two nuclei ^{1}H and ^{19}F because of their high natural abundance and large magnetogyric ratio.

Fluorine has a wide range of chemical shifts. Figure 13-23 shows typical resonance positions of some functional groups and molecules containing fluorine. ^{19}F spectra have been used as indicators in various biological systems. Where it can be introduced by a chemical reaction, there is no interference with its resonance by a large number of other peaks as is true for a hydrogen absorption in an organic molecule. By suitable reactions CF_3CO groups can be placed in macromolecules or in molecules involved in biological reactions, and the CF_3 group used as a probe. For example, certain amino acids in an enzyme can be selectively trifluoroacetylated, and the presence or absence of a chemical shift change when an inhibitor or substrate molecule is bound may indicate whether that particular amino acid is involved in the binding process. Conversely, if an inhibitor or substrate molecule is labeled, its binding state can be ascertained by following spectral changes. In hemoglobin and related compounds, changes in the CF_3 resonance of attached CF_3CO groups have been used as indicators of conformational changes occurring on the uptake of oxygen.

Carbon-13 spectroscopy has recently been developed into a valuable method for structural determinations of organic compounds and natural products. Because of its low natural abundance and small magnetogyric ratio, ^{13}C has a spectral sensitivity about 1600 times less

Figure 13-23
Fluorine-19 chemical shifts for some simple compounds and ranges of chemical shift for some fluoroorganic structural units.

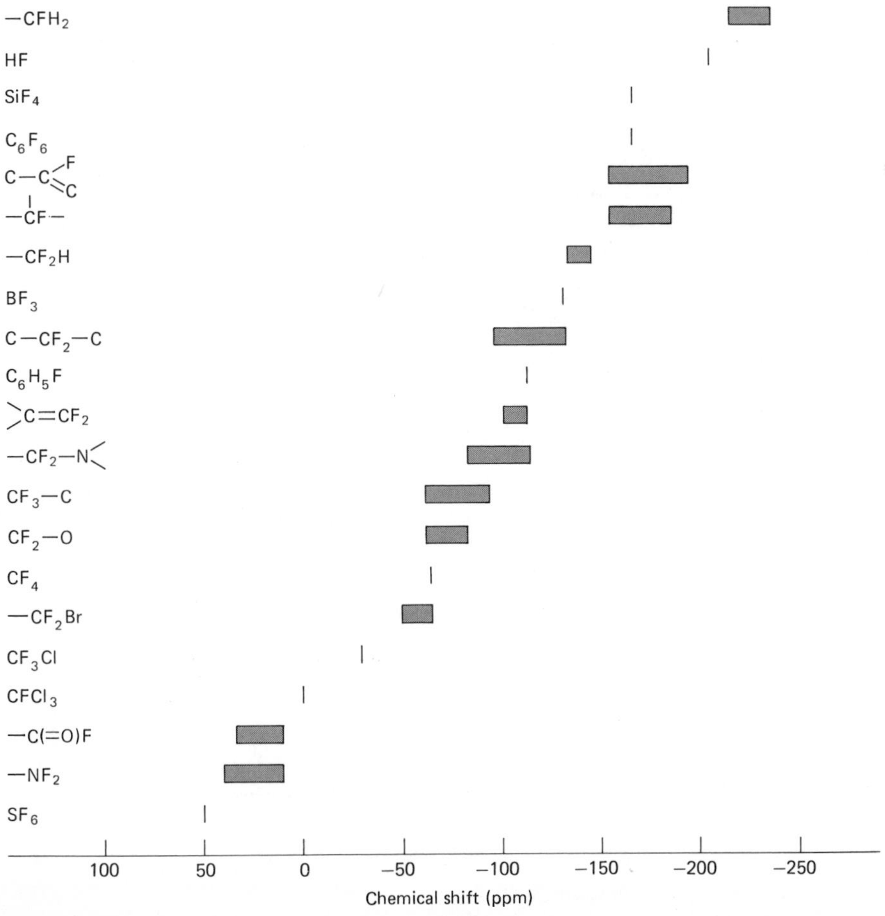

than that of hydrogen. This limitation has been overcome to a considerable degree by methods to be described below. Materials enriched in ^{13}C in one position can be used as tracers to follow a particular group through a reaction sequence. However, molecules unselectively enriched give very complex spectra because of spin–spin coupling between the carbon nuclei, which is almost nonexistent in selectively enriched molecules or in those with only natural abundance of the nuclide, where the probability of finding two ^{13}C atoms adjacent to one another is only about 1 in 10^4.

Like ^{19}F, ^{13}C has a wide range of chemical shifts. The variation in shift with structure is much like that of ^1H, with aliphatic carbons at high field and olefinic and aromatic carbons at low field. Figure 13-24 shows a typical spectrum in which coupling to the hydrogens has been eliminated by "decoupling" which is described below. A particular benefit of carbon spectroscopy of organics compared to hydrogen spectroscopy is that functional groups such as carbonyls, carboxyls, and nitriles, as well as nonprotonated carbons in aromatic systems, are evident in the spectrum. Table 13-2 includes some information about ^{13}C chemical shifts.

Relaxation measurements on ^{13}C have provided extensive information about molecular motion. Thus in phospholipid components of membranes, relaxation times for the individual carbons in the ali-

Figure 13-24
Carbon-13 spectrum of thiamine chloride in water. The spin–spin splitting caused by the hydrogen nuclei has been removed by decoupling or irradiating them. The peaks in the spectrum are labeled according to the designations indicated in the structural formula for the carbon atoms. Spectrum courtesy of Mr. Thomas Baugh, University of Florida.

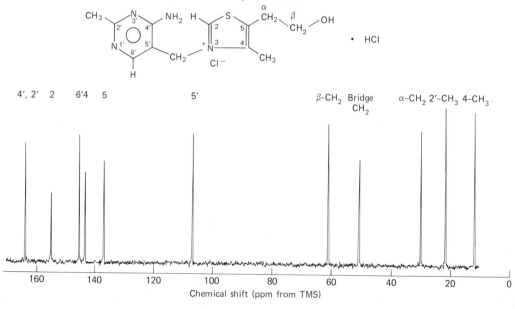

Table 13-2
Carbon-13 chemical shifts and substituent effects[a]

Shift ranges in unsubstituted hydrocarbons

a. Aliphatics		b. Aromatics	122 to 138
CH_3	5 to 32	c. Alkenes	80 to 145
CH_2	16 to 50	$=CH-$	123 to 140
CH	25 to 55	$=CH_2$	115

Shift ranges of functional groups

$>C=O$	150 to 215	$-C\equiv N$	110 to 125

Effect of substituent groups in aliphatics on directly attached carbon

$-OR, -OH, -OCOR$	$+35$ to $+45$	$-Cl$	$+23$
$-NH_2$	$+20$	$-F$	$+60$
$>C=O$	$+10$ to $+15$	$-I$	-6 to $+4$

Effect of substituent groups in benzene on the shifts of directly attached and para carbon atoms

$-C\equiv N$	$-15, +4$	$-F$	$+35, -5$
$-COCF_3$	$-6, +7$	$-NH_2$	$+18, -10$
$-CHO$	$+9, +6$	$-OCH_3$	$+31, -8$
$-NO_2$	$+20, +6$	$-OH$	$+27, -7$

[a]Shifts are given in parts per million and are positive to higher frequency; shift ranges are measured from the resonance of TMS.

phatic chains can be measured, showing that mobility is least near the polar group and greatest near the free end of the chain. An advantage over proton relaxation studies is that in the latter *spin exchange* occurs; that is, the effect of the transition of one hydrogen is felt by a hydrogen in the next group and thus travels along the chain, causing all the hydrogens to have the same, averaged, relaxation time. In contrast, ^{13}C nuclei in natural abundance are very well isolated and relax independently of one another.

Nitrogen-15, like ^{13}C and ^1H, has a spin of $\frac{1}{2}$, as compared to ^{14}N, which has a spin of 1 and therefore gives very broad lines because of the quadrupole moment. The natural abundance of ^{15}N is 0.36 percent, and natural abundance spectroscopy of this nuclide is now just about feasible, although much effort is required. Enriched samples are very promising for biological research.

Phosphorus-31 spectroscopy presents no sensitivity problems, and the nuclear spin is also $\frac{1}{2}$. As an example of its application in biological systems, the various forms of phosphate such as inorganic phosphate, ATP and ADP, creatine phosphate, and so on, can be distinguished by chemical shift differences. In muscles, including a functioning heart of a mouse, removed from the body and kept in the "living" state by bathing in suitable fluids, it has been possible to follow the changes in the relative concentrations of the various forms of phosphate under changing physiological conditions.

Many metal nuclei can be observed with a spectrometer capable of being tuned to the appropriate resonance frequencies. Lithium has been used with enzymes that normally require for function other monovalent cations which are more difficult to observe directly because of quadrupole effects. The magnetic isotopes ^{113}Cd and ^{199}Hg can be substituted for the normally present zinc in human carbonic anhydrase without loss of activity, and their NMR spectra followed as various conditions that influence the enzymic process are altered.

13-5
SPECIAL METHODS
IN NMR SPECTROSCOPY

DOUBLE RESONANCE

The term *double resonance* refers to simultaneous irradiation of another nucleus along with the one under observation. The second variety of nucleus may be of the same element as the one observed, for example, both may be hydrogen nuclei but with different chemical shifts, or it may be of a different element, as when ^{13}C is observed while ^{1}H is irradiated.

Double resonance can be used to unravel the details of or to simplify complex spectra. For instance, if two nuclei are spin–spin coupled, irradiation at the resonance frequency of one may cause changes in the pattern of the other. Irradiation with low-power levels of radio frequency (rf)—spin tickling—results in effects such as intensity changes or splitting of the peaks of coupled nuclei into multiplets.

Irradiation with high-power levels wipes out the coupling effects of a nucleus entirely, and other multiplets arising because of its spin orientation collapse to single peaks or are decoupled. A simple, if not rigorous, description of decoupling is the statement that the nuclear spins of the irradiated type flip so rapidly between upper and lower energy levels that other nuclei see only their average orientation, which is directed perpendicular to the strong magnetic field, rather than parallel or antiparallel.

Many ^{13}C spectra are obtained with all the hydrogen nuclei in the molecule irradiated by a broadband signal, so that they are decoupled. This has two advantages. First, collapse of the spin multiplet to a single peak greatly increases the signal height. Second, spin flips of the protons are coupled to spin flips of the ^{13}C nuclei by the direct dipolar interaction in such a way that the Boltzmann population distribution of the protons is transferred to the ^{13}C nuclei. This is called the nuclear Overhauser effect and results in an intensity increase to as much as three times that otherwise found, because of the larger gyromagnetic ratio of hydrogen.

METHODS FOR
SENSITIVITY
ENHANCEMENT

NMR is not intrinsically a very sensitive spectroscopic method, for the energy separations involved are so small that the Boltzmann factor yields only a very slight excess of population in the ground state. If care is not exercised in keeping down the rf signal from the transmitter during an experiment, the populations of the two states can indeed be equalized, so that the signal becomes saturated.

To extend the technique to lower concentrations, several approaches have been developed. The availability of multichannel systems for data storage has made possible time averaging, in which the spectrum is scanned repeatedly and the data points for each scan are added to the contents of a series of discrete channels in a computer memory. Noise that is random tends to cancel out over a period of time, whereas the signal builds up in intensity. The ratio of spectral signal to noise increases as the square root of the number of scans. Thus doubling the signal intensity requires four times the number of scans, and, although signal averaging is very valuable, after some stage further increase in intensity requires prohibitively long times.

A method of increasing efficiency of spectrometer operation is by use of pulsed Fourier transform (FT) spectroscopy. A powerful rf signal pulse, lasting only from 10 to 100 μsec, is applied somewhere in or near the spectral region. Because the pulse is very short, it contains a whole band of frequencies which can simultaneously excite nuclei with resonance frequencies throughout the region. Nuclei in various environments precess around the external magnetic field at their appropriate frequencies, which are identical with their transition frequencies. The various precession frequencies produce signals in the receiver which interfere with one another, yielding a beat note pattern, called the *free induction decay* (FID), because it occurs after the transmitter pulse has ended. The FID is decomposed into its component frequencies by a mathematical operation called a Fourier transformation, according to the equation

$$f(\nu) = \int_{-\infty}^{+\infty} F(t)e^{2\pi i \nu t}\, dt \qquad (13\text{-}8)$$

The pulsed Fourier transform method has the advantage that spectrometer time is used much more effectively than in continuous scanning of the magnetic field or frequency, in which a large fraction of the spectral range is idle at any instant during the spectral acquisition. Combined with time averaging, the FT method permits proton spectra to be obtained with microgram samples, and ^{13}C spectra with milligram quantities. A parallel method has also been applied to infrared spectroscopy in order to increase its sensitivity and versatility. The mathematical processing of results is similar, but the beat pattern is produced optically by mixing a reference beam of light with a beam that has passed through the sample.

13-6
APPLICATIONS OF NMR

In this section we mention a few of the ways in which NMR can be applied to biological systems. The first is the study of hydrogen bonding. Involvement of a hydrogen atom in a hydrogen bond leads to a substantial change in chemical shift, an unshielding or downfield shift as a result of the restrictions imposed by the electric field of the dipole, which attracts the hydrogen, on the freedom of motion of electrons about the hydrogen nucleus. Thus the hydroxyl resonance of a pure alcohol, almost completely hydrogen-bonded, appears at 6 to 7 ppm from TMS, whereas in dilute solution in an inert solvent the resonance is upfield of a methyl resonance, at about 1 ppm from TMS.

The chemical shift of the hydroxyl hydrogens as observed for a solution is the concentration-weighted average of the shifts of the hydrogen-bonded protons and of the protons not involved in hydrogen bonds. If the limiting shifts can be evaluated, it is then possible to calculate the relative amounts of the monomeric species and of the hydrogen-bonded molecules in solution. If it can be assumed that only one aggregated species, say a dimer, is formed, then the equilibrium constant can be calculated, and from the temperature dependence of the equilibrium constant, the enthalpy change associated with the formation of hydrogen bonds can be determined.

Of course, if the hydrogen bond in question is an intramolecular bond, it is less susceptible to changes in environment. For example, acetylacetone exists in an enol form, with the hydroxyl chemical shift of about 15 ppm from TMS, and with the shift essentially independent of concentration in a solution.

Another type of application involves the use of line widths and relaxation times to follow molecular interactions. For example, the extent of binding of small molecules, which sometimes *intercalate* themselves into the nucleic acid double helix, can be followed by the line broadening their resonance undergoes when their motion is restricted. The line widths of inhibitors bound to enzymes, as well as their chemical shift changes on binding, have been used to follow the binding equilibrium and also the binding rate. An example is the binding of methyl-N-acetyl-D-glucosamine to lysozyme. By investigating the relaxation times of the protons in solvent water when a paramagnetic ion is involved in the active site of an enzyme, and using the fact that the relaxation rate is the weighted average of that of free water and of bound water, it is possible to estimate the degree to which the metal ions are accessible to the solvent.

13-7
PRINCIPLES
OF ELECTRON
PARAMAGNETIC
RESONANCE

In systems with unpaired electrons, the energy levels of the electronic magnetic moments are quantized in an external magnetic field and, as for nuclei, radiation of suitable wavelength can stimulate transitions between these levels. For a convenient magnetic field of 3500 gauss, this radiation is in the microwave X band and has a wavelength of the order of centimeters; for 13,000 G, it is in the K band with millimeter wavelengths. Corresponding frequencies are 10^{10} to 10^{11} Hz, so high that the radiation cannot effectively be transmitted by a metallic conductor; instead, it is passed through hollow rectangular tubes called *waveguides*. The sample is placed in a quartz tube inserted in the waveguide; to obtain a spectrum, the magnetic field is swept while a microwave bridge measures the absorption of energy. For instrumental reasons, the first derivative of the energy absorption, rather than the absorption itself, is usually plotted. The X band happens to coincide with a region where liquid water absorbs energy in the form of energy of rotation of the molecules moving against the frictional force of their neighbors, so that it is difficult to work with aqueous solutions in an X-band spectrometer.

This technique has been called electron paramagnetic resonance (EPR) spectroscopy. It has sometimes been known, less properly, as electron spin resonance (ESR) spectroscopy.

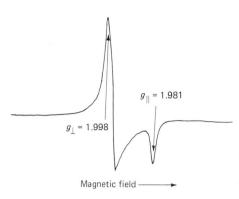

$g_{||} = 1.981$

$g_{\perp} = 1.998$

Magnetic field ⟶

Figure 13-25
The EPR powder spectrum of radicals
formed when CO is adsorbed on
thorium oxide.

g VALUE

The position of absorption of energy in an EPR spectrum is described
by the g value, defined as $h\nu/\beta H$, where β is the *Bohr magneton* and
H is the magnetic field strength. It is essentially a measure of the
frequency/field ratio. For the free electron, the g value g_e is equal to
2.0023, as described in Section 13-1, and for organic radicals such as
the methyl radical, $CH_3\cdot$, its value is very close to this spin-only value.

For transition metal ions and their complexes, the interaction between the electron spin and the electron orbital magnetic moment
prevents complete quenching of the orbital contribution, and the g
value is more variable than for organic radicals. Further, g is anisotropic, so that, under conditions of restricted motion as in a solid or in
a colloidal system, its magnitude depends upon the direction of measurement. Many complexes have axial—that is, cylindrical—symmetry,
which means that their properties in two dimensions, the x and y directions, are identical, and those in the third dimension, the z direction, are different. One can imagine how this arrangement is obtained
by visualizing an octahedral complex in which two diametrically opposed ligands are pulled or pushed symmetrically while the other four
remain undisturbed. The g value when the magnetic field is in the z
direction is then designated $g_{||}$, and that when the field is perpendicular
to the z axis and thus in the xy plane is termed $g_{\perp}$. In a single crystal,
the individual complexes are arrayed in an orderly fashion so that, by
rotating the crystal, g values can be observed to change and the symmetry properties can be determined. In a "powder," a term used to
refer to any unordered system, the paramagnetic species are randomly
arranged. Since for any direction in space there is then twice as great a
probability that $g_{\perp}$ will be encountered as that $g_{||}$ will be found, the
portion of the spectrum corresponding to $g_{\perp}$ is twice as intense as the
portion corresponding to $g_{||}$, as shown in Figure 13-25.

The g value of a complex is often diagnostic of the metal forming the
complex and of the type of complex. In general, for *d* shells less than
half-filled with electrons, g is less than g_e, and for *d* shells more than
half-filled, g is greater than g_e. For copper complexes, $g_{||}$ is usually
about 2.2 and $g_{\perp}$ falls between 2.04 and 2.09. For Fe^{3+}, high-spin complexes have values of about 2.0 to 9.7, but low-spin complex values are
1.4 to 3.1.

For a radical or complex rotating freely in solution, the g value observed is the average of the three g values in the x, y, and z directions.

HYPERFINE SPLITTING

EPR spectra often show splitting resulting from the effects of magnetic nuclei on the electronic energy levels. This multiplicity is referred to as hyperfine splitting, and the mechanism producing it is quite similar to that causing spin–spin splitting in NMR spectra. For example, the EPR spectrum of a hydrogen atom is a doublet with splitting of 1420 MHz, or 507 G at a 3500-G magnetic field. The spectrum for a methyl radical is a 1:3:3:1 quartet because of interaction of the unpaired spin with the three hydrogen atoms, with a hyperfine splitting of 22 G at 3500 G. The spectrum of $\cdot CH_2OH$ is a triplet of doublets, as shown in Figure 13-26.

If the nuclear spin is greater than $\frac{1}{2}$, other rules apply to the multiplicity. A copper nucleus, which has a spin of $\frac{3}{2}$, gives a hyperfine pattern of four lines of equal intensity. In copper complexes such as those of histidine, ethylenediaminetetraacetic acid (EDTA), and carboxypeptidase A, the hyperfine splitting A is 150 to 180 G. In the oxidizing enzyme ceruloplasmin, there are, however, eight copper atoms per molecule with A values of only 77 G, indicating that the odd electron is delocalized more than usual from the copper atom onto the surrounding ligand groups. That this has some relation to the enzymatic activity is indicated by a doubling of A when the enzyme is denatured by urea.

Figure 13-27 shows the spectrum of a radical produced by irradiation of the molecule CF_3CCl_3. The hyperfine splitting confirms the identity of the radical as $CF_3CCl_2\cdot$, for the large quartet splitting is associated with the presence of three fluorine nuclei with a spin of $\frac{1}{2}$, and the seven closely spaced components of each part of the quartet are consistent with the splitting of two chlorine atoms, each having a spin of $\frac{3}{2}$ and thus a maximum total spin of 3.

Unpaired electrons in aromatic radicals, such as benzene negative ion, benzene to which an electron has been added, are usually in π orbitals, which have zero electron density at the carbon and hydrogen positions of the aromatic ring. To a first approximation, the EPR spectrum should show no hyperfine splitting, but the spectrum of $C_6H_6^-$ is a septet with splitting of 3.75 G. The presence of hyperfine splitting, however, can be explained by a mechanism called *spin polarization,* in which the unpaired electron causes magnetic polarization of electrons in other orbitals that do have some s character. In an aromatic

Figure 13-26
Stick form of the EPR spectrum of the radical $\cdot CH_2OH$ showing hyperfine splitting.

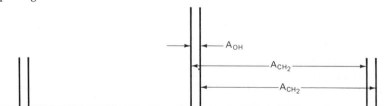

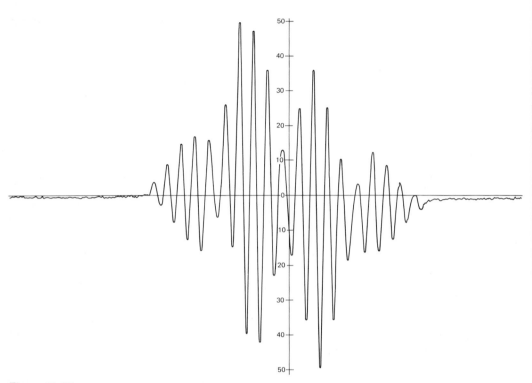

Figure 13-27
Derivative EPR spectrum of $CF_3CCl_2\cdot$ at $-100°$. The large quartet pattern results from the hyperfine splitting of the three fluorine nuclei; each part of the quartet consists of seven lines from the effect of the two chlorine nuclei. Some of the lines in one part of the quartet overlap those in the neighboring quartet component.

molecule, the hyperfine splitting by hydrogen is related to the fraction of an unpaired electron ρ associated with the carbon to which it is attached, by the approximate expression

$$A_H = 25\rho \text{ gauss} \qquad (13\text{-}9)$$

Indeed, the phenomenon of spin polarization must be present in many metal ions in which the unpaired electrons are principally in orbitals with no s character.

In Chapter 14, we shall encounter several examples in which EPR spectroscopy has been applied to the identification of products of radiolysis reactions and to the study of the pathways of processes such as photosynthesis. In Section 13-8 is described the application of EPR to systems in which radicals have been intentionally introduced to serve as probes.

13-8
SPIN LABELING

Spin labels are stable radicals which can serve as reporters of conditions at the location where they are chemically or physically bound to some other system. A type of compound extensively used for this purpose is a nitroxide, of which the two following structures are typical:

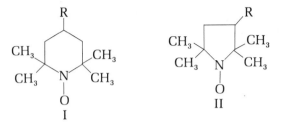

This type of molecule has an odd electron localized on the N—O group, and the EPR pattern shows the hyperfine splitting of the nitrogen nucleus. For a freely rotating spin label the pattern consists of three equally intense lines with an A value of about 15 G. As motion becomes successively more restricted, the high-field line broadens and diminishes in height, the low-field line broadens, and both lines move out from the central line, as shown in Figure 13-28. In the "rigid" pattern, the distance between the two outer peaks is A_z, whereas A_x and A_y, the other anisotropic hyperfine components, are much smaller and are not resolved within the three individual peaks.

From a particular line shape, the mobility of the nitroxide can be estimated. Steroid and fatty acid molecules to which are attached nitroxide groups have been introduced into such systems as liquid crystals and lipid bilayers, and the line shapes observed give an indication of the extent of ordering of the molecules as well as their ability to move about in the structured environment. For example, the degree of mobility varies considerably as the location of the spin label is moved along the hydrocarbon chain in a phospholipid.

In the study of enzyme systems, it is possible to attach the spin label either to the enzyme, or to a ligand molecule—a small molecule, such as an inhibitor, which might be bound to the enzyme. Phosphorylase b is an enzyme that can be specifically labeled with a type I spin label having an iodoacetamide group as R. The EPR spectrum of the bound label shows substantial restriction on mobility. The molecule adenosine monophosphate (AMP) causes the label to become less mobile, indicating that AMP induces a conformational change in the enzyme; this change is apparently a prerequisite to enzyme activity.

In another type of experiment, the effect of the paramagnetic label on the NMR line shape may indicate the approach of the nucleus responsible for the NMR absorption to the vicinity of the label. In some proteins, resonances of the histidine residues can be resolved in the low-field portions of the NMR spectrum and the effect of labeled inhibitors on the resonances may give clues as to where the ligand is bound. Conversely, if the label is attached to the enzyme, broadening of NMR lines of a ligand molecule can be used to determine the binding constant as well as the location of binding. By a combination of spin-label methods and NMR relaxation studies with x-ray diffraction data, rather detailed models of the atomic positions at the active sites of many enzymes have been worked out.

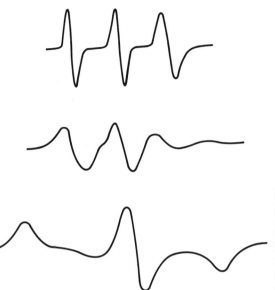

Figure 13-28
Effect of increasing motional
restriction, from top spectrum
to bottom spectrum, on the
EPR pattern of a nitroxide
radical. The spectra are
shown in the usual EPR
derivative form.

As a last example, hemoglobin has been spin-labeled by reaction of the SH group of cysteine 93 in the β chain with both type I and type II labels having R as —NH—C—CH_2I. Oxygenation and deoxygena-

$$\overset{\|}{O}$$

tion at various of the four sites in the hemoglobin unit cause changes in the line shape of the immobilized label. This is consistent with a model in which the cooperative effect of oxygen uptake—the surprising fact that binding of a first oxygen makes a second one bind more easily—travels from one of the four subunits to another by a conformational change which is transmitted by way of the subunit interfaces.

EXERCISES

1. The magnetic susceptibility of a transition-metal complex indicates that the magnetic moment is 3.85 μ_B. Assuming the spin-only approximation applies, how many unpaired electrons are there per complex?

2. Based on Equation (13-3), describe an experiment by which the molar magnetic susceptibility of a substance can be separated into the diamagnetic and the paramagnetic contributions.

3. For a nucleus with spin quantum number of 4, what is the magnitude of the angular momentum in units of $h/2\pi$? Draw a diagram showing the various possible orientations of the spin axis of this nucleus in an external magnetic field and label the orientations with appropriate quantum numbers which distinguish them.

4. The magnetogyric ratio of the ^{23}Na nucleus is 7081 radians/(sec G), and that of ^{1}H is 26,753. What is the approximate resonance frequency of the ^{23}Na nucleus in a field at which the hydrogen resonance frequency is 100 MHz? What magnetic field would be required for the resonance of ^{23}Na at a frequency for which the proton resonance appears at 5000 G? What would that frequency be?

5. In a hydrogen NMR spectrum obtained on a 60-MHz spectrometer, two chemically shifted peaks are 363 Hz apart. What is the

magnitude of the chemical shift difference in parts per million? What would be the separation in hertz between the peaks if the spectrum were obtained on a 100-MHz instrument?

6. Why does the hydrogen NMR resonance for the para hydrogen in aniline appear at higher field than the resonance of the hydrogens in benzene?

7. Sketch the proton NMR spectrum you would expect for $(CH_3)_2CHNO_2$ and that for $CH_3CH_2CH_2NO_2$.

8. In the AA'B . . . X notation, to what kind of system would each of the following correspond? (a) The ring hydrogens in 1,2,3-trichlorobenzene. (b) The ring hydrogens in m-xylene. (c) The ring hydrogens in benzoic acid. (d) The fluorines and hydrogens in 1,1-difluoro-2-methoxycyclopropane.

9. At room temperature, three methyl peaks are observed in the hydrogen NMR spectrum of N,N-dimethylacetamide. As the temperature is increased, two of these peaks merge into a single one of double intensity. Explain these results.

10. The fluorine spectrum of the molecule BF_2Br shows four peaks of equal intensity, with successive peaks separated by 56 Hz. Explain the origin of these peaks and describe the expected appearance of the boron spectrum of the same molecule.

11. Explain why the hydrogen NMR spectrum of 99.5 percent deuterated $(CH_3)_2SO$ shows a five-line multiplet with an intensity ratio of 1:2:3:2:1.

12. The hydrogen NMR spectrum of the substance $CH_3CH=NOH$ shows, along with other peaks, resonances for two different kinds of CH hydrogens. Explain.

13. Silicon has an isotope, ^{29}Si, which has a spin of $\frac{1}{2}$ and comprises 4.7 percent of Si in natural abundance. How does the presence of this isotope affect the hydrogen NMR spectrum of TMS?

14. The following are line widths at various temperatures for the hydrogen resonance of water adsorbed on a protein. From a suitable plot, estimate the activation energy for the motional process which is narrowing the line.

Temperature, °C	Line width at half-maximum intensity
−10	800
0	465
+10	260
+20	168
+30	105

15. State how many resonance peaks you would expect to find in a proton-decoupled ^{13}C spectrum of each of the following molecules. To the extent that you can, arrange the peaks in order of chemical shift: (a) Propylene (b) Toluene (c) 2-Butanone (d) Glucose (e) Propionic anhydride (f) Cyclohexane (g) ATP (h) Cyclobutane (i) Hydroxycyclopentanone.

16. An EPR signal appears at a field of 3500 G for a frequency of 9.8×10^9 Hz. At what field would it appear in a spectrometer of frequency 35×10^9 Hz?

17. A radical has a g value of 4.52. At what field would you look for its resonance in an X-band EPR spectrometer?

18. The proton hyperfine splitting in the methyl radical is 22.3 G. (a) Sketch the EPR spectrum. (b) What is the hyperfine splitting for the $CD_3\cdot$ radical if the magnetogyric ratio for deuterium is 1/6.5 that of hydrogen? (c) What does the EPR spectrum of $CD_3\cdot$ look like?

19. Draw a diagram for the EPR spectrum of each of the following radicals, given the hyperfine couplings specified:

(a) $CF_3CF_2CF_2\cdot$
    $A(CF_2) = 86.2$, $A(CF_2) = 15.1$,
    $A(CF_3) = 3.61$ G
(b) $CH_3CHF\cdot$
    $A(F) = 59.2$, $A(H) = 17.3$,
    $A(CH_3) = 24.5$ G
(c) $CH_3CCl_2\cdot$
    $A(CCl_2) = 4.0$,
    $A(CH_3) = 19.7$ G
(d) The radical cation formed by adding one electron to an antibonding orbital of naphthalene
    A(four alpha hydrogens) $= 4.9$ G,
    A(four beta hydrogens) $= 1.8$ G.
(e) $^{51}V^{2+}$; $A(V) = 94$ G, $I(V) = 7/2$

REFERENCES

Books

Addison Adult and Gerald O. Dudek, *NMR, An Introduction to Proton Nuclear Magnetic Resonance Spectroscopy*, Holden-Day, San Francisco, 1976. An excellent step-by-step introduction to understanding spectra.

Edwin D. Becker, *High Resolution NMR: Theory and Chemical Applications*, Academic Press, New York, 1969. Good general treatment at an intermediate level.

Lawrence J. Berliner, Ed., *Spin Labeling*, Academic Press, New York, 1976. A well-written but advanced account of nitroxide spin labels and their use.

Malcolm Bersohn and James C. Baird, *An Introduction to Electron Paramagnetic Resonance*, W. A. Benjamin, Menlo Park, Calif., 1966. A good introduction.

Frank A. Bovey, *Nuclear Magnetic Resonance Spectroscopy: Principles and Applications in Organic Chemistry*, Academic Press, New York, 1969. An excellent introduction and intermediate-level treatment.

Britton Chance, Takashi Yonetani, and Albert S. Mildvan, Eds., *Probes of Structure and Function of Macromolecules and Membranes*, 2 vols., Academic Press, New York, 1971. Various physical methods, with emphasis on NMR relaxation and EPR spectroscopy.

Raymond A. Dwek, *Nuclear Magnetic Resonance in Biochemistry: Applications to Enzyme Systems*, Oxford University Press, London, 1973. A thorough review at an advanced level.

E. R. Haws, R. R. Hill, and D. J. Mowthorpe, *The Interpretation of Proton Magnetic Resonance Spectra*, Heyden, London, 1973. A programmed introduction, suitable for self-study.

L. M. Jackman and S. Sternhell, *Applications of Nuclear Magnetic Resonance Spectroscopy in Organic Chemistry*, 2nd ed., Pergamon Press, 1969. A thorough coverage on a qualitative level.

Thomas L. James, *Nuclear Magnetic Resonance in Biochemistry*, Academic Press, New York, 1975. Good introductory and intermediate-level overview.

LeRoy F. Johnson and William C. Jankowski, *Carbon-13 NMR Spectra: A Collection of Assigned, Coded, and Indexed Spectra*, Wiley, New York, 1972. Proton-decoupled spectra of a variety of organic compounds.

George C. Levy and Gordon L. Nelson, *Carbon-13 Nuclear Magnetic Resonance for Organic Chemists*, Wiley, New York, 1972. Interpretation of spectra of organic compounds.

J. C. Metcalfe, "NMR Spectroscopy," Chapter 14 in *Physical Principles and Techniques of Protein Chemistry*, S. J. Leach, Ed., Part B, Academic Press, New York, 1970.

F. C. Nachod and J. J. Zuckerman, Eds., *Determination of Organic Structures by Physical Methods*, Vol. 4, Academic Press, New York, 1971. Describes pulse and high-field NMR methods, and spectra of nuclei other than hydrogen.

William W. Paudler, *Nuclear Magnetic Resonance*, Allyn and Bacon, Boston, 1971. Introductory-level treatment of hydrogen spectra of organic compounds.

Henry A. Resing and Charles G. Wade, Eds., *Magnetic Resonance in Colloid and Interface Science*, American Chemical Society, Washington, D.C., 1976. Applications of NMR and EPR to specific systems.

G. C. K. Roberts and Oleg Jardetzky, "Nuclear Magnetic Resonance Spectroscopy of Amino Acids, Peptides, and Proteins," in *Advances in Protein Chemistry*, Vol. 24, Academic Press, New York, 1970. Detailed review of the field.

M. Shporer and M. M. Civan, "The State of Water and Alkali Cations within the Intracellular Fluids," in *Current Topics in Membranes and Transport*, Vol. 9, Academic Press, New York, 1977. Includes information obtained from NMR.

Harold M. Swartz, James R. Bolton, and Donald C. Borg, *Biological Applications of Electron Spin Resonance*, Wiley, New York, 1972. Excellent introduction to the method and an account of some applications.

Brian D. Sykes and Marian D. Scott, "Nuclear Magnetic Resonance Studies of the Dynamic Aspects of Molecular Structure and

Interaction in Biological Systems," in *Annual Review of Biophysics and Bioengineering*, Vol. 1, Annual Reviews, Palo Alto, Calif., 1972.

F. W. Wehrli and T. Wirthlin, *Interpretation of Carbon-13 NMR Spectra*, Heyden, London, 1976. Excellent presentation of practical details in spectral interpretation.

John E. Wertz and James R. Bolton, *Electron Spin Resonance: Elementary Theory and Practical Applications*, McGraw-Hill, New York, 1972. Excellent introduction.

Sidney J. Wyard, Ed., *Solid State Biophysics*, McGraw-Hill, 1969. Includes some biological applications of EPR.

Journal Articles

F. A. L. Anet and G. C. Levy, "Carbon-13 Nuclear Magnetic Resonance Spectroscopy," *Science* **180**, 141 (1973).

N. J. M. Birdsall, J. Feeney, A. G. Lee, Y. K. Levine, and J. C. Metcalfe, "Dipalmitoyl-lecithin: Assignment of the ^{1}H and ^{13}C Nuclear Magnetic Resonance Spectra, and Conformational Studies," *J. Chem. Soc. Perkin II*, 1441 (1972).

F. A. Bovey, A. I. Brewster, D. I. Patel, A. E. Tonelli, and D. A. Torchia, "Determination of the Solution Conformation of Cyclic Polypeptides," *Acc. Chem. Res.* **5**, 193 (1972).

C. Tyler Burt, Thomas Glonek, and Michael Barany, "Analysis of Living Tissue by Phosphorus-31 Magnetic Resonance," *Science* **195**, 145 (1977).

D. Allan Butterfield, "Electron Spin Resonance Studies of Erythrocyte Membranes in Muscular Dystrophy," *Acc. Chem. Res.* **10**, 111 (1977).

Richard L. Carlin, "Paramagnetic Susceptibilities," *J. Chem. Educ.* **43**, 521 (1966).

Jane H. Chin and Dora B. Goldstein, "Drug Tolerance in Biomembranes: A Spin Label Study of the Effects of Ethanol," *Science* **196**, 684 (1977).

Mildred Cohn and Jacques Reuben, "Paramagnetic Probes in Magnetic Resonance Studies of Phosphoryl Transfer Enzymes," *Acc. Chem. Res.* **4**, 214 (1971).

Thomas H. Crawford and John Swanson, "Temperature Dependent Magnetic Measurements and Structural Equilibria in Solution," *J. Chem. Educ.* **48**, 382 (1971).

Charles M. Deber, Vincent Madison, and Elkan R. Blout, "Why Cyclic Peptides? Complementary Approaches to Conformations," *Acc. Chem. Res.* **9**, 106 (1976).

Hermann Dugas, "Spin-Labeled Nucleic Acids," *Acc. Chem. Res.* **10**, 47 (1977).

O. H. Griffith and A. S. Waggoner, "Nitroxide Free Radicals: Spin Labels for Probing Biomolecular Structure," *Acc. Chem. Res.* **2**, 17 (1969).

John J. Grimaldi and Brian B. Sykes, "Stopped Flow Fourier Transform Nuclear Magnetic Resonance Spectroscopy. An Application to the α-Chymotrypsin-Catalyzed Hydrolysis of *tert*-Butyl-L-phenylalanine," *J. Am. Chem. Soc.* **97**, 273 (1975).

J. A. Hamilton, N. J. Oppenheimer, R. Addleman, A. O. Clouse, E. H. Cordes, P. M. Steiner, and C. J. Glueck, "High-Field ^{13}C NMR Studies of Certain Normal and Abnormal Human Plasma Lipoproteins," *Science* **194**, 1424 (1976).

Donald P. Hollis, R. L. Nunnally, G. J. Taylor, M. L. Weisfeldt, and W. E. Jacobus, "Phosphorus Nuclear Magnetic Resonance Studies of Heart Physiology," *J. Magn. Resonance* **29**, 331 (1978).

Edward G. Janzen, "Spin Trapping," *Acc. Chem. Res.* **4**, 31 (1971).

D. R. Kearns and R. G. Schulman, "High-Resolution NMR Studies of the Structure of Transfer Ribonucleic Acid and Other Polynucleotides in Solution," *Acc. Chem. Res.* **7**, 33 (1974).

Roy D. Lapper and Ian C. P. Smith, "A ^{13}C and ^{1}H Nuclear Magnetic Resonance Study of the Conformations of 2',3'-Cyclic Nucleotides," *J. Am. Chem. Soc.* **95**, 2880 (1973).

Y. K. Levine, N. J. M. Birdsall, A. G. Lee, and J. C. Metcalfe, "^{13}C Nuclear Magnetic Resonance Relaxation Measurements of Synthetic Lecithins and the Effect of Spin-Labeled Lipids," *Biochemistry* **11**, 1416 (1972).

A. Gavin McInnes and Jeffrey L. C. Wright, "Use of Carbon-13 Magnetic Resonance Spectroscopy for Biosynthetic Investigations," *Acc. Chem. Res.* **8**, 313 (1975).

C. C. McDonald and W. D. Phillips, "Manifestations of the Tertiary Structure of Proteins in High-Frequency Nuclear

Magnetic Resonance," *J. Am. Chem. Soc.* **89,** 6332 (1967).

C. C. McDonald, W. D. Phillips, and J. D. Glickson, "Nuclear Magnetic Resonance Study of the Mechanism of Reversible Denaturation of Lysozyme," *J. Am. Chem. Soc.* **93,** 235 (1971).

John L. Markley, "Observation of Histidine Residues in Proteins by Means of Nuclear Magnetic Resonance Spectroscopy," *Acc. Chem. Res.* **8,** 70 (1975).

Albert S. Mildvan, "Magnetic Resonance Studies of the Conformations of Enzyme-Bound Substrates," *Acc. Chem. Res.* **10,** 246 (1977).

William A. Pryor, "Free Radicals in Biological Systems," *Sci. Am.* **223,** 70 (August 1970).

G. K. Radda and R. J. P. Williams, "The Study of Enzymes," *Chem. Brit.* **12,** 124 (1976).

R. M. Silverstein and R. G. Silberman, "Troublesome Concepts in NMR Spectrometry," *J. Chem. Educ.* **50,** 484 (1973).

Ian C. P. Smith, Harold J. Jennings, and Roxanne Deslauriers, "Carbon-13 Nuclear Magnetic Resonance and the Conformations of Biological Molecules," *Acc. Chem. Res.* **8,** 306 (1975).

E. Wasserman and R. S. Hutton, "Electron Paramagnetic Resonance of Triplet States: Cyclic 4-Electron Systems, CH_2, and Environmental Effects," *Acc. Chem. Res.* **10,** 27 (1977).

Fourteen
Photochemistry and Radiation Chemistry

In this chapter we are concerned with chemical reactions induced by light, other forms of radiation, and high-energy particles, rather than those in which the thermal energy of molecules—translational, vibrational, and rotational energy—is able to overcome an energy barrier to break a chemical bond. Under the heading *photochemistry* are included reactions in which the sample absorbs light in the visible and ultraviolet spectral regions and undergoes electronic excitation such as was described in Chapter 9. This initial event may be followed by a chemical process consisting of the breaking of a bond, resulting in such processes as the dissociation of a molecule into fragments or the conversion of one olefinic isomer to another. In many cases, more complicated sequences of events may follow: A chain reaction such as those described in Section 10-5 may be initiated, a radical formed may extract an atom from another molecule, a photosynthetic process may occur, or a ring compound may be formed between two molecules or within a molecule, as in the reactions

$$2\ CH_2=CH \atop CH=CH_2 \longrightarrow \quad (14\text{-}1)$$

$$\longrightarrow \quad (14\text{-}2)$$

For many reactants, several reaction pathways are possible, and which one is followed depends upon the state to which the reactant is excited on the absorption of a photon.

Radiation chemistry is concerned with the effects produced either by photons or by material particles having energy sufficiently high to cause ionization. Of electromagnetic radiation, this includes x rays and gamma rays emitted by nuclei. Frequently used in the study of radiochemical processes are electrons with high kinetic energy, which they may have acquired because they were emitted as beta particles in a nuclear disintegration or as a result of intentional acceleration by the experimenter through a large electric potential difference in a linear accelerator or a cyclotron. Other nuclear particles, such as alpha particles or neutrons, also induce radiochemical changes, but usually they act indirectly through the effects of electrons set free as the heavier particles travel through matter.

14-1
GENERAL PRINCIPLES
OF PHOTOCHEMISTRY

Only photons that are absorbed by a sample of matter can produce a change in that sample. According to the principle of Stark and Einstein, each photon is absorbed by a particular atom or molecule, and the energy of that atom or molecule is increased by the amount of energy $h\nu$ in the photon. The energy carried by Avogadro's number of photons is called an *einstein*, and this is the energy given to 1 mol of a substance, each molecule of which absorbs one photon. Of course, photons in the visible and ultraviolet can be absorbed only by a species that has a suitable electronic level to which it can be raised by the excitation process.

For each chemical reaction a molecule can undergo, one can measure a quantum yield ϕ, the ratio of the number of molecules reacting along that particular path to the number of photons absorbed. Since many photochemical reactions proceed through several successive or competing steps, the quantum yield may be greater or less than unity. For a chain reaction, the quantum yield is equal to the average length of the chain, and it may thus be quite large.

Example: The yield in a photochemical reaction is found to correspond to 0.368 mol of product formed for 75 kcal of energy absorbed from a beam of ultraviolet radiation of wavelength 282 nm. Calculate the quantum yield.

Solution: For this radiation, the energy per einstein is

$$E = \frac{hcN}{\lambda}$$

$$= \frac{(6.6 \times 10^{-27} \text{ erg sec/photon})(3 \times 10^{10} \text{ cm/sec})(6.02 \times 10^{23} \text{ photons/einstein})}{(282 \times 10^{-7} \text{cm})(10^{7} \text{ ergs/J})(4.184 \times 10^{3} \text{J/kcal})}$$

$$= 101 \text{ kcal/einstein}$$

Therefore

$$\phi = (0.368 \text{ mol})(101 \text{ kcal/einstein})/75 \text{ kcal} = 0.496 \text{ mol/einstein}$$

In studying photochemical reactions in the laboratory, the most convenient source of radiation in the visible and ultraviolet regions is usually a mercury lamp in which an arc is maintained electrically. Lamps operating at low pressure produce very high light intensities at 253.7 nm, the wavelength corresponding to the transition from the 6s6p triplet state down to the ground state (see Figure 8-23). High-pressure mercury lamps give almost continuous emission throughout the visible and ultraviolet regions. Readily available "germicidal" lamps which operate in an ordinary fluorescent lamp receptacle are sources of moderate-intensity ultraviolet radiation. All ultraviolet sources are very hazardous; even the briefest exposure of the eye to the beam can be damaging, and the skin can be affected by momentary exposure.

Laboratory studies of photochemical reactions are carried out in vessels in which the irradiation beam is directed onto the sample, either by immersing the light source in the sample vessel or by means of an optical train which focuses the light on a window in the vessel. Pyrex glass absorbs radiation below about 300 nm, and so quartz vessels or windows must be used for shorter wavelengths.

Quantitative studies require the measurement of light intensities, either by means of a thermopile, a device in which light falling on a metal strip heats the strip proportionately to its intensity and the temperature of the strip is measured by a battery of thermocouples, or by comparing a reaction of known quantum yield, such as that shown below in Equation (14-12) in the same cell and under the same conditions as the reaction under study.

14-2
PHOTOCHEMICAL
PROCESSES

In this section we examine, in an order roughly paralleling the sequence in which they typically occur, various steps in the overall photochemical reaction, including a variety of photophysical processes that precede the chemical reaction. Figure 14-1 shows a much simplified scheme of the relations between some of the types of events that can occur.

Figure 14-1
Schematic diagram of relationships of various possible photochemical and photophysical processes.

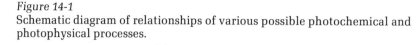
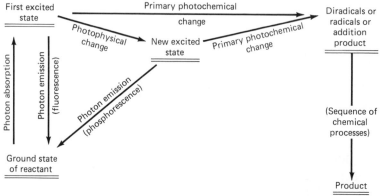

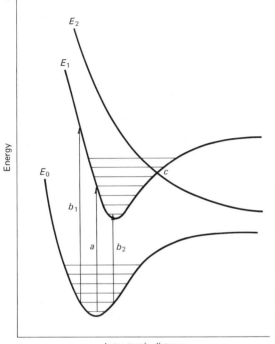

Energy

E_2

E_1

E_0

b_1

a b_2

c

Interatomic distance

Figure 14-2
Potential energy curves for the
ground state and two excited
states of a molecule. Arrow a
corresponds to a transition
from the lowest vibrational
level of the ground state. The b_1
and b_2 arrows represent
possible excitations from the
first excited vibrational level of
the ground state. Transition b_1
leads directly to molecular
dissociation.

ELECTRONIC EXCITATION AND MOLECULAR DISSOCIATION

In Section 9-9, we outlined some typical processes in which molecules
are excited to higher energy levels by the absorption of photons of
visible or ultraviolet radiation. Figure 14-2 shows a potential energy
diagram for several states of a diatomic molecule, with vibrational
levels superimposed. The curves in such a diagram could also represent the variation in potential energy with distance for one bond in
a molecule with more than two atoms, provided that the stretching
of the bond is independent of vibrations in the remainder of the molecule. Curve E_0 is for the electronic ground state. Curve E_1 represents
a stable excited state, stable in the sense that there is a minimum of
potential energy about which the bond distance oscillates. Curve E_2
represents a *repulsive* or unstable state in which there is no genuine
bond.

The vertical arrow a in Figure 14-2 shows the excitation of a molecule from the zeroth vibrational level of the ground electronic state
to the third vibrational level of the excited state E_1. The equilibrium
distance is longer in the excited state than in the ground state, because
the bond is weaker. Now, the electronic excitation occurs so rapidly
that the nuclei can move only a negligible distance while it is hap-
pening. This is a statement of the *Franck–Condon principle*. Further-

more, except for the zeroth vibrational level, in which the molecule spends a lot of time very near the equilibrium position, it tends to be for the longest periods near the extremes of the vibrational "pendulum swing," because there the atoms are moving most slowly. Thus the arrow a, which goes directly upward, in accord with the Franck–Condon principle, and which terminates near the extreme of an excited-state vibration, represents a transition of high probability.

Arrows b_1 and b_2 represent possible excitations from the first excited vibrational level of the ground state. The transition b_2 leads to the lowest vibrational level of the excited electronic state, with the molecule still firmly held together, but b_1 leaves the molecule with energy above the potential energy plateau of state E_1 at large distance, so much energy that the molecule is carried right out of the potential energy well represented by the E_1 curve and dissociates on the first vibration, possibly to form product fragments in electronically excited states. For example, oxygen gas irradiated with ultraviolet light between 129 and 175 nm dissociates into one ^1D atom (an excited atom) and one ^3P atom (a ground-state atom).

Excitation to a level below the plateau in E_1 but above point C where the curves E_1 and E_2 cross may also lead to dissociation. A molecule having for an instant the interatomic distance corresponding to this point may undergo a crossover from one state to the other without a change in energy and therefore without the need for absorption or emission of radiation. Once the repulsive state is reached, the molecule continues to separate along the E_2 curve into two parts. This process is evident by a smearing out of the vibrational structure of the electronic emission bands of the molecule and is termed *predissociation*. When irradiated in the region between 176 and 195 nm, molecular oxygen predissociates by this type of process, producing two ^3P, or electronic ground-state, oxygen atoms.

Ammonia molecules are photolyzed to form NH_2 radicals and hydrogen atoms through excited molecular states which, interestingly, are planar rather than pyramidal like the ground state of ammonia. Hydrogen iodide dissociates photochemically to hydrogen and iodine atoms; the maximum in absorption of hydrogen iodide is near 220 nm, and the reaction occurs chiefly through repulsive states like E_2 in Figure 14-2, rather than by predissociation or through an excited stable state. The extra energy set free in this process is converted to kinetic energy of the product atoms; at 185-nm irradiation wavelength, the hydrogen atom can carry away as much as 84 kcal/mol of kinetic energy, so that it is termed a "hot atom" and is able to enter into a variety of reactions as a result of this energy.

Diatomic halogen atoms have an absorption spectrum consisting of resolved bands at longer wavelengths (804 to 510 nm for I_2), and excitation in this region occurs by predissociation through a triplet state (unexpectedly because the ground state of the molecule is a singlet). At shorter wavelengths, there is a *continuum*, a continuous smooth absorption envelope, corresponding to excitation along arrow b_1 in Figure 14-2, but again the excited state is a triplet state; one ground-state halogen atom and one excited-state halogen atom are produced by this pathway. When photolysis of Cl_2 is conducted in the presence of H_2, a chain reaction ensues with the production of HCl, much as for the H_2–Br_2 reaction described in Chapter 10, a reaction that also can be photochemically initiated. If Cl_2 is photolyzed in the presence of a hydrocarbon, a chain reaction resulting in chlorination of the organic species can occur.

FLUORESCENCE
AND QUENCHING
OF FLUORESCENCE

When a molecule is raised to an excited electronic state but does not dissociate, one of the possible events that may follow is emission of a photon, simply returning the molecule to the ground state. Usually the electronically excited molecule loses its vibrational energy very rapidly by collision or by other radiationless processes and reaches the zeroth vibrational level of the excited electronic state before it is electronically de-excited, so that the emitted photon is of longer wavelength than the photon absorbed. This kind of emission of light of lower energy is called *fluorescence*; the lifetime of an excited state before fluorescence is of the order of 10^{-5} to 10^{-10} sec.

The measurement of intensity of fluorescence can be used as an indirect way of determining the rates at which other, more rapid reactions of the excited molecules occur. Such reactions are said to *quench* the fluorescence, since they reduce the intensity of the light emitted in the process. Quenching may occur by reaction with the solvent or by a change from the initially formed excited state to another excited state, such as from a singlet to a triplet state, effects that are called *internal* quenching. Quenching may also result from the presence of some added species which accepts energy from the excited molecules; this is termed *external* quenching. An example of external quenching is found in the behavior of chlorophyll. On irradiation of a solution of pure chlorophyll a in the laboratory, fluorescence occurs. Addition of oxygen or a quinone eliminates the fluorescence, as does the presence of the electron-transfer enzyme systems with which the chlorophyll is associated in the living plant, indicating that any one of these materials can accept the excitation energy from the chlorophyll faster than the latter can fluoresce.

The effects of quenching agents, Q, can be expressed quantitatively, based on the following set of processes:

Excitation:	$X + h\nu \longrightarrow X^*$	(14-3)
Fluorescence:	$X^* \xrightarrow{k_1} X + h\nu'$	(14-4)
Internal quenching:	$X^* \xrightarrow{k_2} X$	(14-5)
External quenching:	$X^* + Q \xrightarrow{k_3} X + Q'$	(14-6)

The rate of excitation can be expressed as a function of the light intensity, or $f(I)$. The steady-state treatment of $[X^*]$ leads to

$$f(I) = k_1[X^*] + k_2[X^*] + k_3[X^*][Q] \qquad (14\text{-}7)$$

The quantum yield for fluorescence is

$$\phi_Q = \frac{k_1[X^*]}{f(I)} = \frac{k_1}{k_1 + k_2 + k_3[Q]} \tag{14-8}$$

In the absence of external quenching, the fluorescence yield is

$$\phi_0 = \frac{k_1}{k_1 + k_2} \tag{14-9}$$

The ratio of quantum yields for the two cases is

$$\frac{\phi_0}{\phi_Q} = \frac{k_1 + k_2 + k_3[Q]}{k_1 + k_2} = 1 + \frac{k_3}{k_1 + k_2}[Q] = 1 + k_3[Q]\tau \tag{14-10}$$

The quantity τ in this equation is equal to the lifetime of X^* in the absence of external quenching. The equation is known as the Stern–Volmer relation, and it predicts a linear dependence of ϕ_0/ϕ_Q on $[Q]$; the slope of the line can be used to evaluate k_3 if τ is known or can be measured.

ENERGY TRANSFER AND SINGLET–TRIPLET CONVERSION

As we saw in Section 9-9, transitions from the singlet ground state of a molecule usually lead to an excited singlet state, but often the excited singlet can undergo intersystem crossing to a triplet state that lies at a lower energy. The triplet state has a lifetime of the order of 10^{-5} to 10^{-1} sec, much longer than that of the corresponding singlet state, since its transition to the singlet ground state is forbidden. This gives the triplet a greater opportunity to enter into reactions with other species. In the solid state, especially at a low temperature, where there is not very much molecular diffusion or vibration, the triplet may lose its energy by phosphorescence, going back to the ground state.

Reactions of a triplet state are often very different from those of the corresponding singlet state. Thus when anthracene solutions are irradiated in the absence of oxygen, a dimer is produced, and fluorescence measurements indicate that this occurs by way of the excited singlet state. In the presence of oxygen, there is also formed—almost certainly through the triplet state—anthracene peroxide:

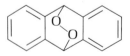

In general, oxygen acts to quench triplet states, as it does here by reacting to give the peroxy species, but it does not quench excited singlet states to nearly so great an extent.

Another example of directive effects in a photochemical reaction is provided by the photolysis of uracil in aqueous solution, which leads both to dimers and to hydration products:

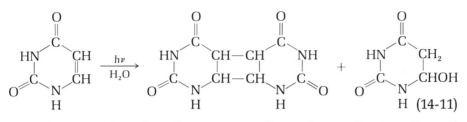

$$(14\text{-}11)$$

Addition of 2,4-hexadienol or oxygen reduces the production of uracil dimers, which are therefore formed by way of the triplet state, but does not affect the yield of photohydration product, which must therefore be produced through an excited singlet state.

It is often possible to induce or at least to favor a particular photochemical reaction by utilizing an effect known as *photosensitization,* which involves the excitation of one species by irradiation, followed by transfer of the excitation energy to a second species which then enters into a reaction. Sensitization most commonly proceeds by way of energy transfer from one triplet state to form a triplet state of another species. The initially excited species can lose energy best if it is a state that is metastable, such as the triplet, so that there is sufficient time for encounter to occur. Rules of spin conservation require that, if the species giving up the energy goes from a triplet state back down to a singlet ground state, the species accepting the energy should undergo the opposite change in spin, from a singlet to a triplet state.

In gas phase photochemistry, mercury atoms have often been used as triplet photosensitizers. Figure 8-23 showed the energy-level diagram of mercury, and there is seen to be a metastable triplet state, with valence electrons in the 6s and 6p orbitals, which lies about 112 kcal/mol above the ground state and which emits photons of 253.7 nm when returning to the ground state. As mentioned earlier, this radiation constitutes the principal part of the output of a low-pressure mercury arc lamp. For example, the ethylene in an ethylene–mercury vapor mixture is sensitized upon irradiation by the 253.7 line, going to the triplet state, from which it can decompose to acetylene and hydrogen. In the complete absence of mercury vapor, ethylene absorbs radiation only at wavelengths shorter than 200 nm.

Another photosensitized reaction, one sometimes used to measure the amount of radiation emitted by a source, is the decomposition of oxalic acid in solution in the liquid phase in the presence of uranyl salts. The UO_2^{2+} ion absorbs light in the blue and ultraviolet regions, and the reaction resulting can be written

$$(UO_2^{2+})^* + (COOH)_2 \longrightarrow UO_2^{2+} + CO_2 + CO + H_2O \qquad (14\text{-}12)$$

In quantitative application of this reaction, the unreacted oxalate can be titrated with permanganate solution.

A typical sensitizer frequently used in organic reactions is benzophenone, for which the n–π^* absorption is centered at about 345 nm and the π–π^* absorption at about 245 nm. The upper singlet state is converted rapidly by internal conversion to the lower n, π^* singlet

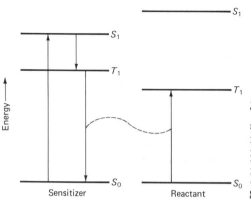

Figure 14-3
Typical relation between energy levels of a sensitizer species and a photochemical reactant, where sensitization occurs via a metastable triplet of the sensitizer. The dotted line links the transition in which the sensitizer loses energy with that in which the reactant gains the energy.

state, and the efficiency of crossing from this state to the first excited triplet state is high, a prerequisite for function as a sensitizer. At the same time, the triplet state lies fairly high in energy, at 69 kcal/mol. Figure 14-3 shows the usual relationship between energy levels of a sensitizer and of the reactant. The triplet state of the reactant must of course lie below the triplet state of the sensitizer so that the energy available is sufficient to raise the reactant molecule to its triplet state.

As molecules for which the course of reaction can be modified by sensitizers, conjugated dienes, such as butadiene, serve as excellent examples. The triplet energy is not too high, only about 50 to 60 kcal/mol above the ground state, but intersystem crossing from the singlet to the triplet is not very efficient, so that products from reaction of the singlet predominate on direct irradiation. Thus butadiene in an inert solvent reacts to give cyclobutene and bicyclobutane by an intramolecular ring-closing process:

$$CH_2=CH \diagdown CH=CH_2 \longrightarrow \begin{matrix} CH-CH_2 \\ \| \quad | \\ CH-CH_2 \end{matrix} + \begin{matrix} CH_2-CH \\ | \diagup | \\ CH-CH_2 \end{matrix} \qquad (14\text{-}13)$$

When a triplet sensitizer is added, the reaction shown in Equation (14-1) leading to the formation of divinylcyclobutane occurs along with the production of 4-vinylcyclohexene. Sensitizers with energies greater than 60 kcal/mol favor formation of the cyclobutane dimers, whereas when the energy lies between 54 and 60 kcal/mol, the yield of vinylcyclohexene is enhanced. It is thought that there are two excited triplet states of butadiene: a cis rotamer which requires less energy to form and which reacts to give the cyclohexene, and a trans rotamer, which requires more than 60 kcal to form and which is not geometrically able to form the six-membered ring. The reaction producing the cyclohexene can be represented as

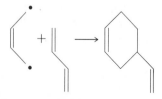

$$(14\text{-}14)$$

SCAVENGING
AND CAGE EFFECTS

Often an investigator wishes to establish the existence of a radical intermediate in a reaction but is unable to obtain direct spectroscopic evidence because of the low concentration in which it is present. One approach is to use a trapping agent, or scavenger, a species that reacts with the intermediate as rapidly as it is formed, so that it has no opportunity to take part in its usual secondary reactions. Many common radical traps have been used principally in gas-phase reactions; these include nitric oxide, oxygen, and propylene. However, there is often produced a complex mixture of products. Molecular iodine seems to be the most generally satisfactory scavenger; it can be used both in the liquid and vapor phases, and the reactions resulting are often quite "clean." Alkyl radicals, for instance, react quite readily with iodine to produce alkyl iodides.

In favorable circumstances, scavenging can be used to determine the rate at which radicals form, in other words, the quantum yield for a particular radical reaction pathway, but it is then necessary that essentially every radical formed react with the scavenger. However, in a reaction in which a molecule is photochemically dissociated in the liquid phase into a pair of radicals, the two radicals tend to be held together in the solvent *cage* in which they are formed. Restrained from diffusing apart by the surrounding solvent molecules, at least for a short period, they have a greater probability of recombining than if the same total number of radicals were uniformly distributed throughout the solution. When the radicals are formed in a cage, the scavenger, no matter how effective a trapping agent it is, may not be able successfully to compete with the recombination reaction. An illustration is the photolysis of molecular iodine in carbon tetrachloride, for which the quantum yield for iodine formation as measured by the reaction of the atoms with other molecules, or scavengers, is only 0.14 at 25°C, indicating that most of the atoms recombine before ever leaving the cage. In the irradiation of azobisisobutyronitrile, some radicals escape the cage and enter into reactions which depend upon the availability of other reagents in the solution, but some remain and recombine to give either of two products:

$$(CH_3)_2\underset{\underset{CN}{|}}{C}-N{=}N-\underset{\underset{CN}{|}}{C}(CH_3)_2 \longrightarrow N_2 + 2\ (CH_3)_2\underset{\underset{CN}{|}}{C}{\bullet} \longrightarrow$$

$$(CH_3)_2C{=}C{=}N-\underset{\underset{CN}{|}}{C}(CH_3)_2 + (CH_3)_2C(CN)C(CN)(CH_3)_2 \quad (14\text{-}15)$$

REACTIONS
OF CARBONYL COMPOUNDS
AND CYCLOADDITION
REACTIONS

Aldehydes and ketones form triplet states in high yield when they are irradiated and thus are able to undergo a variety of photochemical reactions, as well as to serve as energy-transfer agents, as we have seen for benzophenone. Indeed, benzophenone often undergoes other photochemical changes in the same reactions in which it is used as a sensitizer. The final products of the carbonyl photochemical reaction depend upon whether the molecule is an aldehyde or a ketone and upon the exact nature of the groups attached to the carbonyl, in particular whether they are aliphatic or aromatic. The latter difference may be largely a consequence of the fact that, for aliphatic ketones, such as acetone, the lowest singlet and triplet states are primarily of n,π^* character because the π,π^* states are at much higher energy, but for aromatic molecules, the excitation may be partially delocalized into the ring and the excited states thus have mixed n,π^* and π,π^* character. For states that are predominantly n,π^* in nature, the carbon atom is electron-deficient, or electrophilic, and the reactions that result are chiefly those in which the atom seeks to satisfy this deficiency.

The predominant carbonyl photochemical reactions can be classified into several categories:

(a) A bond alpha to the carbonyl group of a ketone is cleaved, forming two radicals. For acetophenone, the reaction appears to go by way of the triplet state [M. Berger and C. Steel, *J. Am. Chem. Soc.* **97**, 4817 (1975)]:

$$\begin{array}{c} C_6H_5 \\ \diagdown\!CO \\ CH_3 \end{array} \longrightarrow CH_3\!\cdot\! + C_6H_5CO\cdot \tag{14-16}$$

This reaction may be followed by elimination of CO by the radical and subsequent recombination of the two hydrocarbon radicals, the end result being the elimination of CO by the carbonyl compound:

$$C_6H_5COCH_3 \longrightarrow C_6H_5CH_3 + CO \tag{14-17}$$

Another example of an alpha-bond cleavage reaction is the ring opening of cyclopentanone:

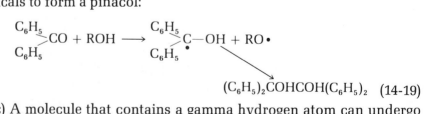

$$\tag{14-18}$$

(b) The oxygen atom may extract a hydrogen atom from another molecule, as illustrated by the reaction of benzophenone with an alcohol; in this case the subsequent reaction is the combination of two radicals to form a pinacol:

$$\begin{array}{c} C_6H_5 \\ \diagdown\!CO + ROH \\ C_6H_5 \end{array} \longrightarrow \begin{array}{c} C_6H_5 \\ \diagdown\!C\!-\!OH + RO\cdot \\ C_6H_5\!\cdot \end{array}$$

$$(C_6H_5)_2COHCOH(C_6H_5)_2 \tag{14-19}$$

(c) A molecule that contains a gamma hydrogen atom can undergo an intramolecular process, called the "Norrish type II" reaction, which leads to the net result of splitting out an olefinic unit from the carbonyl compound:

$$CH_3CH_2CH_2CHO \longrightarrow CH_2\!=\!CH_2 + CH_3CHO \tag{14-20}$$

(d) Aldehydes may split out a molecule of CO if irradiated with light of wavelength less than 300 nm. Thus when benzaldehyde is irradiated at 276 nm so as to cause a $\pi-\pi^*$ transition, there is apparently an inter-system crossing to a vibrationally excited triplet state which decomposes to benzene and CO, provided the pressure is low enough so that the excited state is not deactivated by collision. Since oxygen does not scavenge the intermediates, the reaction evidently does not proceed by way of radicals [M. Berger, I. L. Goldblatt, and C. Steel, *J. Am. Chem. Soc.* **95**, 1717 (1973)].

(e) The carbonyl compound may add to a C=C or C≡C bond, leading to formation of a four-membered ring containing oxygen; the saturated ring compound is called an oxetane. We consider some details of this type of cycloaddition reaction in order to illustrate further some of the considerations involved in the course of photochemical reactions.

The usual intermediate in the cycloaddition reaction seems to be the n,π^* triplet state. In general, the molecules that undergo cycloaddition also abstract hydrogen from suitable solvents such as 2-propanol when the mixture is irradiated, and they are observed to phosphoresce from the n,π^* triplet state. Carbonyl compounds that do not take part in the cycloaddition reaction and do not abstract hydrogen atoms from a suitable donor usually are found to phosphoresce from a π,π^* state.

The structure of the olefin influences the nature of the cycloaddition reaction. If the olefin is electron-rich, the initial step of the reaction is apparently attachment of the electron-deficient oxygen atom to one carbon of the double bond. If a molecule of benzophenone is reacted with *cis*-2-butene or with *trans*-2-butene, the same mixture of isomers is produced, approximately half with the two methyl groups in the oxetane cis and the other half with the two methyl groups trans. This is evidence that the reaction takes place in two successive steps, with a diradical intermediate:

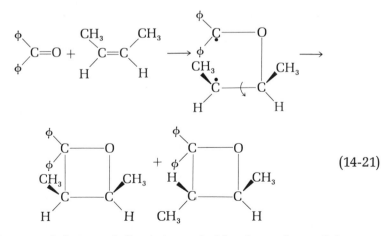

$$(14\text{-}21)$$

With an electron-deficient olefin, it is probably the carbon of the carbonyl that first becomes attached to the olefin. The mechanism is clearly quite different from that above, since the arrangement of groups in the unsymmetrical olefin is retained in the addition product. When *trans*-dicyanoethylene is the reactant, for instance, the oxetane formed has the two CN groups trans to one another:

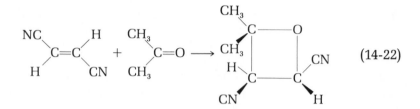

(14-22)

If the triplet energy of the olefin lies below the triplet of the carbonyl, cycloaddition may be prevented because the olefin quenches the carbonyl triplet state, being itself converted into a triplet. This is the case for many conjugated dienes on reaction with most carbonyl compounds, for the diene triplet energy is about 50 to 55 kcal/mol. However, 1,4-benzoquinone, which has a triplet energy of only 50 kcal/mol, can add to a diene because the diene cânnot quench it.

14-3
HIGH-INTENSITY PHOTOCHEMICAL SOURCES

Electronically excited states and unstable intermediates in photochemical reactions can be produced in concentrations high enough to be observed by utilizing intense radiation sources. In *flash photolysis,* an intense pulse of light is applied, usually produced by suddenly passing a high-voltage electric discharge from a bank of electric capacitors across the gap between metal electrodes inserted in a tube containing gas at low pressure. The light intensity may be a million or more times that of the usual continuous source, and the pulse length is of the order of 10^{-5} to 10^{-3} sec. To induce reaction in a liquid phase, the flash tube is coiled around a long, cylindrical quartz cell containing the sample. A second flash, of low intensity, can be sent through the cell along its length and allowed to fall on a detector. If the second flash follows within about 20 μsec after the first flash, intermediates of moderately short lifetime are still present in sufficient concentration to be observed. If the measuring pulse is of white light, the detector can be a monochromator followed by a photographic plate, and a spectral analysis of the mixture is obtained. Alternatively, a continuous beam of monochromatic light at a wavelength absorbed by one component in the mixture is allowed to fall on a photomultiplier tube which is sensitive to blue or ultraviolet radiation, and a time curve of the concentration of that component is obtained by following the variation in output with an oscilloscope or digital memory device, from which it can be read out onto a permanent record at the experimenter's convenience.

In the past few years, lasers have come into use as radiation sources for photochemical processes, as well as for many other technical and scientific purposes. These devices have the advantages that they can produce monochromatic beams of high power and that these beams can be concentrated in very small regions. Pulsed lasers are well suited for flash photolysis experiments, because their peak power can be very high, the pulses very short, and the cutoff at the end of the pulse very sharp.

The term *laser* is an acronym for "light amplification by stimulated emission of radiation." The basis of the laser is the principle that incident radiation having a frequency corresponding to the energy difference between two levels of a molecule or ion or atom can induce transition of an excited particle from the upper level to the lower level with accompanying emission of energy, just as it induces a transition from the lower level to the upper level with absorption of energy.

Consider a molecule with several electronic or vibrational energy levels, or an atom or ion with several electronic levels, as represented in Figure 14-4. Normally practically all the molecules are in the ground state, but some external source of energy can be applied to "pump" a substantial number of the molecules in a sample to an excited state. In Figure 14-4a, level C is the excited state to which transitions are induced; many of the molecules in this state drop back to level B which has a somewhat longer lifetime. Transitions from level B to the ground state are accompanied by the spontaneous emission of photons which travel through the sample, stimulating other molecules to undergo transitions between A and B and between B and A. If more molecules are in state B than in state A, the net effect is a release of energy and an increase in the photon stream. Provided the sample is contained in a cavity with suitable geometry, usually a cylinder with mirrors at each end, the photon beam can be reflected back and forth, continuing to grow in intensity. For laser action, it is also necessary that the rate at which molecules are pumped to the upper level, supplying population to level B, must exceed the rate at which energy losses occur by absorption in the mirrors, emission of light to the side, and diffraction effects.

So long as the *population inversion* between the lower and upper states is maintained, the wave within the laser not only adds photons

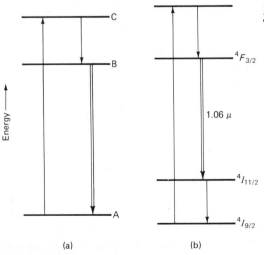

Figure 14-4
Two arrangements of energy levels utilized in lasers. The laser transition is indicated by a double arrow. (a) Three-level system. (b) Four-level system, labeled with term symbols appropriate to Nd^{3+} in glass.

but also becomes more sharply defined in wavelength and direction by repeated passage through the lasing medium. When the intensity of the beam reaches an appropriate level, it is permitted by the arrangement of the device to escape as a pulse or as a continuous beam. A key feature of the light generated in this way is that all parts of the ray are precisely in phase, or coherent. This is a consequence of the fact that the stimulated emission is always exactly in phase with the stimulating photon. Thus all the photons in the laser move as a single wave front, and it is possible to focus the beam very precisely. The beam travels over long distances without diverging, for the normal cause of divergence is interference between waves coming from separate and independent oscillators in a light source.

The ruby laser is a commonly used solid-medium laser. Ruby is a crystal consisting mostly of aluminum oxide with a low concentration of chromium ions substituted for some of the aluminum ions. The energy level arrangement is like that in Figure 14-4a, and the chromium ions are initially excited by a flash of light. The ground state of the chromium ions is conventionally labeled a ^4A state, the level C to which excitation occurs, a ^4F state, and the upper laser level B, a ^2E state. The superscripts have the meaning described in Chapter 8 in terms of electronic spin, and the capital letters are arbitrary designations of the symmetry of the occupied electronic orbitals. The energy difference between the ^2E and ^4A states corresponds to light of 694 nm. The ruby is used in the form of a rod with the ends polished optically flat and parallel, one end silvered to form a mirror and the other end lightly silvered. The beam traveling parallel to the length of the rod grows in intensity until it escapes through the partially silvered mirror. The pumping radiation can be supplied in the form of a flash of light, and the laser beam then emerges as a pulse. With continuous pumping radiation, the output beam can be continuous with a power as high as a watt (W). Alternatively, an arrangement called *Q-switching* can be used, in which power is allowed to build up within the laser and is then allowed to escape in a very short pulse of high intensity. In one method of Q-switching, a rotating mirror is used as the rear reflector, and one pulse is formed each time the mirror comes into precise alignment. Power levels of a megawatt or more have been reached for brief pulses with a ruby laser using Q-switching.

Laser action can be obtained more readily from a four-level system as shown in Figure 4-14b, in which the states are labeled with the normal spectroscopic term symbols for the example of neodymium ion, Nd^{3+}, in glass. In this scheme, the laser transitions are from a higher excited state to a lower excited state, which is normally empty, rather than to the highly populated ground state. Thus a population inversion of the lasing levels requires only a small occupancy of the upper level rather than the excitation of more than half the molecules present as is needed for the three-level scheme. Radiation from the neodymium glass laser has a wavelength of 1.06 μm.

Gases can also act as laser media, although it is natural to expect a lower power output than for solids. One commonly used system is the combination of nitrogen and carbon dioxide. The vibrational levels of N_2 are excited by a pump as shown in Figure 14-5. The en-

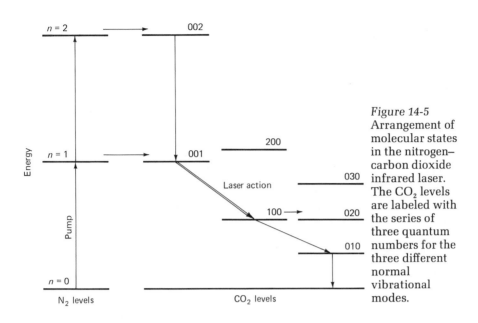

Figure 14-5 Arrangement of molecular states in the nitrogen–carbon dioxide infrared laser. The CO_2 levels are labeled with the series of three quantum numbers for the three different normal vibrational modes.

ergies of these levels are almost the same as those of the asymmetric stretching vibration of CO_2, and energy transfer to this vibrational mode of CO_2 occurs by collisions, populating the level 001 (this label gives in order the vibrational quantum numbers for the symmetric stretching, bending, and asymmetric stretching vibrations of the CO_2 molecule.) Laser action then occurs between the 001 and 100 levels, and the latter is depopulated by transitions to the 020 and 010 states. The emitted radiation has a wavelength of 10.6 μm with power up to 1 kilowatt. In other gas lasers, such as a system involving argon ions, electronic energy levels rather than vibrational levels are involved.

Fluorescent organic molecules in solution form the basis for dye lasers. Light pumps molecules to various vibrational levels of the first excited electronic state, and the molecules return to the lowest vibrational state of this electronic level, losing vibrational energy by collision. Fluorescent emission from this level provides the basis for laser action at a wavelength longer than that of the exciting radiation.

Any one laser system is limited to the generation of radiation at a particular frequency or narrow band of frequencies. To extend the range of radiation to higher frequencies, it is possible to pass the output through nonlinear amplifiers which generate overtones at multiples of the basic frequency, although with an accompanying loss in power. Some lasers can be tuned over limited frequency ranges; dye lasers, for instance, can be tuned by concentration or solvent changes.

The pumping device for a laser can be a flash lamp activated by an electric discharge, another laser, or even a chemical reaction that produces products in vibrationally excited states. With respect to power, the efficiency of conversion of input energy into laser beam energy varies widely among laser systems, but the output obviously cannot be any greater than the input, averaged over time. However, the power can be concentrated into short pulses. For example, 0.1 J for 1 nsec corresponds to a power of 10^8 W. The power can also be concentrated in space: 10^8 W focused on 10^{-4} cm^2 corresponds to 10^{12} W/cm^2 or an electric field strength of 10^6 V/cm.

When the output of an ordinary pulsed laser is closely examined, it is found that what is normally observed as a single "giant" pulse consists of a random sequence of short pulses. Order can be achieved in the output—it can be obtained in the form of a sequence of uniformly spaced, narrow pulses—by the technique of mode locking. A given laser can oscillate at any of a series of slightly different frequencies over the total bandwidth of the laser. Each successive mode corresponds to a difference of one in the number of wavelengths representing one round trip of the radiation through the length of the laser. By inserting a device that resonates at the difference in frequency between adjacent modes, so that its frequency of oscillation corresponds to the beat frequency between neighboring modes, it is possible to lock together all the modes and to produce very short pulses of picosecond (10^{-12} sec) length.

Nanosecond pulse lengths are appropriate for flash photolysis experiments in which it is desired to examine intermediates with lifetimes of 10^{-7} sec or more, but picosecond pulses do even better in terms of finding very short-lived species. Intermediate steps in the processes of vision and of photosynthesis, which are described in Section 14-4, have been investigated by picosecond laser techniques.

There are of course many important uses of lasers other than the study of very fast chemical reactions. Very precise measurements of distance and direction are possible because of the coherence of laser radiation and the associated possibility of focusing to a very narrow beam. The concentration of energy into a small region also makes possible such processes as the melting of glass or steel, as well as medical uses such as the repair of detached retinas.

14-4
LIGHT-INDUCED
BIOLOGICAL PROCESSES

Beginning with the fundamental source of organic materials and the storage process for solar energy, photosynthesis, light is indispensible for several biological processes. In addition to vision and photosynthesis, which are discussed in this section, these include the conversion of certain sterols to vitamin D and photoperiodicity effects, which govern the daily life cycles of many organisms.

VISION

The process by which light striking the eye is converted into an impulse to the brain begins with the absorption of photons by materials in the retina of the eye, but the subsequent mechanism of generating a nerve impulse and restoring the condition of the retina involves a series of events the detailed understanding of which still presents a scientific challenge of some magnitude.

The retina contains a layer of interconnected light-sensitive cells and nerve cells. The light-sensitive or photoreceptor cells in the human eye are of two types, rods and cones. Rods are cylindrical in shape, about 0.002 mm in diameter. Each rod consists of an inner segment, responsible for the normal cell functions, and an outer segment, containing a series of light-sensitive discs, as shown in Figure 14-6. The discs are flattened sacs of bilayer membranes, with which are associated the visual pigments.

Cones are tapered, with a smaller diameter at the top than at the base; in these cells there are no separate discs, but the visual pigments are attached to invaginations of the outer cone membrane. Cones are concentrated at the center of the retina, and their varying spectral characteristics provide sensitivity to different colors. However, to function, they require a higher light intensity than the much more sensitive rods, which are distributed throughout the entire retinal surface but cannot distinguish color.

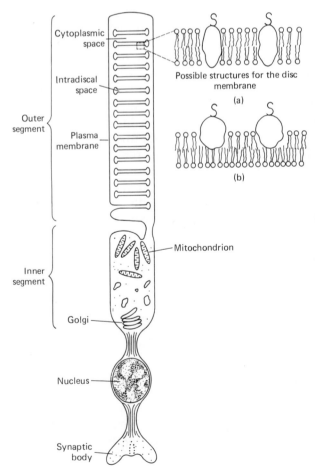

Possible structures for the disc membrane

(a)

(b)

Cytoplasmic space

Intradiscal space

Outer segment

Plasma membrane

Inner segment

Mitochondrion

Golgi

Nucleus

Synaptic body

Figure 14-6
Schematic representation of a visual rod cell. In a bovine cell, the outer segment contains approximately 1500 discs in an outer segment 50 μm in length. The enlarged views show possible membrane structures, with the rhodopsin molecules bearing an attached carbohydrate moiety symbolized by "S." [Adapted from W. L. Hubbell, *Acc. Chem. Res.* **8,** 86 (1975), by permission. Copyright American Chemical Society.]

The molecule of the visual pigment that absorbs light is vitamin A aldehyde, or 11-*cis*-retinal:

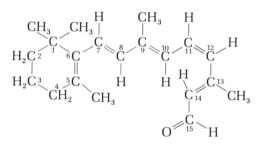

Retinal is attached by a Schiff-base type of linkage to a lysine residue in a glycoprotein called opsin:

$$C_{19}H_{28}C=O + H_2N—opsin \longrightarrow C_{19}H_{28}C=N—opsin + H_2O \quad (14\text{-}23)$$

Retinal is found generally, but different species of animals have different sorts of opsin, and the color discrimination in the cones is associated with different types of opsin. The type found in rods is termed *rhodopsin*, that in cones, *iodopsin*.

Since retinal bound to opsin is less reactive than when free, it is thought to be contained in a sort of pocket in the opsin, somewhat isolated from the environment. X-ray studies of the discs in the outer segments of the rods indicate a disc-to-disc repeat distance of 300 to 320 Å. The disc membranes are about 70 Å thick, with a space between the two membranes of 10Å. One interpretation of the x-ray diffraction results is that the high-density region facing the inside of the disk sac is 28 Å thick, but that the high-density layer facing the interdiscal space is only 23 Å thick. If this interpretation is correct, the membrane is asymmetrical, with most of the mass of the rhodopsin molecules on the inside of the bilayers. It seems likely that the opsin spans the bilayer, perhaps with a globular head on the inside and a tail stretching through to the outside, for certain groups in the protein can be shown to react with suitable reagents in the external solution which are unable to penetrate the membrane.

Retinal can exist in a variety of isomeric forms. Experiment has shown that only the 11-cis form combines with opsin, but there is still a question about the conformation about the 12C—13C bond which has some double-bond character, as do the other formally single bonds in the conjugated chain. In the crystal, groups on 13C are rotated about 39° out of the plane of the remainder of the molecule, but in solution, there are indications that there is a mixture of two structures, a distorted s-cis and a distorted s-trans with respect to the bond between 12C and 13C. Nothing is known about the conformation when it is bound to opsin; binding produces a large shift in the electronic spectral absorption maximum, as indicated in the scheme below, which has been variously attributed to protonation of the Schiff-base linkage by which the retinal is bound, or to distortions from the solution conformation.

Absorption of a photon by rhodopsin results in a $\pi-\pi^*$ transition which evidently leads to cis-trans isomerization of retinal to the all-trans form. If rhodopsin is studied in solution, it is found to be bleached by light, changing from red to yellow, and the chromophore molecule is then found to be detached from the opsin and present as the all-trans form. Various processes in a series following the initial photochemical event have been differentiated, and some intermediates have been characterized by visible, ultraviolet, or laser Raman spectroscopy:

rhodopsin (498 nm) $\longrightarrow$ prelumirhodopsin (or bathorhodopsin, 543 nm) $\longrightarrow$
metarhodopsin I (478 nm) $\longrightarrow$ metarhodopsin II (380 nm) $\longrightarrow$ metarhodopsin
III (or pararhodopsin, 465 nm) $\longrightarrow$ *trans*-retinal (387 nm) + opsin $\longrightarrow$
cis-retinal (387 nm) + opsin $\longrightarrow$ rhodopsin

These reactions were originally studied by cooling the system to very low temperatures, at which they proceed at rates conveniently measurable. Thus on slow warming, prelumirhodopsin is stable below $-140°C$, at $-40°$ it is converted to lumirhodopsin, and at $-15°C$ this in turn is converted to metarhodopsin I, which becomes metarhodopsin II at $0°C$. Recent techniques for following very rapid reactions, including flash photolysis and picosecond pulsed laser experiments, have made it possible to measure some of the rates at ambient temperature. Thus rhodopsin irradiated with a picosecond laser pulse gives an increase in absorption at 560 nm in an interval of less than 6 picoseconds, indicating the formation of prelumirhodopsin. This first process may well be the cis-trans isomerization, but the time interval is almost too short to permit such a change in configuration. M. Karplus and co-workers have suggested that this step is indeed the conversion of 11-cis to 11-trans, and that the conversion from prelumi- to lumirhodopsin involves the change from 12-s-cis to 12-s-trans, which is also necessary to reach an all-trans structure. The conversion of metarhodopsin I to metarhodopsin II has been followed in a flash photolysis experiment, and this may indeed be the key step in which information about light reception is passed on to the nervous system, for it includes the uptake of a proton from the medium. Steps beyond this point are slower than the nerve response to light and thus cannot be involved in it; they are concerned with the regeneration of rhodopsin.

Since the visual pigments are located in the disks in the outer segments of a rod cell and the nerve synapse, or junction, is at the other end of the cell, the signal resulting from light absorption must be transmitted over a considerable distance between these points. Presumably it is carried by an electric current corresponding to a flow of ions. In the dark, there is found to be transport of sodium ions from the inner segment to the outer segment of the rod. In vertebrate eyes, light causes this flow to diminish because the permeability of the membranes decreases. It appears that transfer of Ca^{2+} is also involved in this process, and perhaps the Ca^{2+} concentration controls the permeability of the membrane to sodium ions. In any case, the resulting change in electric conduction at the synapse at the base of the rod initiates the nerve impulse to the brain.

PHOTOSYNTHESIS

Green plants, as well as some bacteria, are able to convert water and carbon dioxide into glucose, utilizing energy from the radiation of the sun. The overall chemical reaction is the opposite of the oxidation of glucose and requires a large input of energy:

$$6CO_2 + 6H_2O \longrightarrow C_6H_{12}O_6 + 6O_2 \qquad \Delta H = +673 \text{ kcal}; \Delta G° = +686 \text{ kcal}$$
$$(14\text{-}24)$$

If this process were to take place by the absorption of a single photon for each reaction of a molecule of carbon dioxide and water, light of about 240-nm wavelength would be needed; nature has instead arranged for the reaction to occur in steps, utilizing the energy of several photons of visible light—most evidence indicates eight photons are required per molecule of carbon dioxide.

Certain bacteria can carry out the photosynthetic process with hydrogen sulfide instead of water, producing elemental sulfur:

$$6CO_2 + 12H_2S \longrightarrow C_6H_{12}O_6 + 12S + 6H_2O \qquad (14\text{-}25)$$

Moreover, tracer studies with ^{18}O have shown that, in the usual photosynthetic reaction, all the gaseous oxygen comes from water. One should then rewrite Equation (14-24) as

$$6CO_2 + 12H_2O^* \longrightarrow C_6H_{12}O_6 + 6H_2O + 6O_2{}^* \qquad (14\text{-}26)$$

where the asterisk indicates oxygen for which water is the source. Thus the overall photosynthetic process can be regarded as the splitting of water or hydrogen sulfide to produce hydrogen, which then "reduces" carbon dioxide to yield glucose plus water. The "hydrogen" is probably produced in the form of hydrogen ions plus electrons; the latter are transferred along an electron-transport chain to bring about reduction. The other product is molecular oxygen or sulfur. The entire sequence of events is quite complex, being by no means understood in full detail, and varies among different organisms, particularly between green plants and bacteria.

In green plants, the photosynthetic centers are localized in units of the cell called chloroplasts. These have a lamellar structure of membranous sacs, termed *thylakoids*, which carry the photosynthetic pigments. The key molecule in photosynthesis is the green pigment chlorophyll, which is a complex containing a magnesium ion surrounded by four nitrogen atoms of a porphyrin ring system:

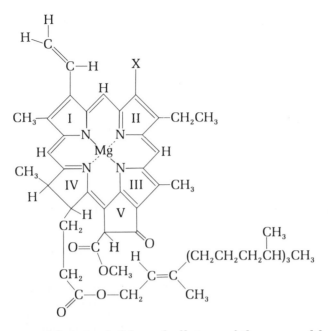

There are several forms of chlorophyll; two of these are chlorophyll a in which X is a methyl group and chlorophyll b in which it is a formyl group. The long hydrocarbon chain which is attached to the ring through an ester linkage and which makes the molecule soluble in fats or organic media is a phytyl group.

Chlorophyll absorbs strongly in the red spectral region from 600 to 700 nm, as well as in the region below 450 nm, and this spectrum pretty well parallels the efficiency of light in photosynthesis. Other pigments, the carotenoids, which are orange-red, supplement the action of chlorophyll by absorbing some radiation in the green portion of the spectrum, but this energy is passed on to chlorophyll molecules.

Chlorophyll molecules are arranged in clusters or arrays of several hundred molecules. It has been suggested that the carbonyl group in ring V of one chlorophyll molecule is complexed to the central magnesium atom in another chlorophyll to form a long string of molecules, the members of which act as light antennae and pass along the light energy, after it has been absorbed, from one molecule to another until it reaches a *reaction center*. Light energy absorbed by carotenoids is also fed to the same chlorophyll reaction center. A center consists of one particular chlorophyll molecule or a pair of chlorophylls, possibly linked together by a water molecule bridge, which is associated with one or more electron carrier systems. Light energy excites an electron in the reaction center chlorophyll molecule, and this electron is then transferred to an oxidizing agent in the electron carrier chain. From the reduced molecule formed by the electron jump, the electron is passed on through a sequence of events in which other molecules are successively reduced. At the same time, the chlorophyll positive ion formed by loss of the electron oxidizes a molecule from another elec-

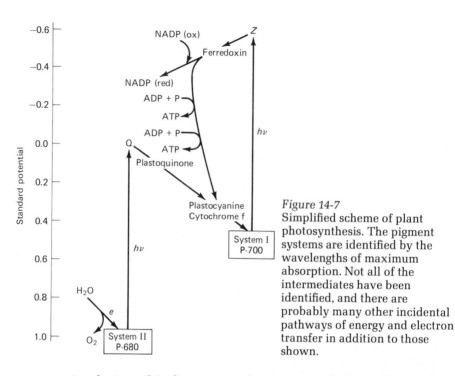

Figure 14-7
Simplified scheme of plant photosynthesis. The pigment systems are identified by the wavelengths of maximum absorption. Not all of the intermediates have been identified, and there are probably many other incidental pathways of energy and electron transfer in addition to those shown.

tron carrier chain and is thus restored to its neutral, ground-state form.

Two different kinds of photoreaction systems, having differing sorts of reaction centers, have been isolated from the chloroplasts of green plants. Their relationship is indicated in Figure 14-7.

System II, which contains both chlorophyll a and chlorophyll b, generates a strong oxidizing agent which is able to oxidize water—probably in the form of OH^-—and the electrons obtained in this process are transferred by a weak reducing agent, Q, to system I. The oxidation of water is known to require Mn^{2+} and Cl^-, and there is evidence that it occurs in four successive steps, each requiring a photon of light and each corresponding to the removal of one of the four electrons required to generate a molecule of O_2, but the identity of the species that transports electrons from water to chlorophyll is unknown.

The identity of the material that is the primary electron acceptor from system II is also uncertain, but a substance called plastoquinone either is this acceptor or is the secondary acceptor to which electrons are next transferred. Apparently the stage beyond plastoquinone is a b-type cytochrome with an optical absorption maximum at 559 nm; one molecule of this has been shown to be associated with each system II unit. An important property of an electron-transfer agent, which places it at a certain point on an electron-transfer chain, is its reduction potential, and one puzzling point about this cytochrome is that it apparently exists in two different forms with potentials of about 350 and 70 mV. Furthermore, it is not clear whether all electrons go to system I via cytochrome b_{559}, or whether some can go directly from plastoquinone. Another possibility is that electrons from cytochrome b_{559}

do not go to system I at all but are used in some other reduction process, following which they return to the chlorophyll of system II.

System I contains mostly chlorophyll a, and the focal point of activity is a reaction center containing chlorophyll for which the intensity of the typical optical absorption with a maximum at 700 nm varies in a regular way as the system accepts or donates an electron. Electrons go from this chlorophyll, often designated P700, to ferredoxin, an iron–sulfur protein. There is evidence that this transfer occurs by way of an even stronger oxidizing agent designated Z, whose identity is unknown, although a substance capable of reducing ferredoxin, and thus having a suitable reduction potential, has been isolated from chloroplasts.

From ferredoxin, some of the electrons pass through a transport chain involving cytochrome b_6, cytochrome f, and possibly plastocyanin, returning to neutralize the chlorophyll positive ion in system I. In the course of this flow, they convert molecules of ADP into ATP at each of two stages, so that the whole process is referred to as *cyclic photophosphorylation*. Somewhere along this chain, possibly at the cytochrome f step or at the plastocyanin step, the electrons arriving from system II enter system I.

Electrons from other ferredoxin molecules serve to reduce a flavoprotein, the molecule nicotinamide adenine dinucleotide phosphate (NADP). Both the reduced NADP and the ATP formed as described above are utilized in the reactions in which CO_2 is consumed and glucose is produced.

Before turning to a brief outline of the CO_2-to-glucose conversion, we summarize some of the methods applied in trying to understand the complexities of the photosynthetic system. First, there are procedures of isolation, in which attempts are made to separate various units or segments of the overall system. This is usually done with the aid of detergents which weaken the hold of the membrane upon the active components. The segments are then examined to determine what substances they contain, the nature of their electronic spectra, and what reactions they can bring about in isolation, if any. Thus it may be possible to correlate particular reactions with particular materials, or to relate spectroscopic changes to specific reactions. A second approach is to follow changes in activities and spectra when an oxidation–reduction titration is carried out, either on the isolated portion or the intact cell, again attempting to relate activities or spectra to the level of the reduction potential. Since many of the intermediates involved are radicals, extensive studies have been carried out in an effort to find characteristic EPR spectra and relate these to the progress of certain stages in the reactions. Of course, there is always the question to be resolved as to whether the observed signals belong to a genuine reaction intermediate or come merely from a by-product of no great significance in the main reaction. Finally, chemical clues can be found by observing the effect of adding or removing a material suspected of being an intermediate; for example, removal of an intermediate may lead to piling up of the product of the previous step.

When light reaching a photosynthesizing system is turned off, the production of carbohydrate continues for several minutes. This *dark* reaction utilizes the ATP and reduced NADP made in the photochemical process. M. Calvin and co-workers used radioactive ^{14}C tracers to show the path of the dark reaction and demonstrated that the initial process is the combination of CO_2 with the 1,5-diphosphate derivative of the five-carbon sugar ribulose, to form two molecules of 3-phosphoglycerate:

$$CO_2 + HO-\begin{matrix} CH_2OPO_3{}^{2-} \\ | \\ C=O \\ | \\ C-H \\ | \\ H-C-OH \\ | \\ CH_2OPO_3{}^{2-} \end{matrix} \longrightarrow 2H-\begin{matrix} CH_2OPO_3{}^{2-} \\ | \\ C-OH \\ | \\ COO^- \end{matrix} \qquad (14\text{-}27)$$

This is followed by the reaction

$$2H-\begin{matrix} CH_2OPO_3{}^{2-} \\ | \\ C-OH \\ | \\ COO^- \end{matrix} + 2NADP(red) + 2ATP \longrightarrow$$

$$2H-\begin{matrix} CH_2OPO_3{}^{2-} \\ | \\ C-OH \\ | \\ CHO \end{matrix} + 2NADP(ox) + 2ADP + 2HPO_4{}^{2-} \quad (14\text{-}28)$$

We have not completely balanced these equations because many of the materials involved ionize to an extent determined by the pH of the medium and may indeed be present as mixtures of ions with different charges. The 3-phosphoglyceraldehyde resulting from the reaction in Equation (14-28) is in part converted through fructose 1,6-diphosphate to glucose by a reaction sequence which is the reverse of glycolysis (Section 4-10), and in part used to regenerate ribulose 1,5-diphosphate by a quite complex enzymatic mechanism.

14-5
PHOTOINACTIVATION
OF BIOLOGICAL SYSTEMS

Along with the important biological functions of light described in Section 14-4, the fact that bacteria are inactivated by light, especially by ultraviolet radiation, losing their ability to divide or being killed, has been known for many years. F. L. Gates in 1928 showed that the dependence of light effectiveness in killing bacteria on frequency parallels the extent of absorption of light of various frequencies by nucleic acids. Such a spectral dependence of biological effects upon radiation frequency, like the one described above for photosynthesis, is often referred to as an *action spectrum*. Analogous deleterious effects are suffered by higher organisms when subjected to ultraviolet light under some conditions.

PHOTOCHEMICAL EFFECTS
ON NUCLEIC ACIDS

Studies of the effect of light on bacteria have indicated that inactivated cells had lost their ability to synthesize DNA (see Section 12-2 for structures of the nucleic acids). The purine and pyrimidine heterocyclic rings are responsible for a maximum in ultraviolet absorption at about 260 nm, in contrast to proteins which have maxima in ultraviolet absorption at about 280 nm for aromatic side-chain groups and at about 200 nm for peptide bonds.

When an aqueous solution of thymine, one of the pyrimidine components of DNA and RNA, is irradiated with ultraviolet light, there is formed a dimeric product which does not revert to monomer on heating or acidification. Formation of the dimer occurs even more readily when thymine is irradiated in the solid state, and it is believed that this is because the relative positions of the molecules in the solid are very favorable for the reaction, which leads to a product containing a cyclobutane ring:

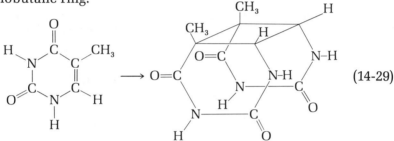

$$\qquad\qquad\qquad\qquad\qquad\qquad\qquad\qquad (14\text{-}29)$$

This equation shows only one of several possible isomeric products. Similar thymine dimers are formed in polynucleotides containing thymine residues adjacent to one another in the chain. Less frequently formed, but also known, are dimers of cytosine, uracil, and the three binary combinations of thymine, cystosine, and uracil. Equation (14-11) represented the formation of uracil dimers.

It is found that the yield of thymine dimers on irradiation of polynucleotides and nucleic acids depends upon the extent to which the macromolecule is in the helical conformation: An increase in temperature, which tends to reduce stacking of the planar purine or pyrimidine rings, or a change to a solvent that is less favorable for stacking, reduces the yield of dimers obtained on irradiation. Thus the proximity of the two molecules is very important in determining whether ring formation can occur.

Ultraviolet removal of the ability of "transforming DNA" to transform the DNA of cells to which it is added has been shown to parallel the amount of dimer formation, and thus the two effects must be closely related. Further evidence that the principal damaging effect of irradiation on cells is that on the nucleic acids is the observation that the incorporation of 5-bromouracil into DNA makes the cell containing it more sensitive to radiation. Apparently this molecule is debrominated by the radiation, yielding a uracil radical that may abstract a hydrogen atom from the adjacent deoxyribose ring, resulting in a break in the DNA chain.

An interesting feature of the behavior of thymine dimers is that the dimerization can be reversed by suitable irradiation of the sample. Apparently an equilibrium is set up between dimer and monomer, and the point of equilibrium depends upon the local geometric situation and upon the wavelength of the radiation. If dimers are formed by irradiation and then the circumstances are changed so that the equilibrium is less favorable for their formation, as by increasing the pH or by adding mercuric ion, dimers are converted back to monomers by further irradiation. Furthermore, irradiation at about 240 nm tends to reverse the dimerization effects achieved at longer wavelengths in the range of 260 to 280 nm.

Some organisms, such as certain strains of the microorganism *E. coli,* are more resistant to killing by ultraviolet light than others. The resistant strains have been shown to be able to repair the radiation-damaged DNA with the aid of further irradiation. One process of repair is an enzyme-catalyzed excision of the damaged region from one strand of double-stranded DNA, followed by replacement of the proper unit to fit the structure of the other strand. Apparently the enzyme binds to the damaged site, and visible light is required for the subsequent step in which the enzyme takes out the damaged section.

Another effect of radiation on cells is the cross-linking of protein to DNA, particularly when cells are irradiated while frozen. However, this change cannot be reversed in the same way as dimer formation.

PHOTOSENSITIZED OXIDATION

It has been known for a long time that light affects living cells in the presence of both oxygen and a dye in ways beyond the changes occurring in the absence of either the oxygen or the dye. The dye acts as a photosensitizer, and the effect has been referred to as *photodynamic action.* By this type of oxidation, viruses can be inactivated and other microorganisms can be killed. Cells suffer damage to nucleic acids, which results in mutation, membrane damage which causes permeability changes, and enzyme inactivation. As we might well expect, the sensitizer dye is usually one that readily forms a metastable triplet state. Most of the molecules which have been found effective are derivatives of the skeleton

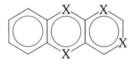

where the X's represent nitrogen, sulfur, or carbon. Methylene blue (page 228), riboflavin (one moiety of the diphosphate FAD on page 234), and acridine orange are examples of such dyes.

Nucleic acids, when subject to photodynamic action, suffer selective destruction of guanine residues. Although direct irradiation of proteins leads to rupture of peptide and disulfide bonds, sensitized oxidation leads to reaction of the side-chain groups in cysteine, histidine, methionine, tryptophan, and tyrosine. As an example of photodynamic effects on carbohydrates, it is found that the mucopolysaccharide hyaluronic acid, which consists of a long chain of D-glucoronic acid and N-acetyl-D-glucosamine units and which occurs in the eyes of mammals, is depolymerized. Dyed cellulose fibers exposed to light and oxygen of the air lose mechanical strength by a similar process.

Assuming that the reaction proceeds through the triplet state of the sensitizer, there are many possible processes by which the subsequent reaction can occur. The first step may be a reaction with oxygen by (1) transfer of an electron to the oxygen molecule, forming O_2^- and D^+ (where D represents the dye), (2) transfer of spin excitation to the oxygen molecule, producing the singlet dye and singlet oxygen, or (3) combination with the oxygen molecule to form a peroxy compound DO_2 which can then react with the substrate S to form SO_2 with regeneration of D. There has been considerable evidence that mechanism 2, involving singlet oxygen, a species described in Section 9-9, occurs frequently. The other path by which photosensitized oxidation can begin is reaction of the triplet dye with the substrate, either (1) by extraction of hydrogen from the substrate to produce a radical form of the latter, (2) by electron transfer to or from the substrate to produce a radical or radical ion, or (3) by excitation of the substrate to the triplet state. To some extent, the choice of whether the sensitizer reacts with oxygen or with the substrate can be controlled for a particular reaction system by varying the conditions, in particular, the partial pressure of oxygen gas in the system.

Probably various reactions proceed by each of the mechanisms mentioned, and there may also be some reactions that involve complexes of the dye with substrate or with both oxygen and substrate that are formed prior to the absorption of a photon, which then serves to excite the complex. Each of the steps described may be followed by any of a number of possible subsequent processes.

14-6
SOURCES OF HIGH-ENERGY RADIATION

In the previous sections of this chapter, we have considered processes in which incident radiation produces electronic excitation but, at least in the initial event, the electron simply moves to a higher-energy orbital of the same molecule. Radiation that comprises a stream of moving particles, carrying mass and considerable momentum, or which consists of extremely high-energy photons such as x rays, may lead to an initial event in which the electron is completely ejected, leaving behind a positive ion and transferring the negative charge of the electron to a new location. In the following parts of the chapter, we describe the sources and nature of ionizing radiation, as well as its effects, and compare these effects with those of visible and ultraviolet light. Some attention is devoted to nuclear processes, both radioactivity that occurs naturally and nuclear changes intentionally induced by humans, since these lead to particles and rays with a wide variety of properties.

X RAYS

Electrons that are accelerated by being allowed to pass through a large electric potential difference, perhaps 15,000 V, upon striking atoms or molecules cause the latter to emit x rays. These rays were discovered in 1895 by William Roentgen while he was studying the behavior of electric discharges in gases. The atoms of each element have a characteristic x-ray emission spectrum with groups of lines, which are designated K, L, M, and so on, in order of decreasing frequency. Within each group, the lines are labeled by Greek subscripts, as K_α, K_β, and K_γ. For any particular line, such as K_α, the square root of the frequency varies from element to element linearly with the atomic number.

The process by which incident electrons produce characteristic x rays begins with the ejection of an electron from the atom, usually one in an inner shell near the nucleus. This is followed by a transition in which an electron in a higher-energy orbital drops into the vacant orbital, with the energy excess carried off as the x-ray photon. K lines correspond to the initial expulsion of electrons from an orbital with principal quantum number equal to 1, L lines to expulsion of an $n = 2$ electron, and so forth. An α line corresponds to filling the vacancy with an electron from the next higher principal shell, a β line to a jump of 2 in value of n, and so on. X-ray tubes contain a target consisting of a particular metal, often copper, and the metal determines the frequency of the radiation produced. Frequently there is also obtained a continuous distribution of x-ray energies arising when fast electrons are decelerated in the very strong electric field existing near an atomic nucleus and called *bremsstrahlung,* from the German word for "braking radiation."

We saw earlier how x-ray diffraction can be used to determine repeat distances in liquids and solids. Extensions of this method have led to complete crystal structure determination and location of all the atoms, except possibly hydrogen, in such molecules as lysozyme and myoglobin. And of course the uses of x rays in medical diagnosis and treatment are well known.

NATURAL RADIOACTIVITY

In 1896, Henri Becquerel was studying the effect of x rays in exciting fluorescence in various materials. One of the substances he investigated was a compound of uranium, potassium uranyl sulfate $K_2UO_2(SO_4)_2 \cdot 2H_2O$, and he found that a sample of this material was capable of darkening a photographic plate even before exposure to x rays and even if the salt had been prepared and kept in complete darkness. Certain uranium ores, however, were found to give an intensity of radiation too high for the proportion of uranium present, and from these ores Pierre and Marie Curie isolated the new elements, radium and polonium.

When the radiation given off by naturally *radioactive* materials is filtered by passage through metallic aluminum, one portion is absorbed in a few hundredths of a centimeter of the metal. This is *alpha radiation* and consists of positive particles with a mass of 4 on the atomic weight scale; alpha particles are converted to helium if given the opportunity to acquire two electrons each and thus correspond to helium nuclei. Alpha particles from a given nuclide have a particular energy value. Since each particle contains two protons and two neutrons, the net effect on a nucleus of emission of such a particle is a loss in mass of four units and a decrease in atomic number of two units. For example, the loss of an alpha particle by uranium-238 gives a nuclide of mass 234 which is an isotope of thorium, atomic number 90:

$$^{238}_{92}\text{U} \longrightarrow {}^{4}_{2}\text{He} + {}^{234}_{90}\text{Th} \qquad (14\text{-}30)$$

A second type, *beta radiation,* is more penetrating, with a range of the order of 100 times that of alpha particles. That the beta radiation consists of high-velocity electrons was shown by electrostatic and magnetic deflection experiments. Emission of a beta particle results in an increase of one unit in the atomic number of the decaying element, indicating that the negatively charged beta particle comes from the nucleus of the atom rather than from among the extranuclear electrons. The isotope of thorium formed by decay of uranium-238, shown in Equation (14-30), decays by beta emission:

$$^{234}_{90}\text{Th} \longrightarrow {}^{0}_{-1}\beta + {}^{234}_{91}\text{Pa} \qquad (14\text{-}31)$$

A still more highly penetrating radiation, *gamma radiation,* consists of photons of electromagnetic radiation rather than of charged particles, for the beam is undeflected by electric or magnetic fields. Emission of a gamma ray by an atom has no effect upon the mass number or atomic number; excess energy is carried off from the nucleus by the emission of a gamma ray, which often accompanies or follows shortly after the emission of an alpha particle or a beta particle. Gamma rays emitted by a particular nuclide have a line spectrum of energies; in contrast to beta particles, which have energies spread over a range from zero up to a maximum characteristic of the nuclide, gammas have only one or a few characteristic energy values.

The naturally occurring radioactive nuclides of the heavy elements fall into three families or series: the uranium, thorium, and actinium families. The name refers in each case to one of the parent elements of the series, and in each series the intermediate members are produced by the preceding member, each decaying in turn to form the next member. Any nuclide in one of the three series is finally converted by successive decays into one of the stable isotopes of lead.

Radioactive disintegration is kinetically a first-order process. Unlike the rate constant for a chemical reaction, however, the decay constant is little affected by changes in temperature, pressure, or the state of combination of the atom of which the radioactive nucleus is a part. As in other first-order reactions, the time required for a given fraction of the initial material to decompose is independent of the amount of material initially present. Radioactive nuclides are often characterized by the time required for one-half of the starting material to decay— the *half-life,* which is inversely proportional to the decay constant.

Radioactive species can be found naturally only if their half-lives are of the order of, or greater than, the age of the material from which the earth is formed, or if they are continuously regenerated by some natural process. The parents of the natural radioactive families have half-lives of about 10^9 to 10^{10} years, while some other radioactive nuclides of light elements which are found naturally have half-lives of this magnitude or greater: ^{40}K, 1.2×10^9 years; ^{87}Rb, 6.2×10^{10} years; and ^{138}La, 2×10^{11} years. Short-lived members of the families are found only because they are continuously regenerated by the long-lived parents, while such species as tritium, the radioactive isotope of hydrogen, and ^{14}C, "radiocarbon," are found in small amounts because they are continuously produced in the atmosphere by the effect of cosmic rays.

PARTICLE ACCELERATORS

Nuclear particles with large kinetic energies are available naturally from radioactive substances which decay by alpha emission, and are also produced, although in rather heterogeneous mixtures, as by-products of atomic reactors. In order to provide more intense and controllable beams of specific types of ions, several kinds of ion accelerators have been devised. These impart kinetic energy, either to electrons or to positively charged particles, by applying a high electric potential difference, which results in a strong force on the ions, causing them to be pulled toward the electrode of opposite sign. With a suitable geometric arrangement, most of the ions can travel right on through the electrode in the form of a beam ready for use in bombarding a target.

One accelerating device is the van de Graaff electrostatic generator, which stores up a large amount of charge on a spherical conductor, thus producing a high potential in the following way: A moving belt enters the sphere through an opening; the electric charge is sprayed onto the belt outside the sphere from a source at 10,000 to 30,000 V dc, and a sharp point connected to the sphere's inner surface and located near the belt picks up the charge from the belt. The mutual repulsion of like charges by one another causes them to move to the outer surface of the sphere, where the potential builds up continuously until leakage equals the rate of transfer of charge to the surface.

Another approach to the problem of imparting large energies to a beam of particles is the application of a moderate electric potential many times in succession. This can be done as a pulse of ions is traveling along a straight tube in a linear accelerator, or as the ions travel in a circular path in a cyclotron. In a linear accelerator, the ion pulse passes through a series of hollow metallic cylinders placed in a row. Every other cylinder is connected together electrically, so that at any given time alternate cylinders are oppositely charged. If positive ions are to be accelerated, the electrode just ahead of a pulse of ions is negatively charged, and as the pulse enters this electrode it is accelerated; before the pulse reaches the next electrode, the signs are reversed, and thus the same type of acceleration is repeated. In a microwave linear accelerator, a klystron tube generates traveling waves of high frequency which pass down an evacuated waveguide. Batches of electrons introduced into the waveguide are accelerated by the moving wave to high energies, and the electrons emerge through a window in pulses of a microsecond or less duration.

The technique of *pulsed radiolysis* utilizes the output of a linear accelerator to study reactions in the same way that flash photolysis is used in photochemistry. Following irradiation of a sample with a short pulse of electrons, spectroscopic methods are used to search for short-lived reaction intermediates. The pulse may trigger a spectrograph with a photographic plate to survey a region of the optical spectrum at a given instant after irradiation, or the absorption at one wavelength can be monitored with time to obtain kinetic results.

NUCLEAR TRANSFORMATIONS AND INDUCED RADIOACTIVITY

Nuclear reactions or "transmutations" may occur when a stream of particles of high-energy radiation impinges upon a sample of matter. Bombarding agencies may be alpha particles, which were the earliest used because of availability from naturally radioactive sources, neutrons, protons, deuterium nuclei, or gamma rays. The result of collision with a nucleus may be either a stable or radioactive nuclide, or almost instantaneous ejection of a second particle, producing another species which again may be stable or radioactive.

The first transmutation reaction was effected by Ernest Rutherford in 1919. Alpha particles striking nitrogen atoms were found to set free protons, in the reaction

$$\text{}^{4}_{2}\text{He} + \text{}^{14}_{7}\text{N} \longrightarrow \text{}^{17}_{8}\text{O} + \text{}^{1}_{1}\text{H} \qquad (14\text{-}32)$$

In a reaction of this sort the sum of the atomic numbers and the sum of the mass numbers must be the same for the products as for the reactants. The reaction cited is referred to as an alpha-particle–proton, or (α,p), reaction on ^{14}N. Using the symbols n for neutron, d for deuteron, and γ for gamma ray, we can represent other common forms of nuclear processes as (α,p), (α,n), (d,p), (d,n), (d,α), (n,p), (n,α), (p,n), (p,α), (n,γ), (p,γ), $(n,2n)$, and (p,d) reactions.

A general requirement for a particular nuclear reaction to occur is that sufficient energy be available for the process. If the sum of the masses of the products is less than the sum of the reactant masses, then there is automatically an excess of energy, for the mass that is lost is converted into energy. For example, in the reaction $^{10}\text{B} \ (n,\alpha) \ ^{7}\text{Li}$ the masses are:

Reactants: B, 10.01290; neutron, 1.00867; total, 11.02157

Products: Li, 7.01600; alpha, 4.00260; total 11.01860

Thus there is a decrease in mass of 0.00297 units, and the reaction can proceed spontaneously. Nuclear reactions requiring an input of energy because of a slight increase in mass can sometimes proceed if the bombarding particle has sufficient kinetic energy to compensate for the mass added.

Let us now consider the characteristics of a few important types of nuclear reactions. All nuclides, except ordinary helium, will react with neutrons of low energy, given the designation *thermal* because they have an energy of the order of magnitude of that they would have if they were in equilibrium with the molecules of a gas by exchanging kinetic energy through collisions. The reaction may be a *radiative capture* process: an absorption of a neutron into the nucleus, followed by emission of a gamma ray. However, in order for this to occur, the energy levels of the product nucleus must include two at the same distance apart as the excess energy carried off by the gamma ray. Thus a particular nucleus will have a greater cross section for certain energies of the incident neutrons or will be *resonant* to those energies.

It is also possible that protons or alpha particles will be emitted as a neutron is captured. Processes in which a positively charged particle is emitted require more energy than those in which other types of radiation are emitted, since an energy barrier at the "surface" of the nucleus must be crossed. Just as the net positive charge on the nucleus repels positively charged particles that approach the nuclear surface from the outside, so it repels positively charged particles approaching the nuclear surface from the inside. The energy barrier at the surface is lower for elements of low atomic number, and therefore proton or alpha-particle emission is easier in these elements, requiring less neutron energy. When particles of matter, instead of gamma rays, are emitted, the process is not resonant, because the emerging particle has a wide latitude in the energy it can carry away as kinetic energy. Some examples of neutron reactions are

$$^{127}_{53}\text{I} + {}^{1}_{0}\text{n} \longrightarrow {}^{128}_{53}\text{I} \tag{14-33}$$

$$^{106}_{46}\text{Pd} + {}^{1}_{0}\text{n} \longrightarrow {}^{106}_{45}\text{Rh} + {}^{1}_{1}\text{H} \tag{14-34}$$

When protons are the bombarding particles, they can be given kinetic energy in an accelerator. Simple capture, with emission of a gamma ray, may occur; resonance effects exist as for neutron capture. The resulting gamma rays appear in good yield and with rather high energies, and this is a useful method for producing gamma rays. The (p,n) process required a certain threshold energy for the proton, since the neutron is heavier than the proton and a less stable nucleus is produced.

Deuteron-induced reactions occur less readily than neutron reactions, because the particle is positively charged and thus on the one hand is repelled by the nucleus if too slow, and on the other hand collides with extranuclear electrons when accelerated. Good yields of neutrons of a controlled energy can be obtained by (d,n) reactions, such as

$$^{9}_{4}\text{Be} + {}^{2}_{1}\text{H} \longrightarrow {}^{10}_{5}\text{B} + {}^{1}_{0}\text{n} \tag{14-35}$$

Artificial radioactivity was discovered in 1934 by Irene Curie and Frederic Joliot, who found that boron, aluminum, and magnesium are converted by bombardment by alpha particles into emitters of positrons, and that the emission continues after cessation of the bombardment. *Positrons* are particles with a mass equal to that of the electron and with a charge equal in magnitude to that of the electron, but positive in sign. For example, the reaction with boron produces a radioactive isotope of nitrogen which decays by the emission of positrons:

$$^{10}_5B + {}^4_2He \longrightarrow {}^{13}_7N + {}^1_0n \tag{14-36}$$

$$^{13}_7N \longrightarrow {}^0_{+1}\beta + {}^{13}_6C \tag{14-37}$$

The decay process is similar in general characteristics to the decay of naturally occurring species, although there exist several additional decay pathways for artificially produced nuclides:

(1) *Position or "positive beta" emission:* The result is a decrease in atomic number by one unit.
(2) *Electron capture:* One of the extranuclear electrons, usually one of the K electrons, is drawn into the nucleus, decreasing the atomic number by one unit. Since the removal of the electron from its K orbit leaves a vacancy, the electrons in higher-energy levels in the atom drop down the energy scale to fill this vacancy, and the energy lost by these electrons leaves the atom in the form of x rays.
(3) *Internal conversion:* A gamma ray leaving the nucleus strikes an extranuclear electron, ejecting the latter from the atom. Again, x rays result as the electrons in higher-energy orbits fill the vacancy.

The energy quantities involved in nuclear reactions are many orders of magnitude greater than the energies involved in ordinary chemical reactions, large enough in fact so that they represent a measurable change in mass. Einstein's equation is the basis for calculation of the energy change from the change in mass, or vice versa:

$$\Delta E = c^2 \Delta m \tag{14-38}$$

On an atomic scale, the *electron volt* (eV) is often used as a unit of energy. It is defined as that amount of energy imparted to an electron moving through a potential difference of 1 V. Electrical work is the product of the charge moved, 1.602×10^{-19} coulomb for the electron, and the potential difference through which it is moved, 1 V, so that the work done is 1.602×10^{-19} joule or 1.602×10^{-12} erg, and this is the magnitude of 1 eV. One atomic mass unit (amu) is equivalent to 932 million electron volts (MeV), as shown by the following calculation:

$$\frac{(2.998 \times 10^{10})^2 \text{ ergs/g}}{6.022 \times 10^{23} \text{ amu/g}} \times \frac{1}{1.602 \times 10^{-12} \text{ ergs/eV}} = 932 \times 10^6 \text{ eV/amu}$$

Upon bombardment with neutrons, certain heavier nuclides such as uranium undergo fission in which, instead of simply emitting an alpha particle or a beta particle and changing one or two units in atomic number, the nucleus breaks into two fragments of about equal mass, plus a number of neutrons. The emission of neutrons is a consequence of the fact that the neutron/proton ratio of elements in the region of the fission products is lower than that for the original uranium atom. Provided the neutrons produced in fission can be caused to initiate other fission processes, the reaction can be made self-sustaining.

14-7
METHODS OF HANDLING IONIZING RADIATION

INSTRUMENTATION FOR DETECTION

Charged particles moving through matter cause ionization of the atoms of the matter, and this ionization is utilized to show the presence of the particles. Several different arrangements of apparatus are available for detecting this ionization.

In an *ionization chamber* two conductors are arranged with a gas space between, and an electric field is applied. When a particle passes through the gas space, ions are formed, the electrons migrate toward the positive electrode, and the heavier, slower positive ions migrate toward the negative electrode. When the charged particles reach the electrode, there may be observed a momentary change in voltage.

An ionization chamber can be modified by constructing it with one electrode as a long, thin cylinder and the other as a wire placed at the axis of the cylinder. This produces a rather large electric field at the surface of the wire. As electrons travel toward the wire, they are accelerated by this strong field, receiving energy sufficient so that they can ionize other molecules in the gas, freeing electrons which in turn are energized by the field and repeat the ionization. This *avalanche* produces a pulse of current that is proportional to the initial ionization, so that this is termed a *proportional counter*. If the voltage applied to a proportional counter is increased to perhaps 1000 V, the proportionality disappears and all pulses become the same size regardless of the initial ionization. Under these conditions, the counter is operating as a *Geiger-Müller counter*.

The production of light in a crystal, or *scintillation* has been utilized for the detection and counting of beta and gamma rays. Anthracene, stilbene, or sodium iodide crystals, or solutions such as p-terphenyl or stilbene in xylene or toluene can be used as the scintillator; the light produced is fed into a photomultiplier tube in which the photons eject electrons from a sensitive cathode surface, and the stream of electrons is multiplied until it becomes a measurable electric current.

The first method of detection of radioactive radiation was the darkening of a photographic film. The general darkening of a piece of film worn on a badge can be used to indicate for an individual the total radiation exposure over a certain time interval. The distribution of a radioactive tracer in a solid sample, such as a metallic surface or a thin section of biological material, can be determined by placing the sample in contact with a photographic plate. Alpha particles, protons, or mesons, which have high ionizing power, can sensitize particles as they travel through nuclear emulsions. Thus the tracks of the radiation are made visible. Radiation of lower ionizing power can be detected by secondary ionizing particles, which are produced upon interaction of the primary radiation with matter.

The *curie* (Ci) of radioactivity is an amount of material that gives 3.7×10^{10} disintegrations per second. The larger the decay constant, the smaller the amount of an element corresponding to a curie. The millicurie, 10^{-3} Ci, and the microcurie, 10^{-6} Ci, are units also in use.

TRACERS

Since the several isotopes of an element behave in an almost identical fashion chemically but can be distinguished by physical measurements, experiments can be carried out in which a certain atom in a molecule, or a certain type of molecule in a mixture, is labeled, the label being in the form of a particular isotope. The labeled material can then be followed or traced through a chemical or physical change.

Stable isotopes, such as deuterium, can be used as tracers. In order to analyze the material for their presence, it may be necessary to destroy the sample. If a deuterium-containing sample is oxidized, the density of the water resulting is a measure of the deuterium present; alternatively, a sample can be analyzed in a mass spectrograph. If sufficient concentration of a stable isotope with a magnetic moment, such as ^2H, ^{13}C, ^{15}N, or ^{17}O, can be introduced, a sample can be analyzed nondestructively by NMR spectroscopy. Stable isotopes do not lead to deleterious side reactions such as might be produced by the radiation from radioactive tracers. However, the latter have the advantage that analysis can be carried out rather simply and with great sensitivity by counting the radiation produced.

Radioactive tracers must have half-lives suitable for the type of experiment in which they are to be used. If they are to be employed in studies of metabolism, for example, they must last long enough so that the physiological processes concerned have a chance to occur. Five of the more commonly used radioactive tracers in work of this sort are ^{14}C with a half-life of 6400 yr, ^{32}P with a half-life of 14.3 days, ^{35}S with a half-life of 87.1 days, ^{131}I with a half-life of 8 days, and ^{24}Na with a half-life of 14.8 hr. All of these are negative beta-particle emitters. The half-life of radiocarbon is so long that the radiation emitted is somewhat difficult to detect, while iodine approaches the lower limit of time for convenient work, and use of the sodium isotope requires special care.

Mechanisms of organic reactions can be followed with the aid of tracers. For example, the hydrolysis of esters with water containing the stable isotope of oxygen, ^{18}O, showed that the oxygen of the water appears only in the acid, so that the oxygen in the alcohol comes from the ester.

The *radioautograph* method is very useful in biological work for qualitative studies of the distribution of radioactive tracers throughout tissues. A thin section of the tissue is taken and placed next to a photographic film. After the film has been exposed for a time, it is developed and the location of radioactivity in the tissue can be noted from the pattern on the film.

Another technique of importance is the method of *isotope dilution*. Suppose that it is desired to determine the amount of a particular amino acid in a mixture of amino acids, and that the acid to be determined can be obtained in pure form from the mixture but cannot be quantitatively separated. In the isotope dilution method, a small, known amount of labeled acid of the sort to be determined is introduced into the mixture. A representative portion of the amino acid is then recovered in pure form from the mixture and the amount of isotopic label determined. The ratio of labeled acid to nonlabeled acid in the recovered portion is the same as the ratio of the labeled acid introduced to the total amount of acid originally present, and so the total amount originally present can be calculated.

14-8
EFFECTS OF IONIZING
RADIATION ON MATTER

In this section we first consider some general effects resulting from the passage of ionizing radiation through matter, and then discuss in more detail processes occurring during the radiolysis of water because these may well occur in the aqueous portion of living cells. Then we examine a few examples of radiolysis processes in biological materials and conclude with examples of the use of electron paramagnetic resonance in identifying radiation products.

GENERAL EFFECTS OF
IONIZING RADIATION

On passage through matter, alpha particles traverse an essentially straight path, interacting primarily with electrons to produce excitation or ionization. An ionizing event occurs every few angstroms, and considerable kinetic energy may be transferred to the products. If the total energy loss in air is divided by the number of ionizations, the result is about 35 eV per event, but probably about half of this energy goes into electronic and vibrational excitation, processes much more difficult to detect than are ionizations. The range in air for 5-MeV alpha particles is about 5 cm, but in a solid it is only about 10 μm. As a result of the very short range in a solid, most alpha radiation is absorbed in the outer layer of dead skin on a human, and thus external exposure is not particularly hazardous. However, ingestion of isotopes such as ^{226}Ra and ^{239}Pu can be very dangerous, because their chemical nature causes them to concentrate in the bone or in body organs, and intense localized damage is produced.

Electrons, or beta particles, because of their much smaller mass, follow a more erratic path, and their range is much greater than that of alpha particles because of their smaller collision probability. They penetrate tissue to millimeter and centimeter depths and can cause severe skin burns. Gamma and x rays are by far the most penetrating, being absorbed nearly completely only by very thick, solid barriers. Their absorption is exponential, with no definite limiting range.

Both beta and gamma rays ionize matter on their passage through it. Indeed, as for alpha particles, it is the *secondary electrons* that are ejected from atoms by these rays which cause most of the observable consequences of irradiation. As a high-energy electron goes through a sample of matter, for example, it leaves behind a track, with pockets of ionization perhaps 1000 Å apart. These pockets are called *spurs*, and each spur contains, within a region of 10 to 20 Å, the ionization products produced by a secondary electron, which may have a kinetic energy up to several hundred electron volts.

Neutrons penetrate quite extensively, because their lack of charge eliminates repulsion by nuclei or electrons. Their direct effects are not as serious as their secondary effects, for upon collision with a nucleus, ionizing particles such as electrons or protons may result. If a neutron with high energy strikes a nucleus, the nucleus recoils for a distance large on the atomic scale, disrupting molecules in its path. Slow neutrons, however, can convert nuclei into radioactive isotopes.

Radiation is often characterized by its *linear energy transfer* (LET) value, the amount of energy deposited in the sample per unit path length in keV per micrometer. Alpha particles of a given energy have a much higher LET value than beta particles of the same energy. The amount of radiation exposure is expressed in terms of the *roentgen* (R), which is defined as the amount of radiation that produces 1 esu of charge of one sign in 1 cm^3 of air at standard conditions, or 2.58×10^{-4} coulomb in 1 kg of air. This is equivalent to the release of 84 ergs/g of air. Since the ionization produced by various types of particles varies with the nature and energy of the particles, there is no simple relation between the number of particles and the exposure in roentgens.

The absorbed dose, the amount of energy taken up by unit weight of the medium, is measured as multiples of the *rad*, which is 100 ergs/g of absorber, or 6.2×10^{13} eV/g, about equal to the effect of one roentgen of medical x rays. Based on the effects of high-energy gamma rays as unity, the *rem* (originally standing for "roentgen equivalent man") is the dose in rads corrected for the LET of the radiation and for the extent to which the energy of the radiation is localized in the sample. For neutrons, 1 rem is perhaps one-fifth to one-twentieth the number of neutrons that produce the same ionization in air as a roentgen of gamma rays, so that neutrons are relatively much more harmful than gamma rays.

RADIOLYSIS OF WATER

When alpha, beta, or gamma radiation passes through water, the initial process is the detachment of electrons:

$$H_2O \longrightarrow H_2O^+ + e^- \qquad (14\text{-}39)$$

The cation left behind can react with another water molecule to produce a hydroxyl radical:

$$H_2O^+ + H_2O \longrightarrow H_3O^+ + OH\cdot \qquad (14\text{-}40)$$

The hydroxyl radical may serve directly as an oxidizing agent, or two of these radicals may combine to produce H_2O_2, a species relatively stable but still capable of acting as an oxidant. The fact that addition of inert salts to the solution to change the ionic strength does not affect the behavior of the oxidizing species indicates that they are neutral.

The electron set free in the initial process usually exists, for a short but finite time, in a rather surprising form: the hydrated electron, symbolized as e_{aq}^-. This species has been identified in pulse radiolysis experiments by its optical absorption at 720 nm, as well as by its EPR spectrum, a single narrow line with a g value of 2.0002. Solvated electrons are also known in liquid ammonia and in organic solvents. The hydrated electron can react directly with a variety of oxidizing agents. If no other reducible species are present, it can combine with one of the positively charged particles, causing dissociation to hydrogen atoms by virtue of the energy it carries:

$$e_{aq}^- + H_2O^+ \longrightarrow H\cdot + OH\cdot \tag{14-41}$$

or

$$e_{aq}^- + H_3O^+ \longrightarrow H\cdot + H_2O \tag{14-42}$$

The resulting hydrogen atoms then become the reducing species in the solution.

The nature of the radiolysis products depends upon the LET values of the radiation. With low LET radiation, the radicals OH· and H· predominate, but with high LET radiation, the molecular species H_2 and H_2O_2 predominate. These results clearly have something to do with the ratio of the concentration of intermediates in spurs to the bulk concentration in the solution, but the quantitative details are by no means established. Some of the molecular species form by combination of the H· and OH· radicals, but it appears that H_2 in particular must also be formed by other reactions.

It turns out that highly purified water, irradiated by low LET radiation under conditions in which gas cannot easily escape from solution, does not appreciably decompose. The various species formed react with each other to regenerate H_2O. Thus the hydrated electron and H_2O_2 combine to form OH^- and $OH\cdot$, the hydroxyl radical reacts with hydrogen atoms to produce water or with molecular hydrogen to produce water and hydrogen atoms, hydrogen atoms pair off, yielding molecular hydrogen, and so on. In contrast, high LET radiation, like alpha particles, produces more H_2O_2 than hydrated electrons, and the H_2O_2 reacts with OH· to form peroxy radicals, $HO_2\cdot$. These combine to form O_2 and H_2O_2.

The presence of oxidizable and reducible solute species substantially modifies the radiolysis process, because the intermediates from water are in part used up by reactions with these other materials. In moderately dilute solutions, the yields of H_2 and H_2O_2 decrease with increasing concentration of solute. For example, dissolved O_2 reacts with reducing radiolysis products to form peroxy radicals:

$$e_{aq}^- + O_2 \longrightarrow O_2^- \tag{14-43}$$

$$O_2^- + H_2O \longrightarrow OH^- + HO_2\cdot \tag{14-44}$$

$$H\cdot + O_2 \longrightarrow HO_2\cdot \tag{14-45}$$

EFFECTS OF IONIZING RADIATION ON BIOLOGICAL MATERIALS

Ionizing radiation produces substantial effects on living cells, and on model compounds of components of cells. The results of irradiation can be studied in simple organisms such as bacteria by following either the decrease in their ability to reproduce or the rate at which death occurs. Effects in higher organisms are more complicated, and most of what has been proposed to explain the observed changes is based upon the study of simple model compounds. In its interaction with matter, ionizing radiation is less selective than ultraviolet light, and the consequences of irradiation are correspondingly more difficult to unravel.

If a model compound is irradiated as a dry solid, the effects observed must be the direct effects of the radiation. However, if studies are made in aqueous solution, as is often done, the principal effects are indirect, resulting from reactions induced by the radiation products of water. Whether radiation effects are direct or indirect can be ascertained by measuring the dependence of product yield upon solute concentration. The situation in the living cell may well be somewhere intermediate between these two limits—a combination of direct and indirect effects.

It seems clearly established that DNA is the principal site of radiation damage in the cell, as it is for damage by ultraviolet. Among the pieces of evidence supporting this is a correlation between sensitivity of various strains of bacteria to killing by x rays and the guanine and cytosine content of their DNA. Events that may occur in irradiated DNA are hydrogen-bond breaking, strand breaking, and destruction of the bases. Radiation-resistant strains of bacteria have also been shown to contain an enzyme that repairs strand breaks, the events that seem to be most closely related to biological inactivation, as described above for ultraviolet damage. However, simultaneous breaks in both strands of the double helix apparently cause irreparable damage; in studies of this effect, the amount of double-strand breakage in DNA can be identified by the molecular-weight statistics obtained from sedimentation measurements on the nucleic acids extracted from the cell.

Extensive studies have been carried out on the radiation processes taking place in the solid state in purines and pyrimidines, as well as in nucleic acids, and a variety of products has been found. ESR spectra have been used to identify the radical products, and one characteristic eight-line spectrum, often found in irradiated nucleic acids, has been assigned to the radical formed from thymidine by the addition of a hydrogen atom to the double bond in the ring:

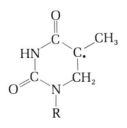

The group of eight lines results from hyperfine interaction of about 20.5 G with the methyl hydrogens and of about 40.5 G with the hydrogens in the CH_2 group.

Of the effects of radiation on proteins, the most significant is the destruction of enzymatic activity. Radiation products from solid amino acids have been extensively investigated as models for protein reactions. In solid glycine, it appears that the principal radical formed is $NH_3^+\dot{C}HCOO^-$. In acetyl alanine, the EPR spectrum shows a quartet with splitting of about 18 G, indicating the radical is $CH_3CONH\!-\!\dot{C}(CH_3)COOH$. Cystine yields $HOOCCH(NH_2)CH_2S\cdot$. Irradiation of solid proteins at 77 K yields a set of overlapping spectra from several radicals. Warming the proteins to room temperature, or direct irradiation at room temperature, produces simply two spectra, one of which is assigned to $RS\cdot$; the other, a doublet of 17 G, may represent the radical $-NH\!-\!\dot{C}H\!-\!CO-$. In the presence of oxygen, peroxy radicals, $ROO\cdot$, are formed instead of the radicals mentioned.

Certain sulfur-containing compounds, including the amino acids cysteine and cystine, are found to be effective in protecting against radiation damage. They may interfere with indirect processes by scavenging intermediates such as the hydrated electron formed from the solvent, since the reaction rate of these sulfur compounds with radicals is much greater than that of other species. It is also possible that they function by repairing damaged molecules because of the easy transfer of hydrogen atoms from the SH group:

$$R'SH + R\cdot \longrightarrow RH + R'S\cdot \tag{14-46}$$

Oxygen enhances the effect of ionizing radiation, primarily because peroxy radicals cannot be readily reconverted to the preirradiation molecular form.

EXERCISES

1. What energy in kilojoules corresponds to 1 einstein of radiation of visible light of wavelength 450 nm, near the region of maximum absorption of chlorophyll?

2. When HI is irradiated by ultraviolet light of 185-nm wavelength, the hydrogen atom is produced as a *hot atom,* able to produce other reactions because it has as much as 84 kcal/mol of kinetic energy. At what irradiation wavelength would the kinetic energy available for the hydrogen atom be 10 kcal/mol?

3. A triplet sensitizer activates a reactant molecule by exciting it from a ground singlet to an excited triplet state. How would you describe, in usual photochemical terminology, the effect of the reactant molecule upon the sensitizer?

4. Many proteins fluoresce as a result of the presence of the tryptophan moiety. For pepsin, the wavelength of the fluorescence emission is 342 nm. If the exciting wavelength is 254 nm, what amount of energy per mole of excited tryptophan is dissipated before the ultraviolet photon is emitted by the protein?

5. The quantum yield in the photochemical decomposition of acetone for light of 300 nm is about 0.2. If an ultraviolet source emitting 10 W of this radiation is used, how many moles per hour of acetone react, assuming 50 percent of the light is absorbed by the sample?

6. In a uranyl oxalate actinometer, light of wavelength 253.7 nm from a mercury source is completely absorbed. The actinometer contains 25 cm^3 of 0.030 M oxalic acid, a quantity that initially requires 79.2 cm^3 of $KMnO_4$ solution for titration. After one hour, 48.7 cm^3 of the same $KMnO_4$ solution is required in the titration. From the known quantum yield of 0.57 for the actinometer reaction, calculate the energy, in joules per hour, falling upon the solution.

7. Show why the EPR spectrum of the thymidine-derived radical described on page 575 appears to consist of eight lines.

8. A quantity of 1.50 mg of glycine labeled with 1000 counts/min of ^{14}C is added to 0.429 g of a mixture of amino acids. From the mixture there is then separated by suitable processes a portion of 4.68 mg of pure glycine, which has a total radioactivity of 150 counts/min. Calculate the percent by weight of glycine in the original mixture of amino acids.

9. The ionization potential of the hydrogen atom is 13.6 V. Calculate the energy in kilojoules per mole required to ionize a hydrogen atom.

10. Calculate the change in mass per mole occurring during a chemical change in which the energy change is 500 kcal/mol.

11. Radium has a half-life of 1590 years. Show that 1 g of radium corresponds to 1 Ci of radioactivity.

12. The nuclide ^{232}Th decays stepwise by emission of a total of six alpha particles and four beta particles, producing a final stable product designated thorium D. Give the mass number of this nuclide and tell of what element it is an isotope.

13. Write equations for the reactions: (n,p) on ^{32}S; (n,α) on ^{28}Si; (d,α) on ^{24}Mg.

14. Specify the product formed by each of the following processes: (a) Emission of a positron by ^{62}Cu; (b) emission of a negative beta particle by ^{40}K; (c) emission of an alpha particle by ^{203}Bi; (d) capture of an electron by ^{123}I.

15. The half-life of ^{38}K is 7.5 min. What time is required for 90 percent of a sample of this isotope to decay?

16. The energy required to dissociate molecular oxygen into two ground-state oxygen atoms (^3P) is 5.08 eV. When oxygen dissociates into one ^3P atom and one atom in the ^1D state, radiation equivalent to 176 nm is required. In kilojoules per mole, how far does the ^1D state lie above the ground state for atomic oxygen.

17. For a continuous-wave laser operating at 10.6 μm with a power output of 1.00 W, how many photons are emitted per second? If a similar laser is operated in a pulsed mode, at 1000 pulses of length 2.50 nsec each, with the same total average power, how many photons are emitted in each pulse?

18. Calculate the energy produced in electron volts per particle in a process in which there is a decrease in mass of 0.0069 amu per particle.

19. When maleic anhydride reacts with acetone, the fluorescence of acetone which otherwise occurs is found to be quenched by the presence of the anhydride. What does this indicate about the form of acetone that enters into the reaction?

20. The photochemical reaction of chlorine and hydrogen leading to HCl has a quantum yield of about 10^5. Propose an explanation.

21. If one could observe K_α x rays for a hydrogen atom, what would be their wavelength?

22. Beta rays are absorbed exponentially in metals. If the thickness of Al metal required to absorb 50 percent of a beam is 3 mm, what thickness is required to absorb 95 percent?

23. Calculate the kinetic energy of an electron accelerated through a potential difference of 15,000 practical volts.

24. The half-life of ^{32}P is 14.2 days. What quantity of this isotope corresponds to 1 mCi?

REFERENCES

Books

D. R. Arnold, N. C. Baird, J. R. Bolton, J. C. D. Brand, P. W. M. Jacobs, P. de Mayo, and W. R. Ware, *Photochemistry: An Introduction,* Academic Press, New York, 1974. An introduction to the subject suitable for students with a fairly good chemical background.

Robert Callender and Barry Honig, "Resonance Raman Studies of Visual Pigments," in *Annual Review of Biophysics and Bioengineering,* Vol. 6, Annual Reviews, Palo Alto, Calif., 1977. A review of recent applications of the technique to the chromophore of rhodopsin and its reaction products.

J. G. Calvert and J. N. Pitts, Jr., *Photochemistry*, Wiley, New York, 1966. The standard textbook, comprehensive and fairly advanced.

Herbert M. Clark, "The Measurement of Radioactivity," in *Techniques of Chemistry*, A. Weissberger and B. W. Rossiter, Eds., Vol. 1, Part IIID, Chapter IX, Wiley-Interscience, New York, 1972.

R. K. Clayton, *Light and Living Matter*, McGraw-Hill, New York, 1971. A fairly elementary introduction.

Dwaine O. Cowan and Ronald L. E. Drisko, *Elements of Organic Photochemistry*, Plenum Press, New York, 1976. An introductory textbook.

J. M. Coxon and B. Halton, *Organic Photochemistry*, Cambridge University Press, 1974. Emphasizes applications to organic reactions.

Malcolm Dole, Ed., *The Radiation Chemistry of Macromolecules*, 2 vols., Academic Press, New York, 1973. Detailed and specialized accounts of radiation effects in polymers; Chapter 17 in Vol. II treats macromolecules of biological interest.

G. E. Fogg, *Photosynthesis*, American Elsevier, New York, 1972. An introductory treatment.

K. W. Foster, "Phototropism in Coprophilous Zygomycetes," in *Annual Review of Biophysics and Bioengineering*, Vol. 6, Annual Reviews, Palo Alto, Calif., 1977. Summary of recent research on the effects of light on certain fungi.

W. A. Hagins, "The Visual Process: Excitatory Mechanisms in the Primary Receptor Cells," in *Annual Review of Biophysics and Bioengineering*, Vol. 1, Annual Reviews, Palo Alto, Calif., 1972.

Richard Henderson, "The Purple Membrane from *Halobacterium halobium*," in *Annual Review of Biophysics and Bioengineering*, Vol. 6, Annual Reviews, Palo Alto, Calif., 1977. An unusual organism which can use the effect of light on retinal as a source of metabolic energy.

Barry Honig and Thomas G. Ebrey, "The Structure and Spectra of the Chromophore of the Visual Pigments," in *Annual Review of Biophysics and Bioengineering*, Vol. 3, Annual Reviews, Palo Alto, Calif., 1974.

D. L. King and D. W. Setser, "Reactions of Electronically Excited-State Atoms," in *Annual Review of Physical Chemistry*, Vol. 27, Annual Reviews, Palo Alto, Calif., 1976. Includes photosensitization and transfer of energy from atoms to molecules.

A. A. Lamola and N. J. Turro, *Energy Transfer and Organic Photochemistry*, Vol. XIV of *Technique of Organic Chemistry*, A. Weissberger, Ed., Wiley-Interscience, New York, 1969. Excellent account of reactions of organic molecules with emphasis on the liquid phase. Includes general principles, examples of reactions, and experimental methods.

A. L. Lehninger, *Bioenergetics*, W. A. Benjamin, Menlo Park, Calif., 1965. Has excellent introduction to photosynthesis.

M. S. Matheson and L. M. Dorfman, *Pulse Radiolysis*, M.I.T. Press, Cambridge, Mass., 1969. Detailed account of the study of reaction mechanisms by this method.

C. B. Moore, Ed., *Chemical and Biological Applications of Lasers*, Academic Press, New York, 1974. Accounts of studies of a wide variety of systems.

R. B. Setlow and J. K. Setlow, "Effects of Radiation on Polynucleotides," in *Annual Review of Biophysics and Bioengineering*, Vol. 1, Annual Reviews, Palo Alto, Calif., 1972.

Kendric C. Smith, Ed., *Aging, Carcinogenesis, and Radiation Biology: The Role of Nucleic Acid Addition Reactions*, Plenum Press, New York, 1976. A collection of research papers on the photochemistry of nucleic acids.

K. C. Smith and P. C. Hanawalt, *Molecular Photobiology, Inactivation and Recovery*, Academic Press, New York, 1969. Describes effects on whole cells and cell components as well as mechanisms for damage repair; intermediate level.

A. J. Swallow, *Radiation Chemistry*, Halsted Press, New York, 1973. A very excellent general account of the field at an introductory level.

Harold M. Swartz, James R. Bolton, and Donald C. Borg, Eds., *Biological Applications of Electron Spin Resonance*, Wiley-Interscience, New York, 1972. Chapter 6 is a good treatment of photosynthesis, and Chapter 10 describes the use of ESR in identifying radiation-produced radicals.

N. J. Turro, *Molecular Photochemistry*, W. A. Benjamin, Menlo Park, Calif., 1965. An advanced account of organic reactions.

C. R. Worthington, "Structures of Photoreceptor Membranes," in *Annual Review of Biophysics and Bioengineering*, Vol. 3, Anual Reviews, Palo Alto, Calif., 1974. Recent structural studies of organs of vision.

Journal Articles

Robert T. Alfano and Stanley L. Shapiro, "Ultrashort Phenomena," *Physics Today*, page 30 (July 1975).

Bernard Alpert and Lars Lindqvist, "Porphyrin Triplet State Probing the Diffusion of Oxygen in Hemoglobin," *Science* **187**, 836 (1975).

V. Balzani, L. Moggi, M. F. Manfrin, F. Bolletta, and M. Gleria, "Solar Energy Conversion by Water Photodissociation," *Science* **189**, 852 (1975).

J. A. Bell and J. D. MacGillivray, "Photosensitized Oxidation by Singlet Oxygen," *J. Chem. Educ.* **51**, 677 (1974).

Michael W. Berns, "Directed Chromosome Loss by Laser Microirradiation," *Science* **186**, 700 (1974).

Robert R. Birge, Michael J. Sullivan, and Bryan E. Kohler, "The Effect of Temperature and Solvent Environment on the Conformational Stability of 11-*cis*-Retinal," *J. Am. Chem. Soc.* **98**, 358 (1976).

Nicolaas Bloembergen, "Lasers: A Renaissance in Optics Research," *Am. Sci.* **63**, 16 (1975).

Warren L. Butler, "Primary Photochemistry of Photosystem II of Photosynthesis," *Acc. Chem. Res.* **6**, 177 (1973).

Melvin Calvin, "Photosynthesis as a Resource for Energy and Materials," *Am. Sci.* **64**, 270 (1976).

A. J. Campillo, V. H. Kollman, and S. L. Shapiro, "Intensity Dependence of the Fluorescent Lifetime of *in Vivo* Chlorophyll Excited by a Picosecond Light Pulse," *Science* **193**, 227 (1976).

Richard H. Clarke, R. E. Connors, T. J. Schaafsma, J. F. Kleibeuker, and R. J. Platenkamp, "The Triplet State of Chlorophylls," *J. Am. Chem. Soc.* **98**, 3674 (1976).

K. B. Eisenthal, "Studies of Chemical and Physical Processes with Picosecond Lasers," *Acc. Chem. Res.* **8**, 118 (1975).

Barbara J. Finlayson and James N. Pitts, Jr., "Photochemistry of the Polluted Troposphere," *Science* **192**, 111 (1976).

Christopher S. Foote, "Mechanisms of Photosensitized Oxidation," *Science* **162**, 963 (1968).

Martin Gibbs, "The Inhibition of Photosynthesis by Oxygen," *Am. Sci.* **58**, 634 (1970).

Govindjee and Rajni Govindjee, "The Absorption of Light in Photosynthesis," *Sci. Am.* **231**, 68 (December 1974).

P. Grutsch and C. Kutal, "Mechanistic Inorganic Photochemistry," *J. Chem. Educ.* **53**, 437 (1976).

Sterling B. Hendricks, "How Light Interacts with Living Matter," *Sci. Am.* **219**, 174 (September 1968).

Hyman Katz, "Orbital Symmetry in Photochemical Transformations," *J. Chem. Educ.* **48**, 84 (1971).

H. H. Jaffe and A. L. Miller, "The Fates of Electronic Excitation Energy," *J. Chem. Educ.* **43**, 469 (1966).

Stephen R. Leone, "Applications of Lasers to Chemical Research," *J. Chem. Educ.* **53**, 13 (1976).

Eva L. Menger, Ed., "Special Issue on the Chemistry of Vision," *Acc. Chem. Res.* **8**, 81 (1975). An excellent series of five articles on various aspects of vision.

Douglas C. Neckers, "Photochemical Reactions of Natural Macromolecules," *J. Chem. Educ.* **50**, 164 (1973).

C. Michael O'Donnell and Thomas S. Spencer, "Some Considerations of Photochemical Reactivity," *J. Chem. Educ.* **49**, 822 (1972).

P. S. Rao and E. Hayon, "Reaction of Hydroxyl Radicals with Oligopeptides in Aqueous Solutions. A Pulse Radiolysis Study," *J. Phys. Chem.* **79**, 109 (1975).

Kenneth J. Rothschild, James R. Andrew, Willem J. De Grip, and H. Eugene Stanley, "Opsin Structure Probed by Raman Spectroscopy of Photoreceptor Membranes," *Science* **191**, 1176 (1976).

Arthur L. Schawlow, "Laser Light," *Sci. Am.* **219**, 120 (September 1968).

Peter Sorokin, "Organic Lasers," *Sci. Am.* **220**, 30 (February 1969).

Walther Stoeckenius, "The Purple Membrane of Salt-Loving Bacteria," *Sci. Am.* **234,** 38 (June 1976).

Nicholas J. Turro, "Triplet-Triplet Excitation Transfer in Fluid Solution—Applications to Organic Photochemistry," *J. Chem. Educ.* **43,** 13 (1966).

Nicholas J. Turro and Gary Schuster, "Photochemical Reactions as a Tool in Organic Synthesis," *Science* **187,** 303 (1975).

W. W. Waddell, A. P. Yudd, and K. Nakanishi, "Micellar Effects on the Photochemistry of Rhodopsin," *J. Am. Chem. Soc.* **98,** 238 (1976).

G. Wald, "Molecular Basis of Visual Excitation," *Science* **162,** 230 (1968).

J. T. Warden and J. R. Bolton, "Light-Induced Paramagnetism in Photosynthetic Systems," *Acc. Chem. Res.* **7,** 189 (1974).

Alvin M. Weinberg, "The Maturity and Future of Nuclear Energy," *Am. Sci.* **64,** 16 (1976).

D. Pan Wong, "Introduction to the Laser in the Physical Chemistry Course," *J. Chem. Educ.* **48,** 654 (1971).

Richard J. Wurtman, "The Effects of Light on the Human Body," *Sci. Am.* **233,** 69 (July 1975).

W. Yang and E. K. C. Lee, "Liquid Scintillation Counting," *J. Chem. Educ.* **46,** 277 (1969).

James T. Yardley, "Tunable Coherent Optical Radiation for Instrumentation," *Science* **190,** 223 (1975).

Richard W. Young, "Visual Cells," *Sci. Am.* **223,** 80 (October 1970).

Howard E. Zimmerman, "Mechanistic and Exploratory Organic Photochemistry," *Science* **191,** 523 (1976).

Table of Symbols and Abbreviations

Constants

c	velocity of light	2.998×10^8 m/sec
e,	electronic charge	1.602×10^{-19} coulombs
$\mathscr{F}$	faraday	96,494 coulombs
g	acceleration of gravity	980.7 cm/sec^2
h	Planck's constant	6.626×10^{-27} erg sec
k	Boltzmann constant	1.381×10^{-16} erg/K
m_e	electronic mass	9.110×10^{-28} g
N	Avogadro's number	6.022×10^{23} mol^{-1}
R	gas constant	8.314 J/(mol K)
		1.987 cal/(mol K)
		82.057 cm^{3}atm/(mol K)

Units

amp	ampere (electric current)
Å	angstrom (length, 10^{-10}m)
atm	atmosphere (pressure, 101,325 N/m^2)
cal	calorie (energy, 4.184 J)
D	debye (dipole moment, 10^{-18} esu cm)
dyn	dyne (force, g cm/sec^2)
eV	electron volt (energy, 1.602×10^{-12} erg)
g	gram
G	gauss (magnetic field)
Hz	hertz (frequency, sec^{-1})
J	joule (energy, kg m)
K	kelvin (temperature, degrees absolute)
m	meter
m	molality (mol/kg solvent)
M	molarity (mol/liter solution)
N	normality (equivalents/liter solution)
N	newton (force, kg m/sec^2)
P	poise (viscosity, dyn sec/cm^2)
sec	second
torr	torr (pressure, 1/760 atm)
V	volt (electric potential)
W	watt (power, J/sec)

Metric multipliers

	Prefix	Symbol
10^9	giga	G
10^6	mega	M
10^3	kilo	k
10^{-1}	deci	d
10^{-2}	centi	c
10^{-3}	milli	m
10^{-6}	micro	μ
10^{-9}	nano	n
10^{-12}	pico	p

Index

A
B 8
C 9
D 0
E 1
F 2
G 3
H 4
I 5
J 6